FATS AND OILS

FATS
and
OILS

Formulating
and Processing
for Applications

Richard D. O'Brien

Fats and Oils Consultant
Plano, Texas
Formerly with Kraft Food Ingredients
and Anderson Clayton Foods

TECHNOMIC
PUBLISHING CO., INC.
LANCASTER · BASEL

Fats and Oils

a **TECHNOMIC** publication

Published in the Western Hemisphere by
Technomic Publishing Company, Inc.
851 New Holland Avenue, Box 3535
Lancaster, Pennsylvania 17604 U.S.A.

Distributed in the Rest of the World by
Technomic Publishing AG
Missionsstrasse 44
CH-4055 Basel, Switzerland

Printed in the United States of America
10 9 8 7 6 5 4 3 2 1

Main entry under title:
    Fats and Oils: Formulating and Processing for Applications

A Technomic Publishing Company book
Bibliography: p.
Includes index p. 667

Library of Congress Catalog Card No. 97-61637
ISBN No. 1-56676-363-0

CONTENTS

Chapter 3. Fats and Oils Analysis   181

Chapter 4. Fats and Oils Formulation   251

Chapter 13. Troubleshooting 559

PREFACE

This book was written as a practical reference for fats and oils processing and formulation for effective food processor, foodservice, and household applications. It is designed to be an information source for personnel and students of fats and oils processing, as well as personnel of the user industries.

A number of books are available that cover various aspects of fats and oils chemistry, technology, and processing. However, the author is not aware of a text devoted to fats and oils processing and formulation from an applications and quality control viewpoint. These are mandatory attributes for the success of a fats and oils processor. Product application influences the design of the finished product. Quality assurance provides consistent, uniform functionality.

Applications development and quality management begins when the customer's requirements are identified and continues through product design, sales, manufacturing, product costing, delivery, and service. This book was designed to delve into the technical aspects that control the functional characteristics of fats and oils products, how these characteristics can be modified to perform as needed, and the processing control necessary to produce these characteristics on a continuing uniform basis at the most economical costs. A thorough understanding of the functions and properties of the various fats and oil products is the basic key to formulation for the desired attributes. Chapters dealing with the raw material or source oils and fats, performance evaluations, formulation attributes, and application were designed to provide the elements of formulation. Control of these processes requires consideration of the problems associated with the properties of the raw materials and each process, as well as customer problems after the sale. Therefore, a quality management chapter was included to help establish when a system is in or out of control, and a troubleshooting chapter was included to identify solutions to problems or at least stimulate the thought process for solving the problem.

Fats and oils user industries have always shown a great deal of interest in the development, processing, and quality control of shortening, margarine, and liquid oil products. Tailored fats and oils products for specific applications, processes, nutritional contributions, and other functionalities have been successful due to the cooperative development efforts between fats and oils processors and the users. Therefore, this text was compiled with the intent of focusing on fats and oils products application for both the processors and the users. The intent was to help both better understand the functionalities, limitations, and potential fats and oils modifications. As stated before, understanding the functions and properties of a shortening, oil, margarine, or other fats-based product are a key element for proper usage and product formulations. Likewise, understanding the requirements of the finished product are necessary to develop fats and oils products that perform more than adequately.

Materials for this book have been gathered over the past 35 years from patents, trade journals, scientific journals, personal experience, and exposure to many of the industry's recognized leaders, as well as many other individuals who took a genuine pride in their work and product produced.

CHAPTER 1

Raw Materials

1. INTRODUCTION

Fats and oils are the raw materials for margarine, shortening, salad oil, and other specialty or tailored products that become essential ingredients in food products prepared in the home, restaurants, and by the food processor. Fats and oils are constituents of all forms of plant and animal life. However, the plants and animals that produce oil in sufficient quantities to be commercially significant are comparatively few.

Oil-bearing fruits, nuts, and seeds have been grown and used for food for many centuries. More than 100 varieties of plants are known to have oil-bearing seeds, but only a few have been commercialized. The largest source of vegetable oil at present is the seeds of annual plants such as soybean, cottonseed, peanut, sunflower, safflower, corn, and canola or rapeseed. Many of these oil-bearing seeds are not only a source of oil, but also protein, and in many cases the protein portion has the most value. Other oil-bearing seeds such as peanuts and corn are by-products of a crop grown primarily for other purposes. A second source of vegetable oils is the oil-bearing fruit and nuts of trees such as coconut, palm, palm kernel, and olive. The oil from the palm and olive are extracted from the fruit rather than the seed of the fruit. All of the oil-bearing trees require a relatively warm climate, two of which are tropical: coconut and palm.

Oil contents for vegetable oil-bearing materials vary between 3 and 70% of the total weight of the seed, nut, kernel, or fruit. The oil contents for the oil-bearing materials used in the United States are [1]

1

Oil Bearing Material	Oil Content, %
Canola	40 to 45
Coconut	65 to 68
Corn	3 to 6
Cottonseed	18 to 20
Olive	25 to 30
Palm	45 to 50
Palm kernel	45 to 50
Peanut	45 to 50
Safflower	30 to 35
Soybean	18 to 20
Sunflower	35 to 45

Meat fats are derived almost entirely from three kinds of domestic animals: hogs, cattle, and sheep. Other animals such as poultry have small carcasses that yield very little fat. Wild animals are no longer considered a source of fats, except in possibly some local situations. Milk fat production approaches that of animal fats. In some regions, fats extracted from the milk of goat, buffalo, reindeer, and so on are locally important, but the bulk of the world's milk fat production consists of butterfat from cow's milk [2].

Fats and oils are a unique class of agricultural product in that a high degree of interchangeability among them is possible for many products and uses. Additional processing and/or blending of one or more source oils may be necessary for a satisfactory substitution. Knowledge of the physical and chemical properties of each individual raw material is necessary to successfully duplicate or improve on the functionality of the original source oil's functionality.

2. U.S. UTILIZATION OF EDIBLE FATS AND OILS

The U.S. market for food fats and oils has expanded sharply during the past 45 years, and dramatic changes have occurred in the structure of the food fat economy. Distinct trends away from animal fats to vegetable oils are recorded in Table 1.1, which reviews source oil utilization in the United States over the past 45 years, as well as the annual per capita consumption [3–6]. Edible fats and oils per capita consumption increased almost 40% during the 35-year period from 1950 to 1985 but then dropped off about 30% in the next 5-year period. The annual average usage increase for the total period was from 45.9 to 63.1 pounds of fat per person. The gains occurred largely in the 25-year period from 1960 through 1985. The changes in the use of food fats reflect the changing eating habits of Amer-

TABLE 1.1. U.S. Edible Fats and Oils Disappearance.

	Year (million of pounds)									
	1950	1955	1960	1965	1970	1975	1980	1985	1990	1995
Canola	129	194	172	272					577	1257
Coconut	223	233	310	427	788	1175	1032	1019	897	999
Corn	1445	1341	1225	1410	445	559	673	863	1149	1334
Cottonseed	79	52	51	44	891	451	523	650	851	990
Olive			1	13	67	63	58	105	213	251
Palm					182	883	299	591	256	240
Palm kernel	26	36	53	80	94	151	NR	333	362	293
Peanut	103	48	62	70	193	237	112	121	197	192
Safflower				51	100	75		51	58	35
Soybean	1446	2309	3011	3750	6253	7906	9114	10553	12164	13450
Sunflower						80	64	144	200	125
Lard	2050	1986	1889	1772	1645	803	1023	810	807	914
Tallow	156	239	328	529	518	506	995	1595	955	1348
Butter	1327	1237	1113	1040	1075	1021	1017	1164	1095	1182
Total	6984	7675	8215	9458	12251	13910	14910	17999	19781	22610
Per capita pounds	45.9	46.3	45.5	48.6	52.5	55.8	57.3	64.1	62.9	65.3

NR = not recorded.

icans in time, place, and frequency of eating. During the periods of high increases, convenience and snack food popularity rose sharply, along with a rapid growth in the fast-food industry, which relies heavily upon deep fat frying.

A definite shift from animal fats to vegetable oils has occurred for food uses. In 1950 the food fat market was split approximately equally between animal fats (lard, tallow, and butter) and edible vegetable oils. Twenty years later in 1970, edible vegetable oils accounted for three-fourths of the total and animal fats only one-fourth. On a per capita basis, as illustrated in Chart 1.1, edible vegetable oil consumption had increased during the four decades from 22 to 54 pounds; meanwhile, animal fats use fell from 23 to 9 pounds. Vegetable oils became dominate mainly because of competitive pricing for soybean oil, increased hydrogenation capacity, consumer preference shifts from butter to margarines, and nutritional concerns regarding cholesterol and saturated fats.

Soybean oil has emerged as the leading vegetable oil in the U.S. edible oil economy, currently accounting for over half of all fats and oils utilized in food products. Soybean oil's growth has more than offset the decline in cottonseed oil, previously the preferred vegetable oil produced and consumed in the United States.

3. FATS AND OILS CHARACTERIZATION

Chemically, fats and oils are a combination of glycerin and fatty acids. The glycerin molecule has three separate points where a fatty acid molecule can be attached, thus the common reference to fats and oils as triglycerides. Physically, fats and oils differ in that fats are a solid at normal room temperatures and oils are liquids at room temperature. The different properties are to a large extent determined by the fatty acid composition and the extent of saturation or unsaturation present. These aspects are identified by the carbon chain length and the number and position of double bonds for the individual fatty acids, and their position on the glycerin. Generally, solid fats are indicated by a dominance of saturated fatty acids, and liquid oils are evidence of a high level of unsaturated fatty acids.

Edible fats and oils carbon chain lengths vary between 4 and 24 carbon atoms with up to three double bonds. Table 1.2 reviews the chain length, common name, saturation, and unsaturation for the most common fatty acids identified in edible fats and oils. The most prevalent saturated fatty acids are lauric (C-12:0), myristic (C-14:0), palmitic (C-16:0), stearic

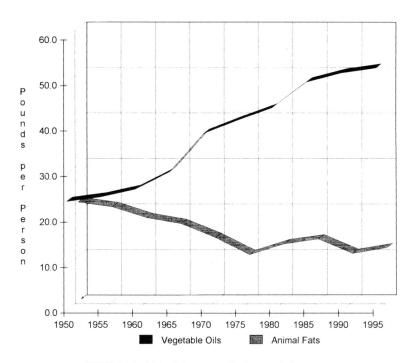

CHART 1.1. United States capita fats and oils usage.

5

TABLE 1.2. Edible Fats and Oils Principal Fatty Acids.

Chain Length	Fatty Acid (double bonds)			
	Saturated (none)	Mono (one)	Di (two)	Tri (three)
C-4	Butyric			
C-6	Caproic			
C-8	Caprylic			
C-10	Capric			
C-12	Lauric			
C-14	Myristic	Myristoleic		
C-15	Pentadecanoic			
C-16	Palmitic	Palmitoleic		
C-17	Margaric	Margaroleic		
C-18	Stearic	Oleic	Linoleic	Linolenic
C-20	Arachidic	Gadoleic	Eicosadienoic	
C-22	Behenic	Erucic		
C-24	Lignoceric			

(C-18:0), arachidic (C-20:0), behenic (C-22:0), and lignoceric (C-24:0). The most important monounsaturated fatty acids are oleic (C-18:1) and erucic (C-22:1). The polyunsaturated fatty acids are linoleic (C-18:2) and linolenic (C-18:3).

Natural fats and oils physical properties vary widely, even though they are composed of the same or similar fatty acids. These differences result from differences in the proportion of the fatty acids and the structure of the individual triglycerides. Among the factors that affect the vegetable oil fatty acid compositions are climate conditions, soil type, growing season, plant maturity, plant health, microbiological seed location within the flower, and the genetic variation of the plant. Animal fats and oils composition varies according to the animal species, diet, health, fat location on the carcass, and maturity [7].

Physical properties of an oil or fat are of critical importance in determining its functional characteristics or use in food products. One fundamental physical property of importance is demonstrated by the terms *fats* and *oils,* which indicate whether a lipid is liquid or solid at ambient temperatures. Some vegetable oils appear to disagree with this designation until the place of origin temperatures are considered. Vegetable oil products that are solid at ambient temperatures in a temperate climate are liquid at the tropical ambient temperatures where they are grown.

4. SOYBEAN OIL

Soybean oil is obtained from soybeans, *Glycina maxima,* which are grown in several countries of the world. Soybeans are native to eastern Asia, where ancient Chinese literature indicates that soybeans have been an important part of their diet for centuries. Even though soybean oil currently fulfills over half of the U.S. edible fats and oils requirements, soybeans are a relatively new food crop for the United States. Soybeans were grown in the United States as early as 1804 but remained an agricultural curiosity and minor crop for over a century. Initially, most of the soybean acreage was used for forage and pasture; only limited quantities were grown and harvested for seed [8]. In 1908 some European countries started importing soybeans from Manchuria to supplement short supplies of cottonseed and flaxseed. The soybeans were processed into oil and meal; the oil was utilized predominately for soap manufacture and the meal for cattle feed. The European successful utilization influenced similar experimentation in the United States, which led to processing of imported soybeans in 1911 and domestic soybeans in 1915. The early ventures were hampered by difficulties in obtaining a suitable supply of soybeans, a lack of processing experience, and the development of a market for the soybean oil and meal.

Growth of the soybean industry in the United States was influenced more by the shortage of oil and its relatively high cost than the need for protein. A bushel of soybeans yielded approximately 10.5 pounds of oil and 47.8 pounds of 44% protein meal with solvent extraction. Since the oil was the most valuable product from soybeans, the meal was considered a by-product, which prompted the soybean producers to urge plant breeders to develop new varieties with high oil contents.

Soybean oil is high in polyunsaturates and is classified as a semidrying oil. In the 1920s, it was used mostly for inedible products like soaps, paints, and varnishes. The use of soybean oil in foods was limited because of its high linolenic fatty acid content, which created a flavor stability problem. This soybean oil flavor problem had to be solved before it would be accepted by the food industry [9].

Soybean oil production grew to 0.3 billion pounds in 1938 and further to 1.3 billion pounds in 1945, despite the flavor and odor deficiencies, predominately because of World War II shortages. Fats and oils processors had incorporated as much soybean oil into formulations as possible to take advantage of the 4- to 9-cent per pound discount over cottonseed oil. In many products, like margarine, 30% soybean oil could be blended into a formulation without a noticeable flavor degradation. However, flavor was still the limiting factor for future soybean acceptance as a food oil. Soybean oil had initially been identified as an industrial oil, but as a paint oil it dried slowly and developed "after tack," while as an edible oil it tasted like paint.

During World War II, German edible oil processors developed a formula or process to cure soybean oil reversion. It involved many water washings and steps contacting the oil with water glass. Objective flavor and odor methods helped determine the reason the German solution worked; the process included the addition of citric acid to the deodorized oil, which complexed trace prooxidant metals. Upon release of this information, edible oil processors almost immediately adopted metal deactivators, of which citric acid is still the most popular.

Objective flavor and odor evaluations substantiated that trace metals were a significant contributor to the soybean oil flavor problem. Other edible fats and oils can tolerate copper and iron in the parts per million (ppm) range, while soybean oil flavor is ruined by as little as 0.5 ppm iron and 0.01 ppm copper. This discovery promoted the removal of brass valves and conversion of cold, rolled steel deodorizers to stainless steel for processing. Another effective precaution was to blanket oils with inert gas at all critical high-temperature processing steps, including packaging.

Investigations still had not identified the cause of the off-flavor development with soybean oil. Circumstantial evidence pointed to the 7 to 8% linolenic fatty acid content. A classic experiment interesterified 9% linolenic fatty acid into cottonseed oil, which typically contains less than 1% of the C-18:3 fatty acid. Subsequent flavor panels identified this modified product as soybean oil. This result presented three alternatives for improving the flavor stability of soybean oil: (1) breed it out, (2) extract it out, or (3) hydrogenate it out. Hydrogenation to reduce the linolenic fatty acid content was chosen as the most practical short-term approach [10].

Soybean salad oil with reduced linolenic fatty acid (3 to 4%) was introduced to the U.S. market in the early 1960s. This product was lightly hydrogenated and subsequently winterized to remove the hard fractions developed during hydrogenation. This soybean salad oil was quickly accepted by the retail salad oil consumers and also industrially as a component of salad dressings, mayonnaise, margarines, and shortenings [11]. In the late 1970s, improvements in soybean oil processing produced a refined bleached and deodorized (RBD) oil that was more acceptable for industrial users and was eventually introduced to the retail market. This RBD soybean oil, promoted as all natural and light, was introduced to the retail market in the late 1980s and rapidly replaced the hydrogenated, winterized, and deodorized salad oil.

Soybean oil has become a popular vegetable oil for foodstuffs due to its nutritional qualities, abundancy, economic value, and wide functionality. Soybean oil utilization has almost tripled in the past 45 years, apparently due to the U.S. food industry's recognition of these attributes. The acceptance of soybean oil is illustrated by the increased use in relation to the other food oils available to the U.S. consumers:

Year	Percent
1950	20.7
1960	36.6
1970	51.3
1980	61.1
1990	61.5
1995	59.5

4.1 Soybean Oil Composition and Physical Properties

Soybean oil is a very versatile oil as far as processing and product formulation are concerned: (1) it refines with a low loss; (2) it is a natural winter or salad oil; (3) it has heat-sensitive color pigments that deodorize to a red color much less than 1.0 Lovibond; (4) it develops large, easily filtered crystals when partially hydrogenated and fractionated; (5) it has a high iodine value that permits hydrogenation of basestocks for a wide variety of products; and (6) it has a high essential fatty acid content [12]. The typical characteristics and fatty acid composition for soybean oil are [13,14]

Specific gravity at 25/25°C		0.917 to 0.921
Refractive index at 25°C		1.470 to 1.476
Iodine value		123.0 to 139.0
Saponification number		189 to 195
Unsaponifiable matter, %		1.5 max.
Melting point, °C		−23.0 to −20.0
Fatty acid composition, %		
Myristic	C-14:0	0.1
Palmitic	C-16:0	10.6
Palmitoleic	C-16:1	0.1
Margaric	C-17:1	0.1
Stearic	C-18:0	4.0
Oleic	C-18:1	23.3
Linoleic	C-18:2	53.7
Linolenic	C-18:3	7.6
Arachidic	C-20:0	0.3
Behenic	C-22:0	0.3

Two deficiencies inherent in soybean oil are that it has a linolenic fatty acid content of 6 to 8% and that the hydrogenated oil is a beta crystal former. Due to the high linolenic fatty acid content, careful handling to prevent oxidation and metal chelating is mandatory to avoid beany, painty, or green flavors. Soybean oil-based shortenings and other plasticized products must have at least 10% of a low iodine value cottonseed or palm oil hardstock added to provide a beta prime crystal form. Plasticized shortenings with 100% soybean oil develop a grainy, nonuniform consistency somewhat similar to products made with 100% lard. Margarine products can be formulated with all soybean oil bases because of the refrigerated storage and shipping conditions exercised in the United States. Soybean oil has been found to be the perfect base oil for liquid shortenings that require a beta crystal form for a stable pourable product [12].

5. COTTONSEED OIL

Cotton is one of the oldest cultivated crops. It was grown in India 4,000 years ago, and Columbus found cotton already growing in America. However, the use of cottonseed oil did not emerge until the nineteenth century when European businessmen began to extract oil from a variety of seeds and nuts to find a more affordable fat source. Several attempts were made to introduce cottonseed oil to the U.S. market, but the first successful venture was as an adulterant for olive oil sold misbranded to immigrants. Its second opportunity was also as a dilutant for another product; it was secretly added to lard. This venture led to lard substitutes or compounds used for bakery products. Several important oil processing developments were responsible for the eventual U.S. consumer's acceptance of cottonseed and other vegetable oils: caustic refining, bleaching, deodorization, and hydrogenation. Cottonseed oil remained the principal raw material for vegetable oils and shortenings in the United States until the mid-twentieth century. This predominance began to erode after World War I, and soon after World War II soybean oil became the principal source of vegetable oil. Today, vegetable oil users have a wide range of raw materials to choose from, but cottonseed oil pioneered the U.S. vegetable oil industry and is still preferred for many applications [15].

5.1 Cottonseed Oil Composition and Physical Properties

The majority of the cottonseed oil in the United States and the world is obtained from seeds of the plant *Gossypium hirsutum*. Crude cottonseed oil is unusual because it contains many kinds of nonglyceride materials, which contribute to the strong red-brown color and odor of the unrefined oil. More than 2% of cottonseed oil is made up of gossypal, phospholipids, sterols, resins, carbohydrates, and related pigments. Almost all of these compounds are removed during the refining process.

A number of factors contribute to minor variations in the properties of cottonseed oil, which include [16]:

- climate—Cotton is very productive at high temperatures when adequate water is available. However, the free fatty acid level of the oil may rise during hot, humid conditions.
- region—Differences in climate and soil conditions cause geographical regions to have a major influence on cottonseed oil composition. For example, oil from the Mississippi Delta region is usually more unsaturated than cottonseed oil from Texas. The iodine value of oils tends to increase in cottonseeds grown farther north.
- varieties—Glandless cotton varieties tend to contain more oil that is slightly more unsaturated and bleaches to a lower red color than the glanced varieties.
- crop fertilizers—High nitrogen levels favor increased cottonseed protein content with less oil, probably due to increased plant maturity.
- seed handling and storage—Oil from improved stored cottonseeds will develop a dark color and free fatty acid levels. Ideally, cottonseeds should be stored at a moisture content of less than 9% with ventilation at ambient temperatures.

The typical characteristics of cottonseed oil are [16,17]

		Typical	Range
Specific gravity at 25/25°C			0.916 to 0.918
Refractive index at 25°C			1.468 to 1.472
Iodine value			99.0 to 113.0
Saponification number			189 to 198
Unsaponifiable matter, %			1.5 max.
Titer, °C			30.0 to 37.0
Melting point, °C			10.0 to 16.0
Cloud point, °C			−1.0 to 3.0
AOM stability, hours		15.0	
Fatty acid composition, %			
Myristic	C-14:0	0.7	0.5 to 2.0
Palmitic	C-16:0	21.6	17.0 to 29.0
Palmitoleic	C-16:1	0.6	0.5 to 1.5
Stearic	C-18:0	2.6	1.0 to 4.0
Oleic	C-18:1	18.6	13.0 to 44.0
Linoleic	C-18:2	54.4	33.0 to 58.0
Linolenic	C-18:3	0.7	0.1 to 2.1
Triglyceride composition,[1] mole %			
Trisaturated (GS_3)		0.1	0 to 0.1
Disaturated (GS_2U)		13.2	14.0
Monosaturated (GSU_2)		58.4	50.0 to 58.0
Triunsaturated (GU_3)		28.3	28.0 to 36.0

[1]G = glyceride, S = saturated, U = unsaturated.

The specific fatty acid profile of the triglycerides in cottonseed oil is dependent upon the variety of cotton grown and the growing conditions, which include temperature and rainfall. The differences are illustrated by fatty acid composition analysis of cottonseed oils grown in California and Texas [18]:

Fatty Acid Composition, %		California	Texas
Myristic	C-14:0	0.7	0.9
Palmitic	C-16:0	22.7	25.2
Palmitoleic	C-16:1	0.6	0.8
Stearic	C-18:0	2.3	2.7
Oleic	C-18:1	17.3	17.5
Linoleic	C-18:2	55.8	52.6
Linolenic	C-18:3	0.3	—

The reverted flavor of cottonseed oil is usually described as nutty or nutlike, which is acceptable at higher degrees of oxidation than other vegetable oils. In some cases, cottonseed oil has been utilized to mask the less desirable flavors from other vegetable oils by blending. This characteristic has made RBD cottonseed oil a favorite for frying snack foods.

Hydrogenated cottonseed oil is stable in the beta-prime crystal form and imparts this property to product blends at levels as low as 10% of the composition. This is an important consideration since the fats and oils processor has a limited source of oils with the stable beta-prime crystal, i.e., cottonseed oil, palm oil, and tallow.

6. PEANUT OIL

The origin of peanuts is unknown, though they were supposed to have been carried to Africa from Brazil or Peru by explorers or missionaries. Slave traders brought them to North America as food while on shipboard because peanuts were inexpensive, had high food value, and did not spoil readily. In North America, peanuts were not used extensively until after the Civil War in 1865, and then the crop was confined to two southern states: Virginia and North Carolina. The greatest factors contributing to the expansion of the U.S. peanut production was the invention of equipment for planting, cultivating, harvesting, separating the nuts from the plants, shelling, cleaning the kernels, roasting, blanching, salting, producing peanut butter, and packaging.

About two-thirds of the world's peanut crop is crushed for oil. However, peanuts in the United States are subject to mandatory price supports and are grown mainly for food delicacies, while other countries grow them for oil and meal. Due to competition from other vegetable oils and the demand for edible nuts, relatively small quantities are crushed for oil. In general, oil stock peanuts are those that are rejected or diverted from the edible nut channels. Rejection may be due to oversupply, variety different from that in demand, improper grading resulting in lowered quality, or inadequate storage resulting in nuts that are rancid, moldy, weathered, or insect infested [19]. It is fortunate that aflatoxin contamination, which can be prevalent and serious with peanuts, can be controlled with respect to the quality of peanut oil with normal processing. Moldy peanuts that have been infected to the extent of 5,500 ppb with aflatoxins yielded a peanut oil with 812 ppb, which was reduced to 10 to 14 ppb after caustic refining and less than 1 ppb after bleaching [20].

6.1 Peanut Oil Composition and Physical Properties

Botanically, the peanut, *Arachis hypogea,* belongs to the same legumes family as the soybean, *Papilionaceous flowers,* but in composition it is more like other nuts than most beans or peas. Peanuts are rich in oil, naturally containing from 47 to 50%. The oil is pale yellow and has the characteristic odor and flavor of peanuts. Compared to other seed oils, particularly cottonseed oil, it is relatively free of phosphatides and nonoil constituents. The typical characteristics of peanut oil are [20–22]

		Typical	Range
Specific gravity at 25/25°C			0.910 to 0.915
Refractive index at 25°C			1.467 to 1.470
Iodine value			84.0 to 100.0
Saponification number			188 to 195
Unsaponifiable matter, %			0.2 to 1.0
Cloud point, °C		40.0	
Pour point, °C		34.0	
Titer, °C			26.0 to 32.0
Melting point, °C		−2.0	
Fatty acid composition, %			
Myristic	C-14:0	0.1	less than 0.1
Palmitic	C-16:0	11.1	6.0 to 15.5
Palmitoleic	C-16:1	0.2	less than 1.0
Margaric	C-17:0	0.1	
Margaroleic	C-17:1	0.1	
Stearic	C-18:0	2.4	1.3 to 6.5
Oleic	C-18:1	46.7	36.0 to 72.0
Linoleic	C-18:2	32.0	13.0 to 45.0
Arachidic	C-20:0	1.3	1.0 to 2.5
Gadoleic	C-20:1	1.6	0.5 to 2.1
Behenic	C-22:0	2.9	1.5 to 4.8
Lignoceric	C-24:0	1.5	1.0 to 2.5

	Typical	Range
Triglyceride composition, mole %		
Trisaturated (GS_3)	—	
Disaturated (GS_2U)	1.0	
Monosaturated (GSU_2)	56.0	
Triunsaturated (GU_3)	43.0	

Peanut oil, with only a trace of linolenic fatty acid, has excellent oxidative stability and is considered a premium cooking and frying oil. Peanut flavor is closely related to the oil, and on separation the flavor goes with the oil rather than the meal. In practically all cases, the flavor is accentuated with heating.

7. CORN OIL

Corn, *Zea mays*, is native to both North and South America and was the staple grain of the Indians for centuries before Europeans reached the New World. Corn, one of the principal crops of the United States, is grown for food and livestock fodder. Only a small fraction is used for obtaining corn or maize oil. Corn oil is a by-product of the two corn milling industries: the starch/sweetener/alcohol industries and the corn meal industry, which produces corn meal, hominy grits, corn flakes, and so on. The germ, which contains about 50% oil, is obtained with a wet degermination process from the sweetener, starch, and alcohol processors. In the corn milling industry, the germ is obtained with a dry degermination process and contains from 10 to 24% oil.

7.1 Corn Oil Composition and Physical Properties

Crude corn oil has a darker reddish amber color than other vegetable oils, which can usually be processed to a light colored oil. Some wet milled oils are more difficult to bleach to a light color, possibly due to the conditions used to flake the germ before extraction. In other cases, corn oil has been deliberately underbleached and deodorized to retain a darker color for appeal to the retail consumer.

Corn oil composition depends upon the seed type, climatic conditions, and growing season. The typical characteristics for corn oil are [23,24]

Specific gravity at 25/25°C	0.915 to 0.920
Refractive index at 25°C	1.470 to 1.474
Iodine value	103.0 to 128.0

Saponification value		187 to 193
Unsaponifiable matter, %		2.0 max.
Titer, °C		14.0 to 20.0
Melting point, °C		-12.0 to -10.0
Fatty acid composition, %		
Myristic	C-14:0	0.1
Palmitic	C-16:0	10.9
Palmitoleic	C-16:1	0.2
Margaric	C-17:0	0.1
Stearic	C-18:0	2.0
Oleic	C-18:1	25.4
Linoleic	C-18:2	59.6
Linolenic	C-18:3	1.2
Arachidic	C-20:0	0.4
Behenic	C-22:0	0.1
Triglyceride composition, mole %		
Trisaturated (GS_3)		0
Disaturated (GS_2U)		2.2
Monosaturated (GSU_2)		40.3
Triunsaturated (GU_3)		57.5

Corn oil contains traces of waxes (0.05%) that are esters of myricil and ceryl alcohols with tetracosanoic acid. The melting point of these waxes is 81 to 82°C. The waxes cause the oil to cloud when cooled to a low temperature unless removed by a dewaxing process. Dewaxed oil is important for retail salad oil production but is not required when corn oil will be hydrogenated for margarine or shortening production.

Corn oil has a relatively high tocopherol content (about 0.1%), along with the presence of a small amount of another antioxidant (ferulic acid) component, probably accounts for its excellent oxidative stability. In addition, the reverted flavor of corn oil is rather pleasant, usually characterized as popcorn-like or possibly musty [25,26].

8. SUNFLOWER OIL

The wild sunflower is believed to have originated in the southern part of the United States and Mexico, where it was found growing as a weed. In 1569, the Spaniards brought the sunflower to Spain where it was an ornamental plant and from where it spread throughout Europe. Peter the Great brought sunflower to Russia, and manufacture of sunflower oil was established around 1830. In the former USSR, seed quality and oil yields were substantially improved from a 29% average yield in 1940 to a 46% oil yield in 1971. In post-war years and more particularly in the late 1960s, efforts were made to establish sunflower as an oilseed crop in other parts of the world. The sunflower is the second largest world source of vegetable oil with the majority of the production still in Russia. In the United States the development of sunflower was fostered first by the export of sunflower seed and later sunflower oil [27,28].

8.1 Sunflower Oil Composition and Physical Properties

Sunflower oil is obtained from the seed of the plant *Helianthus annuus*. Crude sunflower oil is light amber in color; the refined oil is a pale yellow and is similar to other oils. The crude contains some phosphatides and mucilaginious matter but less than cottonseed or corn oils. Sunflower oil has a distinctive, not altogether unpleasant, flavor and odor, which is removed by deodorization. The typical characteristics of sunflower oil are [29]

Specific gravity at 25/25°C		0.915 to 1.474
Refractive index at 25°C		1.472 to 1.474
Iodine value		125.0 to 136.0
Saponification number		188 to 194
Unsaponifiable matter, %		1.5 max.
Wax, %		0.2 to 3.0
Titer, °C		16.0 to 20.0
Melting point, °C		−18.0 to −16.0
Cloud point, °C		−9.5
Fatty acid composition, %		
Myristic	C-14:0	0.1
Palmitic	C-16:0	7.0
Palmitoleic	C-16:1	0.1
Margaric	C-17:0	0.1
Stearic	C-18:0	4.5
Oleic	C-18:1	18.7
Linoleic	C-18:2	67.5
Linolenic	C-18:3	0.8
Arachidic	C-20:0	0.4
Gadoleic	C-20:1	0.1
Behenic	C-22:0	0.7

Few vegetable oils reflect the influence of climate, temperature, genetic factors, and position of seed location in the flower head so significantly in their composition as does sunflower oil. Generally, sunflower grown above the 39th parallel in the United States will be high in linoleic fatty acid, and that grown below will be high in oleic fatty acid. These differences vary with the temperature also. A hot summer will lower the linoleic content of northern sunflower oils. Therefore, fatty acid compositions will vary from year to year and region to region. Average fatty acid compositions from selected northern and southern grown sunflower oil are [30]

		Northern	Southern
Fatty acid composition, %			
Palmitic	C-16:0	5.9	5.2
Stearic	C-18:0	4.7	4.4
Oleic	C-18:1	26.4	50.9
Linoleic	C-18:2	61.5	37.9
Arachidic	C-20:0	0.5	0.5
Behenic	C-22:0	0.7	0.7
Iodine value		129.2	109.4
AOM stability, hours		10.5	17.9

The majority of the sunflower oil produced in the United States could be classified as a natural winter oil that does not require winterization to remove hard fractions that crystallize when chilled. However, most U.S. crushers solvent extract seed and hull together for operation efficiency. The hulls contain waxes that are soluble in the oil portion. The waxes do not affect the emulsion stability of mayonnaise as do stearic hard fractions; however, bottled oil will cloud for an unsightly appearance at cool temperatures. These waxes must be removed with a dewaxing process to prevent the unsightly appearance

9. HIGH-OLEIC SUNFLOWER OIL

Russian scientists used chemical mutagenesis to alter the sunflower plant to create a high oleic variety stable to climate conditions. High-oleic sunflower seed was grown commercially in the United States for the first time in 1984. The U.S. developed progenies from the Russian cultivar varied only 4 to 5% in oleic fatty acid content when grown in the cool climates of Minnesota or the warm conditions in Texas. The U.S. developed sunflower substantially reduced the linoleic fatty acid content in favor of oleic with more uniform results [31].

9.1 High-Oleic Sunflower Oil Physical Properties and Composition

Oxidative stability improvement, a major objective for developing high-oleic sunflower oil, was achieved with the reduction of linoleic fatty acids. Typical physical characteristics and composition for high-oleic sunflower oil are [23,32]

Refractive index at 25°C	1.467 to 1.469
Iodine value	78.0 to 88.0

Saponification value		188 to 194
Unsaponifiable matter, %		1.3 max.
Wax, %		0.03 to 0.10
Mettler dropping point, °C		4.4
Capillary melting point, °C		7.2
AOM stability, hours		40.0
Fatty acid composition, %		
Palmitic	C-16:0	3.7
Palmitoleic	C-16:1	0.1
Stearic	C-18:0	5.4
Oleic	C-18:1	81.3
Linoleic	C-18:2	9.0
Arachidic	C-20:0	0.4
Behenic	C-22:0	0.1

High-oleic sunflower oil should be an adequate replacement for saturated fats for a number of applications due to the high oxidative stability. Some of the applications identified are

- spray oil—protective coatings for cereals, grains, fruits, nuts, and baked products like crackers, cookies, croutons, pie shells, and pizzas
- retail oil—bottled oil for general household applications such as pan and deep fat frying, salads, sauces, gravies, and others where creaming or aeration is not required
- frying—deep fat frying where the characteristics of liquid oil frying are preferred; however, an indication of a reduction of the preferred fried food flavor has been identified with laboratory evaluations [33].
- ingredient—food processor ingredient for prepared foods such as salad dressing, mayonnaise, margarine, sauces, bakery mixes, and so on.

10. SAFFLOWER OIL

Safflower, *Carthamus tinctorius,* is among the oldest crops known to man. The species is believed to be indigenous to Southeastern Asia but has long been cultivated in China, the Near East, and Northern Africa. Safflower was brought to the United States by immigrants primarily from Spain and Portugal. Until recent years, the history of safflower has been concerned mostly with the use of its florets, from which carthamine, a dye, was extracted. The introduction of other more stable dyes replaced this use for the safflower plant.

Safflower was a relatively insignificant oilseed crop until the early 1950s when higher yielding oil-bearing varieties were developed and it was established as a source oil for surface coatings. The composition of safflower oil is largely made up of linoleic fatty acid with a practical absence of linolenic acid, which results in very nearly an ideal drying oil. Interest in the unsaturated liquid oil's ability to lower serum cholesterol levels catalyzed the development of an edible grade of safflower oil [34,35].

10.1 Safflower Oil Composition and Physical Properties

Safflower oil is obtained by pressing the seed or by solvent extraction. The most distinctive characteristic for safflower oil is the high unsaturated fatty acid content of predominately linoleic (C-18:2). The typical characteristics for safflower oil are [36,37]

Specific gravity at 25/25°C		0.919 to 0.924
Refractive index at 25°C		1.748 to 1.4752
Iodine value		140.0 to 150.0
Saponification value		186 to 197
Unsaponifiable matter, %		1.5 max.
Wax, %		ca. 0.5
Melting point, °C		−18.0 to −16.0
Fatty acid composition, %		
Myristic	C-14:0	0.1
Palmitic	C-16:0	6.8
Palmitoleic	C-16:1	0.1
Stearic	C-18:0	2.3
Oleic	C-18:1	12.0
Linoleic	C-18:2	77.7
Linolenic	C-18:3	0.4
Arachidic	C-20:0	0.3
Gadoleic	C-20:1	0.1
Behenic	C-22:0	0.2

The oxidative stability of crude safflower oil precludes storage for indefinite periods before processing. Generally, standard AOM stabilities of crude safflower oil without added antioxidants range from 4 to 8 hours shortly after crushing, which reduces to 1 to 3 hours after 2 to 4 months of normal storage. Thus, it is obvious that safflower oil high in linolenic fatty acid with a minimum of natural antioxidants is not particularly stable.

Safflower oil has been used in food products where a high polyunsaturated fatty acid content is desired. It has been utilized in mayonnaise, salad dressings, and liquid margarine and was the original source oil for the first soft-tub margarine. Flavor stability has been a constant problem with products containing appreciable quantities of safflower oil due to the high linolenic fatty acid content.

Safflower oil is readily hydrogenated in conventional processing equipment. Hydrogenation improves oxidative stability, and the products can be used in margarine or shortening products to replace the usual beta crystal habit basestocks. However, the oxidative stability of hydrogenated safflower oil is significantly less than similar products produced with soybean oil or corn oil hardened to the same degree [38].

11. HIGH-OLEIC SAFFLOWER OIL

A safflower plant natural mutation, discovered by the University of California at Davis, produces an oil in which the normal levels of linoleic and oleic fatty acids are reversed, i.e., high levels of oleic instead of linoleic. Consequently, the oil has a substantially improved oxidative stability over normal safflower oil due to the replacement of the polyunsaturates with monounsaturates. An added benefit derived from the safflower mutation was that the plant and seeds can be produced at the same costs as normal safflower in a wider range of climates [39].

Fatty acid compositions of the oil for both the high-linoleic and -oleic types of safflower have been found to be remarkably uniform at different growing temperature conditions. The slight variations experienced with both types was an increase in oleic at higher growing temperatures and an increase in linoleic at lower temperatures. These results indicate that both safflower oil types should provide very uniform reliable product results for each crop year [40].

11.1 High-Oleic Safflower Oil Composition and Physical Characteristics

High-oleic safflower oil has retained the light color and flavor characteristics of normal safflower oil, but the oxidative stability measured by AOM stability has increased by three and a half times. Typical physical characteristics and fatty acid composition for high-oleic safflower oil are [23,41]

Specific gravity at 25°C		0.910 to 0.920
Refractive index at 25°C		1.467 to 1.469
Iodine value		85.0 to 95.0
Saponification value		186 to 197
Unsaponifiable matter, %		1.5 max.
Setting point, °C		−10.0 to −18.0
Flow point, °C		13.0
AOM stability, hours		35.0 to 45.0
Cold test, hours		24.0 min.
Fatty acid composition, %		
Lauric	C-12:0	0.1
Palmitic	C-16:0	3.6
Palmitoleic	C-16:1	0.1

Stearic	C-18:0	5.2
Oleic	C-18:1	81.5
Linoleic	C-18:2	7.2
Linolenic	C-18:3	0.1
Arachidic	C-20:0	0.4
Gadoleic	C-20:1	0.2
Behenic	C-22:0	1.2
Lignoceric	C-24:1	0.3

The relationship between the degree of unsaturation of lipids and their susceptibility to oxidative deterioration is well documented. Early kinetic studies on oxidation of fatty acids noted the difference in oxidation rates of linoleic and oleic fatty acids. The slower rate for oleic and the saturated fatty acids has been a major reason for the hydrogenation of oils. Elimination of one fatty acid double bond significantly increases the resistance to oxidation [42]. Utilization of oils from nature rich in oleic fatty acid also avoids some of the side effects of hydrogenation such as *trans* acids and hydrogenation flavor.

Frying evaluations have shown that high-oleic safflower is an excellent frying oil. It resisted oxidation and polymerization better than a hydrogenated frying shortening in controlled testing [43] and had acceptable potato chip stability ratings equivalent to hydrogenated frying oil products [42]. Other probable applications would be wherever a high-stability liquid oil is desirable.

12. CANOLA OIL

Canola is the registered trademark of the Canola Council of Canada for the seed, oil, and meal derived from rapeseed cultivars, *Brassica napus* and *Brassica campestris*, low in erucic fatty acid and low in glucosinolates. Rapeseed has been grown for decades as a source of industrial oil, but its edible use has been limited because of high levels of erucic fatty acid (C-22:1). Oils high in erucic fatty acid have been shown to cause heart muscle lesions followed by other cardiac problems in laboratory animals. In response to a petition from Canada, the United States in 1985 affirmed low erucic acid rapeseed oil (LEAR oil) as a food substance generally recognized as safe (GRAS). In 1988, the U.S. FDA agreed that low erucic acid rapeseed oil (2.0% max.) could be identified as canola oil.

12.1 Canola Oil Composition and Physical Properties

Canola seed is flaked and cooked to denature or inactivate the enzyme *myrosinase* to prevent hydrolysis of glucosinolates into undesirable breakdown products. The oil is extracted from the cooked flake by pressing and solvent extraction procedures. Usually, the crude canola oil is degummed to remove the water-hydrated gums to phosphorous levels of approximately 240 ppm with water degumming or approximately 50 ppm with acid degumming procedures. Typical characteristics for degummed canola oil are [23,44]

Specific gravity at 20/20°C		0.914 to 0.920
Refractive index at 25°C		1.470 to 1.474
Iodine value		110.0 to 126.0
Saponification value		182 to 193
Unsaponifiable matter, g/kg		15 max.
Titer, °C		26.0
Pour point, °C		−12.0
Specific heat at 20°C		0.488
Fatty acid composition, %		
Myristic	C-14:0	0.1
Palmitic	C-16:0	4.1
Palmitoleic	C-16:1	0.3
Margaric	C-17:0	0.1
Stearic	C-18:0	1.8
Oleic	C-18:1	60.9
Linoleic	C-18:2	21.0
Linolenic	C-18:3	8.8
Arachidic	C-20:0	0.7
Gadoleic	C-20:1	1.0
Behenic	C-22:0	0.3
Erucic	C-22:1	0.7
Lignoceric	C-24:0	0.2

Distribution of the two fatty acids important for flavor stability, linoleic and linolenic, have been found primarily in the 2-position of the triglyceride similar to high-erucic rapeseed oil. This differs from other oils, which usually have a random distribution for linoleic and linolenic fatty acids, and the somewhat lower total unsaturation indicates better resistance to oxidation than oils with similar linoleic and linolenic fatty acid contents.

Canola oil differs from most other vegetable oils by its content of chlorophyll and sulfur compounds. Removal of these compounds during processing is required for acceptable product quality. Pretreatment of the crude oil with 0.05 to 0.5% phosphoric acid not only helps precipitate phosphatedic materials, but also aids in chlorophyll removal. Bleaching with acid-activated clays is essential for chlorophyll removal. Chlorophyll removal is most important at the prebleach stage. Chlorophyll cannot be heat bleached in hydrogenation or deodorization but must be removed during bleaching. Hydrogenation or deodorization of inadequately bleached canola oil fixes the green color and makes it almost impossible to adsorb on bleaching earth.

The hydrogenation equipment and conditions such as temperature, pressures, and catalyst are essentially the same for canola oil as required for soybean oil. However, a higher catalyst concentration may be necessary due to the presence of low levels of sulfur compounds (3 to 5 ppm) remaining after refining and bleaching, which can poison the hydrogenation catalyst [45].

Canola oil is a natural winter oil that does not require fractionation to remove a hard fraction that could crystallize at refrigerator temperatures. However, most canola crushers solvent extract the seed and hull together for operational efficiency. The seed hulls contain waxes that are soluble in the oil. These waxes solidify after a period of time in retail bottled oil to appear as an unsightly thread or layer of solidified material. The waxes must be removed by a dewaxing process for oils intended for the retail salad oil market.

The formation of large beta crystals limit the level of hydrogenated canola oil that can be utilized in margarine or shortening formulations. Hydrogenated canola oil crystals grow larger over time to produce a sandy or grainy consistency unless the crystal habit is modified by the addition of a beta-prime crystal former.

13. PALM OIL

Palm oil is derived from the fruit of the oil palm tree, *Elaesis guineensis*, which has the appearance of a date palm with a large head of pinnate feathery fronds growing from a sturdy trunk. The fruit grows in bunches usually weighing in excess of 40 pounds, with 400 to 2000 individual fruits. Each individual fruit consists of the outer pulp, which is the source of the crude palm oil; an inner shell which is used for fuel; and two or three kernels that are the source of another oil type—palm kernel.

The quality of the palm oil achieved at the mill and subsequent processing is dependent upon the fruit bunches delivered to the oil mill. Overripe fruit bruises easily, accelerating free fatty acid rise through enzymatic hydrolysis and adversely affecting bleachability of the extracted oil. After receipt at the oil mill, extraction of the oil from the fruit is accomplished in four stages [46]:

(1) Sterilization—The fruit bunches are subjected to live steam under pressure to deactivate the enzymes responsible for hydrolysis and to loosen the fruit from the bunches. Improper deaeration or oversterilization has been found to create bleaching problems.

(2) Digestion—The palm fruit is mashed at high temperatures to break up the cells to allow release of the oil particles.

(3) Pressing—Screw presses with gradual compression and live steam injection are used to extract the oil from the fruit.

(4) Clarification—The oil settled out in a clarification tank is centrifuged to remove moisture and impurities followed by atmospheric or vacuum drying.

13.1 Palm Oil Composition and Physical Properties

Crude palm oil has a deep orange-red color contributed by a high carotene content: 0.03 to 0.15%, of which 90% consists of alpha- and beta-carotene. Caustic refining has little effect upon the color, but it can be heat bleached to low colors if the carotene has not been oxidized and fixed; abused crude palm oil may develop a brown color that is very difficult to remove. Palm oil, carefully processed to retain the carotene content, as much as possible, has been utilized as a natural colorant for margarine and shortening products.

Crude palm oil has a characteristic "nutty" or "fruity" flavor that can be removed easily with refining. Processed palm oil generally develops a slight, distinctive violet-like odor with oxidation. The oxidative stability of palm oil is affected by the presence of high levels of beta-carotene, which acts as a prooxidant even in the presence of high-tocopherol concentrations [48]. Therefore, it is better to inventory RB palm oil rather than crude. Typical characteristics for crude palm oil are [47,48]

	Typical	Range
Specific gravity, 5°C/25°C		0.8919 to 0.8932
Refractive index at 40°C		1.4565 to 1.4585
Iodine value		46.0 to 56.0
Saponification value		196 to 202
Unsaponifiable matter, %		0.2 to 0.5
Titer, °C		43.0 to 47.0
Melting range, °C		36.0 to 45.0
Carotene content, mg/kg		500 to 1600
Solids fat index, %		
at 10.0°C	34.5	30.0 to 39.0
at 21.1°C	14.0	11.5 to 17.0
at 26.7°C	11.0	8.0 to 14.0
at 33.3°C	7.4	4.0 to 11.0
at 37.8°C	5.6	2.5 to 9.0
at 40.0°C	4.7	2.0 to 7.0
Mettler melting point, °C	37.5	35.5 to 39.5

	Typical	Range
Fatty acid composition, %		
Lauric C-12:0	0.1	
Myristic C-14:0	1.0	
Palmitic C-16:0	44.3	
Palmitoleic C-16:1	0.15	
Stearic C-18:0	4.6	
Oleic C-18:1	38.7	
Linoleic C-18:2	10.5	
Linolenic C-18:3	0.3	
Arachidic C-20:0	0.3	
Triglyceride composition, %		
Trisaturate, S_3	7.9	
Monounsaturate, SUS	42.8	
Monounsaturate, SSU	6.6	
Diunsaturate, SU_2	35.7	
Triunsaturate, U_3	6.8	

The structural or physical properties of palm oil are similar to those of tallow and lard; however, based on the crystal habit characteristics, its similarity is limited to tallow. It could be a candidate for replacement of modified or interesterified lard, which changes from a beta to a beta-prime crystal habit with randomization. The physical characteristics for unhardened palm oil are similar to hydrogenated basestocks prepared from various vegetable oils but with mutual crystal habit characteristics limited to hydrogenated cottonseed oil.

Palm oil is distinguished from other oils by a very high level of palmitic fatty acid. This compares with a typical level of 21.6% for cottonseed oil, which is the vegetable oil with the next highest palmitic fatty acid level available in the United States. Hydrogenated hardstocks from both of these source oils, cottonseed and palm, are utilized to produce shortenings with beta-prime crystal habits.

Palm oil appears to have a built-in crystal modifier to effect a slow crystallization behavior confirmed by X-ray studies that showed an unusually long alpha- to beta-prime transition time [49]. The reasons for the long alpha lifetime is induced by the nontriglycerides, the most probable being diglycerides, and the presence of a high percentage of the symmetrical monounsaturated triacyl group or saturate-unsaturate-saturate (SUS) with palmitic-oleic-palmitic (POP) predominating [50]. Hydrogenated palm oil products do not have the problem of slow crystallization. Low iodine value hydrogenated palm oil is an excellent shortening hardstock, which helps extend the plastic range and tolerance to high temperatures.

Malaysia, the leading producer of palm oil is also the major processor and exporter of palm oil products. The palm oil processing industry in

Malaysia offers regular palm oil processed to different points, as well as the olein and stearin fractions of palm oil. Two processing techniques are utilized for palm oil exports, i.e., physical refining and neutralization. The four processing endpoints for all three palm oil products, regular, olein, and stearine, are [51]

- neutralized—caustic refined only (NPO)
- neutralized and bleached—caustic refined and bleached (NB PO)
- neutralized, bleached, and deodorized—caustic refined, bleached and deodorized (NBD PO)
- refined, bleached, and deodorized—physically refined (RBD PO)

Typical physical characteristics for the palm oil olein and stearin fractions are compared to regular palm oil in Table 1.3. The palmitic fatty acid tends to migrate to the stearin fraction. However, the olein fatty acid composition is relatively similar to regular palm oil in spite of fractionation.

The normal arbitrary criterion for palm olein is that cloud point results should be below 10°C (50°F). Cloud point refers to the temperature at which the oil turns cloudy while cooling at a rate of 1°C per minute. Most of the olein products have iodine value results within a relatively narrow

TABLE 1.3. Fractionated Palm Oil Characteristics [52].

	Palm Oil Fraction		
Characteristic	Whole	Olein	Stearin
Softening point, °C	3.0–38.0	19.0–24.0	44.0–56.0
Titer, °C	42.0–46.0		46.0–54.0
Density at 50/25°C	0.892–0.893	0.909–0.903	
Density at 60/25°C			0.882–0.891
Iodine value	51.0–55.0	51.0–61.0	22.0–49.9
Saponification value	190–202	194–202	193–206
Cloud point, °C		6.0–12.0	
Unsaponifiable matter, %			0.1–1.0
Solid fat content (NMR), %			
at 10°C	47–56	28–52	54–91
at 20°C	20–27	3–9	31–87
at 30°C	6–11	0	16–74
at 40°C	1–6		7–57
at 50°C			0–40
Fatty acid composition, %			
Myristic C-14:0	1–1.5	1–1.5	1–2
Palmitic C-16:0	42–47	38–42	47–74
Stearic C-18:0	4–5	4–5	4–6
Oleic C-18:1	37–41	40–44	16–37
Linoleic C-18:2	9–11	10–13	3–10

range, but the stearin fractions fall into a wide range. The three distinct grades of palm stearin fall into a wide range directly related to the fractionation process utilized:

	Fractionation Process		
	Detergent	Slow Dry	Fast Dry
Softening point, °C	53 to 56	50 to 51	46 to 49

Palm oil contains about 94% triglycerides with the remainder made up of minor components. Most of the minor components remain with the olein fraction, i.e., fatty acids, diglycerides, carotene, sterols, tocopherols, peroxides, and oxidized products. The phospholipids and metals like iron migrate predominately to the stearin fraction; the phosphorus concentration in the stearin may show a threefold increase from the original level in the crude palm oil [53]. Therefore, the stability of palm oil stearin is usually lower, especially with physical refining, because of the low tocopherol content and the iron concentration.

14. OLIVE OIL

The olive tree, *Olea europea,* which can survive several hundred years, thrives in a subtropical climate. It probably originates from Mesopotamia and has been cultivated for many centuries in the southern European countries bordering the Mediterranean and in North Africa. The olive tree region stretches from the Mediterranean across Asia Minor, Southern Russia, and Iran to India [54].

The olive oil production process is relatively complicated. Olive fruits are brought directly to the extraction plant after collection. Being a fruit, it cannot be stored as if it were a seed, nor can it be dried to preserve it. Olive oil has to be extracted as soon as possible before the acidity increases to impair quality. The olive is ground to a fine paste after separating the foreign material and washing. Two general procedures are utilized for oil extraction from the paste, i.e., hydraulic presses or continuous centrifuges. Three fractions are separated from the olive paste: (1) oil, (2) wastewater, and (3) husks or residue. The husks are dried and the remaining oil extracted with solvent. Therefore, more than one olive oil type is obtained [55]:

(1) Virgin olive oil—The oil is obtained by pressing or centrifuging with a free fatty acid level below 4.0% and organoleptic evaluation results deemed characteristics.

(2) Refined olive oil—The solvent extracted oil from the husks and the original pressing oil with a high free fatty acid or unacceptable flavor results is neutralized to reduce the free fatty acid content and blended with good quality virgin oil.

The Codex Alimentarius Recommended International Standard for Olive Oil, CAC/RS 33-1970 includes the following definitions for olive oil [56]:

(1) Olive oil is the oil obtained from the fruit of the olive tree (*Olea europaea* L.) without any manipulation or treatment not specified in definitions (2) and (3).

(2) Virgin olive oil is the oil from the fruit of the olive tree obtained by mechanical or other physical means under conditions, particularly thermal, that do not lead to alteration of the oil. Virgin olive oil is suitable for consumption in the natural state.

(3) Refined olive oil is the oil obtained from virgin olive oil with high acid content and/or undesirable organoleptic characteristics by refining methods that do not lead to alterations in the initial glyceride structure.

(4) Refined olive-residue oil is the oil obtained from olive residues (pomace and pits remaining after extraction of the oil from olives by physical means) by solvent extraction and made edible by refining methods that do not lead to alteration in the initial glycerin structure.

The Codex Recommended International Standard also applies to blends of olive oils and notes that refined olive oil and refined olive residue may be marketed alone or blended with virgin olive oil. The Codex standard does not mention esterified oil, which is banned for sale as edible oil in Italy and other Mediterranean countries and probably not considered by Codex as an olive oil product. The U.S. FDA definitions for olive oil products essentially agree with the Codex definitions but have added requirements for identification of the various fractions on the label; i.e., (1) a blend of virgin and refined fractions can be labeled as such or simply olive oil, (2) refined olive residue oil should be identified as such, and (3) refined olive-residue oil must be clearly labeled [57].

14.1 Olive Oil Composition and Physical Properties

Olive oil usually has a greenish yellow color and a characteristic olive flavor and odor. Good grades of olive oil are consumed without the usual edible oil processing after extraction. The high levels of free fatty acid in olive oil may be attributed to the high moisture content in olive fruits, which is favorable to enzyme action. Another cause for high free fatty acid content may be the bruising of the olive fruits at harvest. Typical characteristics for olive oil are [23,58]

		Typical	Range
Specific gravity at 25/25°C			0.909 to 0.915
Refractive index at 20°C			1.4680 to 1.4705
Saponification value			188 to 196
Unsaponifiable matter, %			1.8 max.
Iodine value			80.0 to 88.0
Titer, °C			17.0 to 26.0
Melting point, °C		0	
Cloud point, °C		−5.6	
Pour point, °C		−10.0	
Fatty acid composition, %			
Myristic	C-14:0		0.1 to 1.2
Palmitic	C-16:0	9.0	7.0 to 16.0
Palmitoleic	C-16:1	0.6	—
Stearic	C-18:0	2.7	1.0 to 3.0
Oleic	C-18:1	80.3	65.0 to 85.0
Linoleic	C-18:2	6.3	4.0 to 15.0
Linolenic	C-18:3	0.7	—
Arachidic	C-20:0	0.4	0.1 to 0.3
Triglyceride composition, mole %			
Monosaturated-diolein (GSU_2)		57.2	
Monosaturated-olein-linolein (GSU_2)		4.2	
Linoleo-diolein (GU_3)		34.0	
Triolein (GU_3)		4.6	

The proportions of oleic and linoleic unsaturated fatty acids and palmitic saturated fatty acids vary, depending upon the origin, as indicated by the fatty acid range. Like most other vegetable oils, olive oil develops more unsaturates in cold climates and with advanced maturity of the fruit.

Olive oil is widely used as a table and cooking oil due to its characteristic flavor and stability. Only 10 to 18% of the olive oil fatty acids are saturated, but the majority of the unsaturates are monounsaturates. The high oleic and low linoleic fatty acid content help make olive oil more resistant to oxidation than most liquid oils. Additionally, virgin olive oil has low tocopherol levels, but the chlorophylls degrade into pheophytins for oxidative stability. The polyphenols are usually removed from other liquid oils during processing [58].

Light causes significant deterioration in olive oil quality in the presence of air. The oil will develop an off-flavor from oxidation and become colorless due to the loss of chlorophyll and carotene. Packaging studies have shown that glass provides better protection than polyethylene plastic containers. A storage life in excess of 2 years was demonstrated by olive oil evaluations of hermetically closed glass bottles stored in the dark [59].

15. COCONUT OIL

Coconut palms are traditionally found in coastal regions in Asia and the Pacific islands. The native habitat is unknown, and a popular theory is that coconuts were carried by sea currents and washed ashore where they germinated. Coconuts require a growing temperature range from 75 to 85°F and never less than 68°F with an evenly dispersed rainfall of 60 to 80 inches. The trees start to bear fruit after 5 to 6 years and can continue to do so for as much as 60 years. Coconuts ripen in 9 to 12 months and can be harvested year round, so labor can be spread evenly over the year. The Philippines is the most important producing and exporting country, followed by Indonesia, India, Sri Lanka, Malaysia, and Oceania [60].

Copra is the trade name for dried coconut meat or kernel. The first step in making copra is to remove the husk from mature nuts, usually performed at the orchard site soon after the fruit is harvested. Next, the nut is opened for drying to produce copra. Coconut oil is removed from copra by pressing, solvent extraction, and other patented procedures [61].

15.1 Coconut Oil Composition and Physical Properties

Coconut oil obtained from the nuts of the coconut palm, *Cocos nucifera*, is a commercially important oil in the lauric acid group. Lauric acid oils differ significantly from other fats and oils; they pass abruptly from a brittle solid to a liquid, within a narrow temperature range. Coconut oil is a hard brittle solid at ambient temperatures (70°F/21.1°C) but melts sharply and completely below body temperature.

The color of coconut oil crude varies from a light yellow to brownish yellow; National Institute of Oilseed Products (NIOP) specification limit is 15.0 maximum Lovibond red color. Normal processing techniques will produce deodorized oils with very pale yellow colors; NIOP specification has limits for cochin type coconut oil of 1.0 maximum Lovibond red color [62].

The odor and taste of coconut oil is largely due to the presence of small quantities of lactones—less than 150 ppm. Since coconut oil is low in unsaturated fatty acids, it has a high resistance to oxidation. However, coconut oil will hydrolyze two to ten times as rapidly as normal oils to produce a disagreeable soapy flavor. Coconut oil hydrolysis proceeds slowly in the presence of free moisture alone but rapidly where a enzyme lipase is present in the food product. Pastry doughs and cake batters are examples of such products. The typical physical characteristics for coconut oil are [63,64]

	Typical	Range
Specific gravity at 99/5.5°C		0.869 to 0.874
at 25/15.5°C		0.917 to 0.919
Refractive index at 40°C		1.448 to 1.450
Iodine value		7.5 to 10.5
Saponification value		250 to 264
Unsaponificable matter, %		0.5 max.
Titer, °C		20.0 to 24.0
Setting point, °C		21.8 to 23.0
Melting point, °C		23.0 to 26.0
Solids fat index, %		
at 10.0°C	54.5	48.5 to 60.5
at 21.1°C	26.6	19.0 to 34.0
at 26.7°C	0.3	0.0 to 1.3
Mettler dropping point, °C	26.3	24.5 to 28.0
Fatty acid composition, %		
Caprylic C-8:0	7.6	
Capric C-10:0	7.3	
Lauric C-12:0	48.2	
Myristic C-14:0	16.6	
Palmitic C-16:0	8.0	
Palmitoleic C-16:1	1.0	
Stearic C-18:0	3.8	
Oleic C-18:1	5.0	
Linoleic C-18:2	2.5	

A high percentage of the coconut oil used in the United States is sold as 76 or 92°F coconut oil. The numbers refer to the Wiley melting point of the products. The 76°F oil is unhydrogenated while the 92°F oil has been partially hydrogenated. Typical solid fat indices of natural and fully hydrogenated coconut are shown below [65]:

	Refined	Hydrogenated
Solids fat index		
at 10°C	59	62
at 21.1°C	29	38
at 26.7°C	0	10
at 33.3°C		2.5
at 37.8°C		0
Wiley melting point, °C	24.4	36.7

16. PALM KERNEL OIL

The fruit of the palm tree, *Elaeis guineenis,* is the source of two distinctively different oil types. The outer pulp contains palm oil and the nut in the fruit contains kernels that are the source of palm kernel oil. Palm oil and palm kernel oil differ considerably in their characteristics and properties even though they are derived from the same plant. The kernel oil is similar to coconut oil in that it is light in color, sharp melting, and high in lauric and myristic fatty acids with an excellent oxidative stability contributed by a low level of unsaturates.

Palm kernels are by-products from the palm oil mills. Kernels constitute about 45 to 48% of the palm nut. After sterilization, the palm nuts are separated from the fruit bunch. This process can be responsible for oxidation and discoloration, which affects bleachability of the extracted oil if not properly controlled. After separation, the nuts are dried, cracked, and the kernels separated from the shell. Kernel oil is extracted by mechanical pressure screw pressing, solvent extraction, or preprocessing followed by solvent extraction. Prior to extraction by any of the methods, the kernels are cleaned and flaked to rupture the oil cells. After extraction, any remaining fines, solids, or other impurities are removed with a filter press [66].

16.1 Palm Kernel Oil Composition and Physical Properties

Typical physical properties and characteristics for palm kernel oil are [23,67,68]

Specific gravity at 99/15.5°C	0.860 to 0.873
at 40/20°C	0.899 to 0.914
Refractive index at 40°C	1.448 to 1.452
Iodine value	13.0 to 23.0
Saponification value	230 to 254
Unsaponificable matter, %	0.8 max.
Melting point, °C	24.0 to 26.0
Softening point, °C	26.0 to 28.0
Titer, °C	20.0 to 29.0
Setting point, °C	20.0 to 26.0
Solids fat index (Typical), %	
at 10.0°C	48.0
at 21.1°C	31.0
at 26.7°C	11.0
at 33.3°C	0

		Typical	Range
Fatty acid composition, %			
Caproic	C-6:0	0.2	<0.5
Caprylic	C-8:0	3.3	2.4 to 6.2
Capric	C-10:0	3.4	2.6 to 7.0
Lauric	C-12:0	48.2	41.0 to 55.0
Myristic	C-14:0	16.2	14.0 to 20.0
Palmitic	C-16:0	8.4	6.5 to 11.0
Stearic	C-18:0	2.5	1.3 to 3.5
Oleic	C-18:1	15.3	10.0 to 23.0
Linoleic	C-18:2	2.3	0.7 to 5.4
Arachidic	C-20:0	0.1	—
Gadoleic	C-20:1	0.1	—

Palm kernel and coconut oils are somewhat alike in physical properties. The fatty acid compositions are quite similar; the amount of the principal fatty acid, lauric, is almost equalivent. Nevertheless, the slight differences in their properties have a definite affect. Palm kernel has a lower content of short-chain fatty acids and a slightly higher oleic fatty acid content. The higher monounsaturate level is reflected both in a higher iodine value and melting point.

Lauric fats are among the most stable oils and fats because low unsaturated fatty acid contents present less opportunities for oxidation. However, palm kernel can develop off-flavors characterized as astringent and coarse. The short-chain fatty acids develop unpleasant soapy flavors when split into free fatty acids. Human palates are very sensitive to low levels of free caproic and caprylic fatty acids. Lauric oils soapy flavors and odors can become serious problems with high-moisture foods.

Lauric oils are substantially different from other edible fats and oils. They are solid fats at room temperature but melt sharply and completely below body temperature and provide a pronounced cooling effect in the mouth. The high levels of relatively low molecular weight saturated fatty acids, which are better than 50% of palm kernel oils composition, are the reason for the distinctive melting properties. The sharp melt, low melting point, and the low unsaturates make palm kernel oil and coconut oil particularly suited as fats for low-moisture food products for applications such as confectionery fats, candy centers, cookie fillers, nut roasting, coffee whiteners, and spray oils [69,70].

17. LARD

Meat has been a human food for thousands of years, and the use of fat from land animals in cooking extends back to antiquity. Swine, sheep, and cattle brought to North America from Europe in the fifteenth and sixteenth century were the forebears of our domesticated meat animals. The fatty tissue from meat animals, which is not a part of the carcass, or that which is trimmed off from it in preparing the carcass for market is the raw material from which lard and tallow are rendered.

Rendering is the process to separate fat, water, and protein material from the fatty or adipose tissues of the animal slaughtered for meat. In rendering fatty tissue for edible fats, it is desirable to obtain the maximum yield together with a light colored fat, bland flavor, and a low free fatty acid. The odor of lard is important because much of it is sold "as is" without further processing other than filtration. To preserve flavor and odor, the animal fat must be rendered as soon as possible before free fatty acid development accelerates. Splitting of the triglyceride to fatty acids and glycerin starts the instant the animal is slaughtered and proceeds even at low temperatures. A lipalytic enzyme present in the tissue is responsible for the reaction [71].

Open kettle rendering, which was more than likely the first commercial rendering process, was essentially cooking of the tissues and evaporating the moisture. It was just an enlargement of the open-fired cast iron kettle method used on farms. Most edible rendering plants now are continuous processes with product and quality controls. Maximum fat quality is maintained by a minimum exposure of the fat to heat, usually no more than 200°F (93.3°C) for a period of 30 minutes or less. It is only necessary to melt the fat in the raw material and to raise the temperature to a level sufficient to achieve separation [72]. Inadequate rendering controls can result in quality problems:

- dark color—Heating or cooking too long or at too high temperatures may scorch proteins and discolor the lard.
- flavor—Steam-rendered lards may have a characteristic strong odor and flavor of boiled pork.
- sour—Excessive protein and moisture in the rendered lard can cause souring during storage and/or transit.
- darkening—Colloidal protein material will darken with high-temperature processing, i.e., deodorization.

For many generations lard was the fat of choice for preparing doughs and batters since it has sufficient plasticity at room temperature to cream with sugar and mix with egg or egg yolk. However, periodic supply shortages promoted the development of all-vegetable substitutes. Eventually, the development of these products exceed lard's performance capabilities for creaming and aeration in bakery products.

17.1 Lard Physical Properties and Composition

The composition and physical characteristics of the lard have wide variations related to diet, climate and the overall structure of the animal. Hogs have the ability to assimilate ingested fat with little change in composition. Consequently, the degree of unsaturation of lard depends upon the amount and fatty acid composition of the oils in the feed. Typical physical properties and composition variations in lard are [23,73]

		Typical	Range
Specific gravity at 99/15.5°C			0.858 to 0.864
Refractive number at 40°C (Zeiss)			45 to 52
Iodine value			46.0 to 70.0
Saponification value			195 to 202
Unsaponifiable matter, %			1.0 max.
Titer, °C			36.0 to 42.0
Solids fat index, %			
at 10.0°C		29.0	26.5 to 31.5
at 21.1°C		21.6	19.5 to 23.5
at 26.7°C		15.3	13.0 to 17.5
at 33.3°C		4.5	2.5 to 6.5
at 37.8°C		2.8	2.0 to 4.0
at 40.0°C		2.2	1.5 to 3.0
Mettler dropping point, °C		32.5	31.5 to 33.0
Fatty acid composition, %			
Capric	C-10:0	0.1	
Lauric	C-12:0	0.1	
Myristic	C-14:0	1.5	1.0 to 4.0
Pentadecanoic	C-15:0	0.1	
Palmitic	C-16:0	26.0	20.0 to 28.0
Palmitoleic	C-16:1	3.3	
Margaric	C-17:0	0.4	
Margaroleic	C-17:1	0.2	
Stearic	C-18:0	13.5	5.0 to 14.0
Oleic	C-18:1	43.9	41.0 to 51.0
Linoleic	C-18:2	9.5	2.0 to 15.0
Linolenic	C-18:3	0.4	trace to 0.1
Arachidic	C-20:0	0.2	0.3 to 1.0
Gadoleic	C-20:1	0.7	
Eicosadienoic	C-20:2	0.1	
Triglyceride composition, mole %			
Trisaturated (GS_3)			2.0 to 5.0
Disaturated (GS_2U)			25.0 to 35.0
Monosaturated (GSU_2)			50.0 to 60.0
Triunsaturated (GU_3)			10.0 to 30.0

Even though lard is low in polyunsaturated fatty acids, it has an oxidative stability no better than vegetable oils with high levels of polyunsaturates.

Lard is rancid at a peroxide value of 20 meq/kg as opposed to 70 to 100 meq/kg peroxide value results for most vegetable oils. This is most likely due to the absence of natural antioxidants. Lard responds well to the addition of antioxidants, BHA, BHT, TBHQ, and various tocopherols, with metal chelators such as citric acid.

Lard's structure contains a high percentage of medium melting disaturated monounsaturated triglycerides. These triglycerides are largely in a symmetrical arrangement, which causes lard to crystallize in the beta form. This characteristic has restricted lard to applications requiring low structural properties but high lubricity. The function of lubricity is to impart tenderness and richness while improving eating qualities by providing a feeling of satiety after eating. Major applications for lard due to the lubricity functionality are pie crusts, frying, and yeast-raised doughs.

Ordinary lard is characterized by a translucence and a poor plastic range. Plasticizing with a scraped wall heat exchanger does not improve the consistency but further introduces an unattractive rubberiness. Addition of hardstock overcomes these difficulties and produces an attractive fresh shortening of excellent performance, but with short duration. With age, especially at 80°F, graininess develops from the unique triglyceride composition of lard. Lard has strong beta tendencies that dominate the crystal form of most blends. Consequently, lard is not the basestock of choice to produce a product with a wide plastic range and a smooth consistency characteristic of the beta-prime crystal habit.

Crystallization tendencies for lard can be modified by interesterification with a catalyst, usually sodium methoxide. Lard contains high proportions of palmitic fatty acid in the 2-position of its disaturated (S_2U) triglycerides. The proportion of C-16:0 fatty acid in the 2-position is decreased by about 64 to 24% on randomization along with other changes for lard to crystallize in the beta-prime phase instead of beta [73]. Plasticized product made from randomized lard has a better appearance because it has more and finer crystals, keeps its appearance better during storage, has better creaming and cake making qualities, and other properties superior to those of a corresponding shortening made with natural lard.

18. TALLOW

Tallow is the hard fat of ruminants. In the United States most tallow is obtained from beef cattle and a lesser amount from sheep. All meat fats are by-products of the meat packing industry, which means that availability is related to meat production, rather than a need for fats as a raw material.

The use of meat fats for edible products has diminished due to medical study recommendations of several diet and food modifications, including reductions in saturated fats and cholesterol.

Meat fats are essentially ready for use as is, except for possibly deodorization as received from rendering after a clarification procedure. The two main impurities in meat fats are proteins remaining from the rendering process and free fatty acids. It is necessary to remove the proteinaceous materials before deodorization; with steam distillation, the proteins turn black and require bleaching for removal. Two processes have been employed to remove the proteinaceous materials from meat fats:

- filtration—Diatomaceous earth and/or bleaching earth is added to the meat fat, followed by filtration to remove traces of moisture and impurities.
- water wash—Meat fats can be water washed with about 10% water to remove the protein, instead of filtration. Water washing is not as widely practiced as filtration for clarification because it requires extra centrifuge capacity while most meat fats require bleaching even after water washing.

Caustic refined or neutralized tallow was the frying fat of choice for french fry potatoes by many fast-food restaurants for many years. To produce this product, good quality fresh tallow was caustic refined, double water washed, and vacuum dried. Neutralized tallow was not bleached, had a 0.05% free fatty acid limit, and a characteristic fresh tallow flavor. The sole purpose of the caustic refining was to neutralize the fatty acids. Bleaching and deodorization of this product was avoided to retain the fresh tallow flavor. This end product use has essentially been eliminated by cholesterol concerns [75].

18.1 Tallow Physical Properties and Composition

Tallow contains very little red and yellow pigments but can have a high chlorophyll content. The chlorophyll is easily removed from fresh tallow with activated bleaching earth to achieve a water white appearance. Typical physical characteristics and compositions for tallow are [23,76]

	Typical	Range
Specific gravity at 99/15.5°C		0.860 to 0.870
Refractive number (Zeiss) at 40°C		46.0 to 49.0
Iodine value		38.0 to 48.0
Saponification value		193 to 202
Unsaponifiable matter, %		0.8 max.

		Typical	Range
Titer, °C			40.0 to 46.0
Solids fat index, %			
at 10.0°C		32.6	28.5 to 36.5
at 21.1°C		22.1	18.0 to 26.0
at 26.7°C		19.6	16.5 to 29.0
at 33.3°C		13.8	11.5 to 16.0
at 37.8°C		8.6	7.0 to 10.5
at 40.0°C		6.1	4.5 to 8.0
Mettler dropping point, °C		42.6	41.0 to 44.0
Capillary melting point, °C		44.5	
Fatty acid composition, %			
Lauric	C-12:0	0.1	trace to 0.2
Myristic	C-14:0	3.2	2.0 to 8.0
Myristoleic	C-14:1	0.9	0.4 to 0.6
Pentadecanoic	C-15:0	0.5	
Palmitic	C-16:0	24.3	24.0 to 37.0
Palmitoleic	C-16:1	3.7	1.0 to 2.7
Margaric	C-17:0	1.5	
Margaroleic	C-17:1	0.8	
Stearic	C-18:0	18.6	14.0 to 29.0
Oleic	C-18:1	42.6	40.0 to 50.0
Linoleic	C-18:2	2.6	1.0 to 5.0
Linolenic	C-18:3	0.7	
Arachidic	C-20:0	0.2	trace to 1.2
Gadoleic	C-20:1	0.3	
Triglyceride composition, mole %			
Trisaturated (GS_3)			15.0 to 28.0
Disaturated (GS_2U)			46.0 to 52.0
Monosaturated (GSU_2)			0 to 64.0
Triunsaturated (GU_3)			0 to 2.0

Tallow contains substantially high levels of saturated fatty acids and is solid at room temperature. The solid triglycerides in tallow provide plasticity, which is ideal for functionality in bakery systems. In many cases, vegetable oils are hydrogenated to achieve the same degree of saturation for functionality that naturally occurs in tallow.

The geometric arrangement of the tallow triglycerides is highly assymmetric, which crystallizes in the beta-prime form. The tallow beta-prime crystalline structure provides a good matrix for small air bubble entrapment and retention. The solids content and consistency provide ideal plasticity for rapid incorporation into mixes. Therefore, tallow functionality centers around applications requiring lubricity and structure like cakes, icings, and pastries. The structural properties of tallow can be used advantageously in product blends for margarine, puff pastry, danish, and other roll-in applications. These blends are intended to provide solids fat index profiles that achieve the roll-in properties, flakiness, and expansion desired. Tallow can provide the plastic consistency necessary to achieve workability over the wide temperature range required [77].

Tallow generally has an undesirable flavor and odor for most products and develops a reverted flavor shortly after deodorization. It contains low levels of linolenic and linoleic fatty acids that may contribute to flavor reversion unless protected with an antioxidant. The effective antioxidant systems for lard also apply to tallow, i.e., BHA and BHT or TBHQ with a metal chelator like citric acid.

19. MILK FAT

Since prehistoric times, man has satisfied his need for fat by eating foods from a variety of animal and vegetable sources. Among these foods, milk from animals has been a prime source of nutrients. For example, the average composition of cow's milk in the United States is 3.7% fat, 4.9% lactose, 3.5% protein, 0.7% minerals, and 87.2% water. Traditionally, milk fat has always had the highest economic value of any of the milk constituents, placing it at an economic disadvantage to other edible fats and oils products. The milk fat products of interest to the edible oil processor are

- butter—Legally, butter has been defined as a food product made exclusively from milk or cream, or both, with or without common salt, and with or without additional coloring matter, and contains not less than 80% by weight of milk fat. The nonfat portion of butter is composed of approximately 16% water, 2.5% salt, and 1.5% milk solids. The soluble components in the fat or triglyceride portion are sterols, pigments, fat-soluble vitamins, and phosphatides or lecithin at approximately 0.2%.
- anhydrous butterfat—milk fat produced by removing the fat from milk or cream
- butteroil—milk fat produced by removing the water from butter

Milk fat contains more fatty acids than any other fat of animal or vegetable origin. A number of factors cause variation in the proportions of the different fatty acids found in milk fat. The melting point of the fat can vary depending upon the change in the proportions of the various fatty acids. The factors causing most of the variations in composition are [78]

- feed—Feed influences the milk fat composition the most of all the factors. A change from winter feeding conditions to pasture can increase the proportion of unsaturated fatty acids and decrease the saturated fatty acids; oleic (C-18:1) is increased at the expense of butyric

(C-4:0) and stearic (C-18:0). This change decreases the melting point. Other changes in the fatty acid composition have been effected by formulating the feed with high quantities of different fatty acids. Feed also contributes to the color of milk fat. The reddish yellow plant pigment, carotene, found in green plants and carrots is the principal substance responsible for color.

- nutrition—Underfeeding of cows markedly affects the physical and chemical constants of milk fat with a decline in the volatile fatty acids, butyric (C-4:0) and caproic (C-6:0), and an increase in oleic (C-18:1) fatty acid.
- lactation stage—Butyric (C-4:0) fatty acid gradually declines from the beginning to the end of the period, with an increase in oleic (C-18:1) fatty acid.
- breed—Milk fat from Jerseys has a higher volatile fatty acid content than milk from Holstein, Ayrshires, or Shorthorns. Also, Guerneys and Jerseys convert less carotene to colorless vitamins than Holsteins, to produce a darker colored milk fat.

19.1 Milk Fat Physical Properties and Composition

Milk fat is distinguished from other fats, except those with high lauric fatty acids, by the low average molecular weight of its fatty acids indicated by a high saponification value and a low refractive index. Butterfat differs from the lauric oils by the high content of steam-volatile fatty acids, i.e., butyric through capric. Ruminant milk fat contains relatively low concentrations of polyunsaturated fatty acids as a result of biohydrogenation of dairy lipids in the rumen. The physical properties and composition of milk fat are [23,79]

	Typical	Range
Specific gravity at 40/15°C		0.907 to 0.912
Refractive index at 60°C	1.4465	
Iodine value	31.0	25.0 to 42.0
Saponification value		210 to 250
Unsaponifiable matter, %		0.4 max.
Mettler melting point, °C	35.0	28.0 to 36.0
Titer, °C	34.0	
Solids fat index, %		
at 10.0°C	33.0	
at 21.1°C	14.0	
at 26.7°C	10.0	
at 33.3°C	3.0	
at 40.0°C	0.0	

		Typical	Range
Fatty acid composition, %			
Butyric	C-4:0	3.6	2.8 to 4.0
Caproic	C-6:0	2.2	1.4 to 3.0
Caprylic	C-8:0	1.2	0.5 to 1.7
Capric	C-10:0	2.5	1.7 to 3.2
Decenoic	C-10:1		0.1 to 0.3
Lauric	C-12:0	2.9	2.2 to 4.5
Dodecenoic	C-12:1		0.1 to 0.6
Myristic	C-14:0	10.8	5.4 to 14.6
Myristoleic	C-14:1	0.8	0.6 to 1.6
Pentadecanoic	C-15:0	2.1	
Palmitic	C-16:0	26.9	26.0 to 41.0
Palmitoleic	C-16:1	2.0	2.8 to 5.7
Margaric	C-17:0	0.7	
Stearic	C-18:0	12.1	6.1 to 11.2
Oleic	C-18:1	28.5	18.7 to 33.4
Linoleic	C-18:2	3.2	0.9 to 3.7
Linolenic	C-18:3	0.4	
Gadoleic	C-20:1/C-22:1	0.1	0.8 to 3.0

Actual and potential flavors are among the most important attributes of milk fat. In dairy products like milk, cream, ice cream, and tablespread butter, the aim is to retain the mild delicate flavor associated with the fat of fresh milk. However, in cooking, baking, and many processed food applications, the object is to generate buttery, creamy, cheesy, and caramel-like flavor qualities from milk fat.

Food flavor is a complex of at least three factors, i.e., aroma, taste, and texture. The fatty acid composition of milk fat provides the melting and solidification or textural characteristics as previously reviewed. The aroma and taste of milk fat are diverse. The isolation, separation, and identification of the flavor compounds in butter fat has developed into a highly specialized field. Although present in butterfat in relatively small quantities, aliphatic lactones are responsible for part of the desirable and characteristic flavor of butter. Stored butter experiences a gradual, but regular, increase in total lactone content even under refrigerated storage conditions for stronger flavors. Therefore, the volatile fatty acids, C-4 through C-10, and the lactones generally are considered to be the substances that comprise the pleasant, nonoxidative flavor and odor characteristics of milk fat [78].

Since butter is more expensive than most other fats, its use is restricted to those products where the distinctive flavor makes a significant contribution to the acceptability or in which its use has advertising or marketing value. As a raw material, milk fat opportunities are limited for edible fats and oils processors since most of the processes would destroy or materially change the flavor attributes. The opportunities that have been utilized in the United States are in final blends with margarine and in some cases with other products, primarily as a flavorant.

A process utilized in Europe and Japan can effect consistency changes in milk fat without damaging the flavor. Unlike most conventional techniques to modify a fat's consistency, dry fractionation does not destroy the milk fat flavor or alter the triglyceride. The flavor and beta-prime polymorphic crystal form are maintained in the stearin and olein fractions from this process. The milk fat stearin fraction has application as a roll-in for puff pastry, danish pastry, croissants, and other baked products, as well as coating products. The olein fractions have been used in cookies, dairy products, and blended with other edible oil products to prepare a soft-tub-type tablespread [80].

20. REFERENCES

1. Lanstraat, A. 1976. "Characteristics and Composition of Vegetable Oil-bearing Materials," *J. Am. Oil Chem. Soc.*, 53(6):241–243.
2. Sonntag, N. O. V. 1979. "Sources, Utilization, and Classification of Oils and Fats," in *Bailey's Industrial Oil and Fat Products, Vol. 1*, 4th Edition, D. Swern, ed. New York, NY: A Wiley-Interscience Publication. p. 273.
3. *Fats and Oils Situation,* April, 1971. U.S. Department of Agriculture, p. 25.
4. *Fats and Oils Situation,* July, 1977. U.S. Department of Agriculture, p. 14.
5. *Oil Crops Outlook and Situation Yearbook,* August, 1985. U.S. Department of Agriculture, p. 8.
6. *Oil Crops Yearbook.* October, 1996. U.S. Department of Agriculture. pp. 19 and 54.
7. Sonntag, N. O. V. 1979. "Structure and Composition of Fats and Oils," in *Bailey's Industrial Oil and Fat Products, Vol. 1*, 4th Edition, D. Swern, ed. New York, NY: A Wiley-Interscience Publication. pp. 3–4.
8. Wolf, W. J. and J. C. Cowan. 1971. "Introduction," in *Soybeans as a Food Source*, Cleveland, OH: CRC Press. pp. 9–13.
9. Smith, A. K. and S. J. Circle. 1972. "Historical Background," in *Soybean's Chemistry and Technology, Vol. 1*. Westport, CT: The AVI Publishing Company, Inc. pp. 4–6.
10. Dutton, H. J. 1981. "History of the Development of Soy Oil for Edible Uses," *J. Am. Oil Chem. Soc.*, 58(3): 234–236.
11. Erickson, D. R. 1983. "Soybean Oil: Update on Number One," *J. Am. Oil Chem. Soc.*, 60(2):356.
12. O'Brien, R. D. 1987. "Formulation—Single Feedstock Situation," in *Hydrogenation: Proceedings of an AOCS Colloquium*, R. Hastert, ed. Champaign, IL: American Oil Chemists' Society. p. 153.
13. Sonntag, N. O. V. 1979. "Composition and Characteristics of Individual Fats and Oil," in *Bailey's Industrial Oils and Fat Products, Vol. 1*, 4th Edition, D. Swern, ed. New York, NY: A Wiley-Interscience Publication. pp. 352–362.

14. Weiss, T. J. 1983. "Commercial Oil Sources," in *Food Oils and Their Uses*, Second Edition. Westport, CT: The AVI Publishing Company, Inc. p. 36.

15. Jones, L. A. and C. C. King. 1990. *Cottonseed Oil*. Memphis, TN: National Cottonseed Products Association, Inc. and The Cotton Foundation. pp. 5–7.

16. Sonntag, N. O. V. 1979. "Composition and Characteristics of Individual Fats and Oil," in *Bailey's Industrial Oils and Fat Products, Vol. 1*, 4th Edition, D. Swern, ed. New York, NY: A Wiley-Interscience Publishing Company, Inc. pp. 352–362.

17. Weiss, T. J. 1983. "Commercial Oil Sources," in *Food Oils and Their Uses*, Second Edition. Westport, CT: The AVI Publishing Company, Inc. pp. 37–38.

18. Cheery, J. P. 1983. "Cottonseed Oil," *J. Am. Oil Chem. Soc.*, 60(2):361–362.

19. Woodroof, J. G. 1973. *Peanuts: Production, Processing, Products*. Westport, CT: The AVI Publishing Company, Inc., pp. 1–8 and 247–257.

20. Sonntag, N.O.V. 1979. "Composition and Characteristics of Individual Fats and Oils," in *Bailey's Industrial Oil and Fat Products, Vol. 1*, 4th Edition, D. Swern, ed. New York, NY: A Wiley-Interscience Publication, pp. 363–368.

21. Langstraat, A. 1976. "Characteristics and Composition of Vegetable Oil-bearing Materials," *J. Am. Oil Chem. Soc.*, 53(6):243–244.

22. Branscomb, L. and C. Young. 1972. *Peanut Oil*. Tifton, GA: Georgia Agricultural Commodity Commission for Peanuts.

23. Capital City Products Co. 1989. *Typical Fatty Acid Composition of Selected Edible Fats and Oils*, Capital City Products Company, Columbus, OH.

24. Sonntag, N. O. V. 1979. "Composition and Characteristics of Individual Fats and Oils," in *Bailey's Industrial Oil and Fat Products, Vol. 1*, 4th Edition, D. Swern, ed. New York, NY: A Wiley-Interscience Publication, pp. 389–398.

25. Leibovitz, Z. and C. Ruckenstein. 1983. "Our Experiences in Processing Maize (Corn) Germ Oil," in *J. Am. Oil Chem. Soc.*, 60(2):395–399.

26. Weiss, T. J. 1983. "Commercial Oil Sources," in *Food Oils and Their Uses*, Second Edition, Westport, CT: AVI Publishing Company, Inc. pp. 39–40.

27. Langstraat, A. 1976. "Characteristics and Composition of Vegetable Oil-bearing Materials," *J. Am. Oil Chem. Soc.*, 53(6):242–243.

28. Adams, J. et al. 1982. *Sunflower*, Bismark, ND: The National Sunflower Association. pp. 2–7 and 13.

29. Sonntag, N. O. V. 1979. "Composition and Characteristics of Individual Fats and Oils," in *Bailey's Industrial Oil and Fat Products, Vol. 1*, 4th Edition, D. Swern, ed. New York, NY: A Wiley-Interscience Publication, pp. 382–387.

30. Morrison, W. H. 1975. "Effects of Refining and Bleaching on Oxidation Stability of Sunflower Oil," *J. Am. Oil Chem. Soc.*, 52(12):522.

31. Purdy, R. H. 1985. "Oxidative Stability of High Oleic Sunflower and Safflower Oils," *J. Am. Oil Chem. Soc.*, 62(3):523–525.

32. *National Institute of Oilseed Products Trading Rules 1994-1995*, p. 97.

33. Haumann, B. F. 1994. "Modified Oil May Be Key to Sunflower's Future," *INFORM*, 5(11):1202.

34. Purdy, R. H. et al. 1959. "Safflower, Its Development and Utilization," *J. Am. Oil Chem. Soc.*, 36(9):26–30.

35. Kneeland, J. A. 1966. "The Status of Safflower," *J. Am. Oil Chem. Soc.*, 43(6):403–405.

36. Weiss, T. J. 1983. "Commercial Oil Sources," in *Food Oil and Their Uses,* Second Edition. Westport, CT: AVI Publishing Company, Inc. pp. 45–46.

37. Sonntag, N.O.V. 1979. "Composition and Characteristics of Individual Fats and Oils," in *Bailey's Industrial Oil and Fat Products, Vol. 1,* 4th Edition, D. Swern, ed. New York, NY: A Wiley-Interscience Publication, pp. 398–403.

38. Blum, J. E. 1966. "The Role of Safflower Oil in Edible Oil Applications," *J. Am. Oil Chem. Soc.,* 43(6):416–417.

39. Smith, J. R. 1985. "Safflower: Due for a Rebound?" *J. Am. Oil Chem. Soc.,* 62(9):1286–1291.

40. Knowles, P. F. 1972. "The Plant Geneticist's Contribution Toward Changing Lipid and Amino Acid Composition of Safflower," *J. Am. Oil Chem. Soc.,* 49(1):27–29.

41. *National Institute of Oilseed Products Trading Rules l994–l995,* pp. 95–96.

42. Fuller, G. et al. 1971. "Evaluation of Oleic Safflower in Frying of Potato Chips," *J. Food Science,* 36:43–44.

43. Fuller, G. et al. 1966. "High Oleic Acid Safflower Oil: A New Stable Edible Oil," *J. Am. Oil Chem. Soc.,* 43(7):477–478.

44. Vaisey-Genser, M. et al. 1987. *Canola Oil Properties and Performance,* D. F. G. Harris, ed. Winnipeg, Manitoba, Canada: Canola Council. pp. 1–3 and 12.

45. Mag, T. K. 1983. "Canola Oil Processing in Canada," *J. Am. Oil Chem. Soc.,* 60(2):380–384.

46. Eng, T. G. and M. M. Tat. 1985. "Quality Control in Fruit Processing," *J. Am. Oil Chem. Soc.,* 62(2):278–280.

47. Rossell, J. B. et al. 1985. "Composition of Oil," *J. Am. Oil Chem. Soc.,* 62(2):233.

48. Sonntag, N. O. V. 1979. "Composition and Characteristics of Individual Fats and Oils," in *Bailey's Industrial Oil and Fat Products, Vol. 1,* 4th Edition, D. Swern, ed. New York, NY: A Wiley-Interscience Publication, pp. 374–382.

49. Berger, K. G. 1975. "Uses of Palm Oil in the Food Industry," *Chemistry and Industry.* Nov. 1, 1975, pp. 910–913.

50. Duns, M. L. 1985. "Palm Oil in Margarines and Shortenings," *J. Am. Oil Chem. Soc.,* 62(2):409.

51. Kheiri, M. S. A. 1985. "Present and Prospective Development in the Palm Oil Processing Industry," *J. Am. Oil Chem. Soc.,* 62(2):210–211.

52. Pantzaris, T. P. V. 1988. *Pocketbook of Palm Oil Uses.* Malaysia: The Palm Oil Research Institute of Malaysia, pp. 9–14.

53. Deffense, E. 1985. "Fractionation of Palm Oil," *J. Am. Oil Chem. Soc.,* 62(2):376–385.

54. Langstraat, A. 1976. "Characteristics and Composition of Vegetable Oil-bearing Materials," *J. Am. Oil Chem. Soc.,* 53(6):242–244.

55. Fedeli, E. 1983. "Miscellaneous Exotic Oils," *J. Am. Oil Chem. Soc.,* 60(2):404–406.

56. Recommended International Standard for Olive Oil, Virgin and Refined, and for Refined Olive-Residue Oil, Codex Alimentarius Commission CAC/RS 33-1970. Food and Agriculture Organization, World Health Organization, Rome, Italy, 1970.

57. Firestone, D. et al. 1985. "Detection of Adulterated and Misbranded Olive Oil Products," *J. Am. Oil Chem. Soc.,* 62(2):1558–1561.

58. Sonntag, N. O. V. 1979. "Composition and Characteristics of Individual Fats and Oils," in *Bailey's Industrial Oil and Fat Products, Vol. 1,* 4th Edition, D. Swern, ed. New York, NY: A Wiley-Interscience Publication, pp. 368–374.

59. Kiritsakkis, A. K. 1984. "Effect of Selected Storage Conditions and Packaging Materials on Olive Oil Quality," *J. Am. Oil Chem. Soc.*, 61(12):1868–1870.

60. Langstraat, A. 1976. "Characteristics and Composition of Vegetable Oil-bearing Materials," *J. Am. Oil Chem. Soc.*, 53(6):242.

61. Woodroof, J. G. 1970. *Coconuts: Production, Processing, Products.* Westport, CT: The AVI Publishing Company, Inc., pp. 43–48.

62. *National Institute of Oilseeds Products Trading Rules 1994–1995.* pp. 85–86.

63. Weiss, T. J. 1983. "Commercial Oil Sources," in *Food Oils and Their Uses.* Westport, CT: AVI Publishing Company, Inc. pp. 49–51.

64. Sonntag, N. O. V. 1979. "Composition and Characteristics of Individual Fats and Oils," in *Bailey's Industrial Oil and Fat Products, Vol. 1,* 4th Edition, D. Swern, ed. New York, NY: A Wiley-Interscience Publication, pp. 311–317.

65. Young, F. V. K. 1983. "Palm Kernel and Coconut Oils: Analytical Characteristics, Process Technology and Uses," *J. Am. Oil Chem. Soc.*, 60(2):374–379.

66. Tang, T. S. and P. K. Teoh. 1985. "Palm Kernel Oil Extraction—The Malaysian Experience," *J. Am. Oil Chem. Soc.*, 62(2):254–258.

67. Rossell, J. B. et al. 1985. "Composition of Oil," *J. Am. Oil Chem. Soc.*, 62(2):226–227.

68. Sonntag, N. O. V. 1979. "Composition and Characteristics of Individual Fats and Oils," in *Bailey's Industrial Oil and Fat Products, Vol. 1,* 4th Edition, D. Swern, ed. New York, NY: A Wiley-Interscience Publication, p. 319.

69. Young, F. V. K. 1983. "Palm Kernel and Coconut Oils: Analytical Characteristics, Process Technology and Uses," *J. Am. Oil Chem. Soc.*, 60(2):374–378.

70. Traitler, H. and A Dieffenbacher. 1985. "Palm Oil and Palm Kernel Oil in Food Products," *J. Am. Oil Chem. Soc.*, 62(2):420.

71. Vibrans, F. C. 1949. "Rendering," *J. Am. Oil Chem. Soc.*, 26(10):575–580.

72. Prokop, W. H. 1985. "Rendering Systems for Processing Animal By-Product Materials," *J. Am. Oil Chem. Soc.*, 62(4):805–811.

73. Swern, D. 1964. "Plastic Shortening Agents," in *Bailey's Industrial Oil and Fat Products,* 3rd Edition. New York, NY: John Wiley & Sons, Inc., pp. 189 and 282–283.

74. Sonntag, N. O. V. 1982. "Fat Splitting, Esterification, and Interesterification," in *Bailey's Industrial Oil and Fat Products, Vol. 2,* 4th Edition, D. Swern, ed. New York, NY: A Wiley-Science Publication, p. 159.

75. Latondress, E. G. 1985. "Refining, Bleaching, and Hydrogenating Meat Fats," *J. Am. Oil Chem. Soc.*, 62(4):812–815.

76. Sonntag, N. O. V. 1979. "Composition and Characteristics of Individual Fats and Oils," in *Bailey's Industrial Oil and Fat Products, Vol. 1,* 4th Edition, D. Swern, ed. New York, NY: A Wiley-Science Publication, pp. 342–352.

77. Kincs, F. R. 1985. "Meat Fat Formulation," *J. Am. Oil Chem. Soc.*, 62(4):815–818.

78. Henderson, J. L. 1971. "Composition, Physical and Chemical Properties of Milk," in *The Fluid-Milk Industry,* 3rd Edition. Westport, CT: The AVI Publishing Company, Inc., pp. 29–34.

79. Sonntag, N. O. V. 1979. "Composition and Characteristics of Individual Fats and Oils," in *Bailey's Industrial Oil and Fat Products, Vol. 1,* 4th Edition, D. Swern, ed. New York, NY: A Wiley-Science Publication, pp. 292–311.

80. Deffense, E. 1987. "Multi-Step Butteroil Fractionation and Spreadable Butter," *Fat Science Technology,* pp. 502–507.

CHAPTER 2

Fats and Oils Processing

1. INTRODUCTION

Crude oils can vary from pleasant smelling products that contain few impurities to quite offensive smelling highly impure materials. Quite clearly, very few edible fats and oils can be used for edible purposes until they have been processed in some manner. Fortunately, researchers have developed processes for changing fats and oils to make them increasingly more useful to the food industry. Processing techniques allow us to refine them, make them flavorless and odorless, change the color, harden them, soften them, make them melt more slowly or rapidly, change the crystal habit, rearrange their molecular structure, and literally take them apart and put them back together again to suit our requirements of the moment. Advances in lipid processing technology during this century have resulted in dramatic increases in the consumption of edible fats and oils. Innovations such as deodorization, hydrogenation, fractionation, and interesterification, along with improvements in other processes, have allowed the production of products that can provide demanding functional and nutritional requirements.

Throughout the world, processing of vegetable oils practically always includes neutralization or refining, bleaching, and deodorization, but differences exist in the equipment and techniques used to produce the finished oil products. The choice of processing equipment and/or techniques can depend upon (1) the source oils handled, (2) the quality of the raw materials, (3) available manpower, (4) maintenance capabilities, (5) the daily processed oil requirements, (6) available financial resources, (7) proximity of the crude oils, (8) the product marketing philosophy, and (9) other con-

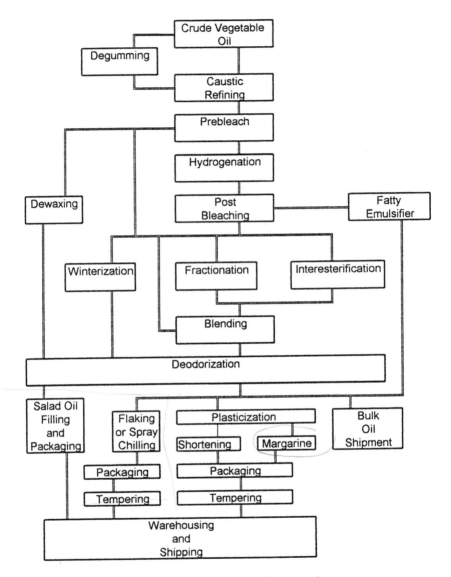

CHART 2.1. Typical fats and oils processing.

siderations. Two general product marketing philosophies practiced in the United States are "crusher/refiner" or "value added." The crusher/refiner plant usually concentrates on a limited number of source oils, which are most likely extracted from the oilseed at the same facility. These plants are usually equipped with continuous, automatic, high-volume systems. The value-added processor usually processes several different source oils into many different specialty or tailormade finished products. The value-added varied and specialized product mix mandates batch or semicontinuous systems with slower throughputs requiring more manpower.

Fats and oils processing consists of a series of unit processes in which both physical and chemical changes are made to the raw materials. Chart 2.1 diagrams the typical flow for vegetable oil processing in the United States. Changes in the food industry requirements have affected how these raw materials are processed to provide fats and oils products with the functionality required for new and improved prepared foods. Processing changes have also been motivated by needs of the industry to (1) improve product quality, (2) improve process efficiencies, (3) reduce capital expenses, and (4) solve or eliminate environmental problems. Consideration of the properties of the raw materials, the measurement methods, and the characteristics of each process or operation is required for process and product control.

2. CRUDE FATS AND OILS

Crude vegetable oils consist primarily of triglycerides or neutral oil with fat-soluble impurities either in true suspension or in colloidal suspension. The quantities of the impurities vary with the oil source, extraction process, season, and geographical source. Purification of the neutral oil portion is the major objective of the refining process. The nontriglyceride portion contains variable amounts of the impurities such as free fatty acid, gums that consist of phospholipids, the crude lecithins, metal complexes (notably iron, copper, calcium, and magnesium in soybean and other oils), peroxides and their breakdown products, color pigments, sterols, tocopherols, meal, waxes, moisture, and dirt. Most of these impurities are detrimental to the finished product's color and flavor stability, as well as their resistance to foaming and smoking, and must be separated from the neutral oil during processing [1]. The primary refining concern is with adequate removal of the free fatty acid and phosphatides.

During refining, the free fatty acid level will be neutralized to a low level, 0.05%. Free fatty acid occurs naturally in crude oils, and the level may be elevated during storage or handling. Some crude oils have much higher initial fatty acid contents due to enzymatic hydrolysis or abuse during harvesting transportation or storage. Free fatty acid content is a good measure of the crude oil quality, as well as that of the refined oil. The level of the free fatty acid in the crude oil will determine the required treatment to neutralize the free fatty acid, which will affect the refining loss.

Phosphatides occur to varying levels in animal and vegetable fats, the latter containing by far the greater amount of phosphatides. The phosphatide content of animal fats is very low. Phosphatides consist of polyhydric alcohols esterified with fatty acids and phosphoric acid, which is combined with a nitrogen-containing compound. Two common phosphatides occurring in vegetable oils are the lecithins and cephalins, which may be considered triglycerides with one fatty acid replaced with phosphoric acid. The position of the phosphoric acid radical is important. When it is attached to an outer carbon link with the glycerol molecule, it is termed an alpha-lipoid and is hydratable. Beta-lipoid compounds with the phosphoric acid radical found in the center position are not hydratable and cannot be removed with a degumming process. Soybean, corn, cottonseed, and canola are the major oils that contain significant quantities of phosphatides. The alkali treatment for free fatty acid neutralization utilized with caustic refining is capable of removing most of these impurities from crude oils [2].

Not all of the impurities in crude oils are undesirable. Tocopherols perform the important function of protecting the oil from oxidation. For this reason, tocopherols are highly desirable constituents of the vegetable oil products. Animal fats do not contain the natural antioxidant protection of tocopherols. Therefore, the objective of refining is to remove the harmful impurities with the least possible damage to the beneficial constituents and with a minimum loss of oil during the process [3]. Chemical refining can remove as much as 10 to 20% of the tocopherols in vegetable oils. Removal is presumably by absorption on the soap that is formed. Removal in steam refining or deodorization depends upon the severity of the process with respect to time, temperature, and stripping steam flow. Drastic conditions have shown a 30 to 60% tocopherol reduction in vegetable oils [4].

Sterols are minor components of all natural fats and oils and comprise most of the unsaponifiable matter, the remainder consisting essentially of hydrocarbons. The sterols are colorless, heat stable, and relatively inert so they do not contribute any important property to the oil. Alkali refining removes a portion of the sterols, but more effective removal requires fractional crystallization, molecular distillation, or high-temperature steam refining or deodorization. The most extensively investigated animal sterol is cholesterol, while the vegetable sterols are known collectively as phytosterols [5].

3. DEGUMMING

Processors have the option of approaching the refining step in two ways. The phosphatides can either be recovered for their by-product value through degumming or treated as impurities that must be removed from the crude oil. Most of the vegetable oils processed in the United States are refined without degumming. It has been estimated that about one-third of the processed soybean oil will satisfy the demand for lecithin [6] and crusher/refiners can add the separated gums back to the meal, but most stand-alone processors lack a market or profitable outlet for the gums and forgo degumming. The decision not to degum oil can be based both on energy conservation and capital savings; a separate combined degumming and refining can be carried out in one step with the primary centrifuge for caustic refining. However, vegetable oil degumming offers several potential advantages:

- It is necessary for lecithin production. The hydrated gums are the raw materials for lecithin processing.
- It satisfies export oil requirements for a product free of impurities that settle out during shipment.
- It reduces the chemical refining neutral oil loss. Gum removal prior to alkali refining often improves yield because the phosphates can act as emulsifiers in a caustic solution to increase the neutral oil entrained in the soapstock.
- It substantially reduces refinery waste load due to the lower neutral oil losses and the reduction of gums discharged.
- It prepares the oil for physical or steam refining. Degummed oil is more suited to physical refining techniques due to the significant reduction in nonvolatile impurities, such as phosphatides and metallic prooxidants.
- It results in improved acidulation performance. The soapstock from alkali refining is easier to acidulate due to a lower emulsifier content, and the acid water has less impact upon the wastewater treatment systems.

Degumming is the treatment of crude oils with water, salt solutions, or dilute acids such as phosphoric, citric, or maleic to remove phosphates, waxes, and other impurities. Degumming converts the phosphatides to hydrated gums, which are insoluble in oil and readily separated as a sludge by settling, filtering, or centrifugal action. The hydrated gums are vacuum dried for crude lecithin processing. The degumming process commonly practiced in the United States, batch water treatment followed by centrifugation, is diagrammed in Chart 2.2. Approximately 2% water,

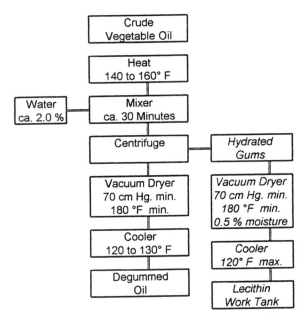

CHART 2.2. Batch water degumming process flow.

by oil volume, is contacted with the crude oil by mechanical agitation in a mix tank. The proper amount of water is normally about 75% of the oil's phosphatide content. Too little water produces dark viscous gums and a hazy oil, while too much water causes excess oil losses through hydrolysis. Complete hydration requires approximately 30 minutes of agitation at 140 to 160°F (60 to 71°C) for batch processing. Temperature is important because degumming is less complete at higher temperatures due to the increased solubility of the phosphatides in the oil, and at lower temperatures the increased oil viscosity makes separation of the phosphatides more difficult. For continuous systems, preheated oil (80°C or 176°F) is treated with water and mixed in a holding tank for approximately 15 minutes or 1 minute maximum with continuous pipeline agitators. The water-oil mixture must be treated very gently to avoid developing an emulsion; high shear stress in the feed pump and at the centrifuge inlet must be avoided. After hydration, centrifuges separate the sludge and degummed oil phases [2,7]. The degummed oil can be vacuum dried and pumped to degummed oil storage or can proceed directly to the refining process.

4. REFINING SYSTEMS

The refining process probably has more impact on the fats and oils quality and the economic performance than any of the other processes during conversion to finished products. Inadequately refined oils will affect the operation of all succeeding processes and the quality of the finished product produced. Additional processing and handling required because of poorly refined oils will increase the costs of a suspect quality finished product beyond that of one produced with a properly refined product that has good quality.

The two refining processes in current commercial use are chemical or caustic soda refining and physical refining. Some oils contain impurities that cannot be removed adequately by the pretreatment steps to enable them to be physically refined to the desired quality. One example of this situation is cottonseed oil, which contains the pigment gossypol, which is not removed with degumming or steam distillation, but can be satisfactorily refined with the caustic soda process. Other oils that contain high levels of nonhydratable phosphatides, trace metals, oxidation products, and red or green pigments have also resulted in unsatisfactory deodorized product with the physical refining process. The shortcomings experienced with physical refining are one reason that alkali refining is the preferred refining technique in the United States. In addition to conventional, other continuous caustic soda refining techniques are also utilized in the United States and other countries: miscella refining, short-mix caustic soda, batch caustic soda refining, and the Zenith process.

4.1 Conventional Caustic Soda Refining

The conventional caustic soda process is the most widely used and best known refining system. The addition of an alkali solution to a crude oil brings about a number of chemical and physical reactions. The alkali combines with the free fatty acid present to form soaps; the phosphatides and gums absorb alkali and are coagulated through hydration or degradation; much of the coloring is degraded, absorbed by the gums, or made water-soluble by the alkali, and the insoluble matter is entrained with the other coagulable material. With heat and time, the excess caustic can also bring about the saponification of a portion of the neutral oil. Therefore, selection of the NaOH strength, mixing time, mixing energy, temperature, and the quantity of excess caustic all have an important part in making the alkali refining process operate effectively and efficiently.

The current alkali refining practices are a result of the gradual application of science to the basic art of batch refining originally performed in open-top, cone-shaped kettles during the first part of this century. Efficient separation of soapstock from neutralized oil is the significant factor in alkali refining, and the technique of using centrifugal separators was conceived in the last century but did not become a commercial reality until 1932. The conventional caustic soda continuous system that evolved has the flexibility to efficiently refine all the crude oils presently utilized in the United States. The system may be outlined as follows:

- crude receipts—Crude or degummed oils are received by railcar, truck, or barge or from on-site extraction or degumming operation.
- sampling—Receipts are sampled, analyzed, and then transferred to the appropriate storage tanks. For optimum performance, degummed soybean oil should have a phosphatide content below 0.3%. If this level is exceeded, nondegummed oil should be blended with the degummed oil to attain a 1.0% phosphatide content.
- crude oil conditioning—As needed, the oils are transferred to the appropriate pretreatment or supply tank. Crude oils with high levels of phosphatides like soybean and canola oils are usually treated with food grade phosphoric acid for 4 hours minimum, 8 hours preferred before refining—300 to 1000 ppm for soybean and 1000 to 3000 ppm for canola. The purpose of the acid pretreatment is to (1) help precipitate phosphatidic materials, (2) precipitate natural calcium and magnesium as insoluble phosphate salts, (3) inactivate trace metals such as iron and copper that may be present in the oil, (4) reduce the neutral oil losses, (5) destabilize and improve the removal of chlorophyll in bleaching, and (6) improve the color and flavor stability of the finished deodorized oil.
- caustic treatment—The degummed or acid conditioned crude oil is continuously mixed with a proportioned stream of dilute caustic soda solution and heated to break the emulsion. Selection of the caustic treatment is determined by the type of crude oil, free fatty acid content, past refining experience with similar oils, and the refining equipment available. In general, the minimum amount of the weakest strength necessary to achieve the desired endpoint should be used to minimize saponification of neutral oil and prevent "three phasing," or emulsions, during separation. Usually, the best results are obtained with relatively weak caustic solutions or lyes on low free fatty acid oils and with stronger lyes on high free fatty acid oils. The strength of the caustic solution is measured in terms of specific gravity expressed in degrees Baumé(Bé). The caustic treat selected for the crude oil will vary with the free fatty acid content, the amount of acid pretreatment, and the level of caustic "excess" over "theoretical" determined for each

oil type from previous experience. The theoretical quantity of caustic is based on the ratio of molecular weights of sodium hydroxide to oleic fatty acid. This factor is determined as follows:

$$\text{Factor} = \frac{\text{NaOH Molecular Weight}}{\text{Oleic Fatty Acid Molecular Weight}} = \frac{40}{282} = 0.142$$

Thus the formula for caustic treatment is:

$$\% \text{ Treat} = \frac{(\% \text{ FFA} \times 0.142 + \% \text{ Excess} + \text{Acid Addition})}{\% \text{NaOH in caustic}} 100$$

The reduction of phosphorus during refining is determined largely by the amount of water present in the caustic solution. Higher excess caustic treatments remove more phosphorus, but the increase in removal is due more to the increased water than NaOH. Comparative studies have shown that more dilute caustic solutions will remove more phosphorus. Therefore, crude oils with high phosphorus levels are best refined with dilute caustic solutions; however, if they become too dilute, difficult emulsion separation characteristics develop. For this reason, dilute caustic solutions or low Bé concentrations are recommended for these oils, i.e., soybean, peanut, safflower, sunflower, and canola oils.

The refining conditions for cottonseed oil are chosen more for the improvement of color because of the presence of the gossypol. This pigment is sensitive to heat and oxidation, forming colored compounds that are difficult to remove from the oil other than by reaction with caustic. Therefore, the caustic treat is a larger excess of a more concentrated NaOH solution.

Palm, palm kernel, and coconut oils require a weaker caustic of approximately 12 Bé to optimize centrifugal separation, reduce saponification of the neutral oil, and minimize emulsions. The diluted caustic for use with the lauric and palm oils is usually preheated to 150°F (65°C) to minimize emulsion formation in the separators. These oils also require only a minimum of excess treatment as well, 0.02%, because they are refined for free fatty acid reduction only. It should be remembered that the free fatty acid for coconut and palm kernel oils is calculated on the basis of lauric fatty acid instead of oleic as for the other vegetable oils.

Suggested caustic concentrations and excess treatments for use with various crude vegetable oils as a starting point before experience is

gained are shown on Table 2.1. A smooth reproducible flow of the
caustic solution into the oil stream is important because pulsating delivery will carry through the mixers and produce varying mixture densities in the centrifuge.

• caustic oil mixing—After the caustic reagent has been proportioned
into the crude oil, it must be adequately blended to insure sufficient
contact with the free fatty acids, phosphatides, and color pigments.
The gums are hydrolyzed by the water in the caustic solution and
become oil-insoluble. The caustic and soft oils are mixed at 30 to 35°C
(86 to 95°F) in a dwell mixer with 5 to 15 minutes residence time.
High oil temperatures during the caustic addition must be avoided
because they can increase the neutral oil saponification and reduce the
refined oil yield. Many refineries will use an inline shear mixer to
obtain the intimate contact time between caustic and oil followed by
a delay period in the dwell mixer prior to centrifugation. After the
caustic mixing phase is complete, the mixture should be delivered to
the centrifuges at a temperature suitable for optimum separation. Most
soft oils are heated to 165°F (74°C) to provide the thermal shock necessary to break the emulsion.

• soap-oil separation—Refining yield efficiency is dependent on the primary separation step. From the caustic-oil mixer, the resultant soap in

TABLE 2.1.

| | Caustic Soda | | |
| | Concentration | | Excess |
	Baumé	NaOH, %*	Treat, %
Soybean oil			
Crude	14 to 18	9.50 to 12.68	0.12
Degummed	14 to 18	9.50 to 12.68	0.10
Mixed	14 to 18	9.50 to 12.68	0.10
Cottonseed oil	19 to 21	13.52 to 15.23	0.16
Corn oil	16 to 20	11.06 to 14.36	0.13
Peanut oil	13 to 15	8.75 to 10.28	0.12
Safflower oil	14 to 18	9.50 to 12.68	0.12
Sunflower oil	14 to 18	9.50 to 12.68	0.12
Canola oil	15 to 16	10.28 to 11.06	0.10
Coconut oil	11 to 12	7.29 to 8.00	0.02
Palm kernel oil	11 to 12	7.29 to 8.00	0.02
Palm oil	11 to 12	7.29 to 8.00	0.02

*At 15°C (59°F).

oil suspension is fed to high-speed centrifuges for separation into light- and heavy-density phases. These separators are designed to divide suspensions of insoluble liquids and solids in suspension with different specific gravities. The light phase discharge is the neutral oil containing traces of moisture and soap. The heavy phase, or soapstock, is primarily insoluble soap, meal, free caustic, phosphatides, and small quantities of neutral oil. Refined oil yield and quality depend upon a uniform feedstock and separation of the heavy phase with the least amount of entrained oil. However, even under the most optimum conditions, complete separation of the two phases cannot be achieved. Therefore, the primary separation is accomplished by allowing a small amount of the soapstock phase to pass along with the refined oil for removal by the water wash centrifuge [8].

There are various types of centrifuges used in vegetable oil refining. However, most centrifuges contain a bowl or hollow cylinder that turns on its axis. The flow of material enters the rotating bowl and is forced outwards to a disc stack. The heavier density soapstock is forced to the outside of the bowl and flows over the top disc and out the discharge port. The lighter neutral oil phase moves to the center of the bowl for discharge from the neck of the top disc. The major factors to consider for improvement of separation completeness are [2] (1) greater differences in specific gravity of the phases, (2) lower viscosities, (3) higher temperatures, (4) shorter travel distance for the heavy particles, (5) increased centrifugal forces, and (6) longer centrifugal dwell times.

- waterwashing—Refined oil from the primary centrifuge is washed with hot softened water or recovered steam condensate proportioned into the oil at a rate of 10 to 20% of the oil flow. Softened water must be used to avoid the formation of insoluble soaps. Sodium soaps remaining from the primary centrifugation phase are readily washable and easily removed from the oil with either a single or double wash. A single wash is usually sufficient. The water-oil mixture passes through a high-speed, in-line mixer to obtain intimate contact for maximum soap transfer from the oil to the water phase. The soapy water-oil mixture continues through to the water wash centrifuge. Similar in action to the refining centrifuge, water-washed oil is discharged as the light phase and the soapy water solution as the heavy phase. The water-washing operation will remove about 90% of the soap content in the refined oil.

Wash water temperature is important for efficient separation in the centrifuge. The water temperature should be 185 to 195°F (85 to 90°C), preferably 10 to 15°F (5 to 8°C) warmer than the oil temperature. The wash water flow rate controls soap removal and affects the oil losses in the wash water. As with the primary centrifuge, a pulsating flow of water must be avoided.

Two things that water washing will not do are remove phosphatides left in the oil after the primary centrifuge and remove unwashable soaps related to the calcium and magnesium content of the crude oil. The metal complexes require phosphoric acid pretreatment or conditioning for their removal in either the degumming of refining steps. Iron soaps are prooxidants while calcium and magnesium result in nonwashable soaps [1].

- vacuum drying—Water-washed oil is usually dried with a vacuum dryer before storage or bleaching. Washed oil at approximately 185°F (85°C) is passed through nozzles into the evacuated section of a continuous vacuum dryer that controls the moisture content of the washed oil to below 0.1%, most often in the range of 0.05%. A typical dryer operates a 70 cm of mercury (Hg) and is equipped with a high-level alarm and automatic shutdown capability. After drying, the dried refined oil should be cooled to approximately 120 to 130°F (49 to 55°C) before storage. For extended storage periods, a nitrogen sparge and/or a nitrogen blanket applied to the surface of the oil will minimize oxidation before use.

4.2 Short-Mix Caustic Soda Refining

Refining practices vary between countries and plants due to the number, quality, and kind of source oils processed. The refining practices in Europe differ from those used in the United States mainly because of the need to process all types of oils and typically poorer quality oils. The European oilseeds or crude oils must be imported, and a typical refinery must be equipped to handle all kinds of oils, dependent upon availability and price. The quality of the oilseeds of crude oils imported is variable, but normally a higher free fatty acid oil is processed. The short-mix process was adopted in Europe after World War II because the relatively high free fatty acid oils made it necessary to avoid the long contact time and the larger excess of caustic used with the conventional caustic soda refining system utilized in the United States. Chart 2.3 compares the differences between the long- and short-mix refining processes.

For the short-mix process, the oil temperature is raised to 80 to 90°C (175 to 195°F) before the addition of the caustic soda. A break between the neutral oil and soapstock takes place immediately, reducing the losses due to emulsification. The contact time between the caustic and oil is reduced to 30 seconds maximum, which helps to reduce the saponification

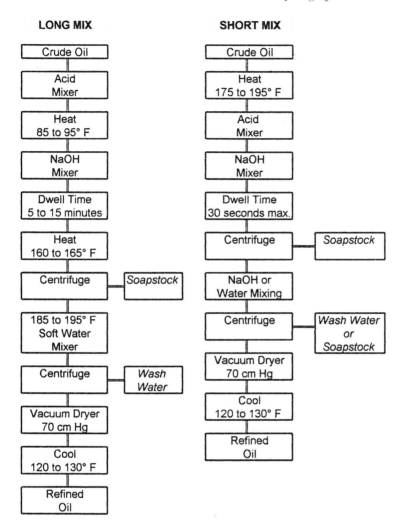

CHART 2.3. Caustic refining process flow: long mix versus short mix.

losses. Because it is standard in Europe to degum solvent-extracted oils and to phosphoric acid condition the oils before refining, the excess caustic treat can be eliminated or reduced substantially. In the case of an oil like cottonseed refined for color removal, a second caustic treatment is used. This additional phase is called re-refining. Since the bulk of the soapstock has already been removed with degumming and the initial caustic treatment, the additional losses with the second caustic treatment are relatively low. The oil is finally washed with demineralized water to help remove the traces of soap remaining in the oil and dried with processes similar to the systems used for the long-mix caustic refining process.

An ultra-short-mix method is used for oils with high acidity to reduce the losses more than the regular short-mix process. For palm and other oils with high acidity levels, the caustic soda is introduced directly into the hollow centrifuge spindle where a special mixing device is located. The very short contact time allows the use of stronger caustic concentrations without excessive saponification of the neutral oil. Comparisons of palm oil refined with the ultra-short-mix contributed considerable yield improvements over the product refined with the short-mix process: 7.0 to 16.5% less loss for one evaluation [10].

4.3 Zenith Process

The Zenith process was developed in Sweden in 1960 to enable better refining of the only oilseed crop grown in that country: rapeseed. The stainless steel continuous process consists of three main refining steps, two of which are semicontinuous to maintain the desired reaction times:

- step one—The oil is treated with concentrated phosphoric acid to remove the nonfatty impurities that influence emulsions. The amount of phosphoric acid depends upon the oil quality, but is normally about 0.2% by oil volume for rapeseed oil. The reaction, performed under a vacuum, requires 20 minutes. The acid sludge formed with the pigments, phosphatides, calcium, magnesium, and other impurities is removed with a sludge separator. Water is introduced in the form of live steam to form liquid crystals of the remaining phosphatides at the interface between the water and the oil.
- step two—Neutralization is performed by introducing the oil at 90°C (194°F) in the form of droplets to the bottom of a vessel almost filled with 0.35 N (2.0 degrees Bé) alkaline solution. The 1.0 to 2.0 mm diameter droplets rise by the difference in specific gravity and are collected in the upper conical part of the vessel forming an oil layer with a typical analysis of 0.05% free fatty acids, 0.2 to 0.3% moisture, and 100 ppm soap.
- step three—The neutralized oil is treated with citric acid to help separate the trace quantities of soap for adsorption by the bleaching earth. The oil is dried and bleaching earth added before it is vacuum bleached for 30 minutes before filtering.

Improved refined oil yields with excellent quality are claimed for the Zenith process [11].

4.4 Miscella Refining

Facilities with an existing oilseed solvent extraction system may find miscella refining to be an advantage by using the same solvent recovery unit for both purposes. Miscella is the solution or mixture that contains the extracted oil. Both continuous and batch miscella refining processes are suitable for the fats and oils available in the United States, such as soybean oil, cottonseed oil, safflower oil, sunflower oil, palm oil, coconut oil, and tallow. This type of refining should be done at a solvent extraction plant as soon as possible, preferably within 6 hours after the oil is extracted from the oil seed or animal. The advantages for miscella refining, as compared to conventional continuous caustic soda refining, are (1) higher oil yield, (2) lighter color oil without bleaching, (3) elimination of the water wash step, and (4) extraction of the color pigments before solvent stripping has set the color [12].

For this purification process, the crude miscella source may be from (a) the pre-evaporator of a direct-solvent extraction plant, (b) a blend of pre-pressed crude oil and solvent-extracted miscella from the press cake, or (c) a reconstituted blend of crude oil with hexane. In the process a mixture of approximately 40 to 58% oil in solvent is heated or cooled to 104°F (40°C) and filtered to remove meal, scale, and other insoluble impurities. Two solvents that have been used commercially for miscella refining are hexane and acetone.

Hydrolysis of phosphatides and pigments in the crude oil miscella requires an acid pretreatment, which usually varies between 100 and 500 ppm by weight of the oil, depending upon the quality of the crude oil. An acid such as phosphoric or glacial acetic has been found effective in improving oil quality and reducing refining losses. Phosphoric acid is used more commonly due to its less corrosive properties and availability. The acid is mixed with the miscella in a static mixer to provide an intimately dispersed acid phase that immediately reacts with the crude miscella.

The pretreated crude miscella is then alkali refined using dilute caustic soda with a 16 to 24 degree Bé and a 0.2 to 0.5% NaOH excess over the theoretical required to neutralize the free fatty acids. The reaction of the caustic soda with the free fatty acids proceeds rapidly at 130 to 135°F (54 to 57°C), using homogenizers with a shear mixing intensity capable of homogenizing milk, hydrolyzing the phosphatides and pigments with the caustic soda to produce a two-phase mixture. The miscella temperature is adjusted to 135°F (57°C) to obtain the best separation of the heavy phase or soapstock from the oil or the light phase with the centrifuge. The neutral oil is then filtered through a diatomaceous earth precoated pressure leaf filter. At this point, the refined and filtered miscella can be stripped of the solvent to produce a neutral yellow oil, or it can further processed as miscella to dewax, fractionate, or hydrogenate the oil [13,14].

Obvious disadvantages for the miscella refining process that may have discouraged many processors from adopting this processing system are [15]

(1) Equipment—All equipment and facilities must be explosion-proof for solvent handling.

(2) Maintenance—The equipment and facilities must be well maintained to avoid excessive solvent losses and accidents.

(3) Laboratory—More elaborate laboratory facilities and staffing is necessary to control this process.

4.5 Batch Caustic Soda Refining

Batch refining is still practiced for some specialty oils, in developing countries, for small production lots, and in most pilot plant operations. Batch refining has some basic advantages: (1) low investment costs, (2) equipment readily available, (3) practical for small lots, (4) suitable for low capacities, and (5) production of a quality refined oil most times. At the same time, batch refining can have some serious drawbacks: (1) high refining loss, (2) high operational costs, and (3) high load on wastewater plants, and (4) it is very time consuming. Batch refining is most likely only installed for special conditions where the refined oil requirements are very low.

Generally, two batch refining procedures are utilized: the dry method and the wet method. The dry method is preferred in the United States for most oils, while the wet method is preferred in Europe. Coincidentally, the batch dry method parameters resemble the long-mix caustic soda continuous refining procedure preferred in the United States and the batch wet method parameters resemble the short-mix continuous procedure preferred in Europe.

DRY METHOD BATCH REFINING

The equipment required for the batch dry refining method is simple, consisting of an open-top, conical bottom tank or kettle equipped with a two-speed agitator and steam coils for heating. The agitator shaft is centered in the vessel and is either suspended from the top or extended to a step bearing on the bottom. The agitator shaft is equipped with sweep arms, each with paddles canted to push the liquid upward during agitation. The usual agitator rates are 30 to 35 rpm maximum and 8 to 10 rpm minimum. The batch dry method consists of the following steps [16,17]:

(1) The first stage is carried out with the oil at ambient temperature or at a temperature just high enough to keep the fat molten and liquid. If

the oil contains occluded air after pumping to the refining kettle, it must be settled long enough to allow the air to escape. The soapstock can entrain enough air to float, which will prevent it from settling to the bottom of the kettle as desired.

(2) The caustic solution or lye is added to the top of the kettle while agitating at high speed. Agitation is continued to thoroughly emulsify the alkali and oil, usually 10 to 15 minutes. Then, the agitation is reduced to slow speed and the oil is heated to 135 to 145°F (57 to 63°C) as rapidly as possible.

(3) A visible break in the emulsion will occur at about 140°F (60°C) where the soapstock separates from the clear oil in the form of small flocculent particles that tend to coalesce with agitation.

(4) After the desired degree of break is obtained, agitation is stopped, heating is discontinued, and the soapstock or foots is allowed to settle to the bottom of the kettle by gravity for 10 to 12 hours minimum.

(5) After the soapstock has settled, the neutral oil can be drawn off the top leaving the soapstock at the bottom of the kettle. For many oils, like soybean canola, sunflower, safflower, and peanut, treated with a dilute caustic (usually 12 to 16 Bé with about 0.25% excess), the settled soapstock should be fluid enough to allow it to be drained from the bottom of the kettle, leaving the refined oil in the kettle.

(6) Traces of moisture and soap remain in the refined oil that should be removed if the oil is going to storage before bleaching. The refined oil can be either filtered through spent bleaching earth or water washed. Kettle water washing consists of adding approximately 15% hot, soft water to the refined oil while agitating for uniform dispersion and then allowing the water to settle for decanting. Water washing may be repeated if warranted.

WET METHOD BATCH REFINING

The preferred European batch refining method heats the crude oil to a relatively high temperature before adding the caustic or lye: 150°F (65°C). A high caustic concentration, 20 Bé, is used for the usually high free fatty acid oils processed with about a 0.10% excess treatment. In many cases, the addition of salt equivalent to about 0.10% sodium chloride per 1.0% free fatty acid is necessary to break the soapstock and oil emulsion. The participated soapstock is washed down with a spray of hot water onto the surface of the oil. Several successive water washes are required to complete the removal of the soap from the oil, with a settling time required between each wash.

The wet method has advantages over the dry method for refining oils with high free fatty acid contents like some palm and olive oils. It has also been used for refining coconut and other lauric oils. Refining equipment for the wet method is not essentially different from that used for the dry process, except that closed tanks that can also be used for vacuum bleaching are usually used to refine and wash the crude oils with the wet method [16].

4.6 Physical Refining

Physical refining was utilized as early as 1930 as a process for the preneutralization of products with a high initial free fatty acid content. In this case, preneutralization was followed by caustic refining. Later, it was found possible to successfully physical refine lauric oils and tallow if the proper pretreatment was applied before steam distillation. Physical refining became a reality in the 1950s for processing palm oil, which typically contains high free fatty acids and low gum contents. The palm oil process subjected the crude feedstock first to pretreatment and then to deacidification. The pretreatment consisted of a degumming step and an earth bleaching step, which together remove certain nonvolatile impurities by filtration. Volatile and thermally liabile components are removed during the conditions of steam distillation under vacuum, which originally gave the process its name of steam refining [18]. However, for vegetable oils like soybean containing relatively low levels of free fatty acids and higher amounts of phosphatides, physical refining became a possibility only recently.

The usual edible oil processing system consists of caustic neutralization, bleaching, and deodorization. Caustic neutralization of vegetable oils with high phosphatide content delivers a soapstock that is a mixture of sodium salts of fatty acids, neutral oil, water, unused caustic, and other compounds resulting from the reactions of the caustic with various impurities in the oil. Disposal of this soapstock or the wastestreams from soapstock processing systems has become increasing more expensive. A second problem associated with chemical neutralization is the loss of neutral oil, which reduces the overall yield from the crude oil. Elimination of the caustic refining step is economically attractive, but it means that degumming or some other pretreatment process, or system must assume all the functions of the alkali refining process, except for free fatty acid removal.

Physical refining can remove the free fatty acid, as well as the unsaponifiable and some other impurities by steam stripping, eliminating the production of soapstock, and keeping neutral oil loss to a minimum. However, degumming and pretreatment of the oil is still required to remove those impurities that darken or otherwise cause a poor quality product when heated to the temperatures required for steam distillation. Crude oil pretreatment is normally a two-step operation—the addition of a chemical to remove any trace quantities of gums remaining after water degumming and bleaching. Following pretreatment, all the free fatty acid and any remaining trace impurities are removed by steam distillation in a single unit. Soapstock acidulation is eliminated with physical refining, and a higher grade distilled fatty acid is recovered directly from the oil without major pollution problems. Chart 2.4 depicts a physical refining process flow.

Vegetable oil refining has to cope with many minor components. After water degumming, a number of impurities must still be removed or converted: carotenoids, chlorophyll, brown pigments, phosphatides, metals, free sugars, free fatty acids, and oxidizing lipids. Steam stripping can convert the carotenoids and remove free fatty acids, most off-flavors, and pesticides, but the other impurities must be handled before the distillation step [19]. Therefore, the pretreatment step is critical to the success of the physical refining process. The major process variables in pretreatment are (a) the pretreatment chemical, concentration, and level; (b) the bleaching clay and level; and (c) the operating conditions. Normally, for a single source oil with a history of consistent quality, the pretreatment process variables can be expected to remain fairly constant, but when more than one source oil is processed, varying conditions and chemical treatments must be considered [20].

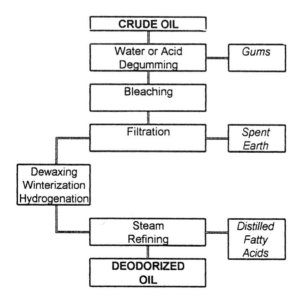

CHART 2.4. Physical refining process flow.

Attempts at pretreating water degummed vegetable oils with levels of phosphorus of approximately 100 ppm have proven costly and the quality is unreliable. An acid degummed feedstock may be the most satisfactory feedstock for physical refining. Also, indications are that a combination of acids, citric and phosphoric, are best for the pretreatment stage, which is critical to the success of the physical refining process. Poor contact between the treatment reagents and the oil does not convert the phosphorus to a bleachable form. A successful approach for good contact has been to mix the oil and reagents together with gentle mixing. A moisture presence in the pretreatment ensures good contact between the phosphorus and the acids because the acids are located in the water phase and the polar phospholipids are attracted to the water surface. After the pretreatment step, bleaching earth is slurried with the oil. The bleaching earth quantity required for this process should be about 115% of the level normally used for bleaching alkali refined oils. The oil should be bleached with a temperature over 212°F (100°C) to ensure that it is thoroughly dried, or the filtration rate will be reduced by slimy phospholipids that blind the filter. When the bleached oil is dry, the phospholipids tend to form a grainy precipitate that is very filterable. The spent bleaching earth must be fully removed from the oil to preserve the oil quality. An increase in free fatty acid of less than 0.2% should be expected, but the final phosphorus content must be reduced to less than 5 ppm.

The pretreated and bleached oil can be hydrogenated before steam refining to strip the fatty acids, flavors, and odors. The operating temperatures of the steam refining deodorizer are the same as those used for deodorization of chemically refined oils, or usually over 440°F (225°C). Deodorizers designed for steam refining of the higher free fatty acid oils should experience no reduction in production rates. However, a rate reduction should be expected with deodorizers designed to remove the lower free fatty acid levels from caustic neutralized oils.

The principal advantage for steam refining a low free fatty acid oil, like soybean, corn, peanut, sunflower, safflower, or canola, is the reduction of plant pollution commonly caused by the acidulation of soapstock produced with conventional caustic refining. The economics for steam refining are usually favorable for high free fatty acid products like palm and the lauric oils. Analyses indicate that there is no operating cost advantages for physical refining the low acidity oils [21], which make up the majority of the oils processed in the United States. Additionally, flavor stability and potential unsatisfactory bleached color concerns still exist. Flavor evaluation work at the USDA Northern Regional Research Center indicated that steam refined soybean oil was equivalent to caustic refined product; however, some of the test results indicated a potential problem with oxidative stability, which was not duplicated with further test work [22]. Another consideration in the United States is that certain oils contain impurities that cannot be adequately removed by the pretreatment process to enable them to be physically refined to the required quality standards. Cottonseed oil falls into this situation due to the gossypol content. This pigment is sensitive to heat and oxidation, forming colored compounds that are difficult to remove from the oil, except by reaction with caustic soda.

4.7 Refining Efficiency

Refining efficiency is generally considered to be the yield of dry neutral refined oil as a percentage of the available neutral oil content of the crude oil. The dry neutral oil is determined by actual weight in a scale tank or volumetric measurement with adjustments as indicated by the product's specific gravity or temperature. The crude neutral oil content is supplied by the laboratory analysis of the incoming crude oil samples or preferably analysis of the feedstock to the refining system. The refining efficiency is expressed as the ratio of neutral oil produced over the analyzed neutral oil in the crude oil, for example,

$$\text{Refining Efficiency} = \frac{\text{Refined Oil Yield}}{\text{Crude Neutral Oil Analysis}} \times 100$$

The specified laboratory loss analysis by trading rules varies with the source oil, which affects the refining efficiency results. Gum containing crude oils, such as soybean and canola, are usually evaluated by the chromatographic method for neutral oil, AOCS Method Ca 9f-57 [23], and the refining efficiency is expressed as the ratio of neutral oil produced to the laboratory determined neutral oil in the crude oil. Corn and cottonseed oils trading rules specify AOCS Method Ca 9a-52 [23], and the refining efficiency is expressed as savings over cup and determined with the following equation:

$$\text{Savings over Cup} = \frac{\%\ \text{Cup Loss} - \%\ \text{Plant Loss}}{\%\ \text{Cup Loss} \times 100}$$

It is customary in Europe to monitor refining efficiency with the "Refining Factor," which is the total loss divided by the fatty acid of the crude oil before refining. This factor permits a direct comparison of the refining efficiency of oils with different free fatty acid starting levels. The refining factor can present a false impression when the amount of gums and other impurities are high in proportion to the free fatty acid content. Experience in European facilities with all types of properly filtered and degummed oils shows that the refining factor varies between 1.4 and 1.7 [10]. In the United States lauric and palm oils are refined against free fatty acid and the efficiency control is measured with the refining factor. The equation for determining the refining factor is [9]

$$\text{Refining Factor} = \frac{\% \text{ Plant Loss}}{\% \text{ Free Fatty Acid}}$$

4.8 Refining By-products

Crude oils contain a number of materials that must be removed to produce neutral, light colored oils. These impurities have been considered waste products constituting a disposal problem; however, they can be valuable by-products when effectively recovered and processed. The two major by-products from the refining processes are soapstock from chemical refining and the hydrated gums from the degumming process prior to caustic refining or the physical refining pretreatment stage.

SOAPSTOCK PROCESSING
Soapstock from alkali refining is a source of fatty acids, but it also presents a handling, storage, and disposal problem. Originally, many years ago, the caustic refining by-products were merely discarded. Then, it became a valuable source of fatty acids for the soapmaker and the fatty acid distiller. Soapstock was shipped from the refiner in the raw form as it was separated from the neutral oil. The growth of synthetic detergents over soaps reduced this market for soapstock considerably, and in the fatty acid field, soapstock utilization was replaced with tall oil, a by-product of the paper industry. These changes turned edible oil refiners to soapstock acidulation to produce "acid oil," which is used as a high-energy ingredient in feeds or provides a more refined product for chemical use [24].

Batch acidulation of the raw soapstock discharged from the caustic refining centrifuges consists of three basic steps [25]:

(1) Acidification with 66 degree Bé sulfuric acid of the highly basic, diluted soapstock to convert the soap into free fatty acids. The acid level required varies with the amount of caustic used in refining to adjust the pH to 1.5 to 2.0.

(2) Breaking the emulsion of fatty acids and foreign materials in water with heat (approximately 195°F or 90°C) and agitation

(3) Separation of the three phases:

- top layer—Fatty acid product is recovered and transported to storage.
- middle layer—This material, which consists of an emulsion of the top and bottom layer material, must be recycled for reprocessing.
- bottom layer—The acidified wastewater is removed, neutralized, and disposed of in the wastewater system.

The problems associated with acidulation of soapstock are mainly the corrosive nature of the process and the fact that the separation of the acid oil phase from the acid water phase is often relatively poor, which leads to high fat losses and wastewater contamination with fatty material. Federal, state, and municipal legislation enacted for pollution abatement mandated more effective processes for clarification of acid water streams from acidulation. Continuous acidulation of soapstock and wash water as it is discharged from the refinery has provided a by-product of acid oil with the required quality standards and an acid wastewater with less and manageable contaminants. Soapstock should be processed as soon as possible after it is produced to minimize fermentation and emulsification. Several different continuous acidulation systems with varying designs are available.

The major reason for acidulating soapstock is for moisture removal so that a smaller volume is obtained for handling or storage. Acid oil is essentially the fatty portions of soapstock with the moisture content reduced to 1.0 or 2.0%. It is traded on a Total Fatty Acid (TFA) basis of 95%, and shipments can be rejected if the TFA falls below 85%. The impurities originally in the crude oil, such as phosphatides, carbohydrates, proteins, pigments, sterols, heavy metals, and so on, are transferred in part or full to the soapstock during refining and then to the acid oil with acidulation.

Probably, the greatest volume usage for acidulated soapstock is for animal feeds. Acid oils have become one of the essential components in many animal feeds. They are high-energy ingredients that provide 9 calories per gram when metabolized as compared to 4 calories per gram from starch or protein. Acid oils act as carriers and protectors for several fat-soluble vitamins and antioxidants and are an excellent source of polyunsaturates in most cases. The main competition for the formulated animal feeds is corn, which provides 3.5 calories per gram. Therefore, acid oils cannot cost more than 2.6 times the price of corn to be economically competitive. Two potential problems for this application of acid oil are that, (1) residual sulfuric acid and its reaction products decrease the palatability to most animals, and (2) deodorizer distillates cannot be a component of the acid oil because of the pesticide contents [27].

An alternative to soapstock acidulation is being used on a limited basis: neutralized dried soapstock process. This alternative process, which converts the soapstock to a neutral pH followed by drum drying, produces a product that has performed well as a fat source in feeds for chickens and cattle, and the only effluent is evaporated water [26]. This process reduces the load on a plant wastewater treatment facility more than any of the alternatives except for shipping raw soapstock.

HYDRATED GUMS PROCESSING

Lecithin is the preferred outlet for the hydrated gums recovered from the degumming stage of refining. Commercial lecithin is one of the most important by-products of the edible oil processing industry because of its functionality and wide application in food systems and industrial utility. However, the gums hydrated from soybean oil alone far exceed the market requirements for lecithin; estimates are that one-third of the soybean oil processed will satisfy the lecithin demands. Another outlet available to the crusher/refiner is to incorporate the recovered gums back into the meal produced. For some operations, this alternative solves a meal dusting problem, as well as adding value to the gums.

Lecithin is the commercial name for a naturally occurring surface active agent made up of a mixture of phospholipids. It can be obtained from a number of vegetable oils, but the major source is soybean oil phospholipids or gums that provide excellent emulsification properties with a good flavor and color. Lecithin production starts with degumming the crude oil with approximately 2% steam or water added with slow agitation to hydrate the lecithin. The hydrated gums are separated from the crude oil and dried carefully below 1.0% moisture to avoid damaging the color. After cooling, ingredients are added to the lecithin to meet the desired specification limits. Soybean oil and fatty acid additions are used to control acetone-insoluble matter, acid value, and viscosity. Lecithin can be chemically bleached with hydrogen peroxide, either before or after drying to control the color. National Soybean Processors Association rules define six common grades of lecithin. In addition, a variety of modified lecithins can be produced for specialty uses [28].

The crude oil degumming process affects the quality and performance of the lecithin products. For example, most additives used to aid degumming are usually deleterious to the lecithin, except for acetic anhydride. It reacts with the lecithin gums to form acylated phoshatidylethanolamine. Other additives, such as phosphoric acid tend to burn and darken the lecithin on drying and pollutes water streams. Oxalic acid does not pollute [29], but renders lecithin toxic, and inorganic salts affect the physical and functional properties [30]. The acid degummed lecithins have also exhibited poor functionality, i.e., inferior instantizing properties in cocoa powders and milk products [31].

5. PREBLEACHING

The purpose of bleaching is not only to provide a lighter colored oil, but also to purify it in preparation for further processing. Refined oil contains traces of a number of undesirable impurities either in solution or as colloidal suspensions. These impurities compete with the color pigments for space on the adsorbent surface. In many cases, the bleaching process is performed more for the removal of the nonpigment materials like soap, gums, and prooxidant metals that hinder filtration, poison hydrogenation catalyst, darken the oils, and affect finished oil flavor. Another function considered primary by many quality processors is the removal of peroxides and secondary oxidation products. The key process parameters for bleaching include (a) procedure, (b) bleaching media and dosage, (c) temperature, (d) time, (e) moisture, and (f) filtration. Each variable must be considered in light of the system used and the oil to be bleached [32–38].

5.1 Procedure

The three most common types of contact bleaching methods used for edible fats and oils are batch atmospheric, batch vacuum, and continuous vacuum. This sequence is also the chronological order in which the different methods were developed. Chart 2.5 diagrams the process flow for the three types of bleaching systems.

BATCH ATMOSPHERIC

Oil at approximately 160°F (71°C) is pumped into an open-top tank equipped with steam coils, or a steam jacket, and a paddle agitator. Bleaching earth is added from the top of the tank with the agitator running; the temperature is raised to bleaching temperature and maintained for a short time. Next, the oil is recirculated through a filter press and back to the bleaching vessel until the oil is clear, and then it is pumped to storage.

BATCH VACUUM

Bleaching earth is added to an agitated slurry tank containing a small portion of the refined oil at 160°F (71°C). This slurry is transferred to the vacuum bleacher, which contains the balance of the oil batch. The bleaching vessel is equipped with coils or steam jacketed, an agitator, and a vacuum system. After the prescribed time at bleaching temperature under vacuum, the bleached oil batch is cooled to 160°F (71°C), the vacuum is broken, and the oil is filtered.

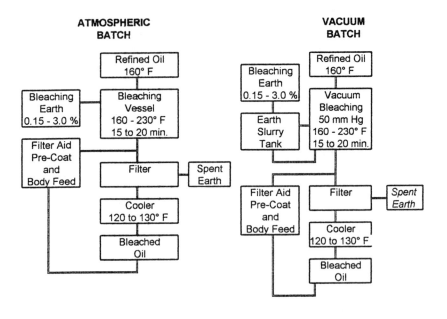

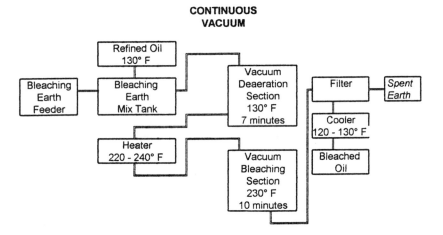

CHART 2.5. Typical bleaching processes flow.

CONTINUOUS VACUUM

Bleaching clay is continuously fed into a stream of oil at 160°F (71°C), and this mixture is sprayed into a vacuum chamber to remove both water and air from the clay and the oil. The product temperature is raised to bleaching temperature with a heat exchanger and then sprayed into a second chamber for bleaching. After the bleaching retention time, it is filtered in a closed-type filter and cooled before the vacuum is broken.

The use of activated clays and higher temperature bleaching led to the need for protection from atmospheric oxidation, which is provided by the vacuum bleaching process. Oxidative reactions during atmospheric bleaching cause a fading of some color pigments to lighten the color but also cause the formation of new nonabsorbent colors and fixation of other colors that darken the oil instead of lightening it. New color formation is believed to be the result of oxidation of red chroman 5,6 quinones from tocopherol, which does not respond to adsorption.

Vacuum bleaching, batch or continuous, is more effective than atmospheric bleaching because it can use less clay, operates at lower bleaching temperatures, effects quicker moisture evacuation for less free fatty acid development from hydrolysis, and does not expose the oil to oxidation at high temperatures. Batch bleaching is preferable to continuous bleaching when a variety of source oils are processed in the same system. However, if only a few different source oils are processed on a volume basis, continuous bleaching is preferable because large volumes can be processed without the necessity of stopping to refill the bleacher, and the bleaching conditions can be adjusted as the operation progresses.

Although vacuum bleaching is preferred, atmospheric bleaching can produce quality bleached oils. Natural clays have been found to perform best with atmospheric bleaching. Natural clay is an oxidation catalyst of low activity because of its low surface area and high pH; consequently, with atmospheric bleaching it is not responsible for forming new color or a fixed color to the extent of color fading. As a result, atmospheric bleaching is superior to vacuum processing where these oxidative effects do not occur [38].

5.2 Bleaching Agents

Chemical agents have been used or proposed for use, but practically all edible oil decoloration and purification is accomplished with absorptive clays or carbons. The basic kinds of absorbents used in edible oil bleaching are neutral clays, activated earths, and activated carbon:

- natural bleaching earth—Bentonite clays that exhibit absorptive properties in their natural state with only physical processing are classed as natural bleaching earths or "fuller's earth." Molecular lattice structure, macropore structure, and particle size all affect the capacity of earths to adsorb water, oil, phosphatides, soap, color bodies, and metals. The better natural earths can absorb 15% of their own weight in pigments and other impurities but also retain about 30% neutral oil. Natural clays perform well with respect to cleaning up residual soap and gums and are employed for easily bleached oils like coconut, lard, and tallow. Additionally, the natural earths do not elevate the free fatty acid content nor isomerize unsaturated fatty acid groups. However, for dark or difficult-to-absorb pigments or impurities, prohibitive levels of the natural earths are required, which make the activated materials more attractive.

- activated bleaching earth—Bentonite clays are also used to produce activated bleaching earths, but a type that contains a high proportion of montmorillonite. This hydrous aluminum silicate has considerable capacity for exchanging part of the aluminum for magnesium, alkalies, and other bases. Interestingly, most bentonites that exhibit high natural bleaching power are not suitable for activation, and most clays used for activated clay products have a poor natural bleaching activity. Treatment to varying degrees with sulfuric or hydrochloric acid, washing, drying, and milling alter the bleaching media's degree of acidity, adsorption capabilities, and particle size distribution [33]. The acid treatment of montmorillonite clay produces a specialty adsorbent from a naturally occurring mineral. During this process, the physical structure and chemical composition are altered in a controlled way to maximize specific properties. An efficient bleaching earth is produced with a surface of the correct chemical composition and pore distribution selectively attractive to the detrimental components of the refined oils.

Particle size is also a major physical parameter affecting bleaching earth performance because all adsorption theory considers adsorption as a surface phenomena. In general, the finest particle size clays have the best bleaching power; however, particles too small create severe filtration problems and oil retention is increased. Therefore, the adsorbent used should have as small a particle size as can be effectively handled by the filter system. In practice, a compromise particle size provides acceptable filtration performance and minimizes oil loss without diminishing bleaching performance [34].

Activated bleaching earths normally contain 10 to 18% moisture, which supports the montmorillonite layers in the clays. If the clay is completely dried prior to bleaching, the layers collapse to decrease the surface area available to adsorb the pigments and other impurities.

Apparent bulk density, weight per unit volume, is dependent upon the amount of void space in the clay; the more void space, the lower the density. Activated clays have a lower bulk density than natural clays, which increases the oil retention. The increased void space and total surface area can retain as much as 70% of the bleaching earths, weight of oil. However, lower activated clay usage level requirements normally result in a lower overall bleached oil loss with a lower bleach color and increased impurity removal [35].

The activated bleaching earths are more likely to split soap residues to elevate free fatty acid, destroy peroxides and secondary oxidation products, and promote isomerization. The latter effect is more pronounced at temperatures above 300°F (150°C), which is well above the optimum bleaching conditions. The modified bleaching earths are especially useful for bleaching the most difficult oils, such as palm, soybean, and canola, or as part of the physical refining pretreatment process for the removal of metals and phosphatides.

- activated carbon—A wide variety of carbonaceous raw materials can be used to form activated carbon by carbonization at high temperatures, combined with the use of activating materials like phosphoric acid, metal salts, and so on. The treated material is washed, dried, and ground to produce activated carbons with various pore sizes, internal specific surface areas, and alkalinity or acidity. Activity is determined by the chemical state and a large specific surface area.

  Carbon is used sparingly by most processors due to problems with filtration, relatively high cost, and high oil retention; it can retain up to 150% of its weight of oil. When utilized, it is normally added in combination with bleaching earths at 5 to 10% of the earth volume. Carbon is effective in adsorbing certain impurities not affected by earths; for example, some aromatic materials that are not volatilized by deodorization can be satisfactorily removed with activated carbon [36].

5.3 Bleaching Earth Dosage

The amount of bleaching earth used depends upon the type of absorbent used and the type of refined oil, as well as the adsorption of color bodies and other impurities required. The percentage of clays used vary in a wide range from 0.15 to 3.0% and only in extreme cases are higher quantities used [37]. Acid treated or activated earths use far exceeds that of the natural clays due to the higher bleaching efficiency, particularly with dark or high chlorophyll oils. On the basis of adsorbent activity, the acid-activated clays are generally one and a half to two times more effective as bleaching agents than the natural earths. The efficiency of an absorbent is measured by the minimum dose required to reduce the concentration of adsorbate to the required level. Therefore, the kind and amount of earth or carbon used need only be enough to clean up the oil preparatory to hydrogenation or deodorization and to remove any undesirable impurities and pigments that will not be removed in later processing. The minimum required bleach is usually best as overbleaching increases oil losses and can lead to flavor, oxidative, and even color instability. The removal of colored pigments is a common, simple visual guide, often used to gauge the overall performance and adjust levels required of a bleaching earth. However, the ability to remove other undesirable impurities is less readily apparent. The choice of the correct bleaching earth and the level to use in any specific application must take into consideration the removal of all the impurities as measured by peroxide value reduction to zero, chlorophyll to less than 1.0 ppm, phosphorus less than 1.0 ppm, negative soap, and the Lovibond red determined for the specific source oil.

5.4 Temperature

It is important to slurry the bleaching earth and oil at relatively low temperatures and then increase the complete mixture up to the final bleach temperature. Experience has shown that final bleach colors are darker when the earth was added to hot oil. Evidently, this effect is due to one or both of two factors:

(1) Adding the earth to hot oil reduces its adsorptive capacity because the moisture in the clay is driven off too rapidly, causing a collapse of the lattice structure, which reduces the effective surface area to adsorb impurities and pigments.

(2) The oil is unprotected against oxidation when heated before the clay is added, which can cause some color fixation or set.

Bleaching clay activity increases as the temperature is increased by reducing the viscosity of the oil, but decoloration declines after the optimum temperature has been reached. The optimum earth/oil contact temperature is dependent upon the oil type and the type of bleaching system. Temperature requirements for vacuum bleaching systems are normally lower than those for atmospheric processing to reach optimum color removal. Temperature also affects other properties of the oil, which dictate that it should be kept as low as possible to minimize product damage, but high enough for adequate adsorbance of the impurities and color pigments [38].

Production of an oil with acceptable oxidative stability requires careful control of the process temperatures. Few problems are encountered when the bleaching temperatures remain below 230°F (110°C) and steps are taken to control air oxidation. Anisidine values begin to rise with bleaching temperatures above 230°F (110°C) indicating damage to the oxidative stability. Nearly all edible oils' optimum bleaching temperatures range between 160 and 230°F (70 to 110°C). The activity of an absorbent in bleaching an edible fat or oil is at a maximum at some particular temperature that varies with oil type and process. Low temperatures favor the retention of the adsorbed pigment on the bleaching media surface, while higher temperatures favor movement into the pores where chemisorption is most likely, which promotes structural changes in the unsaturated fatty acid groups. Extremely high-temperature processing must be avoided to prevent isomerization of the unsaturated fatty acid groups and excessive free fatty acid development.

5.5 Time

In theory, adsorption should be practically instantaneous; however, in practice this is not the case. The rate of color decrease is very rapid during the first few minutes that the adsorbent is in contact with the oil and then decreases to a point where equilibrium is reached and no more color is removed. Time is required for the clay to release all of the bound moisture and take up the color pigments and impurities to maximum capacity. Usually, 15 to 20 minutes contact time is adequate at a bleaching temperature above the boiling point of water. The usual error is to extend bleaching time beyond the optimum. Excessive contact time with atmospheric bleaching usually tends to darken the oil.

Contact time for bleaching is made up of two time periods: (1) the time in the bleaching vessel or continuous stream and (2) the contact time in the filter during recirculation or final filtering. Continued or progressive reduction in peroxides and the other impurities as filtering continues is caused by "press effect," a benefit provided by the earth buildup in the filter with continued use. Some processors take advantage of this effect by decreasing the level of earth used in oils to be filtered with partly filled filters.

5.6 Moisture

The presence of some moisture seems to be essential for good bleaching action. Bleaching earths that have been completely dried before use have been found inactive. The earths normally contain from 10 to 18% moisture, which acts as a structural support to keep the montmorillonite layers apart. During bleaching, it is necessary to remove the moisture in the clay to obtain optimum adsorption capacity; the color bodies and other impurities cannot be adsorbed to maximum capacity until all the water has been removed. The bound moisture is not released until the elevated bleaching temperatures are attained. Refined oil can contain moisture levels from less than 0.1 to as high as 1.0%, which must also be removed for effective adsorption of the traces of soap remaining after refining. Experience has indicated that a slightly wet oil may be beneficial for the removal of color pigments and flavor precursors to provide a lighter, more stable oil. Maximum adsorption is achieved when the bleaching earth is slurried with the oil below the boiling point of water and then gradually increasing the mix to bleaching temperature; addition of the bleaching earth before heating the oil has also been found to inhibit heat darkening.

5.7 Filtration

After an adsorbent has selectively captured the impurities, it must be removed from the oil before it becomes a catalyst for color development and/or other undesirable reactions. Filtration, the separation method most often used for spent bleaching earth removal, is the process of passing a fluid through a permeable filter material to separate particles from the fluid. Examples of the filtration materials utilized are filter paper, filter cloth, filter screen, and membranes. Filter aids such as diatomite, perlite, or cellulose are usually used in conjunction with the permeable filters for surface protection.

There are three steps to filtration: precoating, filtering, and cleaning. The purpose of the precoat is to protect the filter screens, provide immediate clarity, improve the flow rate, and aid in filter cake removal during cleaning. It also helps to prevent blinding, which stops the product flow. Precoating is accomplished by slurring filter aid with previously filtered oil and allowing the oil to carry the filter aid to the filter, deposit it on the filter screen, and return to the precoat slurry tank to pick up more filter aid. The amount of precoat is determined by the filter area, usually 5 to 11 kilograms per square meter. The flow rate during precoating should be the same as during filtration to obtain an even coating on the filter. Uneven coatings result in blinded filters and short filtration cycles.

During filtration "body feed," or the continuous addition of filter aid, can be used to help prevent blinding of the suspended solids on the precoat. The body feed surrounds the suspended solids to provide flow around them. The body feed slurry of filter aid and oil is injected into the system prior to the filter. The suspended solids are ridged or deformable and can elongate under pressure to extrude through the filter cake and slow or block the product flow. Body feed will coat the deformed solids, allowing them to be retained on the filter cake.

Several indicators are utilized to determine the point at which the filter space has been filled with solids from the bleached oil: when the pressure drop across the leaves reaches a predetermined level, a predetermined decrease in flow rate, and/or a calculated load level. Short cycles or premature filter stoppages are usually the result of (a) inadequate body feed; (b) too high a flow rate causing the solids to pack; (c) too low a flow rate, which can allow the solids to settle and block the flow rate; (d) blinded screens that reduce the filter surface area; or (e) a greater solids load than the filter capacity. The perfect circumstance is when the differential pressure is reached and the flow rate is severely reduce at the same time that the calculated filter capacity is exhausted. Once the filter cycle is complete, the filter cake must be removed and the process repeated all over again [39].

Traditionally, either plate and frame filters or pressure leaf filters have been used for bleach clay removal. The sequence of change in usage was approximately as follows: plate and frame, pressure leaf, self-cleaning—closed, and automated filters. Pressure leaf filters began to replace plate and frame presses for several reasons. One of the major ones was that the leaf filters were easier to clean than the plate and frame presses, and labor costs were less. Labor costs have been the impetus for more complete automation of the bleaching operations and all the other processes. Currently, completely self-cleaning closed filters that operate on an automated cycle are available [40].

5.8 Bleaching By-product

The spent bleaching earth removed from the bleached oil with filters represents a substantial amount of waste material. The most common handling procedure is to discard spent bleaching earth to a landfill directly from the filters. The spent bleaching earth oxidizes rapidly when exposed to the air to develop a strong odor, and spontaneous combustion easily occurs, especially with oils high in polyunsaturates. Therefore, it must be covered with soil or sand soon after dumping.

The oil content of the spent bleaching earths may range from 25 to 75% of the earths' weight. Oil retention is affected by the type of filters, the type of refined oil bleached, and the degree of color reduction. It is important to recover as much of this oil as possible, but methods that are too efficient may cause desorption of the impurities adsorbed by the bleaching earth from the refined oil. Since it is possible to remove a substantial portion of the oil from the spent earth, it may become a legal requirement in the future. Oil can be recovered by several methods, some performed on the cake while it is still in the filter and others after it has been removed from the filter [41–44]. Some of the procedures for oil recovery in the filter are

- cake steaming—Blowing steam through the cake in the filter can reduce the oil content as low as 20%. However, the oil content should not be reduced below 25% because the steam wetting may cause desorption of the impurities below this point to lower the quality of the recovered oil. Also, spent earth with a low oil content oxidizes more rapidly when exposed to the atmosphere.
- hot water extraction—Circulation of hot water at 200°F (95°C) through the filter cake while maintaining a pressure of 5 atm at a rapid flow rate can displace as much as 55 to 70% of the oil for collection and separation. Washing time may be extended to 30 minutes, but 90% of the recoverable oil is obtained in the first 10 minutes. After water washing, the filter cake may be partially dried with steam. Drying with air could cause the filter to catch on fire, especially when oils high in unsaturates are processed.
- solvent extraction—Organic solvents can be used to extract the oil from the filter cake in certain enclosed filters. Solvent extraction provides oil yields of over 95% with a quality comparable to the originally filtered oil. Solvent extraction requires explosion-proof environment, buildings, and equipment, which are quite expensive. In most cases, the less efficient hot water extraction will be more practical than solvent extraction, and it may only be feasible for very large processing facilities.

Other potential methods for recouping the oil from bleaching earth after removal from the filters in separate operations are

- solvent extraction—With special precautions to prevent oxidation and selection of the proper solvent, oil with quality equivalent to the original bleached stock can be recovered from the spent bleaching earth. Hexane, a nonpolar solvent, has performed well, but strong polar solvents such as acetone or trichlorethylene may also recover the impurities separated from the refined oil. This recovery process would require a large volume of spent earth to be economically feasible.
- solvent extraction with oilseeds—Extraction of the bleaching earths in a mixture with oilseeds has been practiced by some extraction plants with processing capabilities. However, the potential problems for this recovery may outweigh the savings; i.e., the mineral content of the meal may be increased beyond the acceptable limits, and the recovered oil may decrease the quality of the new oil extracted. The oxidation products and polymers from the recovered oil could contaminate the fresh oil.
- water extraction—Oil can be extracted from the spent bleaching earth by suspending it in double the amount of water and boiling with a

concentrated lye. The oil accumulates on the surface of the slurry for recovery. The remaining slurry can be centrifuged with separations as high as 85% efficiency. The separated bleaching clay has a light gray color, is almost odorless, and does not ignite spontaneously. It can be used as a landfill material to cover other refuse, instead of requiring soil or sand to cover it. The procedure is simple and relatively inexpensive, but a dark colored, low-quality oil suitable only for technical purposes or possibly cattle feed is obtained.

6. HYDROGENATION

In the United States and Northern Europe, animal fats in the form of butter, lard, and tallow were the major source of edible fats until development of the hydrogenation process made it possible for vegetable oils to be converted into plastic fat forms that the people were accustomed to, with greater flavor stability at a lower cost. From the British patent on liquid-phase hydrogenation issued to Norman in 1903 and its introduction in the United States in 1911, few chemical processes have had as great an economic impact on any industry. Hydrogenation opened new markets for vegetable oils processing and provided the means for the development of many specialty fats and oils products.

There are two reasons to hydrogenate oil. One is to change naturally occurring fats and oils into physical forms with consistency and handling characteristics required for functionality. With hydrogenation, edible fats and oils products can be prepared with creaming capabilities, frying stability, sharp melting properties, and the other functional characteristics desired for specific applications. Another reason for hydrogenation is to increase oxidative stability. Flavor stability is necessary to maintain product acceptability for prolonged periods after processing, packaging, and use as an ingredient in a finished product. A wide range of fats and oils products can be produced with the hydrogenation process, depending upon the conditions used, the starting oils, and the degree of saturation or isomerization.

Liquid-phase, catalytic hydrogenation is one of the most important and complex chemical reactions carried out in the processing of edible fats and oils [45–48]. Most chemistry textbooks describe hydrogenation of oils as a simple saturation of double bonds in an unsaturated fat with hydrogen, using nickel as a catalyst. Actually, that is only one of several very complex reactions during hydrogenation. The products of hydrogenation are a very complex mixture because of the simultaneous reactions that occur: (1) saturation of double bonds, (2) *cis/trans* isomerization of double bonds, and (3) shifts of double bond locations, usually to the lower energy conjugated state.

Chemically, fats and oils are a combination of glycerin and fatty acids called triglycerides. The portion of the triglyceride that can be changed with hydrogenation is classified as unsaturated fatty acids. Saturated fatty acids contain only single carbon-to-carbon bonds and are the least reactive chemically. Physically, they have higher melting points and are solid at room temperature. Unsaturated fatty acids contain one or more carbon-to-carbon double bonds and are liquid at room temperature with substantially lower melting points than the saturated fatty acid counterparts. In the process of hydrogenation, it is possible to chemically react hydrogen gas with the double bonds in the carbon chain of the unsaturated fatty acid, converting it to a more saturated fatty acid or a *trans* isomer, both of which increase its melting point. Table 2.2 illustrates the chemical structure of the natural 18 carbon fatty acids and the changes possible with hydrogenation.

Hydrogenation can take place only when the three reactants have been brought together: unsaturated oil, catalyst, and hydrogen gas. The hydrogen gas must be dissolved in the liquid oil before it can diffuse through the liquid to the solid catalyst surface. Each absorbed unsaturated fatty acid can then react with a hydrogen atom to complete the saturation of the double bond, shift it to a new position, or twist it to a higher melting *trans* form. Both positional and geometric or *trans* isomers are very important to the production of partially hydrogenated fats. If the unsaturated oil to hydrogenation contains mono-, di-, and triunsaturates, there may be competition for the catalyst surface. The di- and triunsaturates are preferentially absorbed and partially isomerized and/or hydrogenated to a monounsaturate until their concentration is very low, permitting the monounsaturate to be absorbed and reacted.

Achievement of the desired hydrogenated oil product is usually measured with solids fat index (SFI), which measures the amount of solids present in a fat at different temperatures from below room temperature to above body temperature. Natural fats are not single compounds, and the hydrogenated products are even more complex mixtures due to the simultaneous reactions. Not only are double bonds saturated with hydrogen, but some of the remaining bonds are isomerized: geometric isomerization changes the low melting *cis* form to a higher melting *trans* form, and positional isomers shift the double bond away from its natural position in the carbon chain. Extensive geometrical or *trans* isomerization tends to give products that are hard at low temperatures but soft at high temperatures, which result in steep SFI curves. A lesser, but significant, effect on melting points is contributed by the positional isomerization, since the shift of a double bond in a carbon chain affects the melting point of the hydrogenated oil. Additionally, the bonds that are shifted can be in either the *cis* or

TABLE 2.2. Hydrogenation Reactants and Reaction Results.

Linolenic Fatty Acid (C-18:3) Melting Point 9°F(−13°C)

```
    H  H  H  H  H  H  H  H  H  H  H  H  H  H  H  H  H  H  H
    H-C-C-C=C-C-C=C-C-C=C-C-C-C-C-C-C-C-C-C-H   Tri-unsaturate
    H  H        H        H         H  H  H  H  H  H  H  H   cis isomer
    18 17 16 15 14 13 12 11 10  9  8  7  6  5  4  3  2  1
```

Linoleic Fatty Acid (C-18:2) Melting Point 19°F(−7°C)

```
    H  H  H  H  H  H  H  H  H  H  H  H  H  H  H  H  H  H  H
    H-C-C-C-C-C-C=C-C-C=C-C-C-C-C-C-C-C-C-C-H   Di-unsaturate
    H  H  H  H  H        H         H  H  H  H  H  H  H  H   cis isomer
    18 17 16 15 14 13 12 11 10  9  8  7  6  5  4  3  2  1
```

Oleic Fatty Acid (C-18:1) Melting Point 61°F(16°C)

```
    H  H  H  H  H  H  H  H  H  H  H  H  H  H  H  H  H  H  H
    H-C-C-C-C-C-C-C-C-C-C=C-C-C-C-C-C-C-C-C-H   Mono-unsaturate
    H  H  H  H  H  H  H  H         H  H  H  H  H  H  H  H   cis isomer
    18 17 16 15 14 13 12 11 10  9  8  7  6  5  4  3  2  1
```

Stearic Fatty Acid (C-18:0) Melting Point 158°F(70°C)

```
    H  H  H  H  H  H  H  H  H  H  H  H  H  H  H  H  H  H  H
    H-C-C-C-C-C-C-C-C-C-C-C-C-C-C-C-C-C-C-C-H   Saturated
    H  H  H  H  H  H  H  H  H  H  H  H  H  H  H  H  H  H  H
    18 17 16 15 14 13 12 11 10  9  8  7  6  5  4  3  2  1
```

Trans Isomer—Elaidic Fatty Acid (C-18:1) Melting Point 111°F(44°C)

```
    H  H  H  H  H  H  H  H       H  H  H  H  H  H  H  H  H
    H-C-C-C-C-C-C-C-C=C-C-C-C-C-C-C-C-C-C-C-H   Mono-unsaturate
    H  H  H  H  H  H  H       H  H  H  H  H  H  H  H  H   trans isomer
    18 17 16 15 14 13 12 11 10  9  8  7  6  5  4  3  2  1
```

Positional Isomer—Oleic Fatty Acid (C-18:1) Melting Point 91.4°F(33°C)

```
    H  H  H  H  H  H  H  H  H  H  H  H  H  H  H  H  H  H  H
    H-C-C-C-C-C-C-C-C-C-C-C-C=C-C-C-C-C-C-C-H   Mono-unsaturate
    H  H  H  H  H  H  H  H  H  H  H         H  H  H  H  H   cis isomer
    18 17 16 15 14 13 12 11 10  9  8  7  6  5  4  3  2  1
```

trans form, which further substantiates the complexity of the hydrogenated oil process.

Selective hydrogenation is the tool by which partial hydrogenation can be accomplished in a controlled manner. Selectivity is the saturation with hydrogen of the double bonds in the most unsaturated fatty acid before that of a less unsaturated fatty acid. In a theoretical sense, an oil hardened with perfect preferential selectivity would first have all of its linolenic fatty acids (C-18:3) reduced to linoleic fatty acids (C-18:2) before any linoleic was reduced to oleic (C-18:1); then, all linoleic fatty acids would be reduced to oleic before any oleic were saturated to stearic (C-18:0). Unfortunately, this does not happen in actual practice, but it is possible to vary the hydrogenation rate of linoleic to that of oleic from very selective conditions of 50 to 1 to less selective conditions of 4 linoleic to 1 oleic. The latter is generally described as nonselective.

Formation of the high melting unsaturated fats or isomerization accompanies hydrogenation and appears to be in proportion to the selectivity of the reaction. Therefore, compromises must be made between selectivity and isomer formation when determining the hydrogenation conditions best for the various basestocks. Control of the operating variables that affect the hydrogenation of fats and oils is necessary to produce the desired product functionality.

6.1 Operating Variables

Hydrogenation is a reaction of three components: oil, hydrogen, and catalyst. The reaction takes place on the surface of the catalyst where the oil and gas molecules are adsorbed and brought into close contact. Therefore, any condition that affects the catalyst surface or controls the supply of gas to the catalyst surface will, in turn, affect the course and rate of the reaction. The variables that can affect the results of the hydrogenation are temperature, degree of agitation, hydrogen pressure in the reactor, catalyst amount, type of catalyst, hydrogen gas purity, feedstock source, and feedstock quality. The effects of the variables are as follows:

- temperature—Hydrogenation, like most chemical reactions, proceeds at a faster rate with increased temperatures. An increase in temperature decreases the solubility of the hydrogen gas in the liquid oil while increasing the reaction rate. This causes quicker hydrogen removal from the catalyst to effect a lesser quantity of hydrogen on the catalyst surface, resulting in a high selectivity and isomer formation. Therefore, increased temperature increases selectivity, *trans* isomer development, and the reaction rate that results in a steep SFI curve.

  Since hydrogenation is an exothermic reaction, it will create heat as long as the reaction is active; a decrease of one iodine value increases the reaction temperature by 1.6 to 1.7°C (2.9 to 3.1°F) Temperature increases will increase the reaction rate until an optimum is reached. At this point, cooling of the reaction mixture is required to continue hydrogenation. The optimum temperature varies for different products, but most oils probably reach their maximum temperature at 450 to 500°F (230 to 260°C).

- pressure—Most edible fats and oils hydrogenations are performed at hydrogen pressures ranging from 0.8 to 4.0 atmospheres (10 to 60 psig). At low pressures, the hydrogen gas dissolved in the oil does not cover the catalyst surface, while at high pressure, hydrogen is readily available for saturation of the double bonds. The increased saturation rate results in a decrease in *trans* isomers development and selectivity to produce a flatter SFI curve.

- agitation—The main function of agitation is to supply dissolved hydrogen to the catalyst surface, but the reaction mass must also be agitated for the distribution of heat or cooling for temperature control and suspension of the catalyst throughout the oil mixture for uniformity of reaction. Agitation has a significant effect upon selectivity and isomerization: both are decreased since the catalyst is supplied with sufficient hydrogen to increase the reaction rate.

- catalyst level—Hydrogenation reaction rate will increase as the catalyst concentration is increased up to a point and then levels off. The increase in rate is caused by an increase in active catalyst surface. However, a maximum is reached because at very high levels hydrogen will not dissolve quickly enough to adequately supply the higher catalyst levels. Both selectivity and *trans* isomer formation are increased with catalyst concentration increases but only slightly.

- catalyst type—The choice of catalysts has a strong influence on the reaction rate, preferential selectivity, and geometric isomerization. Nickel catalysts are used almost exclusively for edible fats and oils hydrogenation. Catalysts are prepared by a variety of techniques, some proprietary to the catalyst supplier. However, the usual nickel catalyst is prepared by the reduction of a nickel salt and is usually supported on an inert solid or flaked in hard fat or a combination of the two. Different catalysts will give different selectivity rates. The activity of a catalyst depends on the number of active sites available for hydrogenation. These active sites may be located on the surface of the catalyst or deep inside the pores. High-selectivity catalysts enable the processor to reduce the linolenic fatty acid without producing excessive amounts of stearic fatty acid, thus producing a product with good oxidative stability and a low melting point. The selectivity characteristics of a catalyst are unrelated to the catalyst's ability to form *trans* because the catalyst may have a very low or very high selectivity, but all common nickel catalysts appear to produce the same level of *trans* at the same conditions. However, catalysts may be treated with other materials such as sulfur, which increases the amount of *trans* unsaturation.

Sulfur-poisoned catalysts produce larger quantities of *trans* isomers in hydrogenated oils. Reaction with sulfur inhibits the capacity of nickel to adsorb and dissociate hydrogen, reducing the total activity of the catalyst. As the ability of the nickel to hydrogenate is reduced, its tendency to promote isomerization is enhanced. Hydrogenated oils with a relatively high melting point at a high iodine value, which results in very steep SFI slopes, are the result of the high *trans* isomers content. Commercially, sulfur-treated catalysts have been found to provide more uniform performance than products that are sulfur poisoned during processing.

Copper-chromite catalysts have been used for selective hydrogenation of linolenic fatty acid to linoleic fatty acid in soybean oil for a more flavor-stable salad oil with higher winterization yields. The selectivity offered by these catalysts is excellent, but the activity is poor and they are more sensitive to catalyst poisons.

Precious metals have been investigated and found effective as hydrogenation catalysts. Evaluations have shown that basestocks hydrogenated with 0.0005% palladium modified with silver and bismuth were exceedingly more active and slightly more selective with more *trans* development than equivalent stocks prepared with nickel catalyst. Subsequent evaluations have shown that the precious metals are more active at lower temperatures than nickel. Oils have been hydrogenated at 60°C (140°F) with precious metals while temperatures above 130 to 140°C (265 to 285°F) are required with nickel catalysts. *Trans* isomer development is increased as the hydrogenation temperature is increased. Therefore, less *trans* development should be obtained with the precious metal utilization at low temperatures [50]. Palladium has been found some 30 times as active as nickel since only 6 ppm is required to replace 200 ppm nickel. The principal deterrent to the use of palladium has been economics, both in the initial costs and recovery problems associated with the minute quantities required.

- catalyst poisons—Refined oils and the hydrogen gas can contain impurities that modify or poison the catalyst. Catalyst poisons are a factor that can have a significant effect upon the product. The poisons effectively reduce catalyst concentration with a consequent change in the selectivity, isomerization, and rate of reaction. Impurities present in both the feedstock oil and hydrogen gas are known to have a deleterious effect upon nickel catalysts. Hydrogen gas may contain carbon monoxide, hydrogen sulfide, or ammonia. Refined oil can contain soaps, sulfur compounds, phosphatides, moisture, free fatty acids, mineral acids, and a host of other materials that can change the catalyst. Studies have determined that 1 ppm sulfur poisons 0.004% nickel, 1 ppm phosphorus poisons 0.0008% nickel, 1 ppm bromine poisons 0.00125% nickel, and 1 ppm nitrogen poisons 0.0014% nickel. Sulfur primarily affects the activity to promote isomerization by inhibiting the capacity of the nickel catalyst to absorb and dissociate hydrogen. Phosphorus in the form of phosphatides and soaps affects selectivity by residing at the catalyst pore entrance to hinder the triglyceride exit for a higher degree of saturation [49]. Water or moisture and free fatty acids are deactivators that decrease the hydrogenation rate by reacting chemically with the catalyst to form nickel soaps.

- catalyst reuse—Hydrogenation with used catalysts offers economic advantages when sufficient activity remains after the previous use. However, at some point there are diminishing returns due to filtration problems, as well as changes in reaction rate and selectivity. The filtration problems are that (1) free fatty acids in the oils react with the catalyst to form nickel soaps that can blind filters; (2) decreased particle size from mechanical attrition results in colloidal nickel, which will pass through the filter screens; and (3) excessive levels are required to maintain the desired nickel content and activity due to catalyst poisons and dilution with filter aid and oil. Selectivity will decrease with each use while *trans* isomerization increases with each catalyst reuse. It has also been observed that *trans* isomers increased more rapidly when the catalysts was exposed to air after use. Usually, fats and oils processors specify new catalysts for the production of critical basestocks and reuse catalysts for those products where selectivity is not as important. In a catalyst reuse program, the once used catalysts are segregated to permit identification and are used by lot, with care taken to maintain the identity of each. The lots are graded down on the basis of the activity during the last use. Each succeeding grade is used for products requiring a lower degree of selectivity until finally it is used to harden low iodine value hard fat. In the latter usage, selectivity is of no consequence since the hardstock is substantially completely saturated and it makes no difference if the catalyst is selective or nonselective.

- source oils—Hydrogenation selectivity depends upon the type of unsaturated fatty acids available and the number of unsaturated fatty acids per triglyceride. Those oils with high linolenic and/or linoleic fatty acids levels hydrogenate more rapidly and to higher melting points than oils with high-oleic fatty acid levels. The relative hydrogenation reactivity for the 18 carbon fatty acids are [51]

Fatty Acid		Relative Reactivity
Linolenic	C-18:3	40
Linoleic	C-18:2	20
Oleic	C-18:1	1

6.2 Hydrogenation Systems

Batch hydrogenation is most commonly used in the edible oil industry, primarily because of the simplicity and flexibility for use with different source oils. Essentially all that is required is a reaction vessel, usually referred to as a converter, that can withstand 7 to 10 bar (150 psig) pressure, with an agitator, heating and cooling coils, a hydrogen gas inlet, piping and pumps to move the oil in and out, and a sample port for process control of the reaction. The converter provides the means to control three reaction variables: pressure, temperature, and rate of agitation.

Two different batch converter designs utilized for the partial hydrogenation of edible fats and oils are recirculation and "dead-end." In the recirculation system, hydrogen gas is introduced at the bottom of the vessel, and nonreacted hydrogen gas is withdrawn from the headspace, purified, and returned to the converter. The converter is almost always filled with hydrogen under pressure in the operation of the recirculation system. Hydrogenation begins immediately when the catalyst is added with the oil charge during the heating period and thereafter until the endpoint is attained when recirculation is discontinued. Reaction temperature is controlled by circulating water through the cooling coils to carry away the heat of reaction. The hydrogenated oil is pumped out of the converter through an external cooler to a filter for catalyst removal.

A dead-end hydrogenation system is outlined in Chart 2.6. The converter is loaded with oil from a scale tank or metering device. Converter vacuum is utilized to deaerate, dry, and prevent any hydrogenation while heating with steam to reaction temperature. Catalyst, slurried in a portion of the feedstock, is added during the heating period. When the oil reaches reaction temperature, the vacuum is discontinued, and hydrogen is added until the specified pressure is attained. This pressure is maintained during the hydrogenation. An agitator designed to provide efficient hydrogen dispersion is necessary to create a vortex to draw hydrogen from the headspace back into the oil. When the exothermic reaction has raised the oil temperature close to the maximum specified temperature, cooling water is introduced into the coils. Samples are drawn from the converter via the sample port as the reaction proceeds to measure the hydrogenation progress. Agitation is suspended whenever waiting for laboratory analysis to confirm that the endpoint has been reached. When the endpoint has been attained, the hydrogen is vented to the atmosphere through the vacuum system, and the oil is cooled in the converter, in a drop tank, or with a heat exchanger. After cooling to 150°F (65°C), the oil is filtered through a black press to separate the catalyst from the oil. Hydrogenation black presses traditionally have been of the plate and frame or pressure leaf variety.

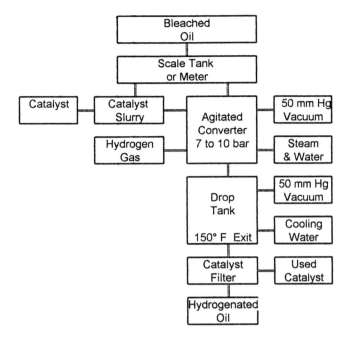

CHART 2.6. Dead-end batch hydrogenation process flow.

From an operations standpoint, the two types of converters do not differ very much. In general, the dead-end type is preferred by many processors because it (1) requires less energy, (2) offers more versatility, (3) requires less capital and operating costs, and (4) is safer than the recirculation system. Quality- and performance-wise, the advantages for the dead-end system are (1) oxidation and hydrolysis prevention through deaeration and dehydration provided by the vacuum during heat up and cooling, (2) more positive control of the reaction for product uniformity, and (3) the ability to vary the hydrogen pressure as well as temperature.

Most hydrogenations of edible fats and oils are performed both in the United States and the rest of the world in batch converters. Continuous hydrogenation systems have been available for quite some time, but their commercial usage has been limited for several reasons. The maximum value for any continuous operation is realized when it is used to produce large quantities of the same product. Considerable out-of-specification product can be produced during a change from one product to another. Since most fats and oils processors produce a variety of products, several different basestocks are routinely required that can be produced more uniformly with batch hydrogenation systems.

6.3 Hydrogenation Control

In any hydrogenation operation, except those carried out to make low iodine value hard fats, the ultimate aim is to produce a partially hydrogenated basestock with a definite preconceived consistency or a basestock suitable for blending with other basestocks or oils to produce the desired finished product functionality. Batch-to-batch variation in consistency is encountered even when the same hydrogenation conditions are maintained due to differences in the feedstock oil, catalyst activity, and selectivity, as well as the other minor variables. It is therefore important to identify controls that will permit the reaction to be stopped at a point that will provide the desired consistency. These controls are usually exercised at the end of the hydrogenation but can be used throughout the reaction to follow the hydrogenation progress. Physical consistency of most finished shortenings, margarines, and other fats and oils products is identified by analytical methods such as solids fat index (SFI), iodine value (IV), and/or melting points. However, time restraints during hydrogenation require more rapid controls. Hydrogenation controls used to determine basestock endpoints are

- refractive index—Hydrogenation reduces both the iodine value and the refractive index of oils. The relationship between the iodine value and refractive index depends upon the molecular weights of the glycerides, which is very nearly the same for most oils. The exceptions are the oils high in either lauric or erucic fatty acids. Correlation between iodine value and refractive index is not precise but will be within one or two units, which should be adequate to monitor the hydrogenation reaction and indicate when to interrupt the reaction for more precise evaluations.
- Mettler dropping point—The relationship of iodine value to melting point can be changed by varying hydrogenation conditions, catalyst types, and levels. Therefore, it is necessary to measure both refractive index and melting characteristics for most basestocks with iodine values below 90. Mettler dropping point analysis can provide a reliable result in less than 30 minutes for these basestocks. Usually, the oil is hydrogenated to a refractive index before determining the Mettler dropping point, which is the controlling analysis. If the melting point (dropping point) is lower than desired, hydrogenation is continued and the process is repeated until the specified melting point is obtained.
- quick titer—Refractive indices are rarely used for low iodine value hard fat hydrogenation control. The refractometers are generally kept at 40.0 1 0.1°C and the hard fats would solidify on the prism at this temperature. The hard fat is too hard for dropping point determinations and

iodine value or official titer determinations are too time consuming. A nonstandardized "quick titer" evaluation is usually used for endpoint control for the hardfats. In this evaluation, a titer thermometer is dipped into a hot sample directly from the converter and rotated in the air until the fat clouds on the thermometer bulb. The correlation between iodine value and quick titer results is different for each source oil; therefore, quick titer limits must be predetermined for each product.

6.4 Hydrogenated Basestock System

Most prepared foods are formulated with ingredients designed for their application or, in many cases, specifically for the particular product and/or processing technique employed by the producer. These customer-tailored products have expanded the product base for fats and oils processors from a few basic products to literally hundreds. Each of these products could be formulated to require a different hydrogenated product for each different product. This practice, with the ever-increasing number of finished products, would result in a scheduling nightmare with a large number of product heels tying up tank space and inventory. Basestock systems with a limited number of hydrogenated stock products for blending to meet the finished product requirements are utilized by most fats and oils processors. The advantages provided by a well-designed basestock system are basically control and efficiency [52]. The control advantages are

- hydrogenated oil batch blending to average minor variations
- increased uniformity by the production of the same product more often
- reduced contamination afforded by the ability to schedule compatible products together
- elimination of product deviations generated from attempts to use product heels
- elimination of rework generated by heel deterioration before use

The efficiency advantages contributed by a basestock system are

- hydrogenation scheduling to maintain basestock inventories, rather than reacting to customer orders
- hydrogenation of full batches, instead of producing some partial batches to meet demands
- better reaction time to meet customer requirements

Basestock requirements will vary with each processor, depending upon the customer requirements, which dictate the finished products produced. The basestock systems can include several source oils or be limited to almost a single oil type. In either case, the basestock inventories usually consist of a few hydrogenated products that cover a wide range for blending to the desired consistencies:

- brush hydrogenated basestocks—For many edible fat ingredient specifications, a liquid oil is required. To guarantee an acceptable shelf life, the level of polyunsaturates should be low, with an absence or severely reduced level of linolenic fatty acids (C-18:3). This can be achieved by a light and highly selective hydrogenation of an oil within the oleic-linoleic fatty acid group like soybean, sunflower, or canola. During hydrogenation the iodine value drop is kept to a minimum to reduce the formation of saturated fatty acids and the *trans* isomers formation is largely suppressed. To reduce the formation of *trans* isomers, the hydrogenation should be performed at a low temperature. A pressure of 3 to 4 bar (30 to 40 psig), in combination with new catalysts with high activity, selectivity, and poison resistance should be used. Optimum conditions will vary considerably, depending upon the geometry of the converter, agitator, hydrogen gas purity, and the other hydrogenation variables. After hydrogenation, this basestock can be winterized or fractionated to produce a flavor-stable salad oil or a high stability liquid oil depending upon the extent of the hydrogenation. This basestock class is also very useful in margarine oil blends, snack frying oil, and in specialty product formulations.

  partially hydrogenated flat basestocks—Many food products require fats and oils products that have an extended plastic range with good oxidative stability. The products must be soft and plastic at room temperature and still possess some body at temperatures of 100°F (38°C), with melting points only slightly above body temperature. Stability is important because of the probable exposure to baking or frying temperatures and/or long shelf life expectancy. These basestock requirements can impose a conflicting set of operating conditions. Highly selective conditions are desirable to convert all linolenic and as much linoleic fatty acids as possible to oleic fatty acids for maximum stability. However, highly selective conditions also favor the formation of *trans* isomers, which are undesirable for this application. The *trans* isomers have higher melting points than the normal oleic fatty acid without the stability improvement. Further, the *trans* isomers restrict the amount of saturated fatty acids that would increase the stability and serve to extend the plastic range while providing the high-temperature body desired. Usually, moderately selective conditions are utilized to produce these flat SFI basestocks, i.e. , relatively low temperatures to 300 to 350°F (150 to 175°C) with high pressures of 20 to 30 psig (2 to 3 bar) with a selective catalyst that has *trans* isomer suppressant qualities. Reuse catalysts should be avoided, which enhance *trans* isomer formation.

- partially hydrogenated steep basestocks—The physical properties of these basestocks are characterized by steep SFI curves or high solids contents at the lower measuring temperatures with an absence of solids at temperatures higher than body temperature. Hydrogenation of these basestocks should be carried out with highly selective conditions or high temperature and low pressure. Used catalysts, sometimes enhanced with new very selective catalysts, can be utilized to help achieve the desired high selectivity and *trans* isomer formation. These basestocks are beneficial in blends for margarine oils, high-stability frying shortenings, nondairy products, fillings, and other products requiring a sharp melting point with good flavor stability while providing the required firmness at room temperature.
- low iodine value hard fats—These basestocks are many times referred to as fully hydrogenated hard fats, or stearins; however, regulations require a zero iodine value for the fully saturated designation. Since catalyst activity is the only criterion with these hydrogenations, used catalysts can be utilized. In general, high-pressure, 5 bar (60 psig) or higher, and high temperature, 450°F (230°C), conditions are used for these basestocks to make the reaction progress as rapidly as possible.

Table 2.3 outlines a soybean oil basestock system with seven hydrogenated stock oils ranging from a lightly hydrogenated 109 iodine value to a saturated hard fat with an 8.0 maximum iodine value. Utilization of a similar basestock system designed for the required product mix should enable fats and oils processors to meet most shortening specifications by blending two or more basestocks, except for some specialty products, which can only be made with special hydrogenation conditions.

The solids fat index (SFI) is one of the most important consistency measurements that also indicates the selectivity of the conditions used to prepare the individual basestocks. It measures the amount of solids present in a fat at different temperatures from below room temperature to above normal body temperature. A fat can appear to be a solid but really exists as a semisolid and does not have a distinct melting point. Natural and hydrogenated fats and oils melt over a wide range of temperatures. Solids fat index analysis determines the solid or unmelted portion of a fat over a measured temperature range. These results relate to the consistency of the fat and oil product in terms of its softness, plasticity, organoleptic, and other physical properties important for its use as an ingredient in prepared foods. The slope of the SFI curve shows the effects of hydrogenation selectivity

TABLE 2.3. Soybean Oil Basestock System.

Basestock Type	Brush	Flat		Steep			Saturated
Iodine value	109	85	80	74	66	60	>8
Solids fat index, %							
at 10.0°C / 50°F	4.0 max.	15-21	22-28	38-44	59-65	65-71	*
at 21.2°C / 70°F	2.0 max.	6-10	9-15	21-27	47-53	56-62	*
at 26.7°C / 80°F		2-4	4-6	13-19	42-48	51-57	*
at 33.3°C / 92°F				3.5 max.	23-29	37-43	*
at 40.0°C / 104°F					3-9	14-18	*
Mettler dropping point, °C	**	28-32	31-35	34-37	41-45	45-48	*
Quick titer, °C	**	**	**	**	**	**	50-54
Hydrogenation conditions	†	†	†	†	†	†	†
Gassing temperature, °F	300	300	300	300	300	300	300
Hydrogenation temp, °F	325	350	350	440	440	440	450
Pressure, bar	3.5	2.0-3.0		0.7-1.0			5
Catalyst, % nickel	0.02	0.02		0.04-0.08			0.04-0.08
Agitation	Fixed	Fixed		Fixed			Fixed

* Too hard to analyze.
** Too soft.
† Optimum conditions will vary considerably dependent upon the converter, agitation, hydrogen gas purity, etc.

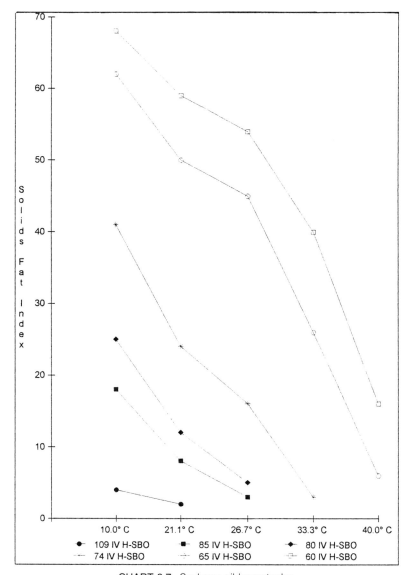

CHART 2.7. Soybean oil basestocks.

as it affects consistency. The SFI curve slope becomes steeper as the hydrogenation conditions are made more selective, that is, the highest temperature, lowest pressure, and the highest level of a selective catalyst. The slope of the SFI curve becomes flatter as the hydrogenation reaction conditions are made less selective with lower temperatures, higher pressure, and low catalyst levels. These effects are illustrated for the soybean oil basestocks graphically in Chart 2.7.

7. POSTBLEACHING

A separate bleaching operation, immediately following the hydrogenation process, has three general purposes: (1) to remove traces of nickel that escape the catalyst recovery filtration, (2) to remove undesirable colors generally of greenish hue, and (3) to remove peroxides and secondary oxidation products. This bleaching process generally employs a bleaching earth and a metal chelating acid to reduce the residual nickel content to the lowest possible level. As much as 50 ppm nickel, mostly in colloidal form, can remain in the hydrogenated oil after the black press filtration. Trace amounts of nickel remaining in the oil adversely affect the stability of the oil by accelerating the oxidation process. After postbleaching, the trace metal levels in the oils should have been reduced to <0.1 ppm nickel and/or <0.02 ppm copper for oxidative stability.

Green colors can emerge in the hydrogenated oils because of the heat bleaching of yellow and red masking pigments during hydrogenation. In the course of hydrogenation, the carotenoid pigments can be reduced to a colorless form while the chlorphyllic pigments merely have their absorption maxima shifted from 660 to 640 μm. After the removal of the masking reddish pigments, the greenish pigments predominate, resulting in green appearing oils. The green pigments can be removed by adsorption on acid-activated clays with some difficulty, depending upon the severity. Green colors are more easily removed in the prebleach process before the color has been set by heating during hydrogenation.

Postbleach systems can be exact duplicates of the prebleach process. However, most fats and oils processors prefer batch systems over continuous due to the production of a wide variety of hydrogenated basestocks from several different source oils. Vacuum batch systems are normally selected over atmospheric processes for the oxidative protection afforded the oils. The bleaching conditions are normally 0.1 to 0.2% activated bleaching earth, with approximately 10 ppm phosphoric acid or 50 ppm citric acid added as a chelating agent and bleached at 180°F under a 25" Hg minimum vacuum. After bleaching, the spent earth and adsorbed impurities are

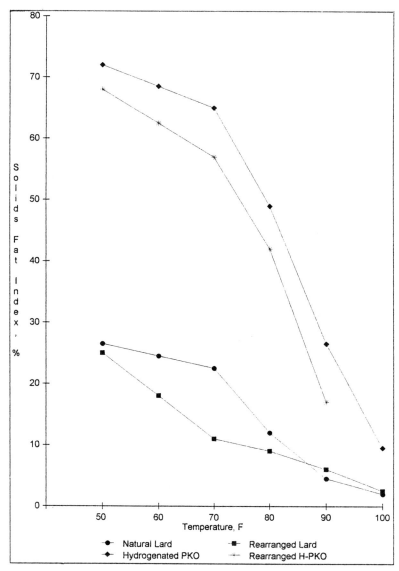

CHART 2.8. Rearrangement effect upon SFI.

removed with a pressure filter and the oil is cooled to 130°F before further processing or inventoried for basestock blending. The process flow is diagrammed on Chart 2.8.

An alternative to postbleaching practiced by some processors is the addition of activated carbon to the oil with the catalyst during hydrogenation. Activated carbon is very effective in removing chlorophyll pigments. After filtering, the oil is treated with citric or phosphoric acid and polish filtered to capture the colloidal trace metals remaining in the oil and any traces of the carbon that has escaped the catalyst filter.

8. INTERESTERIFICATION

The term *interesterification* refers to the fats and oils reaction in which fatty acid esters react with other esters or fatty acids to produce new esters by an interchange of fatty acid groups. More simply stated, interesterification can be visualized as a breakup of a specific glyceride, removal of a fatty acid at random, shuffling it among the rest of the fatty acid pool, and replacement at random by another fatty acid. The process is called interesterification, but because of the random rearrangement of the fatty acids of the natural oil, it is also commonly referred to as randomization, rearrangement, or modification.

Natural fats and oils are mixtures of mixed triglycerides. Their functional properties as ingredients in prepared foods are directly related to the type of triglycerides in the fats and oils. Triglyceride type is determined by the fatty acid composition of the triglyceride and the distribution of the fatty acids on the individual triglyceride molecules. Distribution of the fatty acids on the individual triglyceride molecule and the quantity of each triglyceride type depend on the proportions of the individual fatty acids, the fat or oil source, and the products processing history. Nature provides each fat and oil with a selective fatty acid distribution among the glycerides, which affects the product's consistency as either a solid or a liquid. Trisaturated triglycerides provide structure, disaturated monounsaturated triglycerides provide both structure and lubricity, and the lower melting unsaturated triglycerides provide lubricity only. Table 2.4 shows the melting points of the common triglycerides found in most oleic-linoleic fatty acid group fats and oils [53]. The interesterification processes can alter the original order of distribution of the fatty acids in the triglyceride-producing products with melting and crystallization characteristics different from the original oil or fat. Unlike hydrogenation, interesterification neither affects the degree of saturation nor causes isomerization of the fatty acid double bond. It does not change the fatty acid composition of the starting material but rearranges

TABLE 2.4. Common Triglyceride Melting Characteristics.

	Degree F	Degree C
Trisaturate		
stearic-stearic-stearic	149	65
stearic-stearic-palmitic	142	61.1
stearic-palmitic-palmitic	140	60
palmitic-palmitic-palmitic	133	56.1
Disaturate-Monosaturate		
stearic-stearic-oleic	107	41.7
stearic-palmitic-oleic	100	37.8
palmitic-palmitic-oleic	95	35
stearic-stearic-linoleic	91	32.8
stearic-palmitic-linoleic	86	30
palmitic-palmitic-linoleic	81	27.2
Diunsaturate-Saturate		
stearic-oleic-oleic	73	22.8
oleic-oleic-palmitic	60	15.6
stearic-oleic-linoleic	43	6.1
stearic-linoleic-linoleic	34	1.1
palmitic-linoleic-oleic	27	−2.8
palmitic-linoleic-linoleic	22	−5.6
Triunsaturate		
oleic-oleic-oleic	42	5.6
oleic-oleic-linoleic	30	−1.1
oleic-linoleic-linoleic	20	−6.7
linoleic-linoleic-linoleic	8	−13.3

the fatty acids on the glycerol molecule. The process of interesterification can be considered as the removal of fatty acids at random from the glyceride molecules, shuffling of these acids, and replacement of them on the glyceride molecules at random. This change in the distribution of the fatty acids among the glycerides affects the physical nature and behavior of the fats.

Commercially, the interesterification process is utilized for processing edible fats and oils to produce confectionery or coating fats, margarine oils, cooking oils, frying fats, shortenings, and other special application products. Interesterification has not been a preferred process in the United States, except for some very specific applications: (a) modification of lard to function as a plastic shortening and (b) random rearrangement of lauric oil-based confectionery or coating fats. Interesterification of lard was used to obtain a different triglyceride composition for more desirable physical properties than the original fat. Plasticized shortening made from randomized lard has a smoother appearance and texture because the crystal structure has been changed from beta to beta-prime, which also helps it retain its appearance during storage with better creaming properties. Currently, the market for premium lard shortenings no longer exists due to the reduced use of animal fats and the competitive pricing of vegetable oils. The primary, but limited, use of interesterification in the United States is for the processing of value-added specialty fats such as confectionery or coating fats. These fats, called hard butters, are used as substitutes for cocoa butter in coatings and other applications characterized by a relatively high solid fat content at room temperature with a sharp melting point. The interesterified hard butters are typically composed of rearranged, hydrogenated palm kernel or coconut oils with lesser quantities of other vegetable oils to adjust the melting point. Outside the United States, interesterification has a much wider usage to process basestocks or finished products for margarine oils, cooking oils, cocoa butter equivalents (CBEs), and other specialty products [54].

Two types of chemical interesterification or rearrangement processes are practiced: random or directed. In random rearrangement, the fatty acid radicals freely move from one position to another in a single glyceride or from one glyceride to another. As the fatty acids rearrange, they reach an equilibrium that is based on the composition of the starting material and is predictable from the laws of probability. Directed rearrangement modifies the fatty acid randomization by upsetting the equilibration mixture. This process is carried out at low temperatures to allow crystallization of a portion of the mixture while the interchange of fatty acids is continuing in the liquid portion. This produces a different composition composed of larger proportions of high-melting glycerides and a corresponding larger proportion of very low-melting glycerides. The degree of difference depends upon the temperature, time, and other conditions of the reaction.

A third rearrangement process, enzymatic interesterification, is now used to produce high-value-added products, such as structured triglycerides for confectionery use, in Europe. The major advantages for the enzymatic interesterification over chemical processes are the specificity available in lipase catalysts and the greater degree of reaction control. Oil modification by lipases is performed under anhydrous conditions at temperatures up to 160°F (70°C). Two types of enzyme catalyst are currently available commercially: a random lipase that produces products similar to chemical randomization and a 1,3 fatty acid specificity lipase that allows the production of specific triglycerides at high yields [50].

8.1 Chemical Rearrangement Catalyst

Fatty acid rearrangement may occur without the use of a catalyst at a temperature of 475°F (250°C) or higher, but most processors use alkali metal alkylates or alkali metals to speed up the reaction. Fats reacted at high temperatures without the assistance of a catalyst proceed slowly to equilibrium and have other undesirable changes as well, i.e., isomerization, polymerization, and decomposition. Some of the chemical rearrangement catalysts commercially utilized are [55,56]:

- Sodium methylate, a alkali metal alkylate, is the most widely used low-temperature interesterification catalyst. It is active at a lower temperature, speeds up the reaction, has a relatively low cost, does not require a vacuum during processing, and is easily dispersed in fat. Sodium methylate is either used as a powder or as a dispersion in solvents, such as xylene at very low levels, 0.1% if the starting material has a low free fatty acid level and is dry. However, the average usage range is 0.2 to 0.4%.
- Sodium potassium alloy, which has been used as an interesterification catalyst at 0.05 to 0.1% is liquid at ambient temperature and does not require dispersing in a solvent for introduction to the reaction. It can catalyze low-temperature reactions at faster rates than other catalysts but requires high sheer agitation and is typically more expensive but effects a low oil loss. The interesterification reaction starts almost instantaneously with the addition of the catalyst and is complete in as short a time as 5 minutes. The feedstock oil or fat must be very dry before the addition of this catalyst. It will react with moisture to liberate hydrogen gas, which can inactivate the catalyst. Additionally, this creates an explosion potential from the heat and hydrogen gas generated during deactivation.
- Sodium or potassium hydroxide are the lowest cost rearrangement catalyst, but they must be used in combination with glycerol and require a two-stage reaction under vacuum at high temperatures to effect a reaction. The first stage is conducted at 140°F (60°C) to neutralize any free fatty acids, dry the oil, and disperse the catalyst. The reaction mixture is heated to 285 to 320°F (140 to 160°F) during the second stage to effect rearrangement. Glycerol is a necessary component of this catalyst for the reaction to occur and usually forms small amounts of mono- and diglycerides.

The compounds described above are probably not the real chemical interesterification catalyst but serve as initiators in the process of forming the true interesterification catalyst. Most likely, intermediates like sodium glycerate formed in the fat are the active catalyst. When the catalyst is dispersed in a previously dried oil maintained at about 140 to 175°F (60 to 80°C), a white slurry develops. After heating, a characteristic brown color develops, indicating that the reaction has begun. The color change is associated with the active catalyst formation, which is probably an intermediate glycerate anion.

Rearrangement catalyst must be inactivated and removed at the end of the process. Interesterification is a reversible reaction. Most chemical catalysts can be removed by washing the reaction mixture with water to separate a salt or soap-rich aqueous phase. An alternative method is to terminate the reaction with phosphoric acid and remove the solid phosphate salts by filtration. Either technique results in product loss:

(1) Phosphoric acid termination losses:
- catalyst—One gram of sodium methylate catalyst yields: the interesterified product, 5.519 grams of methyl esters, 5.67 grams of sodium soap, and 2.13 grams of diglycerides. The loss for the catalyst level used can be calculated as 11.2 pounds product loss for each pound of catalyst used.
- bleaching earth oil retention—It may be assumed that the bleaching earth will retain its weight in neutral oil or a 0.1% bleaching earth will remove the phosphate salts and retain 0.1% neutral oil.
- deodorizer losses—Deodorization to a 0.05% free fatty acid would necessitate a loss of the free fatty acid content above this level.
(2) Water termination losses—This neutralization process adds the entrained oil losses experienced with water washing to those for the phosphoric acid termination. The only savings for this procedure are the phosphoric acid costs. However, the oil quality improvement with water termination may justify the added expense.

8.2 Endpoint Control

Visually, a brown color develops when the chemical reaction begins and deepens as the reaction continues. In most operations, the reaction is allowed to proceed for a fixed time period after the appearance of the brown color before sampling to determine if the reaction has been completed. The reaction is most often confirmed by monitoring changes for a characteristic particular to the source oil or mixture of oils processed, which usually involve evaluations for specific product changes. The effect of interesterification of glyceride mixtures differs in different cases, depending upon the composition of the original fat, mixture of fats or prior processing.

- melting point—Interesterification may raise, lower, or have no effect upon the melting point, depending upon the starting fat and oil glyceride composition. A high melting mixture of completely hydrogenated fat with a large proportion of a liquid oil will experience a decrease in the proportion of trisaturated glycerides to effect a lower melting point with interesterification. Applied to an oil like cottonseed oil, with a substantial proportion of solid fatty acids but hardly any trisaturated glycerides, the randomization process will raise the melting point because it increases the proportion of fully saturated glycerides. These changes are illustrated by the melting point results before and after random rearrangement listed in Table 2.5 [55,57]. Specific melting point limits must be established to identify the expected change and the suitability of this analytical method for endpoint control. In some cases, there is not a melting point change, or it may be so slight that it will be within the range of normal analytical error. Even though melting point is a rapid reproducible method, it cannot be used as an interesterification endpoint in all cases.

- solids fat index—SFI analysis control requires measurement at several temperatures for definite results to identify a change, and it is very time consuming. However, small changes in melting point evaluations may be accompanied by more significant changes in the SFI content and slope for the curve throughout the range of functionality important temperatures. The changes in trisaturate and disaturate glycerides with interesterification are reflected in the SFI contents before and after the reaction. Chart 2.8 shows the effects of interesterification on two different fats and oils sources—natural lard and hydrogenated palm kernel oil. These SFI results [58,59] verify that the rearrangement effects are dependent upon the composition of the original fat or oil product. Rearranged lard has a flatter SFI slope caused by a higher trisaturate level, which is a more desirable shortening base. Rearranged hydrogenated palm kernel oil results in a steeper SFI slope with a lower melting point for improved eating characteristics. SFI

TABLE 2.5. *Random Rearrangement Melting Point Changes.*

Fat and Oils Products	Original		Rearranged	
	Degree F	Degree C	Degree F	Degree C
Soybean oil	19.4	−7	41.9	5.5
Cottonseed oil	50.9	10.5	93.2	34
Lard	109.4	43	109.4	43
Palm oil	102.9	39.4	108.9	42.7
Palm kernel oil	82.9	28.3	80.4	26.9
Coconut oil	77.9	25.5	82.8	28.2
H–Palm kernel oil	113	45	93.9	34.4
H–Coconut oil	100	37.8	88.9	31.6

analyses are useful for formulation and to confirm that the predetermined results have been attained but are probably too time consuming for interesterification endpoint control.

- differential scanning calorimetry—DSC is most useful for studying the kinetics of crystallization and melting of triglyceride mixtures under dynamic conditions. The heating and cooling thermograms resulting from DSC show distinct differences between some nonrandomized and randomized fats and oils. The basis of the cooling curve is that crystals of fat give off heat on solidifying from liquid oils and absorb heat on melting. Large crystals give up heat so rapidly during formation that the temperature of the fat may rise rapidly during the chilling cycle. The endpoint of lard interesterification is best determined by a cooling curve analysis that indicates the absence of the heat of crystallization associated with untreated lard. The endpoint of interesterification for lauric hard butters or coating fats might be determined by a loss of rapid solidification of the fully saturated glyceride component on cooling.
- glyceride compositional analysis—The basic change that occurs due to interesterification is in the glyceride composition. Therefore, analysis that can identify the glyceride composition should be the most definitive endpoint possible. High-performance liquid chromatography (HPLC) methods can separate triglycerides according to their level of saturation or on the basis of molecular weights.

8.3 Random Chemical Interesterification Process

Random chemical rearrangement of fats and oils can be accomplished using either a batch or continuous process. Both perform the three important rearrangement steps: (1) pretreatment of the oil, (2) reaction with the catalyst, and (3) deactivation of the catalyst. A typical batch rearrangement reaction vessel is equipped with an agitator, coils for heating and cooling, nitrogen sparging, and vacuum capabilities. The process steps for batch rearrangement are [55,56,58]:

(1) Heat the fat to 250 to 300°F (120 to 150°C) in the reaction vessel under a vacuum to dry the oil. Drying is critical since moisture deactivates the catalyst. Moisture levels in excess of 0.01% will require more catalyst to complete the reaction. Additional catalyst usage results in higher product losses. A rule of thumb is that each 0.1% of sodium methylate catalyst results in 1.1% neutral oil loss.

(2) After drying, the fat is cooled to a reaction temperature that ranges from 160 to 210°F (70 to 100°C), depending upon the product and

desired processing conditions. Sodium methylate powder is sucked into the reaction vessel with the vacuum. The amount of catalyst necessary is the amount required to neutralize the free fatty acid, plus a slight excess to catalyze the random rearrangement. Therefore, since one part of sodium methylate will neutralize 5.26 parts of stearic fatty acid and 0.06% excess is enough to catalyze the reaction, the catalyst requirement can be calculated as: free fatty acid times 0.19 plus 0.06 equals the percent sodium methylate catalyst required for the reaction. This mixture is agitated, to form a white slurry indicating good dispersion, for 30 to 60 minutes or until formation of the distinctive brown color indicates randomization. At this point, the mixture is sampled for laboratory analysis to determine if the reaction is complete or requires additional catalyst and time to attain the predetermined endpoint.

(3) When reaction completion is confirmed by the laboratory results, the catalyst is neutralized in the reaction vessel. Neutralization may include the addition of phosphoric acid or carbon dioxide prior to water washing to deactivate the catalyst. Water combines with sodium methylate to form sodium hydroxide and methyl alcohol, which react with the neutral oil to form soap and methyl esters. Product losses are kept to a minimum by neutralizing with phosphoric acid or CO_2 prior to water washing.

Continuous interesterification processes follow the same cycle as the batch process but utilizes different equipment. The process flow for one continuous system is as follows. The oil is heated with a heat exchanger and flash dried with a vacuum oil dryer to bring the moisture level to 0.01% or less. The catalyst is introduced into the hot oil stream and homogenized for dispersion. The homogenized mixture is then passed through a tubular reactor. The reactor residence time can be adjusted by changing the length of the tube. The catalyst is deactivated with water and centrifuged to separate the soap and oil. After separation, the product is vacuum dried to remove the remaining traces of moisture.

8.4 Directed Chemical Interesterification Process

In directed rearrangement processes, one or more of the triglyceride products of the interesterification reaction is selectively removed from the ongoing reaction. Trisaturated glycerides crystallize and are separated from the reaction when the mixture is cooled below its melting point. This selective crystallization upsets the equilibrium and the reaction will produce more trisaturated glycerides to reestablish equilibrium. Theoretically, this process could continue until all the saturated fatty acids are converted into trisaturated glycerides and separated from the reaction. Since this reaction is directed to produce a particular type of glyceride, it is referred to as directed interesterification.

In directed interesterification, only catalysts active at low temperatures are effective, and the rate of random rearrangement is important since the trisaturated glycerides can only precipate as fast as they are formed in the liquid phase. Sodium-potassium alloy (NaK) is more suitable for directed interesterification than either sodium or sodium methylate because of a more rapid activity at low temperatures.

Continuous processes are normally used for directed interesterification because the batch process is difficult to control and would require a number of extra tanks. The process flow for continuous directed rearrangement [55,58,60] is as follows:

(1) The oil is vacuum dried to 0.01% moisture or less.

(2) After drying, the oil is cooled to a temperature just above its melting point with a heat exchanger.

(3) A carefully metered stream of NaK catalyst is added to the product stream and mixed or homogenized to suspend the catalyst throughout the product.

(4) The homogeneous mixture is quick-chilled with a scraped wall heat exchanger to a predetermined point to initiate crystallization of the trisaturated glycerides.

(5) The cooled mixture is transferred to an agitated vessel, where interesterification proceeds under carefully controlled agitation. At this stage trisaturated glycerides are crystallizing while interesterification of the liquid phase is continuing to form more trisaturated glycerides.

(6) Crystallization of the trisaturated glycerides liberates a considerable amount of heat due to fusion, which can increase the reaction temperature beyond the desired point, which necessitates a second chilling step with a scraped wall heat exchanger.

(7) After the second cooling, the product is transferred to another vessel with controlled agitation, where the precipitation of trisaturated glycerides and interesterification continues to the desired endpoint. Crystallization slows as the trisaturates diminish so that this stage requires more time for reaction completion. The level of trisaturated glycerides in the final product can be adjusted by varying the time in the crystallizer, the crystallization temperature, or a combination of the two.

(8) After the desired endpoint has been reached, the catalyst is "killed" by adding water. The amount of water is calculated to provide the desired fluidity for centrifuging to remove the soap phase. Saponification of the

fat can be minimized somewhat by the addition of carbon dioxide with the water to buffer the caustic to a lower pH.

(9) After neutralization of the catalyst, the product can be heated to melt the trisaturated glyceride crystals for centrifugation followed by vacuum drying.

8.5 Enzymatic Interesterification

Enzymes have been used for many years to modify the structure and composition of foods but only recently have been available on a large enough scale for industrial applications. Enzymatic interesterification is now used to produce high value-added structured fats and oils products [50,61,62]. Useful glyceride mixtures that cannot be obtained by chemical interesterification processes are possible by exploitation of the specificity of lipases. In all glyceride reactions, lipases catalyze either the removal or exchange of fatty acid groups on the glycerol backbone. Different lipases can show preferences for both the position of the fatty acid group on the triglyceride and the nature of the fatty acid. Two types of lipase catalyst identified by application specificity have been identified:

(1) Random lipases catalyze reactions at all three positions on the glyceride randomly.

(2) 1,3 Specific lipases catalyze reactions only at the outer 1- and 3-positions of glycerides.

Random lipase interesterification has very little advantages over standard chemical techniques. Nonspecific catalyzed reactions with triglycerides produce products similar to those obtained by chemical interesterification. However, with a 1,3- specific lipase as a catalyst, fatty acid migration is confined to the 1- and 3-positions to produce a mixture of triglycerides not possible with chemical interesterification. These specific lipases' allow the production of a limited range of glycerides to be produced, which can be separated using physical processes.

Lipases are manufactured by fermentation of selected microorganisms followed by a purification process. The enzymatic interesterification catalysts are prepared by the addition of a solvent like acetone, ethanol, or methanol to a slurry of an inorganic particulate material in buffered lipase solution. The precipitated enzyme coats the inorganic material, and the lipase-coated particles are recovered by filtration and dried. A variety of support materials have been used to immobilize lipases. Generally, porous particulate materials with high surface areas are preferred. Typical examples of the support materials are ion exchange resins, silicas, macroporous polymers, clays, and so on. Effective support functionality requirements are (a) the lipase must adsorb irreversibly with a suitable structure for functionality; (b) pore sizes must not restrict reaction rates; (c) it must not contaminate the finished product; (d) it must be thermal stable; and (e) it must be economical. The dried particles are almost inactive as an interesterification catalyst until hydrated with up to 10% water prior to use.

Lipase-catalyzed interesterification of fats and oils can be accomplished using either a stirred batch reactor or with continuous processing using a fixed-bed reactor. The latter is the preferred process, with the advantage of minimized reaction times due to the high catalyst substrate ratio along with the other advantages: (a) catalyst recovery, (b) reduced catalyst damage, and (c) improved operability. The continuous fixed-bed interesterification process begins by dissolving the feedstock in a solvent followed by treatment to remove enzyme catalyst inhibitors, poisons, and particulate materials. This solution is then partially saturated with water prior to pumping through a bed of hydrated catalyst particles. The reaction products are a mixture of triglycerides and free fatty acids. After the reaction, the acids are removed by evaporation and processed for recovery. The acid-free triglyceride is then solvent fractionated to yield the desired triglyceride composition.

8.6 Interesterified Fats and Oils Applications

Interesterification processes can be used to produce fats and oils products with different physical and nutritional properties. Rearrangement of the fatty acids on the glycerol backbone affects the structural properties or melting behavior of the fats and oils products. Often, interesterification is combined with other specialized processing techniques such as fractionation and/or hydrogenation. By combining interesterification with these and other more sophisticated techniques, the fatty acid and glyceride composition can be manipulated to produce the desired physical and functional properties. These products may also be utilized as basestocks for blending with other selected fats and oils products to produce the desired functional properties.

Chemical and enzymatic interesterification processes can affect physical properties by changing the melting properties and in some cases the crystal behavior. The use of these triglycerides in food products can allow the development of specialty fats and oils products more suitable for the desired application performance, nutritional requirements, or both. Interesterification can be utilized to formulate products with less saturated and/or isomerized fatty acids for the production of products with *trans* acids. Margarine and shortening finished products have been produced using interesterified fats and oils instead of the traditional hydrogenated basestocks.

9. WINTERIZATION

The descriptive term of *winterization* evolved from the observation that refined cottonseed oil stored in outside tanks during the winter months physically separated into a hard and clear fraction. Topping or decanting the clear oil from the top of the tanks provided an oil that remained liquid without clouding for long periods at cool temperatures. In fact, some cottonseed salad oils routinely had cold test results of 100 plus hours when topped from outside storage tanks. The clear oil portion became known as winterized salad oil [63]. The hard fraction from the bottom of the tanks was identified as stearin which is the solid portion of any fat.

A need for a liquid oil with these characteristics was created by the use of refrigerators in the home and the requirements of the mayonnaise and salad dressing industry. Mayonnaise could not be made from oils that would crystallize in the refrigerator and cause the emulsion to break. New terminology emerged because of this association with mayonnaise. Winterized oil became known as "salad oil." Summer oils or oil not subjected to winterization became known as "cooking oils." An increased demand for salad oils made it impossible to rely on long-term storage of refined oils for the winterized oil requirement. Processors recognized the obvious solution and created winter conditions indoors.

9.1 Classic Winterization Process

The indoors process developed to simulate the natural winter process consisted of a chilled room held at 42°F (5.6°C) with deep, narrow, rectangular tanks to provide the maximum surface exposure to cooling. Warm, dry, refined, and bleached oil pumped into the chill room tanks began to cool and crystallize out stearin immediately but slowly. Convection heat transfer simulated the outside storage conditions. Agitation was avoided because it fractured the crystal, causing formation of small, soft crystals that were difficult to filter. Cooling with the 42°F room temperature simulated southern U.S. winter conditions closely, requiring 2 to 3 days to produce the desired large crystals for filtering. After the oil temperature equated with the room temperature, it was held for several hours to allow the stearin or hard fraction to precipitate more fully. The stearin was separated from the liquid oil by filtering with plate and frame presses. Early installations relied on gravity feed to the presses, but later, compressed air or positive displacement pumps were utilized to exert 5 to 20 psig pressure to increase the filtration rate. Care was exercised to avoid breaking up the crystals excessively. A slow filtration rate was necessitated by the high oil viscosity, and excessive pressure pressed the stearin into the filter cloths, causing a blockage that stopped the oil flow. A large filter area on the order of two to three pounds of oil per hour per square foot was the general guideline. The stearin cake was melted with hot fat for removal after the filter press was full.

Winterization is still performed with the classic techniques outlined above, but many processors have made equipment and process modifications to improve efficiency. Jacketed enclosed tanks equipped with programmable cooling and agitation have evolved as crystallization cells to replace the open-top, narrow, rectangular tanks cooled by the chill room temperature. However, attempts at forcing crystallization, by means of an excessively cold coolant and rapid agitation, result in small crystals that are virtually unfilterable. Recessed plate and frame or pressure leaf filters have been used in winterization since these filters have the cake holding capacity that the process requires. Obviously, when 15% or more of the feed is removed in the form of stearin, a substantial solids retention capacity is needed. Separation of the stearin from the liquid oil by means of a centrifuge has had some success. The main problems encountered with centrifugal separation is liquid oil yield since the stearin tends to trap excessive amounts of oil.

9.2 Winterization Principle

Winterization is a thermomechanical separation process where component triglycerides of fats and oils are crystallized from a melt. The two-component fractional crystallization is accomplished with partial solidification and separation of the higher melting triglyceride components. The complex fats and oils triglycerides may have one, two, or all three fatty acids, either all the same or different in any of the possible configurations, depending upon the source oil and prior processing.

Fat crystallization occurs in two steps: the first step is called nucleation and the second is crystal growth. The rate of nucleation depends upon the triglyceride composition of the oil being winterized, the cooling rate of the oil, the temperature of the nucleation, and the mechanical power input or agitation. Growth rate is dependent on the crystallization temperature, time, and mechanical input or agitation. A careful selection of the process variables for a particular oil is very important. The ideal is to produce a small number of nuclei around which the crystals grow larger in size with cooling. A large mass of hard-to-filter small crystals result when a large number of nuclei are formed. Poor separation and yield are also the result when crystals group together in clumps that trap large quantities of the liquid phase. The effect of the major processing variables upon winterization performance are [64]

- source oil composition—Nucleation and crystal growth depend upon the composition of the oil being winterized. The various triglycerides in a particular oil will fractionate in the following order: (1) trisaturate—S_3, (2) disaturate monounsaturate—S_2U, (3) monosaturate diunsaturate—SU_2, and (4) triunsaturate—U_3. A portion of the higher melting glycerides will be found with the lower melting liquid oils as a result of eutectic formation and equilibrium solubility.

Since the mixture of triglycerides of an oil is too complex to predict their phase behavior, a given set of winterization conditions are applicable only for the particular feed oil. For partially hydrogenated oils, the composition of the oil and the hydrogenation conditions affect the winterization yield and quality. Hydrogenation conditions should be selected that produce the lowest level of saturates and *trans* fatty acids but still affect the desired iodine value endpoint.

- cooling rate—An essential requirement of the winterization process is a slow rate of chilling. Rapid cooling of the oil results in (1) a mass of very small alpha crystals and (2) a high nucleation rate that increases the viscosity, which in turn restricts crystal growth. Slow controlled cooling rates produce stable beta or beta-prime crystals, depending upon the dominant crystal habit for the source oil winterized, and the viscosity remains low enough to permit nucleii movement to allow crystal growth. Therefore, the cooling rate is dependent upon the source oil and prior processing.
- crystallization temperature—The crystal growth rate is affected by the temperature of crystallization. A high viscosity resulting from too low a temperature reduces the crystal growth rate. Control of the temperature after crystallization begins is important for transformation from the alpha to the stable beta-prime or beta crystal habit. If the process is not properly controlled at this stage, an unstable crystal will develop.

  A temperature differential between the coolant and the oil must be maintained to avoid shock chilling. A 25°F (14°C) differential has been found appropriate for oil at the beginning of the process. The differential can be reduced to 10°F (5.6°C) by the time the oil reaches 45°F (7.2°C). If the coolant is allowed to become too cold in relation to the oil, a heavy layer of stearin will build up on the surfaces to insulate the oil from the coolant.
- agitation rate—Crystal formation is hastened by stirring to bring the first crystals into contact with more of the liquid. However, mild agitation rates are recommended because high shear rates fragment the crystal during the growth stage to produce more smaller crystals instead of the desirable large crystal.
- crystallization time—Crystallization is inseparably linked to two ele-

ments of time: (1) the time it takes to lower the temperature of the material to the point where crystallization will occur and (2) the time for the crystal to become fully grown. The rate of cooling is a primary factor for determining the size, amount, and stability of the crystals formed. In general, crystals assume their most highly developed and characteristic forms when grown slowly from a melt or solution only slightly supercooled, in which the liquid freely circulates around the crystal. A typical time–temperature sequence for winterization of cottonseed oil is [65]

(1) Refined and bleached cottonseed oil is transferred to the chilling units at 70 to 89°F (21.1 to 26.7°C).

(2) It is cooled to 55°F (12.8°C) in 6 to 12 hours, when the first crystals usually appear.

(3) It is cooled to 45°F (7.2°C) in 12 to 18 hours with a reduced cooling rate. At this point, a 2 to 4°F (1.1 to 2.2°C) heat of crystallization temperature increase should be observed.

(4) After the oil temperature drops slightly below the previous low, approximately 42°F (5.6°C), it is maintained at this temperature for approximately 12 hours. This period is critical for the effectiveness of the process. Since the oil is viscous and molecular movement is slow, crystals continue to grow after the minimum temperature is reached.

9.3 Solvent Winterization

Salad oil production with the traditional winterization procedure is a slow process. Two to three days of chilling time is required for good filtration and yield. Most vegetable oils that cloud at refrigerator temperatures can be solvent winterized for better yields and a salad oil of better quality in less time than by the conventional process. Comparison of the two procedures indicates many similarities. The major advantages of a solvent winterization system are (1) considerably lower viscosity, which allows a faster crystal growth for more rapid stearin separation; (2) the salad oil produced has a better resistance to clouding at cool temperatures for longer cold tests; and (3) less liquid oil is trapped in the stearin component for higher salad oil yields.

An operational continuous solvent process was described by Cavanagh [66,67] for winterization of cottonseed oil. Miscella containing 30 to 60% by weight of oil in hexane with a 50% solution preferred is cooled rapidly with a heat exchanger to either 20 to 26°F (−6.6 to −3.3°C) or 8 to 12°F

TABLE 2.6. Cottonseed Salad Oil Stearin Analysis.

Winterization Process	Conventional	Solvent
Iodine value	95.5	71.6
Solids fats index:		
% at 10.0°C	21.6	52.3
% at 21.1°C	1.3	33.7
% at 26.7°C		1.2
% at 33.3°C		0.1
Fatty acid composition, %		
Mysristic C–14:0	0.7	0.6
Palmitic C–16:0	34.6	52.1
Palmitoleic C–16:1	0.6	0.8
Stearic C–18:0	2.1	1.9
Oleic C–18:1	15.8	9.1
Linoleic C–18:2	46.2	35.5

(-13.3 to -11.1°C). After cooling, the miscella passes through a continuous winterizing column that cools with a series of agitated trays over a 40- to 60-minute period to temperatures as low as -40 F. (-20°C). A continuous solids discharge centrifuge separates the solid stearin from the liquid miscella. The solvent is removed from the liquid oil portion with an evaporator system before deodorization. The solid discharge from the centrifuge is filtered to remove any foreign material before the residual 10 to 15% hexane solvent is removed with an evaporator system.

Controlled agitation of 1 to 10 rpm and a controlled temperature drop to 0°F produces harder, firmer, more compact stearin crystals in solvent, and less oil is entrapped than with conventional winterization systems. A comparison of cottonseed salad oil stearin analytical characteristics from a conventional process and a solvent process shows the differences produced by the two systems (see Table 2.6).

9.4 Winterization Process Control Procedures

The acceptability of a winterized oil is almost always determined by cold test analysis. This method measures the ability of the oil to resist crystallization. The cold test result is the number of hours at 32°F (0°C) for an oil to become cloudy. AOCS Method Cc 11-53 indicates that an oil has passed the test if it is clear and free of any cloud at 5.5 hours [68]. However, most processors and customers have more stringent requirements for cold test hours. Cottonseed and soybean winterized oil products normally have a minimum cold test limit of 10 hours, and some are as high as 20 hours for special products.

Processors have investigated many different potential process control evaluations, procedures, and methods to determine that the winterization process is in control on a timely basis. However, cold test is still the most definitive evaluation, even though the results are not available until a lengthy period after the oil has been winterized. Usually, the winterized oil production is segregated in separate tanks until the cold test results are available. If the oil fails to meet the specific number of hours, it must be rewinterized. Oils that meet the requirements are transferred to salad oil storage for subsequent deodorization, packaging, or shipment as required. The after-the-fact analytical results to determine the acceptability of the winterized oil should place more emphasis on the process techniques to assure that all of the best practices are continually observed.

9.5 Winterization Applications

Historically, winterization has always been associated with cottonseed oil. It and other liquid oils that contain fractions that solidify when chilled must be winterized or fractionated to remain clear at cool temperatures. Oil that is to be refrigerated or stored in cool warehouses must resist clouding for a period of time to be acceptable aesthetically or for performance. Winterized cottonseed oil was the standard salad oil for the retail trade food processors to produce mayonnaise and other salad dressing products because of its pleasing flavor and flavor stability.

Soybean oil was rejected as a salad oil both at the retail level and by food processors until the flavor stability problem was remedied with partial hydrogenation to reduce the linolenic (C-18:3) fatty acid content. Hydrogenation to improve flavor stability also produced a hard fraction in the soybean oil, which crystallized at cool temperatures similar to cottonseed oil. Winterization was employed to separate the hard and liquid fractions. Supply and demand economics and performance elevated partially hydrogenated winterized soybean oil to the leading winterized salad oil product in the United States. A comparison of the two winterized oil products is presented in Table 2.7.

Winterization of hydrogenated soybean oil is very similar to that of cottonseed oil, except that less time is required for crystallization and filtration. The inherent crystallization tendencies for the two source oils are different; the stable crystal form for soybean oil is beta, but beta-prime for cottonseed oil. Beta crystals are large, course, and self-occluding, while the beta-prime crystals are small, needle-like shaped crystals that pack together to form dense, fine structures.

TABLE 2.7. *Typical Cottonseed Oil and Partially Hydrogenated Soybean Oil Winterized Salad Oil Components.*

Source Oil:	Cottonseed Oil			Hydrogenated Soybean Oil		
Winterized Oil Component:	Whole Oil	Salad Oil	Stearin	Base Stock	Salad Oil	Stearin
Iodine value	109	113.5	90.6	108.7	111.4	95.7
Fatty acid composition, %						
Myristic C–14:0	0.9	0.9	0.7			
Palmitic C–16:0	24.8	21.3	38.2	11.2	10.2	15
Palmitoleic C–16:1	0.5	0.5	0.3			
Stearic C–18:0	2.6	2.9	2.3	4.8	4.1	7.9
Oleic C–18:1	16.9	18	13.8	45.4	45.2	46
Linoleic C–18:2	53.7	55.8	44.2	35.4	37.5	29.1
Linolenic C–18:3	0.2	0.2	0.1	3	2.8	1.9
Arachidic C–20:0	0.2	0.2	0.2			
Fraction, %	100	84.6	15.4	100	82.9	17.1
Cold test hours		24			10.5	

10. DEWAXING

An increased demand for salad oils high in unsaturates has resulted in the marketing of source oils that must be dewaxed to maintain clarity during storage, on the retail store shelf, and at refrigerator temperatures. Many vegetable oils have small quantities of waxes that solidify and cause cloudy oil. Most vegetable oils are solvent extracted with the seed and hull together for operational efficiency. The seed hulls can contain waxes that are soluble in oil. Waxes are high-melting esters of fatty alcohols and fatty acids with low solubility in oils. These waxes solidify after a period of time to give the oil a cloudy appearance, an unsightly thread, or a layer of solidified material. The quantity of wax in the various vegetable oils can vary from a few hundred parts per million (ppm) to over 2,000 ppm. The wax content must be reduced to less than 10 ppm to ensure that the oil will not cloud or develop a wisp.

The mechanisms for wax removal from oils is different than those applicable to winterization even though the same equipment can be utilized. The classical dewaxing process, usually performed after prebleaching and prior to deodorization, consists of carefully cooling the oil to crystallize the waxes for removal by filtration. The cooling must be done slowly under controlled conditions. A body feed approximately equal to the wax content of the oils is used to prevent blinding of the filter leaves. Without a body feed, the waxes slime over and blind the screens almost immediately [69].

A continuous dewaxing process that operates efficiently with low wax oils, 500 ppm or less, has the following process flow [70]:

- The oil is continuously cooled with heat exchangers and a crystallizer to 43 to 46°F (6 to 8°C).
- Filter aid at equal quantities to the wax content is added to the crystallizer to facilitate crystallization and filtration.
- Crystallization time is 4 hours minimum, followed by a holding period of 6 hours to develop the wax crystals.
- The oil is carefully heated to 64°F (18°C) before filtering to separate the wax crystals from the liquid oil.

The typical dewaxing process performs well with low wax oils, but some vegetable oils have higher wax contents. The filtration time is increased and higher product losses are experienced with the higher wax content oils. Some of the procedures in use to improve the dewaxing economics are [71,72]:

- simultaneous dewaxing and degumming—The crude oil is cooled to approximately 77°F (25°C) and held at this temperature for 24 hours before water degumming. This process usually reduces the wax content to 200 to 400 ppm.
- wet dewaxing—The phosphatides are first removed from the oil by degumming. The oil is then cooled to 46°F (8°C) and 5% water with sodium lauryl sulfate is added and agitated for 4 hours minimum to crystallize the waxes. The wax crystals should disperse in the water phase for separation with a centrifuge.
- simultaneous dewaxing and chemical refining—The oil is treated with phosphoric acid and neutralized with sodium hydroxide and centrifuged using normal refining techniques. Before water washing, the oil is cooled to 46°F (8°C) and held for 4 to 5 hours with gentle agitation. Then, 4 to 6% water is added and the mixture heated to 64°F (18°C) with agitation. During this mixing, the wax crystals are wetted and suspended in the soapy water phase. This mixture is centrifuged to separate the water and oil phases. Usually, a second water wash is required to complete removal of the wax and soap traces from the oil.
- solvent dewaxing—This procedure is performed after prebleaching and prior to deodorization if it is not an integral part of a miscella refining process. Dewaxing in solvent consists of mixing the oil with a fixed volume of solvent and, after chilling, to promote crystallization of the waxes for separation by either filtration or with a centrifuge.

10.1 Dewaxing Process Control

Currently, the analytical method to determine if an oil is adequately dewaxed is the same as for products winterized to remove large quantities of stearin: cold test, AOCS Method Cc 11-53 [68]. However, this evaluation may be very misleading. Dewaxed oil, which remains clear and brilliant for 5.5 hours, generally remains so for 24 hours in the ice bath, but the same oil can become opaque after only a few hours at room temperature due to the reappearance of waxes as well as glycerides. Therefore, to determine that an oil has been adequately dewaxed, a chill test and a cold test of 24 hours minimum should be required. The chill test consists of drying and filtering the test sample before hermetically sealing it in a 4-ounce bottle. The sample is held at 70°F (21.1°C) and examined after 24 hours for clarity. Any indication of a cloud or wisp indicates the presence of a wax or hard oil contamination.

10.2 Dewaxing Applications

The highly unsaturated vegetable oils marketed in the United States as salad oils, which can cloud due to a wax content, are listed below with a typical wax content:

Source Oil	Typical Wax Content, %
Sunflower	0.2 to 3.0
Safflower	ca. 0.5
Corn	0.5 to 1.0
Canola	ca. 0.2

11. FRACTIONATION

Edible fats and oils are fractionated to provide new materials more useful than the natural product. Edible fats and oils are complex multicomponent mixtures of different triglycerides with different melting points. The melting behavior and/or the clear point of fats are important properties for functionality in the various prepared food products. Fractionation processes separate fats and oils into fractions with different melting points. Fractionation may be practiced merely to remove an undesirable component, which is the case with dewaxing and winterization processes to produce liquid oils that resist clouding at cool temperatures. Separation of a fat or oil into fractions can also provide two or more functional products from the same original product. The production of cocoa butter equivalents or substitutes is a well known application for this type of fractionation.

Separation of fats and oils fractions is based on the solubility of the component triglycerides. The solubility differences are directly related to the type of triglycerides in the fat system. The triglyceride types are determined by their fatty acid composition and the distribution of the fatty acids on the individual triglyceride molecule. Components of a fat or oil that differ considerably in melting point can be separated by crystallization and subsequent filtration for removal of the higher melting portions. In any practical process of fractional crystallization the potential for efficient separation of crystals from the liquid is dependent upon the mechanics of separation as well as the phase behavior of the system. The successive stages of fractionation can be distinguished as

(1) Cooling of the oil to supersaturation to form the nuclei for crystallization

(2) Progressive growth of the crystalline and liquid phases

(3) Separation of the crystalline and liquid phases

Separation efficiency of the liquid and solid fractions depend primarily upon the cooling method, which determines the crystal form and size. Fats and oils can crystallize in several polymorphic forms, specifically alpha, beta-prime, and beta in that order of stability, melting point, heat of fusion, and density. The rate of crystallization for the alpha form is higher than for the beta-prime form, which crystallizes quicker than the beta form. Rapid cooling causes heavy supersaturation that forms many small, shapeless, soft crystals of mixed crystal types with poor filtration properties. Gradual cooling of the oil results in stable beta and beta-prime crystals that are easily filtered from the liquid phase.

Three distinct unit processes for the fractionation of triglycerides, which couple crystallization and separation process, are practiced commercially to produce value-added fractionated fats and oils: (1) dry fractionation, (2) solvent fractionation, and (3) aqueous detergent fractionation [73]. Dry fractionation processes include winterization, dewaxing, hydraulic pressing, and crystal fractionation. It is the most widely practiced form of fractionation in which crystallization takes place without the aid of a solvent. The winterization process is effective for the removal of small quantities of solid fat from a large quantity of liquid oil. Dewaxing can be a variation of the winterization process to remove small quantities of waxes from certain vegetable oils rich in unsaturates. Hydraulic pressing effectively removes small quantities of liquid oil from a large quantity of solid fat. Some oils like palm oil, containing high levels of both liquid and solid fractions, can be separated by dry fractionation, but not as efficiently as with other processes. Solvent or aqueous detergent fractionation processes provide better separation of specific fractions of the feedstock materials.

11.1 Dry Crystal Fractionation

The advantages of fractionation were accidentally revealed to European companies with coconut oil imported from Sri Lanka in long wooden barrels called "Ceylon Pipes." The barrels were filled with warm coconut oil, which cooled slowly during the sea voyage to the cooler temperatures in Europe. The slow cooling, coupled with agitation from the ships' movement, caused the coconut oil to crystallize and separate into a hard and soft fraction. Customers receiving the fractionated product evaluated the properties of the components and realized that the fractions were more useful for some applications than the whole or natural products [74].

Dry crystal fractionation procedures are commonly used for separation of hard stearin and soft olein fractions from natural products that contain high levels of each, like palm oil and the lauric fats. The principal of this fractionation procedure is based on slowly cooling the oil under controlled conditions without the aid of a solvent. The stearin and olein fractions can be separated by various processes, i.e., filtration, centrifugation, hydraulic pressing, rotary drum, or one of the patented processes. In the dry process, large crystals are generally required for efficient separation. The large crystals tend to group together in clumps that can trap part of the liquid olein phase. This results in a soft stearin or a low olein yield due to the poorer separation. A low olein yield can also be experienced from the formation of mixed crystals. Controlled slow cooling of the natural oil will diminish these problems to provide cleaner separation of the olein and stearin fractions.

One unique labor-intensive dry crystal fractionation process still utilized on a limited basis to fractionate lauric oils is to slowly cool the oil until it has a plastic consistency. It is then poured into canvas bags and allowed to cool further to the fractionation temperature, and the crystal is allowed to stabilize or mature. Then it is hydraulic pressed to squeeze the oil portion out of the stearin cake. This procedure can produce a very acceptable confectionery coating fat from palm kernel oil [74]. Previously, this process was utilized to fractionate lard and tallow to produce hard fractions for use in stabilizing compound shortenings and a liquid oleo oil.

11.2 Detergent Fractionation

The aqueous detergent fractionation procedure utilizes the same basic principals of dry fractionation; i.e., crystallization is induced by cooling the oil under controlled conditions without the addition of a solvent [74,75]. The difference is that an aqueous detergent solution is added to the crystallized material to assist in the separation of the liquid olein and the solid stearin. The aqueous solution contains about 5% of a detergent such as sodium lauryl sulphate, which preferentially wets the surface of the crystals displacing the liquid oil. About 2% of an electrolyte such as magnesium or aluminum sulphate is added to the solution to assist in uniting the liquid olein droplets. Separation is then effected with a centrifuge. The heavier phase containing the stearin is heated to melt the stearin and effect a separation of the oil and water phases. Complete separation of these phases is accomplished with a second centrifuge. Aqueous detergent fractionation is more expensive than dry fractionation, but more complete separation of the soft and hard fractions can be effected to produce a higher olein yield and a harder stearin fraction.

11.3 Solvent Fractionation

Solvent fractionation is an expensive process that can only be justified for preparation of value-added, high-quality products. The ultimate objective for the use of solvent fractionation technology is commercial production of fats and oils products with unique properties. Fractional crystallization from dilute solution results in more efficient separation with improved yields, reduced processing times, and increased purity than fractionation carried out without a solvent. The attractive benefits are partially, and for some products completely, offset by high capital costs for the handling and recovery of the solvents, as well as increased cooling capacity requirements. Some of the product categories that have utilized solvent fractionation technology to produce products with unique functional characteristics are

- cocoa butter equivalents (CBEs)—CBEs are fats with chemical compositions similar to cocoa butter, which are capable of replacing it in any proportion in chocolates.
- lauric cocoa butter replacers (CBRs)—These are fractionated coconut or palm kernel oils with physical properties closely resembling cocoa butter.
- non lauric cocoa butter replacers—The most widely used products in this classification are based on solvent fractionated, hydrogenated liquid oils, such as soybean, cottonseed, and palm oils.
- confectionery products—Fat systems with a low melting point but a high solids fat index (SFI) content at room temperature provide products with a quick melt at body temperature, which results in a cooling sensation when eaten. High-quality candy centers and whipped toppings are two specific applications for fractionated products.
- medium chain triglycerides—These are lauric oil fractions containing C-6:0, C-8:0, and C-10:0 saturated fatty acids, which are soluble in

both oil and water systems for quick absorption by the body and transported via the portal system.

- high stability liquid oils—Modification of oils by utilization of hydrogenation with fractionation has permitted the development of liquid oils with high resistance to oxidative degradation. Liquid oils with AOM stability of 350 hours are available commercially.

Commercially, solvent fractionation is carried out by a number of different processes that may be batch, semicontinuous, or continuous [75,76]. Crystallizers, filters, and solvent recovery systems can differ in design, and one of several organic solvents may be employed. Solvents that have been used include acetone, hexane, and 2-nitropropane. Some of the processes are protected by patent and are proprietary, but all require control of certain process parameters:

- Feedstock selection, which can include natural oils and blends, hydrogenated oils and blends, randomly rearranged oils and blends, and blends of any or all of these
- concentration of the fat in the solvent
- fractionation temperature
- cooling rate and residence time in the crystallizer
- separation conditions

A typical solvent fractionation process flow usually begins with heating the feedstock oil to a temperature above the completed melting point and blending with warm solvent in the ratio of one part oil to between three and five parts solvent by weight. The solution is then cooled to crystallize the hard fractions. Crystallization temperatures vary, depending on the nature of the solvent, the concentration of oil in the solution, and the characteristics needed in the final fractions. For example, for lauric oils in acetone solvent, temperatures of 28 to 68°F (-2 to 20°C) have been used to obtain stearin iodine values of 1.8 to 8.3. The crystallized slurry is separated by filtration or a settling technique. The solid material is then stripped of solvent for one fraction. Removal of the solvent from the filtrate yields another fraction. Further fractionation may then be achieved by redissolving either fraction and repeating the process.

12. ESTERIFICATION OR ALCOHOLYSIS

Emulsifiers are usually made by either alcoholysis or direct esterification. In direct esterification, fatty acids and polyalcohols are reacted together under controlled conditions to form esters. In alcoholysis, fats are reacted with polyalcohols to make the surfactants. For example, the production of mono- and diglycerides from fat is an alcoholysis reaction with glycerine as the alcohol.

Mono- and diglycerides were the first fatty emulsifiers to be added to foods. These emulsifiers were first used in margarine for Danish pastry and puff pastry shortening. The first U.S. patents for mono- and diglycerides granted in 1938 disclose the surfactant's usefulness in emulsions and margarine [77]. Monoglycerides with only one fatty acid attached to a glycerol molecule and two free hydroxyl groups on the glycerol take on the properties of both fats and water. The fatty acid portion of the molecule acts like any other fat and readily mixes with the fatty materials, while the two hydroxyl groups mix or dissolve in water; thus, monoglycerides tend to hold fats and water together. Cake shortenings that produced increased aerating, creaming and moisture retention properties contributed by the addition of mono- and diglycerides were introduced in 1933, with patents granted in 1938 as well. Shelf life extension properties for yeast-raised products with mono- and diglyceride addition was introduced soon after to the bread bakers [77]. Since that time, the uses and demand for food emulsifiers has grown dramatically and several other emulsifiers have been developed. Most of the fatty emulsifiers produced are either monoglyceride derivatives or utilize an alcohol other than glycerine, like propylene glycol monoesters (PGME) or polyglycerol esters (PGE). Examples of the monoglyceride derivatives are ethoxylated monoglycerides, acetylated monoglycerides, succinylated monoglycerides, and others.

Emulsifier production is generally performed in versatile batch equipment that is used for a variety of different surfactants. Acid-resistant construction for the tankage and reactors is one or another variant of the 300 series stainless steels. The use of these alloys essentially eliminates the possibility of iron and other metal contamination that may either degrade the product or catalyze undesirable oxidation and other side reactions. Typically, type 304 stainless steel is used for fatty chemical processes where the temperature does not exceed 300°F (150°C) and type 316 is used for temperatures above this level [78]. Internal reactor coils capable of handling high-pressure steam and cooling water, as well as a vacuum system with 0.1 mm Hg capabilities, are necessary for emulsifier reactors. A condenser for the recovery of the excess glycerine, glycol, or other alcohols should be sized for the largest volume of the polyalcohol to be recovered.

12.1 Mono- and Diglycerides

Mono- and diglycerides are the most dominate food emulsifier both from the standpoint of total use and breadth of use. The mono- and diglycerides consist of glycerol esters of various edible fatty acids and fat blends. Three types of standard mono- and diglycerides are manufactured: 40 to 46% alpha monoglyceride content, 52 to 56% alpha monoglyceride content and distilled monoglycerides with 90% alpha monoglyceride content. There are three basic physical forms of the mono- and diglycerides: hard, soft, and intermediate. The differences in the emulsifier consistency is determined by the physical characteristics of the edible fat or oil used in the reaction. The consistency of the mono- and diglyceride, as indicated by the iodine value and/or melting point, determines the functionality of the emulsifier. Generally, a higher melting point emulsifier is preferred when a tight emulsion is desired, while the lower melting or softer products provide better aeration qualities. Intermediate hardness mono- and diglycerides are compromise products suitable to a degree for both application ranges but not specific for either.

Most of the mono- and diglyceride requirements are produced by glycerolysis of triglycerides or fats and oils. In this process, fats and oils of the desired hardness are mixed with an excess of glycerine at elevated temperatures in the presence of an alkaline catalyst, usually either sodium or calcium hydroxide. The reaction mixture is kept at an elevated temperature until the fatty acid radicals of the triglycerides are redistributed at random among the available hydroxyl groups of the glycerine. The reaction mixture is cooled after equilibrium has been attained and the catalyst is "killed" or deactivated by the addition of a food acid, normally phosphoric acid. The phosphate salts that result from the catalyst neutralization must be removed by filtration. The excess glycerine will separate as a lower layer upon cooling and can be partially removed by decanting. The glycerine that remains dissolved in the reaction mixture must be removed by vacuum distillation while steam stripping to reduce the free fatty acid content and remove the oxidation materials that contribute to undesirable flavors and odors.

This process yields substantial amounts of monoglycerides in addition to diglycerides and triglycerides with altered or rearranged structures and free glycerine. The proportion of monoglyceride to diglyceride may be controlled, depending on the relative proportion of reactants, temperature, time, catalyst, and use of stripping steam or inert gas. The normal distribution is 50% monoglycerides, 40% diglycerides, and 10% triglycerides; approximately 85% of the monoglycerides are esterified on the alpha position of the glyceride, the balance being esterified on the center, or beta, hydroxyl group. The composition of this mixture can be changed by reducing the amount of glycerine added or by changing reaction temperature and catalyst level. Most of the mono- and diglyceride emulsifiers made contain about 50% monoglycerides, but 60% monoglyceride levels can be attained. Higher monoglyceride contents can be obtained by distillation of the mono- and diglyceride products.

The amount of fat that will react with the specified amount of glycerine varies with the molecular weight of the fat involved. A product with a high molecular weight, such as soybean oil, requires more fat per pound of mono- and diglycerides produced than does a lower molecular weight product such as tallow or cottonseed oil. Based on typical fatty acid compositions, the average molecular weight for hydrogenated soybean oil containing 3% oleic acid is 882.6, while a corresponding molecular weight for hydrogenated tallow triglyceride is 858.7. Theoretically, it would require 89.37 pounds of hydrogenated soybean oil to produce 100 pounds of glycerine-free mono- and diglycerides while it would only require 88.15 pounds of hydrogenated tallow for 100 pounds of mono- and diglycerides. Since it is impossible to strip all of the glycerine from the finished emulsifier, the above fat requirements become 88.9 pounds of hydrogenated soybean oil for a mono- and diglyceride containing 0.5% free glycerine and 87.7 pounds of hydrogenated tallow to yield 100 pounds of mono- and diglycerides again containing 0.5 free glycerine.

Mono- and diglycerides can be manufactured with either a batch or continuous process. However, a large proportion of the U.S. demand is still produced with batch systems where the reaction time, temperature, and catalyst may be varied. A typical batch process flow to produce a 40 to 46% alpha monoglyceride emulsifier is detailed in Chart 2.9. For continuous processing, the process time is considerably less than required for batch systems, probably less than 30 minutes total, which is promoted in part by the higher reaction temperatures employed; however, the total pounds per man-hour may be equivalent to those produced by a batch system.

Concentrations of monoglycerides to produce the 90% alpha monoglyceride products is achieved by distillation of the mono- and diglyceride emulsifiers. Prior to the actual concentration, the free glycerine content has to be removed to preserve the monoglyceride content. The reaction between triglycerides and glycerine is reversible, especially in the presence of catalyst and at elevated temperatures. Additionally, monoglycerides are not heat-stable and have only limited heat tolerance, so thermal damage must be avoided. Molecular distillation is evaporative distillation where a compound in the liquid state evaporates without boiling, because the high vacuum removes the effect of atmospheric pressure. Monoglycerides vaporize, leaving the heavier di- and triglycerides behind in the distillation residue [77]. The residue is recycled to produce additional mono- and diglycerides for subsequent distillation until the product color becomes too dark.

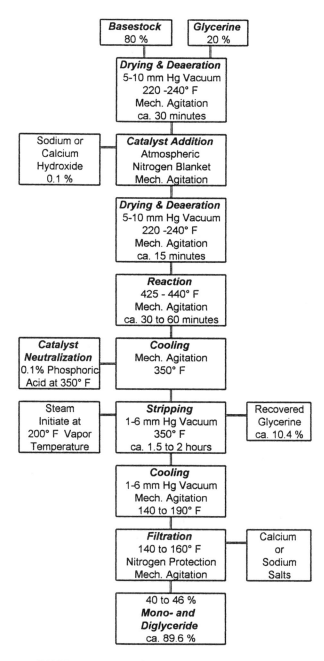

CHART 2.9. Mono- and diglyceride batch process flow.

125

12.2 Monoglyceride Derivatives

Monoglycerides not only are used as surfactants as produced, but can also be further modified to produce other surface-active products suitable for use in prepared foods. The multiple reactive groups of the monoglycerides allow the formation of other functional emulsifiers. One or both of the hydroxyl groups can be replaced by different groups to form esters with specialized functionality characteristics. Modified monoglyceride surfactants can be obtained by treating monoglycerides with an acid, acid anhydrides, acid chloride, or another ester. Some of the monoglyceride derivatives produced are

(1) Lactylated monoglycerides—The lactoglycerides comprise an important series of edible emulsifiers, which are derived from monoglycerides or mixtures of mono- and diglycerides. Lactostearin can be made by subjecting low iodine value hydrogenated soybean oil to glycerolysis, using sodium or calcium hydroxide as a catalyst and then esterifying this mixture directly with lactic acid, followed by water washing to remove the bitter-tasting triglycerides of lactic acid. The resultant product can be identified as glyceryl lactostearate and mono- and diglycerides [79]. The lactylated monoglyceride surfactants are utilized primarily for aeration in cakes and whipped toppings.

(2) Acetylated monoglycerides—Acetylated monoglycerides are characterized by sharp melting points with a waxy, rather than a greasy, feel. The physical and functional properties of a particular acetylated monoglyceride depends on the triglyceride from which it is made and the degree of acetylation. Monoacetylated or partially acetylated monoglyceride molecules each contain one long-chain fatty acid plus one acetyl group and one hydroxyl group. Diacetylated or fully acetylated monoglyceride molecules each contain one long-chain fatty acid and two acetyl groups. Acetylated monoglycerides may be produced by either of two procedures:
- direct acetylation of monoglycerides with acetic anhydride without a catalyst or molecular distillation with the acetic acid, acetic anhydride, and triacetin removed by vacuum distillation
- interesterification of edible fats and oils with triacetin in the presence of a catalyst, followed by molecular distillation or steam stripping

Acetylated monoglycerides are stable in the alpha crystal form. These alpha tending emulsifiers promote the agglomeration of the dispersed fat globules, which induce clumping of fat globules into clusters to form structural networks in whipped toppings and other foams. Saturated monoglycerides can be stabilized in the active alpha crystalline form with the addition of acetylated monoglycerides.

(3) Ethoxylated mono- and diglycerides—Ethoxylated mono- and diglycerides, labeled as "Polyglycerate 60," are formed when 20 moles of ethylene oxide are reacted with a mono- and diglyceride. The reaction is carried out at 265 to 325°F (130 to 165°C) under pressure because ethylene oxide is a gas, catalyzed by a base like sodium hydroxide. The polymerized ethylene oxide chains combine with the mono- and diglycerides to produce an effective emulsifier with strong hydrophilic characteristics. The presence of many ether oxygen atoms in the polyoxyethylene chains offers sites for hydrogen bonding to water molecules and proteins [77]. The physical form contributed by the triglyceride basestock has a direct affect upon functionality; i.e., firm consistencies are preferred for yeast-raised products, and soft consistencies provide better aeration properties for cakes and icings.

(4) Succinylated monoglycerides—Succinylation of monoglycerides will change the properties of the surfactant from one with essentially only bread softening properties to one with both softening and dough strengthening properties. This product is manufactured by reacting monoglyceride with succinic anhydride at temperatures ranging from 175 to 350°F (80 to 180°C). At high temperatures, no catalyst is required, but at reaction temperatures below 200°F (95°C), a low level of an alkaline catalyst may be needed to accelerate the reaction [80].

12.3 Polyglycerol Esters

Polyglycerol esters have unique physical and chemical properties that provide a broad range of functional properties [81,82]. Polyglycerols are polymers that are formed by the dehydration of glycerine. For each molecule of glycerine added to the polymer, there is an increase of one hydroxyl group. The potential number of different polyglycerol esters possible is almost limitless, depending upon the degree of glycerine polymerization. FDA regulations have approved these surfactants, ranging up to and including decaglycerol esters of edible fats and oils from corn, cottonseed, peanut, safflower, sesame, soybean, lard, palm, and tallow. The polyglycerol esters are functional in a very broad range of applications, either singly or blended with other emulsifiers to include crystal inhibitors; aeration of batters, icings, fillings, and toppings; antispattering agent for cooking oils and margarine; low-fat products; ice cream and mellorine; confectionery antibloom agent; viscosity control; nondairy creamers; cake mixes; lubricants; and more.

The esters are made by reacting fat with glycerine that has been polymerized into polyglycerols at high temperatures. To a degree, the polymerization can be controlled to produce varying amounts of diglycerol, triglycerol, tetraglycerol, and so on. Two molecules of glycerol will combine to form diglycerol and a molecule of water. The diglycerol is then transformed into triglycerol as another glycerol molecule is added. The water released by the polymerization distills from the reaction mixture at the high temperature used and is lost.

Similarly, tetraglycerol and the higher polymers can be formed from triglycerol. During polymerization of glycerol, these reactions are all occurring at the same time, but at different rates. Thus, the polymerized product is a mixture of glycerol and various polymers with the actual composition depending on polymerizing conditions and its extent. The average molecular weight of the polyglycerol mixture can be either low or high, depending on the extent of the polymerization, with the number of hydroxyl groups per molecule or sites for esterification with fatty acids increasing as the molecular weight increases. Variations in molecular weight of the polyglycerol or the fatty acids and in the fatty acid/polyglycerol ratio will change the solubility characteristics and emulsification properties of the reaction product.

The following procedure has been utilized to polymerize glycerine and esterify the polyglycerol and triglyceride basestock in the same reaction vessel:

(1) Charge the reactor with glycerine.

(2) Add 1.0% sodium hydroxide (dry basis) to the glycerine while agitating.

(3) Start heating and sparge nitrogen through the mixture to provide a positive nitrogen flow through the reactor outlet through a condenser. Care must be exercised to exclude air from the system. Traces of oxygen lead to the formation of acrolein as well as a dark color that is difficult to bleach.

(4) Heat to 480 to 500°F (250 to 260°C) and hold while continuing the nitrogen sparge to carry off the water liberated.

(5) The polymerization process endpoint may be determined by refractive index and/or measurement of the liberated water removed.

(6) Cool the polyglycerol mixture.

(7) Add the triglyceride basestock at equal proportions to the polyglycerol weight while sparging with nitrogen.

(8) Stop the nitrogen sparge and initiate a vacuum on the reactor.

(9) Heat and maintain the mixture at 410 to 425°F (210 to 220°C) for 2 hours. Approximately 20% of the glycerol should be removed and recovered during the reaction.

(10) Cool the polyglycerol ester before exposing it to the atmosphere. From this point, the polyglycerol esters can be treated much the same as triglycerides. They can be bleached and deodorized by the usual procedures used with fats and oils.

12.4 Propylene Glycol Monoester (PGME)

Propylene glycol has the same carbon chain length as glycerine, but it only has two hydroxyl groups. Commonly used propylene glycol monoesters are generally more lipophilic than many glycerol monoesters because they have fewer free hydroxyl groups. Alcoholysis of triglycerides with propylene glycol leads to mixed partial glycol-glycerol esters or propylene glycol mono- and diesters and mono- and diglycerides. Propylene glycol mono esters are alpha-crystalline tending surface active compounds that have proved to be especially effective emulsifiers for cake baking and whipped toppings. The emulsifier forms an alpha-crystalline film around entrapped air bubbles to stabilize the food systems.

The process for the preparation of PGME emulsifiers is similar to that used for the manufacture of mono- and diglyceride emulsifiers, except propylene glycol is added to the fat base. Mono- and diesters of propylene glycol are formed and glycerine is liberated from the triglyceride. Some of the liberated glycerine reacts with some of the fat present, forming mono- and diglycerides, following the same reaction used for mono- and diglyceride preparation. The propylene glycol monoesters composition will vary, depending upon the mixture to reaction. A reaction mixture of basefat and pure propylene glycol should yield approximately 70% propylene glycol monoesters, 10% monoglyceride, and the balance consisting of propylene glycol diesters, diglycerides, and a small amount of triglyceride.

13. BLENDING

Basestocks are blended to produce the specified composition, consistency, and stability requirements for the various edible fats and oils products, such as shortenings, frying fats, margarine oils, specialty products, and even some salad oils. The basestocks may be composed of hydrogenated fats and oils, interesterified products, refined and bleached oils, and/or fractions from winterization, dewaxing, or fractionation. The products are blended to meet both the composition and analytical consistency controls identified by the product developers and quality assurance. The consistency controls frequently include analytical evaluations for solids fat index, iodine value, melting points, fatty acid composition, and so on. Basestocks should not be blended with a disregard for either specified composition or analytical requirements. It is often possible to meet the specified analytical requirements with several different basestock compositions; however, only the specified composition will perform properly, have the required oxidative stability, or conform to the nutritional and ingredient statements.

The blending process requires scale tanks and meters to proportion the basestocks accurately for each different product. The blend tanks should be equipped with mechanical agitators and heating coils to assure a uniform blend for consistency control. Capacities of the blend tanks should be sized to accommodate the next process, probably deodorization. Nitrogen protection should be provided for the long holding times required to perform some of the analytical evaluations. A typical blending process sequence is:

(1) Determine the proportions for each basestock for the product to be blended.

(2) Add the basestocks to the blend tank at the identified proportions.

(3) Heat the blend if necessary to the specified temperature and agitate for 20 minutes.

(4) Sample and submit to the laboratory for the specified analysis.

(5) Transfer to the next process if the analytical characteristics meet the specified limits.

(6) Blends outside the specified limits require adjustments to bring them into the allowed consistency control ranges; however, only specified basestocks within the allowed ranges should be used for any adjustments.

(7) Resample the adjusted blend after agitating for 20 minutes to achieve a uniform product and resubmit to the laboratory for consistency analysis.

(8) Transfer to the next process after the specified analytical results have been met.

14. DEODORIZATION

Most of the major edible oils retain certain undesirable odors and flavors after refining. Normal bleaching imparts an "earthy" flavor and odor, while hydrogenation adds an odor and flavor that can only be described as typical and certainly undesirable. To provide the bland flavor and odor now required by consumers, it is necessary that these undesirable impurities be removed by deodorization. Deodorization is primarily a high-temperature, high-vacuum, steam distillation process to remove volatile, odoriferous materials present in edible fats and oils. Deodorization is the last major processing step through which the flavor and odor and many of the stability qualities of an oil can be controlled. From this point forward, effort is directed only toward retaining the quality that the deodorized oil possesses. Therefore, considerable care must be given to the selection, operation, and maintenance of the deodorizer equipment and the operating conditions.

In the early stages of the development of the edible fats and oils industry, there was little need for deodorization. Lard and butter were consumed in the same form as produced, and the natural flavors were considered an asset. Olive oil, one of the earliest known vegetable oils, was and still is used for its distinctive flavor. However, the rapid expansion of cotton acreage at the end of the nineteenth century resulted in large quantities of cottonseed oil, which presented an economic incentive to use this vegetable oil. Blends of cottonseed oil with harder naturally occurring fats such as tallow and olein stearin were marketed as lard substitutes. Also, refined and bleached cottonseed oil was offered as a cooking and salad oil. These products enjoyed a price advantage over lard and olive oil, but the unpleasant flavor was so strong that acceptance was poor. In addition, the hydrogenation process was being developed to harden vegetable oils because of the high lard and tallow costs, and the hydrogenated oils had a more disagreeable flavor and odor than the oils in their natural state. Attempts to remove the flavor and odors chemically or to mask them with spices or flavors were unsuccessful. The first successful attempt at removing the disagreeable odors and flavors from a fat and oil consisted of injecting live steam into an oil at high temperatures. It was discovered in England, but this flavor improvement process was soon adapted by most American fats and oils processors. The advantages from treating oils with steam to remove offensive flavors and odors was recognized in the early 1890s by Henry Eckstein. David Wesson improved the process by using higher temperatures and maintaining the oil under vacuum while blowing with superheated steam [83,84].

The basic principal of the deodorization process is essentially steam stripping distillation of minute levels of odoriferous materials from the oil with a minimum of injury to the oil [84–91]. The odoriferous substances are generally considered to be free fatty acids, aldehydes, ketones, peroxides, alcohols, and other organic compounds. Theoretically, any inert gas could be used for deodorization; however, steam is relatively inexpensive, inert, and easily removed from the system by condensation. Therefore, deodorization has universally become a steam distillation process that removes undesirable volatile materials from edible fats and oils.

14.1 Operating Variables

Experience has shown that edible fats and oils flavor and odor removal correlates well with the reduction of free fatty acid (FFA). The odor and flavor of an oil with a 0.1% FFA will be eliminated when the FFA is reduced to 0.01 to 0.03% assuming a zero peroxide value. Therefore, all commercial deodorization consists of steam stripping the oil for FFA removal. The four interrelated operating variables that influence deodorizer design are vacuum, temperature, stripping steam rate, and holding time at deodorization temperatures:

- vacuum—The absolute pressure in the deodorizer must be low enough to permit boiling of the trace impurities from the exposed oil surface. The practical pressure requirements can be kept reasonably low and consequently, the deodorization temperature can also be kept low by maintaining a low absolute pressure in the deodorization system and by employing stripping steam. A low absolute pressure directly affects the vapor pressure of the fatty acid. Therefore, an improved vacuum will permit lower deodorizing temperatures, and conversely, higher temperatures can compensate for a poorer vacuum.

  Vacuum is supplied almost invariably by multistage steam injectors. A well-designed three- or four-stage vacuum system is capable of producing a vacuum in the range of 1 to 6 millimeters of mercury absolute pressure, which is the usual operating range. The available data does not substantiate any finished product quality advantages for 1 mm Hg as opposed to 6 mm Hg operation; however, the most economical steam consumption will be obtained with a four-stage ejector system designed for 6 mm Hg operation.

- temperature—Deodorization temperatures must be high enough to make the vapor pressure of the volatile impurities in the oil conveniently high. The vapor pressure of the odoriferous materials increases rapidly as the temperature of the fat is increased. For example, the vapor pressure of palmitic fatty acid is 1.8 mm at 350°F (176.7°C), 7.4 mm at 400°F. (204.4°C), 25.0 mm at 450°F (232.2°C), and 72 mm at 500°F (260°C). Assuming that the vapor pressure–temperature relationship for all the odoriferous materials is similar to that of palmitic fatty acid, each 50°F (27.8°C) deodorizer temperature increase would triple the odoriferous material removal rate. Or stated another way, it would take nine times as long to deodorize an oil at 350°F than at

450°F. Higher deodorizer temperatures definitely provide shorter deodorization times. However, there are temperature limits due to the development of the unwanted polymers, isomerization to produce *trans* acids, and distillation of the natural antioxidants or tocopherols. Therefore, a compromise must be determined between time and temperature for the particular fat and oil product deodorized. It has been determined that twice as many tocopherols and sterols are stripped out at 525°F (275°C) as at 465°F (240°C) and that pressure variations of 2 to 6 mm Hg had only a slight effect upon tocopherol/sterol stripping.

Deodorizer operation at elevated temperatures can also promote thermal decomposition of some constituents naturally present in oils, such as pigments and some trace metal–prooxidant complexes. The carotenoid pigments can be decomposed and removed by deodorization at approximately 500°F (260°C).

Optimum deodorizer operating temperatures vary from product to product. In general, animal fats require less stringent conditions than the vegetable oil products. Among the vegetable oils, those containing relatively short-chain fatty acids such as coconut and palm kernel oils require lower deodorization temperatures than the domestic oils composed of longer chain fatty acids. Hydrogenated oils are usually more difficult to deodorize because of higher free fatty acid contents and the distinctive odor imparted by the hydrogenation reaction. In general, deodorization temperatures will vary from 400 to 475°F (204 to 246°C) and in some cases will be as high as 525°F (274°C).

- stripping steam—Adequate stripping steam, consistent with the temperature and pressure in the deodorizer, is required. The amount of stripping steam required is both a function of the absolute operating pressure and the mixing efficiency of the equipment design. Agitation of the oil, necessary to constantly expose new oil surfaces to the low absolute pressure, is accomplished by the use of carefully distributed stripping steam. Therefore, oil depth is a primary factor for establishing both the stripping steam requirement and the deodorizing or holding time. The quantity of fatty acids distilled with each pound of steam is directly proportional to the vapor pressure of the fatty acids. Effective steam stripping is dependent upon volume; i.e., a 1 mm Hg operation will require a lower weight percent of stripping steam than a 6 mm Hg operation. Typical stripping steam deodorization conditions are 5 to 15% wt % of oil for batch systems and 1 to 5% for continuous and semicontinuous deodorizer systems.
- holding time—Deodorizer holding time is the period during which the fat or oil is at deodorizing temperature and subjected to stripping

steam. Stripping time for efficient deodorization has to be long enough to reduce the fats and oil products odoriferous components to the required level. This time will vary with the equipment design. For example, a batch deodorizer with an 8- to 10-foot depth of fat above the sparging steam distributor will require a longer deodorization time than a continuous or semicontinuous system that treats shallow layers of fat. Typically, the holding time at elevated temperatures for batch deodorizer systems is 3 to 8 hours, while the holding time for continuous and semicontinuous systems varies from 15 to 120 minutes. Additionally, certain reactions with the oils deodorized are not FFA removal related but help to provide a stable oil after deodorization. These reactions and heat bleaching are time and temperature dependent. Therefore, deodorizing systems provide a retention period at deodorizing temperatures to allow these reactions and the heat bleaching to occur.

14.2 Deodorization Systems

Deodorization equipment in current use can be classified into three principal groups: batch, continuous, and semicontinuous.

- batch—This is basically the simplest type of deodorization system that can be installed. The principal component parts consist of a vessel in the form of a vertical cylinder with dished or cone heads. The vessel is fabricated from type 304 stainless steel to avoid the deleterious catalytic activity of copper and iron on oils, welded to prevent air leaks, and well insulated to minimize heat loss. The usual range of capacity is 10,000 to 40,000 pounds although the preference appears to be batch sizes of 15,000 to 30,000 pounds. Vessel diameters are usually chosen to give a depth of 8 to 10 feet of oil and have a similar amount of headspace above the surface of the oil. It is necessary to allow sufficient head space to avoid excessive entrainment losses from the rolling and splashing of the oil caused by the injected steam. Stripping steam is injected into the bottom of the vessel through a distributor. In addition to the steam ejector system, means for heating, cooling, pumping, and filtering the oil are required. The batch system controls include a device for indicating oil temperature and a pressure gauge designed to indicate accurately low pressures within the deodorizer.

  Equipment operating at a high temperature and 6 to 12 mm Hg pressure requires about 8 hours for a complete deodorization cycle of charging, heating, deodorizing, cooling, and discharging the oil. Some systems operating at higher pressures and/or lower deodorization temperatures may require as long as 10 to 12 hours for a deodorization cycle. The total amount of stripping steam required may vary from approximately 10 to 50 pounds/100 pounds of oil, with the average usage probably about 25 pounds/100 pounds oil. The stripping steam is ordinarily injected at 3 pounds/100 of oil per hour at 6 mm Hg pressure. The oil must be cooled to as low a temperature as practicable after deodorization before it is discharged to atmospheric conditions to minimize oxidation. A temperature of 100 to 120°F (38 to 49°C) is preferred for liquid oils, with higher temperatures necessary for higher melting products, but still maintained as low as possible.

Batch deodorization has the advantage of simplicity of design, flexibility, and ease of operation. It can be operated for as long or as short a period as required with frequent product and even deodorization condition changes. Mechanically, batch deodorizer systems require very little maintenance; however, the cost of utilities for batch deodorization is considerably higher than for the continuous or semicontinuous systems. Batch systems do not provide a convenient means of recovering any substantial portion of the heat required, have a high stripping steam consumption, and require large vacuum systems with high steam and water requirements. But the lower labor and capitalization costs for the original installation may offset a portion of the higher utility costs.

- semicontinuous—These systems operate on the basis of handling finite batches of oil in a timed sequence of deaeration-heating, holding-steam stripping, and cooling such that each quantum of oil is completely subjected to each condition before proceeding to the next step. The semicontinuous deodorizer consists principally of a tall cylindrical shell of carbon steel construction with five or more type 304 stainless steel trays stacked inside of, but not quite contacting, the outer shell. Each tray is fitted with a steam sparge and is capable of holding a measured batch of oil. By means of a measuring tank, oil is charged to the top tray where it is deaerated while being heated with steam to about 320 to 330°F (160 to 166°C). At the end of the heating period, the charge is automatically dropped to the second tray, and the top tray is refilled. In the second tray the oil is heated to the operating temperature and, again after a timed period, is automatically dropped to the tray below. When the oil reaches the bottom tray, it is cooled to 100 to 130°F (38 to 54°C) and discharged to a drop tank from which it is pumped through a polishing filter to storage. Semicontinuous deodorizers are usually automated and controlled from a central panel with time cycle controller and interlocks such that the sequence steps are interrupted in the event of insufficient batch size, improper drop valve opening or closing, or the oil not reaching the preset heating or cooling temperatures in the allotted time.

One of the principal advantages of the semicontinuous deodorization system derives from the fact that all of the trays are under the same relatively high vacuum. All oil receives substantially identical treatment, and the annular space between the trays and the shell provide some insurance against oxidation due to inward leakage of air. The deodorizer arrangement avoids refluxing of once-distilled undesirable materials back into the oil. This reflux, plus any mechanical carryover is permitted to drain from the bottom of the deodorizer shell. The ability to accommodate frequent stock changes with a minimum of lost production and practically no intermixing is an important advantage for the semicontinuous systems over the continuous deodorization systems.

- continuous—Continuous deodorizers operate so that the oil flows continuously through the heating, holding, and cooling phases of deodorization with the retention times controlled by overflow weirs and pipes. Various designs are utilized in attempting to ensure that all of the oil is subjected to the same deodorization conditions of time, temperature, and stripping steam.

An early continuous deodorization system resembled the countercurrent stripping tower utilized in the petroleum industry. The deodorizer consisted of a tower 3 to 6 feet in diameter, depending upon the throughput desired; a 6-foot diameter had a nominal capacity of 5,000 pounds per hour. The tower contained a series of trays fitted with bubble caps. As the oil descended in the tower, it met an ascending stripping steam current. The oil was stripped of volatiles with fresher and fresher steam as it descended to the bottom of the tower. The total pressure drop in the tower rarely exceeded 2.5 mm Hg due to the thin oil layers on each tray. The countercurrent principal introduced efficiencies through more efficient utilization of the injected steam to reduce the quantity required, smaller vacuum requirements due to the small volume of oil treated at a time, and a large percentage of the heat recovered by preheating the feedstock oil by passing countercurrently through a heat exchanger opposite the other oil flow.

Two approaches for continuous deodorization processes have evolved: (1) to perform the heating, cooling, and heat recovery with external heat exchangers or (2) to perform all the functions within the deodorizer. The internal approach is somewhat less efficient for heat recovery but provides a simpler, more reliable method for changing from one product to another. In either case, continuous deodorizers provide uniform utility consumption by not being subject to the peak loads attendant with batch-type heating and cooling of semicontinuous operations. This permits smaller heating and cooling auxiliaries and the optimum in heat recovery through interchange between incoming and outgoing oils. Processors requiring very infrequent product changes can benefit from continuous deodorization; however, processors requiring multiple stock changes will not realize the benefits. Continuous deodorization benefits are lost with as few as three or four stock changes in a 24-hour period due to loss of production (30 to 60 minutes for each stock change) and the likelihood of commingling product.

14.3 Deodorizer Heating Systems

Practical, as well as theoretical, considerations indicate that oils should be heated in the range of 410 to 525°F (210 to 274°C) for reasonably rapid and effective deodorization. The Wesson system utilized a direct heat method that circulated the oil through a direct fired tubular heater. The disadvantages for this system were poor temperature control and localized overheating, which caused hot spots. A more satisfactory system, known as the Merrill system, was used for many years, which heated mineral oil in a direct fired furnace and circulated it through coils in the deodorization vessel. Pump maintenance, care of the tubes in the direct fired furnace, and a tendency of the mineral oil to decompose under the heating system conditions were the major shortcomings of the Merrill system. Electrical heating has been used to heat deodorizers by either strip heaters immersed in the oil or by placing the elements around the outside of the tank and heating tubes. Automatic control was achieved with electrical heating but the danger of localized overheating, as well as high power costs, were problems. The use of condensing "Dowtherm"™ became the most popular method of heating edible oils shortly after its introduction in 1932.

"Dowtherm A" is a tradename of The Dow Chemical Company for a eutectic mixture of 26.5% diphenyl and 73.5% diphenyl oxide. This liquid boils with negligible fractionation at 495.8°F (257.7°C), under atmospheric pressure. Dowtherm A is referred to as a liquid since it is used in liquid form although it is a solid below 53.6°F (12°C). Dowtherm A is relatively stable and has the advantage of low vapor pressure when boiling at high temperature. For example, at 522°F (272°C) the vapor pressure of Dowtherm A is only 4 psig. Dowtherm A has a distinct odor that is immediately evident in case of a leak. A typical Dowtherm heating system consists of a boiler (vaporizer), a burner complete with safety controls required by insurance regulations, and a gravity return system for the condensate from the deodorizer.

In 1973, contamination of a rapeseed oil with a heat transfer medium in Japan resulted in illness and deaths, which were originally blamed on Dowtherm A. However, it was concluded from the investigation following this incident that usual deodorization conditions are sufficient to remove Dowtherm A from the oil [92]. Nevertheless, Dowtherm has been banned in Japan. Chlorinated biphenyl compounds have been prohibited for use in the United States but Dowtherm usage is still permitted. However, interest worldwide seems to favor a shift to high-pressure steam as the heat transfer media. Many European processors use high-pressure steam. Their usual deodorization temperatures of 410 to 465°F (210 to 240°C) permit the use of 40 to 60 kg/sq cm steam, while the U.S. usual deodorization temperatures of up to 525°F (274°C) require 80 to 90 kg/sq cm steam. High-pressure steam generators are available to handle these requirements but at considerably higher capital costs than for a Dowtherm vaporizer.

14.4 Deodorization Process Control

Deodorization is the last major processing step through which flavor, odor, and many of the stability qualities of an edible fat and oil product can be controlled. From this point forward, all of the efforts are directed toward retaining the quality of the freshly deodorized product. In order to produce a quality deodorized product, attention must be focused on all the factors involved with the process. The various factors that influence the quality of the finished deodorized oil are

(1) Undeodorized oil preparation—Preparation of the oil before deodorization has a significant effect upon the finished deodorized product; therefore, the first process control requirement is to assure that the processing steps prior to deodorization have been performed properly as specified. For example, deodorization of high peroxide oils will thermally decompose the peroxides, but the rate of peroxide formation in the oil during subsequent storage will probably increase and the flavor stability will be compromised. Proper handling of the oil would be to rebleach it prior to deodorization. Steam distillation does not remove secondary oxidation products, soap, or phosphatides that are adsorbed in bleaching.

(2) Air elimination—Oil must be scrupulously protected from air throughout the deodorization process. At deodorization temperature, the oil reacts instantly with oxygen for a decidedly bad effect upon the product's flavor and oxidative stability. Potential air sources are as follows:
  - Deaeration of the feedstock is essential because the oil may contain dissolved oxygen from previous exposure to the atmosphere.

- Air leaks at deodorizer fittings below the oil level and in external pumps, heaters, and coolers.
- Stripping steam must be generated from deaerated water to be oxygen-free.

(3) Deodorization construction materials—Heavy metals, particularly those possessing two or more valency states, generally increase the rate of oxidation. Of all metal, copper is the most potent catalyzing one. A concentration high enough to produce a noticeable oxidative effect lies in the proximity of analytical detection limits, probably 0.005 ppm. The corresponding content for iron is 0.03 ppm. Other metals have exhibited varying catalytic powers; for example, the contents of metal ions required to decrease lard keeping time by 50% at 208°F (98°C) are 0.06 ppm manganese, 1.2 ppm chromium, 2.2 ppm nickel, 3.0 ppm vanadium, 19.6 ppm zinc, and 50.0 ppm aluminum [93]. This comparison stresses the importance of not using copper, iron, or some other alloy if the highest possible flavor stability is required. Therefore, deodorizers are fabricated from type 304 stainless steel at points contacted by the oil.

(4) Metal chelating—Fats and oils obtain metal contents from the soils where plants are grown and later from contact during crushing, processing, and storage. Many of the metals promote autooxidation, which results in off-flavors and odors accompanied by color development in the finished fats and oils products. Studies have identified copper as the most harmful metal with iron, manganese, chromium, and nickel following. The effects of prooxidants can be diminished by using chelating agents before and after deodorization. The most commonly used chelating agents are citric acid, phosphoric acid, and lecithin. Citric acid is metered into the oil as an aqueous or alcoholic solution at levels of 50 to 100 ppm. Citric acid decomposes at 347°F (175°C) and the usual practice is to add it during the cooling stage in deodorization. Citric added prior to deodorization decomposes at deodorization temperatures but still affords a degree of protection from trace quantities of oxygen present during preheating. Phosphoric acid, when used, is added to the deodorized in aqueous solution at a concentration of no more than 10 ppm because a slight overaddition can lead to off-flavors in some oils, i.e., watermelon flavor in soybean oil. Lecithin has been used to chelate metals at 5 ppm.

(5) Oil polishing—The final stage of deodorization should be filtration of the oil. The deodorized oil is normally pumped through an enclosed polishing filter to remove any fine particles of soaps, metallic salts, rust, filter aid, polymerized oil, or any other solid impurities. Horizontal

plate filters have long been used as the polishing filter of choice for deodorized oils. These filters are well adapted to this service since oil clarity is excellent and the amount of solids to be removed from the oil is minimal. The disadvantages for this type of filter are the labor requirements to clean and redress the filter plus the space requirements. Therefore, small cartridge or bag filters have become popular for this purpose. These oil polishing filters are relatively inexpensive, require a minimum of space, and are much less labor-intensive, and the bags are relatively inexpensive.

(6) Operating conditions—Operating variables, such as temperature, pressure, stripping steam rate, and time of steaming affect the quality of the finished product. The temperature of the steam required is proportional to the absolute pressure. The time required for efficient deodorization depends upon the rate at which steam can be passed through the oil and is limited by the point at which appreciable mechanical entrainment occurs. The lower the system pressure at a fixed vapor pressure or temperature and sparge steam rate, the greater the FFA reduction. Since the vapor pressure of the FFA and the other volatiles is directly proportional to the temperature, both an increase in temperature and sparge steam rate will increase FFA reduction. However, the maximum temperature that can be used is limited because of the detrimental effects upon oil stability.

Typical deodorizer conditions practiced in the United States are shown in Table 2.8 for the three types of deodorization systems [88]. Most deodorized oil specifications allow a 0.05% maximum FFA; however, most processors will deodorize to a 0.03% maximum limit internally. The higher published limits allow for the addition of antioxidants or other additives that raise the FFA level.

14.5 Deodorizer Distillate

Typically, deodorizers utilize a three- or four-stage steam ejector system with direct contact in the condensers to maintain a vacuum within the deodorization vessel at 1 to 6 mm Hg absolute pressure, and stripping steam is injected into the oil at a temperature of 410 to 525°F (210 to 274°C). The odoriferous materials, free fatty acids, other organic materials, and small quantities of triglycerides are carried off by the stripping steam through the primary steam ejector of the vacuum generating system and are finally condensed along with the stripping and motive steam within the barometric condenser. Scrub coolers are often installed before the barometric condenser to remove as much of the distillate as possible to keep it

TABLE 2.8. Typical Deodorization Conditions.

Deodorization Conditions	Range
Vacuum, absolute pressure, mm Hg	1 to 6
Deodorization temperature, °F	410 to 525
Deodorization temperature, °C	210 to 274
Holding time at deodorization temperature	
Batch deodorizers, hours	3 to 8
Continuous deodorizers, minutes	15 to 120
Semicontinuous deodorizers, minutes	15 to 120
Stripping steam, weight percent of oil	
Batch deodorizers	5 to 15
Continuous deodorizers	1 to 5
Semicontinuous deodorizers	1 to 5
Drop temperature	
Liquid oils, °F	100 to 120
Liquid oils, °C	37.8 to 48.9
Higher melting products:	
°F above melt point	10 to 15
°C above melt point	5.5 to 8.5
Product free fatty acid, %	
Feedstock	0.05 to 6.0
Deodorized product	0.02 to 0.03
Product peroxide value, meq/kg	
Feedstock	2.0 maximum
Deodorized product	Zero
Product flavor, minimum	Bland

out of the recirculating cooling water. The distillate stripped from edible oils can be divided into three component groups. The first group will condense between the deodorization and their solidification temperatures, the second group includes those that are condensed and solidify when cooled to a lower temperature by contact with the vacuum system condensing water, and the final group consists of those that remain volatile even at the lower temperature [94].

Initially, deodorizers used once through cooling water systems that discharged the wastewater into rivers, lakes, reservoirs, or to a water treatment plant. Closed-circuit water systems were introduced when the discharge of organic materials was restricted by government agencies. These systems recirculated the same water through cooling towers and back to the condensers. A rapid buildup on all the surfaces contacted, such as piping, pumps, values, cooling tower surfaces, spray nozzles, and so on, fowled or plugged the equipment for high maintenance costs and created an offensive odor, which resulted in complaints from the surrounding communities.

A direct contact cooling process was developed in the late 1950s, where the deodorizer discharge vapor and vacuum booster steam flows into a tower where it is cooled by direct contact with a stream of circulating distillate. The distillate recovery tower's purpose was to cool the deodorizer discharge to condense as much of the distillate as possible. The location of the scrub cooler within the vacuum system and the operating conditions were selected to maximize distillate recovery, to maintain a pumpable liquid in the tower, and to prevent condensing of the vacuum system steam. The distillate recovery systems were very successful in reducing maintenance and odor problems; however, some organic material still escaped and ended up in the hot well, which contributed to cooling tower problems. This problem was solved with condenser water recycling systems. The basic concept of these systems was to keep the cooling tower clean by cooling the dirty hot well water in a heat exchanger with tower water before recycling to the vacuum system barometric condensers. In this system, there is no direct contact between the two water streams, and therefore, the cooling tower water remains clear [90].

The odoriferous low-boiling compounds that remain volatile at low temperatures continue in the vapor phase through the barometric condenser and exit with the final vacuum stage discharge. The volatile compounds dissolve in the hot well water and are either reintroduced into the air in the cooling tower or build up in the hotwell to cause odor emissions around the deodorizer. Three techniques that have been used for odor control of the noncondensable materials are wet scrubbers, carbon bed systems, and thermal incineration. Carbon beds have not been successful with edible oil systems because of the heat load in the steam being discharged from the deodorizer vacuum system and regeneration problems encountered with the carbon bed. Thermal incineration has not been adopted for deodorization odor control because of the low cost and availability of energy. Wet scrubbers offer the best solution for eliminating noncondensable odor compounds. A wet scrubber is a device where a liquid is used to contact the gas and absorb the soluble components or capture any solid particles. Wet scrubbers normally consist of two components; the first section is a contacting zone where the vapor or particle is captured, and the second section is a disengaging zone where the liquid is eliminated from the cleaned gas [94].

Disposal of deodorizer distillates can be somewhat difficult. As a general rule 0.5% of the deodorizer feedstock would approximate the amount of distillate produced by a typical edible fats and oils processor. The composition of the distillate is generally equal quantities of fatty acids, fat, and unsaponifiables. The distillates contain sterol and tocopherol compounds that are sources for natural vitamin E, natural antioxidants, and other pharmaceuticals. However, the market is limited and the availability of substitutes has substantially decreased the value of deodorizer distillates. The use of deodorizer distillates in animal feeds is forbidden by U.S. regulations because any insecticides in the source oils are codistilled with the other organic compounds. One method of disposal that can realize some value is to utilize the energy value of the dried distillate by mixing it with the fuel oil used to fire the steam boilers; up to 10% of the distillate has been successfully used in fuel oil. The remaining disposal alternatives are to deposit it in a refuse dump, if permitted, or combustion in a fluidized-bed incinerator [95,96].

15. FINISHED FATS AND OILS HANDLING

The last major process that can control the odor and flavor stability of an edible fat and oil product is deodorization. If the conditions of the preceding processing technique and raw material quality have been satisfactory, the result is a tasteless, odorless, and light colored fat and oil ingredient, free from peroxides and contaminates. Processes for handling and storage of these products must protect the achieved quality prior to packaging or use in a prepared food product. Deodorized fats and oils products must be protected from contamination with foreign flavors, foreign odors, impurities, other fats and oils products, microorganisms, and hydrolysis, as well as thermal decomposition and oxidative deterioration.

15.1 Protection against Oxidative Deterioration

Fats and oils containing unsaturated fatty acids are subject to oxidation, a chemical reaction, which can occur with exposure to air, to make them unacceptable to consumers. The double bonds found in unsaturated fatty acids are the site of this chemical activity. The oxidation rate is roughly proportional to the degree of unsaturation: linolenic (C-18:3), with three double bonds, is more susceptible than oleic (C-18:1), with only one. Linoleic (C-18:2) is intermediate but twice as susceptible as oleic. Oxidative deterioration results in the formation of hydroperoxides, which decompose into carbonyls, dimerization, and polymerized materials. It is accelerated by temperature increase, oxygen pressure increase, a higher concentration of oxidation products (peroxides and aldehydes), metal catalyst, lipoxidases, hematin compounds, antioxidant reductions, absence of metal deactivators, storage time, and ultraviolet or visible light [97]. Extensive oxidation would eventually destroy the carotenoids (vitamin A), the essential fatty acids (linoleic and linolenic), and the tocopherols (vitamin E) contained in most oils.

Prevention of oxidation is a major factor that must be designed into all edible fats and oils processes.

NITROGEN BLANKETING

Many of the current edible fats and oils storage, handling, and stabilization practices stem from the results of flavor stability research. Determinations that a correlation exists between peroxide value results and flavor determinations indicate that flavor deterioration can be prevented by excluding air from fat and oil products. The usual procedure involves replacing oxygen with nitrogen. Gas-free finished oil is delivered from the deodorizer to a storage tank under a complete nitrogen blanket supplied by a nitrogen gas generation unit or from liquid nitrogen. A nitrogen blanket is maintained by a pressure system controlled by a regulator. As the tank is filled with oil, the pressure builds and the gas is vented to the atmosphere. Conversely, as the oil is pumped from the tank, the pressure drops and replacement gas enters the tank. It is good practice to install a relief valve or rupture disc for each tank to avoid collapsing a tank with a vacuum. A typical system would be kept under nitrogen pressure of 1 to 15 psi (1 to 2 kg/sq cm). This pressure varies as the tanks are loaded or unloaded, and the pressure relief valves are set to relieve pressures above 15 psi (2 kg/sq cm).

Another nitrogen protective practice utilized is to "sparge" oils at the exit of the deodorizer or a nitrogen blanketed vessel. The principal is to saturate the oil with nitrogen while it is completely deaerated. The spargers discharge tiny bubbles of nitrogen directly into the oil stream. As the oil flows into the unprotected vessel, the effusing gas sweeps the headspace, sweeping most of the oxygen from the vessel. This technique is particularly useful for protecting bulk oil shipments in tank cars and trucks [98].

Oxygen contact can be reduced by keeping the entire edible fats and oils handling process protected with an inert gas. In finished oil processing, the product is preferably protected by nitrogen gas in the storage tanks and bulk transports as well as in packaging. Effective protection against oxidation for packaged products requires that the air in the oil and the package headspace be evacuated and then replaced by a protective gas such as nitrogen.

TEMPERATURE CONTROL

Fats and oils temperatures should be maintained as low as possible during handling because heat accelerates the oxidation reaction. Before exiting the deodorizer, while still protected by vacuum, the product should be cooled to 100 to 120°F (37.8 to 48.0°C) for liquid oils and 10 to 15°F above the melting point of higher melting products. After the polish filter, the liquid oils can be further cooled. The speed of oxidation is doubled for each 27°F (15°C) temperature increase in the 68 to 140°F (20 to 60°C) range. Therefore, an oil can be held four times as long at 68°F than at 122°F, before the same degree of oxidation occurs.

Hydrogenated and higher melting fats and oils must be kept liquid for handling with pumps. Control of the product handling temperature is necessary to prevent oxidation and because the pumping rates plus any volumetric metering systems are dependent upon product viscosity, which in turn is temperature-dependent. A good practice, as outlined above, is to keep the products no warmer than necessary to pump conveniently or 10 to 15°F (5.6 to 8.3°C) above the product's melting point. Bulk handling systems should be designed to accommodate low-temperature product handling and minimize localized overheating with short insulated lines, tank agitators, automatic temperature controllers, hot water or low-pressure steam heating, properly sized tanks, proper placement of temperature probes, and so on.

LIGHT CONTROL

Light exposure contributes to the flavor deterioration of all edible fats and oils products. Only limited exposure of an oil product to sunlight or ultraviolet rays from fluorescent lighting will increase peroxide value results and decrease flavor ratings. Normally, light-induced deterioration is not a processing concern because all processing, handling, and storage is accomplished in closed systems. However, two areas of concern are (1) the samples drawn and submitted to the laboratory for analysis and (2) the liquid oils packaged in clear glass or plastic bottles for the retail market. Transparent or open, large-mouth containers should not be used to obtain samples for laboratory testing of organoleptical properties and peroxide value determinations. Oils stored in brown glass bottles have experienced only slightly more deterioration than those stored in metal or otherwise kept away from light, but the U.S. consumer has shown a preference for salad oils packaged in clear containers.

STORAGE TIME

All fat or oil products experience deterioration even when handled and stored under ideal conditions. Oils that do not require heating to remain liquid resist deterioration more so than the higher melting products. These products should not be allowed to solidify and then reheated for use unless for a down period of extended duration. Most shortenings and other similar products will maintain an acceptable flavor and oxidative stability for 2 to 3 weeks in melted form with adequate controls. Therefore, edible fats and oils storage systems should be designed for turnovers of less than two to three weeks or within the established storage life of the product. Additionally, fresh product should not be added to heels of previous lots. The older product will contain concentrations of oxidation products that will accelerate oxidation of the fresh product.

ANTIOXIDANT ADDITION

Antioxidants are chemical compounds that provide greater oxidative stability and longer shelf life for edible fats and oils by delaying the onset of oxidative rancidity. Oxidation occurs in a series of steps often referred to as free radical oxidation because the initial step is the formation of a free radical on the fatty acid portion of the fat molecule. The free radical is highly reactive and forms peroxides and hydroperoxides by reaction with oxygen. These free radicals also initiate further oxidation by propagating other free radicals. Finally, the hydroperoxides split into smaller organic compounds such as aldehydes, ketones, alcohols, and acids. These compounds provide the offensive odor and flavor characteristic of oxidized oils [99]. Antioxidants function by inhibiting or interrupting the free radical mechanism of glyceride autoxidation. Their ability to do this is based on their phenolic structure or the phenolic configuration within their molecular structure. Antioxidants or phenolic substances function as a free radical acceptor, thereby terminating oxidation at the initial step. Fortunately, the antioxidant free radical that forms is stable and does not split into other compounds that provide off-flavor and odors, nor does it propagate further oxidation of the glyceride [100].

Food products in interstate commerce in the United States are subject to the regulations under the Food, Drug and Cosmetic Act; the Meat Inspection Act; and the Poultry Inspection Act. These regulations establish limitations on the use of antioxidants and other food additives in food products. Antioxidants cleared under these regulations include [99–106]

- tocopherols—Nature's fat-soluble antioxidants exist in four forms: alpha-tocopherol, beta-tocopherol, gamma-tocopherol, and delta-tocopherol. Alpha-tocopherol (vitamin E) is widely used as a nutritional supplement. Beta-tocopherol exists in concentrations too low to have any practical significance. Gamma- and delta-tocopherols are known primarily for their antioxidant properties and are used where consumers require a natural antioxidant. Interestingly, synthetic antioxidants contain the same phenolic structure that makes the tocopherols effective antioxidants.

  In most applications, tocopherol addition levels of 0.02 to 0.06% are sufficient to provide good antioxidant properties. The USDA limits the addition level to 0.03% for lard or rendered pork and rendered poultry fat, but the FDA allows the amount required for the intended technical effect for most products. The antioxidant properties of tocopherols can be enhanced by the addition of an acid synergist such as citric or ascorbic acid. Tocopherol additions have the most application for stabilizing edible fats of animal origin because of the absence of natural antioxidants in these products. Vegetable oils are the source of the tocopherols, and enough of the natural antioxidant survives refining, bleaching, and deodorization to provide the optimum stability available from tocopherols. Performance testing has shown that tocopherols added to vegetable oils above the normal survival level did not improve the oxidative stability and that high enough levels had a negative effect upon flavor stability.

- propyl gallate—*n*-propyl ester of 3,4,5-tri-hydroxybenzoic acid or propyl gallate is an effective antioxidant for shelf life improvement of vegetable oils at usage levels of 100 to 200 ppm, or the FDA and USDA permitted levels. However, its usage is hampered by solubility problems, discoloration, and poor heat stability. Propyl gallate has significant water solubility and seeks the water phase in water-fat systems, which causes a reduction in antioxidant effectiveness and allows complexing with iron to cause iron-gallate discoloration. Darkening of scrambled eggs to a blue-black color when prepared in an iron skillet with oil stabilized with propyl gallate is an example of the deficiency. Additionally, edible oil products stabilized with propyl gallate can darken while stored in black iron vessels, packaged in metal containers, or in contact with metal processing equipment. Finally, propyl gallate may be inactivated readily in alkaline systems and particularly at elevated temperatures. These deficiencies have limited the use of propyl gallate in edible fats and oils in the United States, where an alternative effective vegetable oil antioxidant is approved—TBHQ.
- BHA—Butylated hydroxyanisole (BHA) is outstanding among the antioxidants for its carrythrough effect; i.e., it can substantially withstand food processing temperatures like those experienced in baking and frying. BHA is referred to as a hindered phenol because of the tertiarybutyl group, *ortho-* or *meta-*, to the hydroxyl group. This steric hindrance is probably responsible for the ineffectiveness of BHA in vegetable oils. However, this same steric hindrance is probably responsible for the carrythrough effect for baked or fried foods. BHA has a strong phenolic odor that is particularly noticeable with the initial heating of frying shortenings or oils stabilized with BHA and described as a chemical odor. Another concern is the development of a pink color, which can occur when BHA comes in contact with fairly high concentrations of alkaline metal ions, such as sodium or potassium.

- BHT—Butylated hydroxytoluene (BHT) is another hindered phenol with a molecular structure similar to BHA. BHT is also similar to BHA in performance in regard to the relative weakness to stabilize vegetable oils and the ability to survive baking and frying conditions to carry through into the finished food products. Both BHT and BHA are extremely soluble in edible fats and oils and have practically no water solubility. BHT can experience some darkening in the presence of iron, but the degree is not very serious.

- TBHQ—Tertiary butylhdroquinone (TBHQ) is a relatively recent addition to the approved antioxidant list in the United States and has not gained approval in many other countries. TBHQ has been found to be the most effective antioxidant for the unsaturated vegetable oils along with several other advantages: (1) no discoloration when used in the presence of iron; (2) no discernible odor or flavor imparted to fats and oils; (3) good solubility in fats and oils; (4) effective in poultry and animal fats as well as vegetable oils; (5) carrythrough protection in baked and fried products; and (6) a stabilizing effect upon tocopherols. The use of TBHQ to protect edible fats and oils from crude to deodorized has been found effective for preventing secondary oxidation products as determined by anisidine value evaluations. This protection is completely removed during steam distillation and additional TBHQ must be added to the deodorized products for protection of the finished oils. TBHQ, BHA, and BHT are all volatilized at approximately the same rate, but oils stabilized with TBHQ antioxidant consistently have a higher oxidative stability and a better carrythrough into the fried foods. One concern with TBHQ utilization is the pink color that can develop with an alkaline pH, certain proteins, or sodium salts.

- synergistic antioxidant mixtures—Much of the success of antioxidants depends upon their being in chemical contact with the product they are protecting. Therefore, antioxidant formulations containing various combinations of different antioxidant and chelating agents are generally used, rather than individual antioxidants compounds, in most food applications. Not only does the use of such formulations provide a convenience, in that it is easier to handle the diluted antioxidants, but it also permits the processor to take advantage of the synergistic properties of the different antioxidant compounds. For example, BHA and BHT used in combination provide a greater antioxidant effect than when either is used alone. Propylene glycol and vegetable oils usually serve as solvents for the antioxidant mixtures. Lecithin, citrate, monoglyceride citrate, and mono- and diglycerides are included in the formulations as emulsifiers.

Synergism is a characteristic common to many antioxidant mixtures. A mixture is designated as synergistic when the effect of the mixture is greater than the effect produced by the sum of the individual components. A synergist, such as citric acid, has two important functions in antioxidant formulations: (1) it increases the antioxidant effectiveness of the combination and (2) it ties up or sequesters the trace metals, which are fat oxidizing catalysts, by forming complex, stable compounds (chelates). Other compounds that function as synergists and chelating agents include isoproyl citrate, stearyl citrate, orthophosphoric acid, sodium monohydrogen phosphate, pyrophosphoric acid and its salts, metaphosphoric acid and its salts, calcium disodium EDTA, and disodium EDTA.

15.2 Protection against Contamination

A contaminate is any undesirable material that may taint, infect, corrupt, modify, or degrade by contact or association. Edible fats and oils must be protected from contamination at all stages of processing and especially after deodorization because it was the last opportunity to remove undesirable flavors, odors, colors, and other compounds. The contaminates most likely to affect the deodorized edible fats and oils products to the degree that would require reprocessing or downgrading to a lesser value product include

- moisture—Maintenance of a low moisture content after deodorization will ensure that hydrolysis does not occur. Hydrolysis is the reaction of water with a fat or triglyceride to break it down into a diglyceride and a free fatty acid. This hydrolytic reaction of fats and oils can be prevented by maintaining a moisture-free environment for the product. Some potential sources of moisture contamination during storage and handling are ruptured heating or cooling coils, leaking coolers, inadequate drying of washed tanks, condensation, and steaming of lines.
- impurities—Foreign materials in finished products are usually caused by a malfunctioning polish filter, precipitant of metal salts, polymerized oil from dead spaces in lines, inadequate tank cleaning, and so on. It is a good policy to polish filter all finished products during each transfer, preferably as near to the destination as possible.
- commingling—Inadvertent mixing of two different fats and oils products together is serious contamination. Every finished fats and oils product has specific properties designed for the intended application.

Commingling with another product can change the composition, consistency, performance, and oxidative stability of the contaminated product. The resultant product in most cases requires reprocessing to another product due to potential changes in labeling, religious constraints, and performance. Commingling is usually the result of human error or malfunction of a valve. Murphy's Law dictates that the highest value product is always mixed with the least expensive product handled.

- odors and flavors—Fats and oils easily absorb odors and flavors from other foods, spices, solvents, gases, chemicals, paints, and any other odoriferous or flavorful material. Therefore, extreme care must be exercised to protect the fats and oils products at all times after deodorization, including storage, handling, and transportation of bulk liquid or packaged products.

16. PLASTICIZATION

Considerably more is involved in preparing shortening and margarine for packaging and eventual use as an ingredient or spread than simply lowering the temperature to cause solidification. A grainy, pasty, nonuniform mass is produced when edible fats and oils are allowed to cool slowly. The more saturated triglycerides crystallize first and grow in size to produce an unsightly, difficult to handle, mushy product that lacks many of the basic qualities necessary for shortening or margarine performance. Development of the desired edibility, appearance, stability, texture, functionality, uniformity, and reproducibility in solidified fats and oils products are a function of controlled crystallization or plasticization.

16.1 Plasticity of Edible Fats and Oils

Edible fats and oils products appear to be soft homogenous solids; however, microscopic examination shows a mass of very small interlocked crystals that trap and hold by surface tension a high percentage of liquid oil. The crystals are separate discrete particles capable of moving independently of each other when a sufficient shearing force is applied to the mass. Therefore, shortening, margarine, and other solidified fats and oils products possess the characteristic structure of a plastic solid.

The distinguishing feature of a plastic substance is the property of behaving as a solid by completely resisting small stresses, but yielding at once and flowing like a liquid when subjected to deforming stresses above a minimal value. A firm, plastic material will not flow or deform from its own weight. However, it may be easily molded by slight pressure into any desired form. Plastic solids derive their functionality from their unique plastic nature. Three conditions are essential for plasticity [107]:

- It must consist of two phases, one a solid and the other a liquid.
- The solid phase must be dispersed finely enough to hold the mass together by internal cohesive forces.
- There must be proper proportions of the two phases. The solid portion must be capable of holding the liquid while the liquid portion is sufficient to allow flow when stress is applied.

Plasticity and consistency of an edible fat and oil product depend upon the amount, size, shape, and distribution of the solid material, as well as the development of crystal nuclei capable of surviving high-temperature abuse to serve as starting points for new desirable crystal growth. The factors that influence these characteristics are [108,109]:

- product composition—The factor most directly and obviously influencing the consistency of a plastic shortening is the amount of material in the solid phase; the product becomes firmer as the solids content increases. The solids contents are determined by oil source and prior processing, i.e., the degree of hydrogenation, interesterification, fractionation, and/or naturally solid source oils such as lard, tallow, palm oil, and coconut oil. Solidified fats and oils begin to have enough body to hold their shape well at a solids content as low as 5% and become rigid, losing elasticity as the solids contents reach 40 to 50%. A typical, all-purpose bakery shortening formulated for creaming properties and spreadability or workability attempts to maintain a solids fat index of 15 to 25% over the widest temperature range possible. However, a satisfactory plastic range should be defined as the range of temperature at which the fat and oil product may be used with the intended results.
- crystal size—At elevated temperatures, fats retain enough molecular motion to preclude organization into stable crystal structures. However, edible fats and oils go through a series of increasingly organized crystal phases with cooling until a final stable crystal form is achieved. All fats crystallize from the liquid phase in the alpha form, transform more or less rapidly to beta-prime, and subsequently to the intermediate and/or the beta modifications if they are likely to exhibit these higher polymorphs. This sequence is irreversible; once transformation to the more stable forms has occurred, lower polymorphs can only be obtained by melting the product and repeating the process. This process can occur in fractions of a second or months. The crystal types formed define the texture and function properties of most fat-based

products. Each crystal form possesses its own specific physical properties. They differ in melting point, solubility, specific heat, dielectric constants, and more. The crystal lattice that is formed when the molecules solidify is a relatively loose arrangement. As crystallization proceeds, the molecules tend to pack more closely together. With time, the molecules in the crystal lattice will pack together as closely as their structure permits; therefore, the molecules in the most stable crystal lattice will require the least space.

The body and functionality of plastic fats and oils products is influenced significantly by the size of the crystals formed during solidification. A product becomes progressively firmer as the average size of the crystals decreases and softer as the crystal size increases. A fat that has been melted and allowed to crystallize slowly under static conditions will contain many large crystals plainly visible to the eye. Crystals formed in the same fat by rapid chilling methods will be microscopic in size. Quickly chilled product with very small crystals will be firmer and will have a consistency range much wider than that of a fat slowly crystallized. The slowly crystallized product will also be softer than the rapidly chilled fat.

- supercooling—A very critical and complicating factor in plasticizing of edible fat and oil products is the supercooling properties of triglycerides. Fats can remain liquid when chilled below their melting point. Because of this fact and because fats are polymorphic and can crystallize in two or more forms, the solidification and plasticization process requires careful control. The degree of supercooling and the temperature at which the supercooled product is allowed to reach crystal equilibrium is directly related to the temperature range over which the product will be workable. In practice, the temperature to which the product is supercooled, worked, and packaged is controlled to produce the widest plastic range for the individual product formulation. The extent to which a fat is supercooled can affect not only the consistency, but also the melting point of the solidified product.

- mechanical working—Solidification of the supercooled product without working or agitation will produce a firm consistency and a narrow plastic range. The product will also lack smoothness of texture and have a nonuniform appearance. Solidification without working allows the fat crystals to grow together to form a crystal lattice with greater strength than the same product with the crystals broken into smaller discrete particles. Therefore, for optimum plasticity, the supercooled product must be mechanically worked during this crystal formation period until substantially all of the latent heat of crystallization has been dissipated.

The degree of work applied to shortening and margarine differs due to the finished product consistency desired. Traditional tablegrade margarine, packaged in quarter-pound sticks and one-pound solids, must be firm enough to be handled with print-forming and wrapping equipment. Shortening and soft tub margarines are preferred in a semifluid condition at the filling station. Traditional stick margarines must be allowed to reach crystallization equilibrium with the minimum amount of work for a homogenous mixture. An important precaution for both stick and tub-type margarine plasticization is to avoid too fine a dispersion of the aqueous phase while inducing the larger crystal formation. This prevents a "waxy" mouth sensation and hastens the liberation of the salty aqueous phase which contributes some of the flavors and complements others.

- gas incorporation—Creaming gas, preferably an inert gas like nitrogen instead of air, is incorporated into most standard shortenings at 12 to 14%, regular soft tub margarines at 4 to 8%, whipped tub margarines at 30 to 35%, and precreamed household shortenings at 18 to 25%. The reasons for the addition of creaming gas to these products are (1) white, creamy appearance; (2) bright surface sheen; (3) easier handling—less dense product; (4) texture improvement; (5) homogeneity; (6) increased volume; (7) reduced calories per serving; and (8) reduced saturated fat grams per serving. Stick, liquid, and most industrial margarines do not have creaming gas added during crystallization. The aqueous phase of a margarine emulsion has the same effect on appearance as gas incorporation.

16.2 Solidification Apparatus Evolution [110–113]

Some of the first vegetable oil shortenings were grainy and nonuniform, with an appearance similar to rendered lard. These products were either filled into containers in a molten state and allowed to cool and solidify or were chilled in tanks until crystallization began and then poured into containers where crystallization and solidification were completed. With this type of processing, the rate and degree of chilling that affected the degree of crystallization before packaging were variable, which resulted in products with a nonuniform texture and consistency.

Probably, the earliest device in general use for achieving rapid cooling of fats and oils products was the chill roll, which was a hollow cast iron, closed-end cylinder, mounted on bearings so it could be revolved. The roll was chilled either by pumping brine through it or by direct expansion of ammonia into its interior. In operation, the roll was rotated, and during rotation it passed through or contacted a trough containing molten fat. A film of fat adhered to the roll surface where it was chilled to form a solid layer. The solidified sheet of fat was scraped off the roll surface by a doctor blade just prior to the melted fat pickup point after one revolution. The solidified fat dropped from the roll into a horizontal trough or picker box with a longitudinal shaft fitted with a number of paddles or spikes. The purpose of the picker box was threefold: (1) to beat air into the product, which improved the product appearance and texture; (2) as a holding vessel while crystallization continued; and (3) to dissipate the heat of crystallization. While the fat sheet from the roll was usually solid, it was in a super-cooled condition, and much of the crystallization remained to take place after it left the roll surface. From the picker box, the viscous fat was pumped under pressure through a throttling valve to the container. The pressure and throttling improved the texture and appearance of the short-ening by uniformly distributing the air bubbles within the fat and by break-ing up any large agglomerates of crystals. Use of the chill roll allowed packaging of a reasonably uniform product. By varying the speed of the roll and the temperature of the refrigerant, it was possible to control the rate of chilling and the degree to which the fat was cooled.

In the early 1930s, development work aimed at the development of im-proved heat-transfer equipment for freezing ice cream led to the perfection of a continuous internal chilling machine that was soon applied to the plas-ticization of shortenings. Liquid fat to be chilled is pumped into a relatively small annular space between a large mutator shaft and an outer refrigerated jacket. As the shortening contacts the cold jacket wall, it congeals and is instantly scraped off by scraper blades attached to the rapidly rotating mu-tator. This repeated high-speed congealing scraping sequence provides ex-tremely high heat transfer rates and a homogeneous product. The mutator may be hollow and heated slightly with circulating warm water to prevent the formation of adhering masses of crystallized fat.

As the molten shortening is pumped to the scraped wall heat exchanger, nitrogen, an inert gas, is injected into the stream, which dissolves in the liquid fat under normal pump pressures. As the product leaves the heat exchanger, it flows into a working or tempering unit that simulates, to some extent, the picker box with the chill roll procedure. The worker unit is a closed cylinder with stationary pins attached to the walls, which intermesh with spirally located pins affixed to a helical rotating shaft inside the unit. The purpose of the worker unit is to dissipate the heat of crystallization while working the shortening to provide fine crystals. The worked short-ening passes through an extrusion or throttling valve to make a homoge-neous product and pumped to the filling station where it again passes through another extrusion valve and into the package.

Margarine processing went through somewhat the same developments as shortening processing, except that in the beginning it was chilled by continuously pouring or spraying the molten emulsion into a vat of running cold water or brine. The emulsion entered at one end of the vat and solidified before it reached the other end. The solidified material floated to the surface in the form of a flaky mass that was skimmed off. The water was drained from the solidified emulsion, and it was worked to give the desired texture and to incorporate flavoring ingredients. This procedure gave way to chill roll processing, which was replaced with the scraped wall heat exchanger process in the United States. With this processing, it is customary to make an emulsion containing the fat, moisture, salt, flavoring materials, emulsifiers, color, and so on, which is then chilled with the scrapped-wall heat exchanger. The other differences from shortening plasticization are lower chilling temperatures, the fact that creaming gas or nitrogen is not incorporated for many margarines, and that a quiescent aging tube is utilized, instead of the picking or working unit to provide a firm body suitable for printing into quarters or one-pound solids. Also, margarine handling equipment has to be constructed of corrosion-resistant metals due to the presence of salt, moisture, and acids in the emulsions.

As the continuous closed system for plasticizing margarine was developing in the United States, an intermittent method was becoming popular in Europe. Crystallization in this process is accomplished by delivering the emulsion to internally refrigerated rotating drums or rolls. After crystallization, the product is allowed to temper before it is conveyed through a series of working rolls. It is then transferred to a worker or compactor, cut into slabs, and conveyed to blending machines where the moisture is adjusted. After blending, it is placed in large troughs for chilling to 40°F (4.4°C). The chilled product is then printed and packaged. The advocates of the European process maintain that it provides a preferred product consistency and texture because it more closely approximates butter. Margarine crystallization structure with the continuous process preferred in the United States is finer with a tighter emulsion, which does not experience moisture weeping but probably masks flavor more so than the European process.

16.3 Shortening Plasticization Process

The ultimate polymorphic form of plastic shortening is determined by the triglyceride composition, but the rate at which the most stable form is reached can be influenced by mechanical and thermal energy. Thus, it is customary to process plastic shortenings through various heat exchanger working configurations to remove heat of crystallization and heat of transformation.

A typical U.S. shortening plasticization process, depicted in Chart 2.10, begins when the deodorized shortening blend has been transferred to the

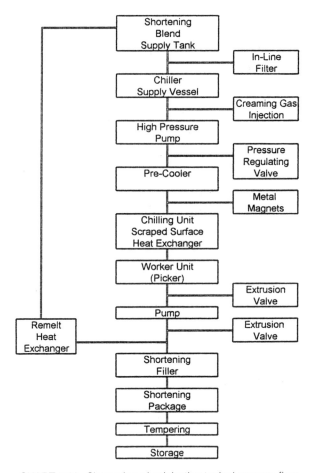

CHART 2.10. Shortening plasticization typical process flow.

packaging department storage vessel, has had all the specified additive materials incorporated, and has met all of the process quality control requirements. The sequence of operations for the plasticization of a shortening are [108–115]

(1) The deodorized shortening blend is transferred from the packaging department storage tank to a small float controlled supply tank adjacent to the chilling unit.

(2) The product is picked up by a gear pump, which maintains pressures of 300 to 400 psi for some systems and up to 1,000 psi for others, on the entire chilling system.

(3) Nitrogen, an inert gas, is introduced into the suction side of the gear pump, usually at 13.0 ± 1.0% for standard shortenings. Creaming gas levels can range from 0 to 30.0%, depending upon the product requirements.

(4) Pressure throughout the entire solidification system is maintained by the use of a pressure regulating valve placed in the product line between the pump and the precooler.

(5) The oil and nitrogen mixture is precooled to 10 to 15°F (5.5 to 8.3°C) above the product's melting point. It is important that the shortening blend remain completely melted to prevent precrystallization before rapid cooling is commenced, as this will result in the formation of large crystals that do not provide a good homogeneous crystal structure. The purpose of the precooler is to
- insure that the shortening blend entering the scraped wall heat exchange is at a constant temperature
- reduce the load on the scraped wall heat exchanger
- insure the presence of a large number of crystal nuclei in the product as it is chilled in the scraped wall heat exchanger

(6) The precooled shortening blend enters the scraped wall heat exchanger, where it is rapidly chilled, usually less than a 30-second residence time, to temperatures ranging from 60 to 78°F (15 to 25.5°C), depending upon the product type and desired firmness.

(7) The supercooled product then passes through one or more worker tubes where the fat crystals are subjected to a shearing action while the heat of crystallization dissipates. The shaft in the worker unit revolves at approximately 125 rpm and the residence time of the chilled product in this unit is usually about 3 minutes. During this time, the temperature of the product rises approximately 10 to 15°F (5.5 to 8.3°C) due to the heat of crystallization.

(8) The worked product is then forced through an extrusion valve that contains a slot or other form of constriction to aid in making the product homogeneous by breaking up any remaining crystal aggregates with an intense shearing action.

(9) A rotary pump delivers the substantially solidified product at pressures in the range of 300 to 400 psi to a second extrusion valve located near the filling station.

(10) The solidified shortening can now be filled. The temperature rise in the container should not be in excess of 1 or 2°F (0.6 to 1.1°C). Increases above this level are indicative of substantial crystallization under static conditions and will cause the consistency to be firmer than desired.

16.4 Liquid Shortening Crystallization

The major attribute of liquid shortenings is fluidity at room temperature. Liquid shortenings are easily poured, pumped, and metered under normal atmospheric conditions, which reduces handling problems for the consumer. Properly processed liquid shortenings do not require agitation to ensure uniformity. Also, oxidative stability is prolonged because no heat is required for fluidity as low as 50°F (10°C) for most liquid shortenings. The products are milky-white in appearance due to the dispersion of hard fats in the form of microcrystalline particles, which do not settle out because of the crystallization process. Fluid shortenings are composed of components that are stable in the beta crystalline form. Low-iodine value, beta crystal-forming hard fats seed crystallization for liquid shortenings. The hard fat level that can be added is limited by the desired fluidity of the shortening and the eating quality requirements of the finished products produced. A stable fluid system in the beta crystalline form will not increase in viscosity or gel once it is properly processed. Hard fats with beta-prime crystal habits are unacceptable for liquid shortenings because the tight knit crystal lattice structure initiates a viscosity change with crystallization to a nonfluid product. Aeration properties normally associated with beta-prime small crystals are achieved by the addition of appropriate emulsifiers. Emulsifiers are also included in some formulations to retard staling of yeast-raised breads and rolls to increase shelf life.

The rate at which a fat transforms into its stable crystal form is important in liquid shortening processing since it must be in the stable form before packaging to avoid solidification in the package. Therefore, the transformation of a liquid shortening into the stable form must be accomplished in a few hours. The quickest transformation of a fat to its stable crystal form can be attained by the following:

(1) Heat the fat until completely melted.

(2) Rapidly cool the fat to just below the alpha crystal melting point. Theoretically, the alpha melting point is very nearly the lower limit to which fatty materials can be cooled without forming any crystals. AOCS Method Cc 6-25 [23] determines the temperature at which a cloud is induced in a fat caused by the first stage of crystallization.

(3) Heat to just above the beta-prime crystal melting point but below the beta crystal melting point. The beta crystal melting point or the highest melting form may be estimated with a capillary melting point, AOCS Method Cc 1-25 [23].

Many different processes have been proposed and patented for preparing liquid shortenings. Most of the methods require rapid chilling of the product with a scraped surface heat exchanger followed by a crystallization period in an agitated vessel. Two process procedures that have produced acceptable liquid shortenings are diagrammed in Chart 2.11. Process "A" has been found acceptable for liquid shortenings with hard fat levels below 5% and Process "B," which requires more process time, has been found acceptable for products with more than 5% hard fat and/or emulsifier additions.

16.5 Margarine Plasticization Process

Several different types of margarine and spreads are produced for the North American retail, foodservice, and food processor markets. Processing parameters like formulations should be tailored to produce the appropriate finished product characteristics. The presence of moisture and salt necessitates the use of sanitary piping and fittings throughout the margarine process, and the material-contacting parts of the equipment are stainless steel or commercially pure nickel, chrome plated. Four basic margarine types produced are stick, soft-tub, liquid, and industrial products. The process flow for these margarine types is illustrated in Chart 2.12.

Processing for all of the margarine types begins with the preparation of the water-in-oil emulsion. The loose emulsion may be produced with either a batch or a continuous process. With batch processing, the emulsion mixing tank or vat is called a churn although it has no resemblance to a butter churn. It is fabricated with stainless steel like all of the margarine equipment to resist corrosion, jacketed for temperature control, and agitated with high-speed counterrotating propellers. Emulsions are prepared by weighing or metering warm margarine oil into the churn and individually weighing and adding the oil-soluble ingredients while under agitation. Concurrently, the pasteurized, aqueous phase is prepared by mixing all the water-soluble ingredients together in a separate vat. Since margarine is a water-in-oil emulsion, the water phase is added to the oil phase. The loose emulsion is agitated, and enough heat is applied to maintain the emulsion temperature at 10 to 15°F (5.6 to 8.3°C) above the melting point of the margarine oil base. The emulsion must be continuously agitated because it is not stable in the melted form and will begin to separate within seconds if not constantly mixed. After the emulsion is well mixed to ensure uniformity, it is transferred to the chiller supply tank. Agitation must be continued and a uniform temperature maintained to supply a uniform product to the crystallization process.

An alternative continuous emulsion preparation process consists of proportioning pump systems capable of metering individual ingredients of the aqueous phase together and also the oil phase components. In-line static mixers are utilized to blend the separate phases, which are then mixed

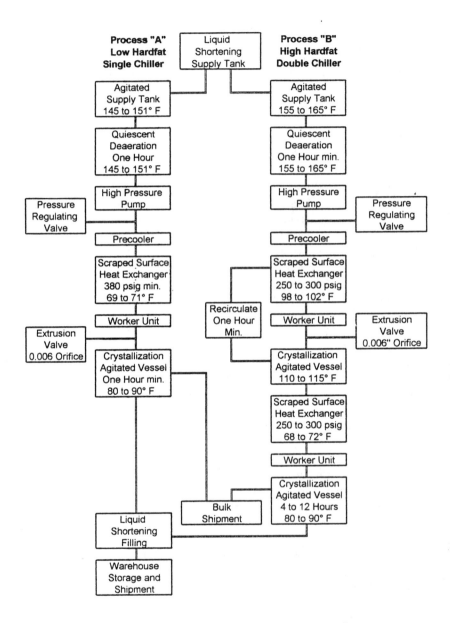

Process "A"
Low Hardfat
Single Chiller

Liquid
Shortening
Supply Tank

Process "B"
High Hardfat
Double Chiller

Agitated
Supply Tank
145 to 151° F

Agitated
Supply Tank
155 to 165° F

Quiescent
Deaeration
One Hour
145 to 151° F

Quiescent
Deaeration
One Hour min.
155 to 165° F

Pressure
Regulating
Valve

High Pressure
Pump

High Pressure
Pump

Pressure
Regulating
Valve

Precooler

Precooler

Scraped Surface
Heat Exchanger
380 psig min.
69 to 71° F

Scraped Surface
Heat Exchanger
250 to 300 psig
98 to 102° F

Worker Unit

Recirculate
One Hour
Min.

Worker Unit

Extrusion
Valve
0.006" Orifice

Extrusion
Valve
0.006 Orifice

Crystallization
Agitated Vessel
One Hour min.
80 to 90° F

Crystallization
Agitated Vessel
110 to 115° F

Scraped Surface
Heat Exchanger
250 to 300 psig
68 to 72° F

Worker Unit

Crystallization
Agitated Vessel
4 to 12 Hours
80 to 90° F

Bulk
Shipment

Liquid
Shortening
Filling

Warehouse
Storage and
Shipment

CHART 2.11. Liquid shortening crystallization process flow.

160

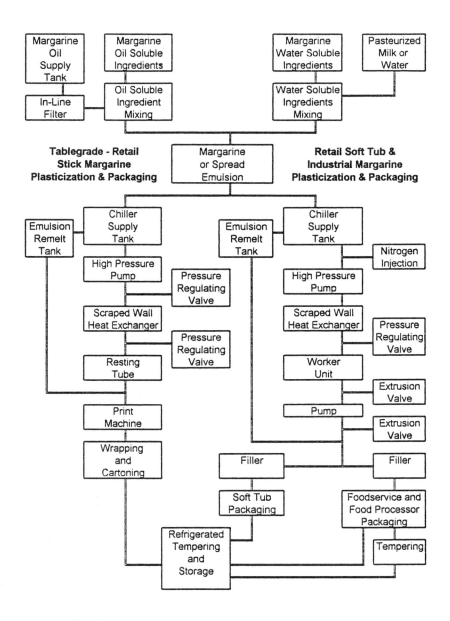

| Margarine Oil Supply Tank | Margarine Oil Soluble Ingredients | | Margarine Water Soluble Ingredients | Pasteurized Milk or Water |
| In-Line Filter | Oil Soluble Ingredient Mixing | | Water Soluble Ingredients Mixing | |

| **Tablegrade - Retail Stick Margarine Plasticization & Packaging** | Margarine or Spread Emulsion | **Retail Soft Tub & Industrial Margarine Plasticization & Packaging** |

Emulsion Remelt Tank — Chiller Supply Tank — High Pressure Pump — Pressure Regulating Valve — Scraped Wall Heat Exchanger — Pressure Regulating Valve — Resting Tube — Print Machine — Wrapping and Cartoning

Emulsion Remelt Tank — Chiller Supply Tank — Nitrogen Injection — High Pressure Pump — Scraped Wall Heat Exchanger — Pressure Regulating Valve — Worker Unit — Extrusion Valve — Pump — Extrusion Valve — Filler — Filler — Soft Tub Packaging — Foodservice and Food Processor Packaging — Refrigerated Tempering and Storage — Tempering

CHART 2.12. Margarine or spread plasticization typical process flow.

161

in-line and emulsified with another static mixer. After in-line blending, the loose emulsion is continuously fed to the crystallization system.

The solidification process for the different margarine types all employ a scraped surface heat exchanger for rapid chilling, but the other steps are somewhat different than for shortening and the other margarine types. The solidification or plasticization process for the four basic margarine types are

- stick margarine—The temperature of the stick margarine emulsion is adjusted and maintained for most tablegrade products that melt below body temperature at 100 to 105°F (37.8 to 40.6°C) before pumping to the scraped surface heat exchanger. It is rapidly chilled to 40 to 45°F (4.4 to 7.2°C) in less than 30 seconds. Stick margarine requires a stiffer consistency than shortening, which is accomplished with the use of a quiescent tube immediately after the chilling unit. This is a warm water-jacketed cylinder that can contain baffles or perforated plates to prevent the product from channeling through the center of the cylinder. The length of this tube may have to be varied to increase or decrease crystallization time, depending upon the product formulation. The supercooled mixture passes directly to a quiescent resting or aging tube for molded print forming equipment. For filled print equipment, a small blender may be utilized prior to the resting tube to achieve the proper consistency for packaging and a slightly softer finished product. A remelt line is necessary because, in all closed filler systems, some overfeeding must be maintained for adequate product to the filler for weight control. The excess is pumped to a remelt tank and then reintroduced into the product line.

  Two types of stick margarine forming and wrapping equipment are in use in the United States, i.e., molded and filled print. The molded print system initially used an open hopper into which the product was forced from an aging tube through a perforated plate in the form of noodles. The margarine noodles were screw-fed into a forming head and then discharged into the parchment paper wrapping chamber and finally cartoned. Closed molded stick systems now use a crystallization chamber instead of the aging tube and open hopper arrangement, which fills the mold cavity by line pressure. The filled print system accepts margarine from the quiescent tube with a semifluid consistency. It is filled into a cavity prelined with the parchment or foil interwrap. The interwrap is then folded before ejecting from the mold into the cartoning equipment.

- soft-tub margarine—The margarine oil blends for soft-tub margarines are formulated with lower solids to liquid ratios than the stick margarine products to produce a spreadable product directly out of the refrigerator or freezer. Crystallization technique contributes to the desirable consistency as well, but the products are too soft to print into sticks. Therefore, packaging in plastic tubs or cups with snap-on lids is utilized.

To fill the container properly, the soft margarine consistency must be semifluid like shortening. Therefore, the crystallization process resembles shortening plasticization more closely than stick margarine processing. The temperature of a typical soft margarine emulsion would be adjusted and maintained at 95 to 105°F (35.0 to 40.6°C) before pumping with a high-pressure pump to the scraped surface heat exchanger. Creaming gas or nitrogen, added to further improve spreadability, is injected at the suction side of the pump at 8.0% for the most spreadable product and at lower levels for a firmer product. The product is rapidly chilled to an exit temperature of 48 to 52°F (8.9 to 11.1°C). The supercooled margarine mixture then passes through a worker unit to dissipate the heat of crystallization. The shaft in the worker unit revolves at about 35 to 50 rpm, with approximately 3 minutes residence time. Worked product is then delivered to the filler where it is forced through an extrusion valve at pressures in the range of 300 to 400 psi Either a rotary or straight line filler may be used to fill the tubs with margarine. The excess product necessary for a uniform supply to the filler is transferred to a remelt tank and eventually reenters the solidification system.

- whipped-tub margarine—The same equipment used to prepare, crystallize, and package regular soft-tub margarine can be utilized for whipped-tub margarines. The difference during crystallization is the addition of 33% nitrogen gas by volume for a 50% overrun. The nitrogen is injected in-line through a flow meter into the suction side of the pump. Larger tubs, required for the increased volume, necessitate change parts for the filling, lidding, and packaging equipment.
- liquid margarine—Both retail and commercial liquid margarines can be crystallized with the same equipment and process flow used for soft-tub margarines, depending upon the formulation and suspension stability requirements. Liquid margarine oil formulations normally consist of a liquid vegetable oil stabilized with either a beta or beta-prime forming hard fat. Beta-prime stabilized liquid margarine can be prepared using the same rapid crystallization process used to prepare soft-tub margarines, omitting the addition of creaming gas. These finished products require refrigerated storage for suspension stability.

Liquid margarines formulated with beta forming hard fats and some processed with beta-prime formulations incorporate a crystallization or tempering step to increase fluidity and suspension stability. This crystallization step consists of a holding period in an agitated jacketed vessel to dissipate the heat of crystallization. The product may be filled into containers after the product temperature has stabilized. Some products with higher solids to liquid ratios are further processed to stabilize the fluid suspension with either homogenization or a second pass through the scraped surface heat exchanger before filling. The additional processing for more stable crystallization will increase fluidity and suspension stability, but the increased production costs may not be justifiable.

- industrial margarines—Foodservice and food processor margarines may be either duplicates of the retail products in larger packages or designed for a specific use, either product- or process-related. Among the specific use margarines, puff paste or danish pastry applications are the most difficult with regard to crystallization. The characteristic features of a roll-in margarine are plasticity and firmness. Plasticity is necessary because the margarines should remain as unbroken layers during repeated folding and rolling operations. Firmness is equally important as soft and oily margarine is partly absorbed by the dough destroying its role as a barrier between the dough layers. As with shortenings, the ultimate polymorphic form for roll-in margarines is determined by the triglyceride composition, but the rate at which the most stable form is reached can be influenced by mechanical and thermal energy. Therefore, the customary crystallization process for roll-in and/ or baker's margarines is a duplicate of the shortening plasticization process depicted in Chart 2.10, except that margarines containing water or milk in emulsion form are normally not aerated. The aqueous phase of a margarine emulsion has the same effect as gas incorporation on appearance and performance. Commercial margarine products are usually packaged in 50-pound corrugated fiberboard cartons, 5-gallon plastic pails, 55-gallon drums, or special packaging designed for each specific use.

16.6 Tempering

Tempering conditions used depend upon the type of product and are done to control the consistency and plasticity of the fat product. Table and kitchen use margarines do not require any heat treatment after packaging and are tempered at refrigerated temperatures to achieve a loosely, structured and brittle consistency; however, a plasticized shortening or margarine requiring a plastic consistency should be tempered for 40 hours or more in a quiescent state at a temperature slightly above the fill temperature immediately after packaging. During tempering, the crystals transform to the polymorphic form in which they normally exist under ordinary conditions. In practice, holding at 85°F (29.4°C) for 24 to 72 hours or until a stable crystal form is reached is an acceptable compromise. The primary purpose of tempering is to condition the solidified shortening so that it will withstand wide temperature variations in subsequent storage and still have a uniform consistency when brought back to 70 to 75°F (21.1 to 23.9°C), which is the use temperature for a majority of the plasticized shortening and margarines [108]. Slow crystallization during tempering favors crystal growth, which extends plasticity for improved creaming properties and baking performance.

Tempering involves a relationship between temperature and time. During the tempering period, the heat of transformation must be dissipated as rapidly as possible. If the fats are allowed to retain this heat by virtue of their normal thermal-insulating capacity, an appreciable portion of the crystal might be melted, and the subsequent gradual cooling under normal storage conditions will tend to promote transformation of the remaining crystal to the undesirable beta form.

A plasticized product is tempered when the crystal structure of the hard fraction reaches equilibrium by forming a stable crystal matrix. The crystal structure entraps the liquid portion of the shortening or margarine. The mixture of low- and high-melting components of the solids undergoes a transformation in which the low-melting fractions remelt and then recrystallize into a higher melting, more stable form. This practice can take from 1 to 10 days, depending on the product formulation and package size. Small packages temper quicker than the larger packages, which have less surface area exposed to the conditioning temperature. After a plasticized product takes an initial set, some alpha crystals are still present. These crystals remelt and slowly recrystallize into the beta-prime form during tempering. Beta-prime crystal form is preferred for most plastic shortenings and margarines, especially those designed for creaming or roll-in applications [114]. Therefore, beta forming products like soybean oil requiring a wide plastic range are formulated with 5 to 20% of a beta-prime tending hard fat like cottonseed oil, palm oil, or tallow. The beta-prime hard fat must have a higher melting point than the soybean oil basestock in order for the entire product to crystallize in the stable beta-prime form.

The effect of tempering on a plasticized product can be demonstrated best by performance testing; tempered products will have superior creaming properties and an extended working range enabling use at both high and low temperatures. In some cases, penetration values undergo some change during tempering, showing a softening of the conditioned shortening versus a nontempered one. The effect of tempering can also be identified by the product feel; a tempered product will be more smooth with a better body while the untempered product will be more brittle and break down sooner when worked. Beta-prime crystal forming products transferred to a cool temperature environment immediately after chilling and working become permanently hard and brittle, and attempts to recondition these products by subjecting them to tempering conditions have not been successful [115]. Once transformation to the more stable crystal form has occurred, lower polymorphs can be obtained again only by melting the product, crystallizing the alpha form, and repeating the sequence of transformations.

QUICK TEMPERING

The expense and logistical problems associated with constant temperature rooms for tempering have led several equipment manufacturers to develop mechanical systems in attempts to eliminate tempering. Most of these systems do not claim complete elimination, but a 50% or more reduction in tempering time. Most of the quick tempering processes add a postcooling and kneading, or working, unit to the conventional type chilling and working systems utilized with tempering. The theory postulated for these systems is that liquid fat is forced to crystallize individually and rapidly, thus creating smaller, more stable crystals, rather than crystallizing onto existing crystals, causing an increase in crystal size or agglomeration as happens in normal tempering. Two basic systems utilizing additional chilling to reduce tempering have been developed to more completely crystallize the product into a stable polymorphic form mechanically. The two systems are compared in Chart 2.13. In both systems the viscous crystalline material that has been supercooled and worked is re-cooled to remove the heat of crystallization developed in the working unit. One of the systems employs another working unit after the second chilling cycle, while the other is filled immediately after the second chilling unit. The best product quality with postchilling temper is achieved when the product discharge temperature from the second set of chilling units is equivalent to or slightly lower than the exit temperature from the first chilling units. Most shortening and margarine products using a postchilling temper process are filled within a temperature range of 65 to 85°F (18.3 to 29.4°C).

Performance characteristics equivalent to well-tempered shortening with only 24 hours conditioning have been claimed for the postchilled quick temper products [116]. These systems have received acceptance from many edible oil processors, but the standard tempering procedures are still practiced by other processors for plastic shortenings and margarines. In many cases, versions of both conventional and quick temper systems are being

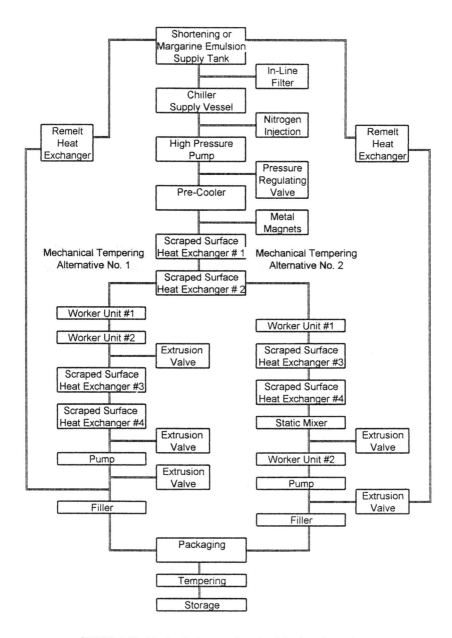

CHART 2.13. Mechanical tempering plasticization alternatives.

167

utilized to produce the various specialized shortening and margarine products, for example,

- Frying shortenings that do not require a plastic consistency are produced with the conventional process with a rapid throughput and tempered for solidification purposes only.
- All-purpose shortenings requiring a wide plastic range are either plasticized conventionally and tempered or utilize a quick temper process.
- Pastry or high solids products are plasticized with one of the two specialized systems and also employ a short conventional quiescent tempering stage.

Another procedure developed to instantaneously crystallize fat products to the desired final form utilizes energy instead of cooling. J. G. Enders et al. found that tempering of shortening with microwave energy produces a substantially improved product with 0.25 to 10 minutes exposure. U.S. patent 3,469,996 indicates that the microwave tempering could be performed continuously before packaging or after packaging on a batch basis. The typical microwave exposure time was 2 minutes with a product exit temperature of 95 to 103°F (35.0 to 39.4°C) [117].

17. FLAKING

Fat flakes describes the higher melting edible oil products solidified into a thin flake form for ease of handling, quicker remelting, or for a specific function in a food product. Flaking rolls, utilized for the chilling of shortening and margarine prior to the introduction of scraped wall heat exchangers, are still used for the production of fat flakes. Chill rolls have been adapted to produce several different flaked products used to provide distinctive performance characteristics in specialty formulated foods as well as the traditional melting point adjustment function. Consumer demands have created the need for such specialty fat products. Specialty high melting fat flakes have been developed for specific applications with varied melting points:

	Melting Range	
Product	°F	°C
Low-iodine value hard fats	125–150	52–66
Hard emulsifiers	140–150	60–66
Icing stabilizers	110–130	43–54
Shortening chips	110–115	43–46
Confectioners fats	97–112	36–44

17.1 Chill Rolls

Chill rolls are available in different sizes, configurations, surface treatments, feeding mechanisms, and so on, but most that are used to flake fats are 4-foot diameter hollow metal cylinders either 9 or 12 feet long, with a surface machined and ground smooth to true cylindrical form. Flaking rolls, internally refrigerated with either flooded or spray systems, turn slowly on longitudinal and horizontal axes. Several options exist for feeding the melted oil product to the chill roll: (1) a trough arrangement positioned at varying locations on the top quarter section of the roll, (2) a dip pan at the bottom of the roll, (3) overhead feeding between the chill roll and a smaller applicator roll, and (4) a double or twin drum arrangement operating together with a very narrow space between them where the melted fat product is sprayed for application to both rolls. A thin film of liquid fat is carried over the roll, and as the revolution of the roll continues, the fat is partially solidified. The solidified fat is cleanly scraped from the roll by a doctor blade, positioned ahead of the feed mechanism with all of the designs.

17.2 Flake Crystallization

During chilling, a portion of the fat is supercooled sufficiently to cause very rapid crystallization. The latent heat released by fat crystallization is absorbed by the cooling medium in the roll. In the crystallization of hydrogenated edible oil products, the sensible heat of the liquid is removed until the temperature of the product is equal to the melting point. At the melting point, heat must be removed to allow crystallization of the product. The quantity of heat associated with phenomenon is called heat of crystallization. Sensible heat or specific heat of the most common hard fat products is equal to 27.8 calories per gram (50 Btu/lb). The amount of heat that must be removed to crystallize low-iodine value hardened oil is 100 times the amount of heat that must be removed to lower the product temperature. Typical coolant temperature requirements for flaked products are

	Coolant Temperature	
Product	°F	°C
Low IV hard fats	70 max.	21.1 max.
Hard emulsifiers	70 max.	21.1 max.
Icing stabilizers	30 max.	−1.1 max.
Shortening chips	5 max.	−15 max.
Confectioners fats	30 max.	−1.1 max.

17.3 Flaking Conditions

The desired flake product dictates the chill roll operating conditions and additional treatment necessary before and after packaging. However, some generalizations relative to chill roll operations and product quality can be made:

- crystal structure—Each flaked product has crystallization requirements dependent upon the source oil, melting point, degree of saturation, and the physical characteristics desired.
- flake thickness—Four controllable variables help determine flake thickness: (1) oil temperature to the roll, (2) chill roll temperature, (3) speed of the chill roll, and (4) the feed mechanism. Normal flake thickness for most flaked products is 0.03 inches but increased for most shortening chips to 0.05 inches.
- in-package temperature—Crystallization of the fat flakes is not complete before removal from the chill roll and heat continues to be released. Heat of crystallization will cause a product's temperature to rise after packaging if not dissipated prior to packaging. The product temperature can increase to the point where partial melting, coupled with pressure from stacking, will cause the product to fuse together into a large lump.
- flake condition—Glossy or wet flakes are caused by a film of liquid oil on the flake surface due to incomplete solidification. Either too warm or too low chill roll temperatures can cause this condition. High roll temperatures may not provide sufficient cooling to completely solidify the flake. Low roll temperatures may shock the oil film, causing the flake to pull away from the surface before completely solidified. Wet flakes from either cause will lump in the package.

18. POWDERED AND BEADED FATS

Powdered and beaded fats are specialized products developed for ease of incorporation, handling, melting efficiency, uniform delivery with vibrator addition systems, and so on. These products may be produced using a hydrogenated fat only, a hard emulsifier only, a blend of hydrogenated fats and emulsifiers, or fats and/or emulsifiers incorporated with other ingredients such as skim milk, corn syrup solids, sodium caseinate, powdered eggs, starch, and other carriers. The blended products are formulated for specialized functions in dairy systems, fillings, prepared mixes, candies, sauces, and other prepared food products. Powdered hard emulsifiers serve the same function as flaked product but will melt quicker and can be incorporated into some finished products "as is." The hard emulsifiers, hydrogenated fats, or blends of the two serve as stabilizers for peanut butter and other products and act as crystallization promoters, lubricants for breading mixes, pan release agents, pharmaceuticals, cosmetics, and melting point adjustment.

Three principal methods of forming powders or beaded fats are practiced in the United States: spray cooling, grinding flaked product, or spray flaking and grinding. Formulated products with fats, emulsifiers, milk solids, and so on for specialized uses are usually spray chilled powders. Hard fats and hard emulsifiers are also spray chilled but may have a disadvantage for feeding or blending accuracy. The spherical shape of the spray chilled powders may act as roller bearings to give erratic feeding rates with vibratory feeding systems or stratify in blends of dry materials. Beaded products produced by grinding flakes or spray flaking and grinding have granular shapes that can be metered at uniform rates with vibratory or screw feeders and resist stratification or separation in mixes with other granular materials.

18.1 Spray Cooling

Generally, this process consists of atomizing a molten fat in a crystallization zone maintained under temperature conditions where a very fine mist of the melted fat is contacted with cooled air or gas to cause crystallization of the fat without marked supercooling. By controlling the temperature conditions and residence time in the cooling chamber, all of the sensible heat and virtually all of the latent heat of crystallization are removed with the formation of many crystallization centers by rapid nucleation, accompanied by an optimum of crystal growth of these nuclei. By controlling the crystal growth with conditions of no appreciable supercooling and almost complete dissipation of the heat of crystallization, the equilibrium condition is achieved almost instantly. This equilibrium condition achieves homogeneity in both inter- and intra-crystal composition of the fat to produce free-flowing, tempered particles.

18.2 Flake Grinding

Flaked, low-iodine value hard fats and hard mono- and diglycerides can be ground with attrition-type mills. This type mill involves impact of the particles both with each other in the air and against a plate. Air flow can be used to move the powdered fats through the equipment. The required air flow is less than involved in spray cooling, but the powder must be separated from the airstream. Precooling of the flakes or dry ice addition must be used to lower the temperature of the product below 40°F (4.4°C) before grinding at a sustained rate is possible. The grinding process causes a rise in temperature, which results in material that is too gummy to grind properly. Flakes can be ground successfully if they are well cooled before the grinding operation begins, with the addition of dry ice, or with air circulation at low enough temperatures to dissipate the heat generated.

18.3 Spray Flaking and Grinding

This patented continuous process allows the manufacture of powdered fats utilizing a chill roll and permits the immediate grinding of the fat into powder without excessive refrigeration in the flaking roll or further cooling of the product. This process consists of [118]

(1) Spraying liquid droplets of molten fat onto the cool chill roll surface and solidified rapidly

(2) Removal of the solidified droplets from the chill roll with a doctor blade

(3) Classification of the solidified fat droplets by size with a vibrating sizing screen, the larger particles being diverted to the grinder while the smaller acceptable size particles proceed directly to the packaging operation

(4) Ground material returning to the vibrating screen for resizing until an acceptable particle size is obtained for packaging

(5) Tempering at an elevated temperature to transform the crystals to the stable polymorphic form

19. SALAD AND COOKING OILS PACKAGING

Clear salad and cooking oils do not require any further processing after deodorization for packaging except for temperature control for flavor, consistency, and weight control plus nitrogen protection for oxidative stability. Liquid oils are currently packaged in 8-, 16-, 24-, 32-, 38-, 48-, and 64-ounce clear plastic containers, as well as 1-gallon opaque plastic containers for the retail market. The same oils and some additional products with additives, such as antifoamers and antioxidants, are packaged in 35-pound or 5-gallon plastic jugs and 425 closed head drums for the foodservice and food processor customers. Plastic containers have replaced glass and metal containers due to improved economics. Both lower package costs and lighter weight packaging contribute to lower product costs.

Salad and cooking oils are sensitive to light, which catalyzes oxidation to produce off-flavors. Sensitizers such as chlorophyll may promote photooxidation. Artificial lighting, as well as sunlight, causes this deterioration in oils. Industrial packaging protects the oils, but the clear plastic retail container like glass offers little protection after removal from the case at the store level.

Oxygen contact contributes to the degradation of an oil and is the most critical factor affecting flavor stability. It may gain access to packaged oil in several ways:

- Atmospheric oxygen may be entrained in the oil at packaging.
- Oxygen may be available in the container headspace.
- Oxygen may permeate the walls of plastic containers.
- Impure nitrogen may contain oxygen.

Liquid oils should be protected from both heat and refrigerated temperatures. High temperatures excellerate flavor degradation and cold temperatures cause crystallization and clouding. Therefore, salad and cooking oils should not be stored in shortening tempering rooms or margarine vaults. Storage and shipment should be at 70 to 75°F (21.1 to 23.9°C).

20. BULK OIL SHIPMENTS

Food processors that use edible fats and oils in large quantities frequently purchase their requirements in bulk. All of the packaged fats and oils products can be shipped liquid bulk in tank cars or tank trucks, except complete margarines and spreads. In this case, the margarine base oil can be shipped to the customers. These products must be handled properly during loading, transit, unloading, and storage in the customers' tanks to ensure acceptable quality at the time of use and consumption.

Bulk handling systems for shipping fats and oils products must be designed and operated with three primary considerations for the maintenance of quality: (1) avoid contamination, (2) avoid overheating, and (3) minimize exposure to air.

(1) Contamination—A contaminate is any undesirable material that may taint, infect, corrupt, modify, or degrade by contact or association. The contaminants most likely to affect bulk shipments are:
- moisture—Precautions must be taken to protect the fat and oil product from hydrolysis and development of free fatty acids. Moisture sources may be wet tank cars or trucks, steamed lines, ruptured tankcar coils, leaking coolers, condensation, rain during loading, and so on.

- impurities—Foreign materials in bulk oil shipments are usually caused by a malfunction in the in-line load out filter, inadequate tank cleaning or rinsing, open hatch covers, or similar. Rail tankcar cleaning is somewhat difficult because of the black iron construction and heating coils, which tend to become coated with oxidized and polymerized oil.
- commingling—Inadvertent mixing of two different fat and oil products is a serious contamination. Every fat and oil product has its own specific properties, and depending upon its field of application, the tolerance toward admixture of other fats is usually very low. Product mixing is usually the result of a mis-pumping, failure to remove a returned heel in a tank car before loading fresh product, or a malfunctioning valve.

(2) Overheating—Most bulk shipped fats and oils must be heated for pumping to tank cars and trucks and again at the customer's location with railcars. Product temperatures should never be at a higher level than necessary. The oxidation rate for fats and oils increases by factor of three for each 20°F (11.1°C); normally, a temperature 10 to 15°F (5.6 to 8.3°C) above the melting point is adequate to keep a fat and oil product liquid for pumping.

Products received at the customer's location in a solid or semisolid state should be heated slowly so that they are liquid and homogenous before pumping. Rail tankcar overheating can occur with high-pressure steam usage or even with low-pressure steam handled improperly. Heating should start at a time calculated to provide the required temperature without exceeding the maximum rate of 10°F (5.6°C) per 24 hours. When steam is used, the steam pressure should not exceed 1.5 kg/sq cm to prevent localized overheating.

(3) Air exposure—Oxidation results when fats and oils are exposed to air, which decreases stability and produces poor flavors. Air can be almost completely excluded by maintaining a nitrogen atmosphere at all stages after deodorization. However, most tank cars and trucks cannot be pressurized to maintain a complete nitrogen blanket. Nitrogen sparging into the stream of oil as it is loading will saturate the oil plus an excess. This excess is released when the oil is loaded into the tanker to displace air in the headspace. This practice requires about 5 cubic feet of nitrogen per 1,000 pounds of oil.

The loading lines should discharge near the bottom of the tanker to minimize aeration of the product. Allowing heated fats and oils to cascade or fall through the air into the tanker allows the product to splash and aerate.

21. REFERENCES

1. Weidermann, L. H. 1981. "Degumming, Refining and Bleaching Soybean Oil," *J. Am. Oil Chem. Soc.,* 58(3):159–165.

2. Carr, R. A. 1978. "Refining and Degumming Systems for Edible Fats and Oils," *J. Am. Oil Chem. Soc.,* 55(11):765–771.

3. Norris, F. A. 1964. "Refining and Bleaching," in *Bailey's Industrial Oil and Fat Products,* Third Edition, Daniel Swern, ed. New York, NY: John Wiley & Sons, Inc., p. 719.

4. Norris, F. A. 1982. "Refining and Bleaching," in *Bailey's Industrial Oil and Fat Products, Volume 2,* Fourth Edition, Daniel Swern, ed. New York, NY: John Wiley & Sons, Inc., p. 259.

5. Swern, D. 1964. "Structure and Composition of Fats and Oils," in *Bailey's Industrial Oil and Fat Products,* Third Edition, Daniel Swern, ed. New York, NY: John Wiley & Sons, Inc., pp. 33–34.

6. Brian, R. 1976. "Soybean Lecithin Processing Unit Operations," *J. Am. Oil Chem. Soc.,* 53(1):27.

7. Brekke, O. L. 1980. "Oil Degumming," in *Handbook of Soy Oil Processing and Utilization,* D. R. Erickson, E. H. Pryde, O. L. Brekke, T. L. Mounts, and R. A. Falb, ed. Champaign, IL: American Oil Chemists Society and American Soybean Association, pp. 71–75.

8. Sullivan, F. E. 1968. "Refining of Oils and Fats," *J. Am. Oil Chem. Soc.,* 45(10):564A–582A.

9. Carr, R. A. 1976. "Degumming and Refining Practices in the U.S.," *J. Am. Oil Chem. Soc.,* 53(6):347–352.

10. Braae, B. 1976. "Degumming and Refining Practices in Europe," *J. Am. Oil Chem. Soc.,* 53(6):353–357.

11. Hoffmann, Y. 1973. "Experiences in Refining Rapeseed Oil by the Zenith Process," *J. Am. Oil Chem. Soc.,* 50(7):260A–267A.

12. Crauer, L. S. 1964. "Continuous Refining of Crude Cottonseed Miscella," *J. Am. Oil Chem. Soc.,* 41(10):656–661.

13. Cavanagh, G. C. 1976. "Miscella Refining," *J. Am. Oil Chem. Soc.,* 53(6):361–363.

14. Hendrix, W. B. 1984. "Current Practices in Continuous Cottonseed Miscella Refining," *J. Am. Oil Chem. Soc.,* 61(8):1369–1372.

15. Norris, F. A. 1982. "Refining and Bleaching," in *Bailey's Industrial Oil and Fat Products, Volume 2,* Fourth Edition, Daniel Swern, ed. New York, NY: John Wiley & Sons, Inc., p. 285.

16. Norris, F. A. 1982. "Refining and Bleaching," in *Bailey's Industrial Oil and Fat products, Volume 2,* Fourth Edition, Daniel Swern, ed. New York, NY: John Wiley & Sons, Inc., pp. 273–276.

17. Erickson, D. R. 1995. "Batch or Kettle Refining," in *Practical Handbook of Soybean Processing and Utilization,* D. R. Erickson, ed. Champaign, IL: AOCS Press and the United Soybean Board, pp. 186–187.

18. Swoboda, P. A. T. 1985. "Chemistry of Refining," *J. Am. Oil Chem. Soc.,* 62(2):287–291.

19. Segers, J. C. 1983. "Pretreatment of Edible Oils for Physical Refining," *J. Am. Oil Chem. Soc.,* 60(2):262–264.

20. Tandy, D. C. and W. J. McPherson, 1984. "Physical Refining of Edible Oil," *J. Am. Oil Chem. Soc.*, 62(7):1253–1261.

21. Carlson, K. 1993. "Acid and Alkali Refining of Canola Oil," *INFORM*, 4(3):273–281.

22. List, G. R., T. L. Mounts, K. Warner, and A. J. Warner. 1978. "Steam Refined Soybean Oil: Effect of Refining and Degumming Methods on Oil Quality," *J. Am. Oil Chem. Soc.*, 55(2):277–279.

23. *The Official Methods and Recommended Practices of the American Oil Chemists' Society*, 1994, 4th ed. Champaign, IL: American Oil Chemists Society.

24. Norris, F. A. 1982. "Refining and Bleaching," in *Bailey's Industrial Oil and Fat Products, Volume 2*, Forth Edition, Daniel Swern, ed. New York, NY: John Wiley & Sons, Inc., p. 290.

25. Levin, H. and J. S. Swearingen. 1953. "An Inexpensive Soap Stock Conversion Plant," *J. Am. Oil Chem. Soc.*, 30(2):85–86.

26. Beal, P. E. et al. 1972. "Treatment of Soybean Oil Soapstock to Reduce Pollution," *J. Am. Oil Chem. Soc.*, 49(8):447–449.

27. Hong, W. M. 1983. "Quality of By Products from Chemical and Physical Refining of Palm Oil and Other Oils," *J. Am. Oil Chem. Soc.*, 60(2):316–321.

28. Woerfel, J. B. 1981. " Processing and Utilization of By-Products from Soy Oil Processing," *J. Am. Oil Chem. Soc.*, 58(3):188–191.

29. Ohlson, R. and C. Swensson. 1976. "Comparison of Oxalic Acid and Phosphoric Acid as Degumming Agents for Vegetable Oils," *J. Am. Oil Chem. Soc.*, 53(1):8–11.

30. Szuhaj, B. F. 1983. " Lecithin Production and Utilization," *J. Am. Oil Chem. Soc.*, 60(2): 308.

31. Ziegelitz, R. 1995. "Lecithin Processing Possibilities," *INFORM*, 6(11):1226.

32. Rich, A. D. 1967. "Major Factors That Influence Bleaching Performance," *J. Am. Oil Chem. Soc.*, 44(7):298A–300A and 323A–324A.

33. Richardson, L. L. 1978. "Use of Bleaching Clays in Processing Edible Oils," *J. Am. Oil Chem. Soc.*, 55(11):777.

34. Morgan, D. A., D. B. Shaw, M. J. Sidebottom, T. C. Soon, and R. S. Taylor. 1985. "The Function of Bleaching Earths in the Processing of Palm, Palm Kernel, and Coconut Oils," *J. Am. Oil Chem. Soc.*, 62(2):292–299.

35. Wiedermann, L. H. 1981. "Degumming, Refining and Bleaching Soybean Oil," *J. Am. Oil Chem. Soc.*, 58(3):163–164.

36. Rini, S. J. 1960. "Refining, Bleaching, Stabilization, Deodorization, and Plasticization of Fats, Oils, and Shortening," *J. Am. Oil Chem. Soc.*, 37(10):515–516.

37. Goebel, E. H. 1976. "Bleaching Practices in the U.S.," *J. Am. Oil Chem. Soc.*, 53(6): 342–343.

38. Rich, A. D. 1964. "Some Basic Factors in the Bleaching of Fatty Oils," *J. Am. Oil Chem. Soc.*, 41(4):315–321.

39. Butterworth, E. R. 1978. "Separation Methods in Processing Edible Oils," *J. Am. Oil Chem. Soc.*, 55(11):781–782.

40. Latondress, E. G. 1983. "Oil-Solids Separation in Edible Oil Processing," *J. Am. Oil Chem. Soc.*, 60(2):259–260.

41. Patterson, H. B. W. 1976. "Bleaching Practices in Europe," *J. Am. Oil Chem. Soc.*, 53(6): 341.

42. Ong, J. T. 1983. "Oil Recovery from Spent Bleaching Earth and Disposal of the Extracted Material," *J. Am. Oil Chem. Soc.*, 60(2):314–315.

43. Hong, W. M. 1983. "Quality of Byproducts from Chemical and Physical Refining of Palm Oil and Other Oils," *J. Am. Oil Chem. Soc.*, 60(2):316–318.

44. Svensson, C. 1976. "Use or Disposal of By-Products and Spent Material from Vegetable Oil Processing Industry in Europe," *J. Am. Oil Chem. Soc.*, 53(6):443–444.

45. Allen, R. R. 1960. "Hydrogenation," *J. Am. Oil Chem. Soc.*, 37(10):521–523.

46. Allen, R. R. 1968. "Hydrogenation: Principals and Catalyst," *J. Am. Oil Chem. Soc.*, 45(6): 312A–314A and 340A–341A.

47. Allen, R. R. 1978. "Principals and Catalyst for Hydrogenation of Fats and Oils," *J. Am. Oil Chem. Soc.*, 55(11):792–795.

48. Allen, R. R. 1987. "Theory of Hydrogenation and Isomerization," in *Hydrogenation: Proceedings of an AOCS Colloquium*, R. Hastert, ed. Champaign, IL: American Oil Chemists' Society, pp. 1–10.

49. Beckmann, H. J. 1983. "Hydrogenation Process," *J. Am. Oil Chem. Soc.*, 60(2):286–288.

50. Haumann, B. F. 1994. "Tools: Hydrogenation, Interesterification," *INFORM*, 5(6):668–678.

51. Hastert, R. C. 1988. "The Partial Hydrogenation of Edible Oils," AOCS Short Course, Phoenix, Arizona.

52. Lantondress, E. G. 1981. "Formulation of Products from Soybean Oil," *J. Am. Oil Chem. Soc.*, 58(3):185.

53. Wiederman, L. H. 1975. "Versatility of Fats and Oils Utilization," in *Military Food and Package Systems Activities Report*, 27(2):100–103.

54. Haumann, B. F. 1994. "Tools: Hydrogenation, Interesterification," *INFORM*, 5(6):672–678.

55. Lanning, S. J. 1985. "Chemical Interesterification of Palm, Palm Kernel, and Coconut Oils," *J. Am. Oil Chem. Soc.*, 62(2):400–404.

56. Sreenivasan, B. 1978. "Interesterification of Fats," *J. Am. Oil Chem. Soc.*, 55(11):796–805.

57. Norris, F. A. 1947. "A New Approach to the Glyceride Structure of Natural Fats," *J. Am. Oil Chem. Soc.*, 24(8):275.

58. Going, L. H. 1967. "Interesterification Products and Processes," *J. Am. Oil Chem. Soc.*, 44(9):414A–422A and 454A–456A.

59. Young, F. V. K. 1983. "Palm Kernel and Coconut Oils: Analytical Characteristics, Process Technology and Uses," *J. Am. Oil Chem. Soc.*, 60(2):378.

60. Hawley, H. K. and G. W. Holman. 1956. "Directed Interesterification as a New Processing Tool for Lard," *J. Am. Oil Chem. Soc.*, 33(1):29–35.

61. Macrae, A. R. 1983. "Lipase-Catalyzed Interesterification of Oils and Fats," *J. Am. Oil Chem. Soc.*, 60(2):291–294.

62. Quinlan, P. and S. Moore. 1993. "Modification of Triglycerides by Lipases: Process Technology and Its Application to the Production of Nutritionally Improved Fats," *INFORM*, 4(5):580–585.

63. Weiss, T. J. 1967. "Salad Oil Manufacture and Control," *J. Am. Oil Chem. Soc.*, 44(4): 146A–148A and 186A.

64. Puri, P. S. 1980. "Winterization of Oils and Fats," *J. Am. Oil Chem. Soc.*, 57(11): 848A–850A.

65. Stirton, A. J. 1964. "Fractionation of Fats and Fatty Acids," in *Bailey's Industrial Oil and Fat Products,* Third Edition, Daniel Swern, ed. New York, NY: Interscience Publishers, a Division of John Wiley & Sons, pp. 1008–1009.

66. Neumunz, G. M. 1978. "Old and New in Winterizing," *J. Am. Oil Chem. Soc.,* 55(5): 396A–398A.

67. Cavanagh, C. C. 1961. "Seed to Salad Oil in 18 Hours," *Food Processing,* 22(4):38–45.

68. *The Official Methods and Recommended Practices of the American Oil Chemists' Society,* 1994. 4th ed. Champaign, IL: American Oil Chemists' Society.

69. Latondress, E. G. 1983. "Oil-Solid Separation," *J. Am. Oil Chem. Soc.,* 60(2):259.

70. Haraldsson, G. 1983. "Degumming, Dewaxing and Refining," *J. Am. Oil Chem. Soc.,* 60(2):259.

71. Leibovitz, Z. and C. Ruckenstein. 1984. "Winterization of Sunflower Oil," *J. Am. Oil Chem. Soc.,* 61(5):870–872.

72. Neumunz, G. M. 1978. "Old and New in Winterizing," *J. Am. Oil Chem. Soc.,* 55(5): 396A–398A.

73. Kreulen, H. P. 1976. "Fractionation and Winterization of Edible Fats and Oils," *J. Am. Oil Chem. Soc.,* 53(6):393–396.

74. Rossell, J. B. 1985. "Fractionation of Lauric Oils," *J. Am. Oil Chem. Soc.,* 62(2):385–390.

75. Defense, E. 1985. "Fractionation of Palm Oil," *J. Am. Oil Chem. Soc.,* 62(2):376–385.

76. Thomas, A. E. and F. R. Paulicka. 1976. "Solvent Fractionated Fats," *Chemistry and Industry,* Sept., p. 774–779.

77. Birnbaum, H. 1981. "The Monoglycerides: Manufacture, Concentration, Derivatives and Applications," *Bakers Digest,* 55(12):6–16.

78. Rice, E. E. 1979. "Materials of Construction in the Fatty Acid Industry," *J. Am. Oil Chem. Soc.,* 56(11):754A–758A.

79. Feuge, R. O. 1962. "Derivatives of Fats for Use as Foods," *J. Am. Oil Chem. Soc.,* 39(12):525–526.

80. Van Haften, J. L. 1979. "Fat-Based Emulsifiers," *J. Am. Oil Chem. Soc.,* 56(11):833A.

81. Nash, N. H. and G. S. Knight. 1967. "Polyfunctional Quality Improvers . . . Polyglycerol Esters," *Food Engineering Magazine,* May, 1967, pp. 79–82.

82. McIntyre, R. T. 1979. "Polyglycerol Esters," *J. Am. Oil Chem. Soc.,* 56(11):835A–840A.

83. Wrenn, L. B. 1995. "Pioneer Oil Chemists: Allbright, Wesson," *INFORM,* 6(1):98–99.

84. Morris, C. E. 1949. "Mechanics of Deodorization," *J. Am. Oil Chem. Soc.,* 26(10): 607–610.

85. White, F. B. 1953. "Deodorization," *J. Am. Oil Chem. Soc.,* 30(11):515–526.

86. White, F. B. 1956. "Deodorization," *J. Am. Oil Chem. Soc.,* 33(10):495–506.

87. Zehnder, C. T. and C. E. McMichael. 1967. "Deodorization, Principals and Practices," *J. Am. Oil Chem. Soc.,* 44(10):478A–480A.

88. Zehnder, C. T. 1976. "Deodorization 1975," *J. Am. Oil Chem. Soc.,* 53(6):364–369.

89. Gavin, A. M. 1978. "Edible Oil Deodorization," *J. Am. Oil Chem. Soc.,* 55(11):783–791.

90. Gavin, A. M. 1981. "Deodorization and Finished Oil Handling," *J. Am. Oil Chem. Soc.,* 58(3):175–184.

91. Durdow, F. A. 1983. "Deodorization of Edible Oils," *J. Am. Oil Chem. Soc.,* 60(2): 272–274.

92. Imai, C. et al. 1974. "Detection of Heat Transfer Media in Edible Oil," *J. Am. Oil Chem. Soc.,* 51(11):495–501.

93. Johansson, G. M. R. 1976. "Finished Oil Handling and Storage in Europe," *J. Am. Oil Chem. Soc.,* 53(6):410–413.

94. Gilbert, W. J. and D. C. Tandy. 1979. "Odor Control in Edible Oil Processing," *J. Am. Oil Chem. Soc.,* 56(10):654A–658A.

95. Watson, K. S. and C. H. Meiehoefer. 1976. "Use or Disposal of By-Products and Spent Material from the Vegetable Oil Processing Industry in the U.S.," *J. Am. Oil Chem. Soc.,* 53(6):437–442.

96. Svensson, C. 1976. "Use of Disposal of By-Products and Spent Material from the Vegetable Oil Processing Industry in Europe," *J. Am. Oil Chem. Soc.,* 53(6):443–445.

97. Johansson, G. M. R. 1976. "Finished Oil Handling and Storage in Europe," *J. Am. Oil Chem. Soc.,* 53(6):410–413.

98. Wright, L. M. 1976. "Finished Oil Handling and Storage in the U.S.," *J. Am. Oil Chem. Soc.,* 53(6):408–409.

99. Buck, D. F. 1981. "Antioxidants in Soya Oil," *J. Am. Oil Chem. Soc.,* 58(3):275–278.

100. Sherwin, E. R. 1976. "Antioxidants for Vegetable Oils," *J. Am. Oil Chem. Soc.,* 53(6): 430–436.

101. Eastman Chemical Company. 1990. Tenox Natural Tocopherols, Technical Data Bulletin ZG–263.

102. Chahine, M. H. and R. F. Macneill. 1974. "Effect of Stabilization of Crude Whale Oil with Tertiary Butylhydroquinone and Other Antioxidants upon Keeping Quality of Resultant Deodorized Oil," *J. Am. Oil Chem. Soc.,* 51(3):37–41.

103. Sherwin, E. R. 1972. "Antioxidants for Food Fats and Oils," *J. Am. Oil Chem. Soc.,* 49(8):468–472.

104. Luckadoo, B. M. and E. R. Sherwin. 1972. "Tertiary Butylhydroquinone as Antioxidant for Crude Sunflower Seed Oil," *J. Am. Oil Chem. Soc.,* 49(2):95–97.

105. Sherwin, E. R. and B. M. Luckadoo. 1970. "Studies on Antioxidant Treatment of Crude Vegetable Oils," *J. Am. Oil Chem. Soc.,* 47(1):19–23.

106. Kraybill, H. R. et al. 1949. "Butylated Hydroxyanisole as an Antioxidant for Animal Fats," *J. Am. Oil Chem. Soc.,* 26(8):449–453.

107. Mattil, K. F. 1964. "Plastic Shortening Agents," in *Bailey's Industrial Oil and Fat Products,* Third Edition, D. Swern, ed. New York, NY: Interscience Publishers, a Division of John Wiley & Sons, pp. 272–281.

108. Joyner, N. T. 1953. "The Plasticization of Edible Fats," *J. Am. Oil Chem. Soc.,* 30(11): 526–538.

109. McMichael, C. E. 1956. "Finishing and Packaging of Edible Fats," *J. Am. Oil Chem. Soc.,* 33(10):512–516.

110. Slaughter, Jr., J. E. 1948. "Plasticizing and Packaging," *Proceedings of a Six-Day Short Course in Vegetable Oils,* The American Oil Chemists' Society, Conducted by The University of Illinois, Urbana, IL: pp. 119–131.

111. Brown, L. C. 1949. "Emulsion Food Products," *J. Am. Oil Chem. Soc.,* 26(10):634.

112. Fincher, H. D. 1953. "General Discussion of Processing Edible Oil Seeds and Edible Oils," *J. Am. Oil Chem. Soc.,* 30(11):479–481.

113. Rini, S. J. 1960. "Refining, Bleaching, Stabilization, Deodorization, and Plasticization of Fats, Oils, and Shortening," *J. Am. Oil Chem. Soc.*, 37(10):519–520.

114. Hoerr, C. W. and J. V. Ziemba. 1965. "Fat Crystallography Points Way to Quality," *Food Engineering*, 37(5):90.

115. Weiss, T. J. 1983. "Basic Processing of Fats and Oils," in *Food Oils and Their Uses*, T. J. Weiss, ed. Westport, CT: AVI Publishing Co. Inc., pp. 96–97.

116. Reigel, G. W. and C. E. McMichael. 1966. "The Production of Quick-Tempered Shortenings," *J. Am. Oil Chem. Soc.*, 43(12):687–689.

117. Endes, J. G., R. J. Wrobel and R. B. Rendek. September 30, 1969. U.S. patent 3,469,996.

118. Campbell, Jr., R. L., C. H. Wood and A. E. Brust. May 26, 1970. U.S. patent 3,514,297.

CHAPTER 3

Fats and Oils Analysis

1. INTRODUCTION

The use of fats and oils prior to the beginning of the nineteenth century was based on practical knowledge that had been accumulated slowly over many centuries. Today, fats and oils products are developed and subsequent production controlled with a knowledge of their composition, structural and functional properties, and the expected reactions obtained through the application of scientific research. Progress in the utilization of fats and oils for the production of useful products is dependent upon a thorough knowledge of the characteristics of the raw materials, the changes effected by each process, and the requirements of the individual prepared food product. The physical, chemical, and performance analyses are tools available to the fats and oils processors for the purchase of raw materials, development of new products, and evaluation of the products produced.

Analysis of fats and oils are required for a number of purposes, beginning with commodity trading. In every fat and oil processing plant, there are analytical requirements for process quality control. In refining, for example, free fatty acid content of the oil is necessary to determine the caustic treat and, again, to determine if the oil has been properly deodorized, as well as being a quality indicator in other areas. Melting points, solids fat index, and other physical evaluations indicate that the product will function as developed. On final edible oil products, organoleptic evaluations, peroxide value, free fatty acid, and other analyses are utilized for assurance that the product has the required bland flavor with predictive analysis like AOM stability, which is utilized to assure the proper shelf life.

Nutritional listing of saturates, polyunsaturates, cholesterol, vitamins, and other product characteristics on food product labels requires accurate analysis for identification of the original values and control to ensure compliance with any claims. Also of importance are analyses for trace constituents, such as pesticides or trace metals, to ensure compliance with governmental regulations.

Investigative analyses are frequently required in fats and oils processing. Identification of the oils and physical characteristics of a somewhat mystery product as a result of a mis-pumping or some other mishap is required to identify the opportunities to salvage the product. Analyses are also needed to determine the cause of product failures either identified by quality evaluations of finished products or customer complaints.

Specialty and tailored fats have individual analytical requirements in terms of chemical, physical, consistency, nutritional, performance, and other properties that must be measured and maintained to assure that the products function as designed. The diversity of the products and applications amplifies the needs for analytical information and the ability to interpret the meaning of the results.

The analytical methods employed should be those universally accepted in the industry, both by the suppliers of the raw materials and the purchasers of the finished products. Historically, development of analytical methods for fats and oils has been undertaken by the processors, professional societies, end users, the academic community, and government-sponsored research. Development of analytical procedures was one of the main reasons for forming the Society of Cotton Products Analyst, which became the American Oil Chemists' Society (AOCS). Methods standardization remains a major interest for AOCS, and procedures with wide interest and application become a part of the *Official Methods and Recommended Practices of the American Oil Chemists' Society.* The AOCS methods are continually updated with unused or less satisfactory methods dropped and new procedures added. Other standardized methods with potential application are published by the Association of Official Analytical Chemists (AOAC), American Society for Testing Materials (ASTM), and American Association of Cereal Chemists (AACC).

2. NONFATTY IMPURITIES

Impurities present in fats and oils are contaminates that must be removed during processing to prevent an adverse reaction or an undesirable appearance. Most of the trading rules for edible oil products have specific limits for the various nonfatty materials. Analysis for these materials during processing is part of good process control, and quality processors have stringent as-shipped requirements for finished products to prevent these contaminates from reaching the customer. The nonfatty impurities analytical methods are identified by the predominate contaminate in most cases.

2.1 Moisture Analysis

Moisture can contaminate a fat or oil in many different ways: condensation, broken coils, intentionally added during processing, and so on. Continued presence of moisture will induce hydrolysis with a resultant free fatty acid increase and off-flavors. Some of the analytical methods to identify and quantify moisture are

- hot plate method—One of the most common methods of determining moisture in a fat or oil is AOCS Method Ca 2b-38 for Moisture and Volatile Matter [1]. Approximately 10 grams of a representative sample is heated in a beaker with gentle agitation on a hot plate until foaming stops and incipient smoking begins. The loss in weight between the beginning and ending sample is the moisture and volatile matter.
- air oven method—The results of the air oven method, AOCS Ca 2c-25 [1], are more accurate and reliable but also more time-consuming than the hot plate method. For the air oven method, approximately 5 grams of a representative sample is weighed into a dried, tared moisture dish and dried in the oven for 30 minutes at $101 \pm 1°C$ and repeated until a constant weight is determined. The loss in weight is calculated as the moisture and volatile matter.
- vacuum oven method—AOCS Ca 2d-25 [1] provides more accurate results than the hot plate or air oven procedures and is especially applicable where moisture is deep-seated and must diffuse largely through the capillaries. The same procedure as for the air oven method is observed, except for a vacuum not to exceed 100 mm of mercury at a temperature of 20 to 25°C above the boiling point of water at the operating pressure until a constant weight is obtained in successive 1-hour drying periods. Again, the weight loss is calculated as the moisture and volatile matter.
- Karl Fisher method—This method is adaptable for determining moisture in a wide variety of materials and has been adapted by many laboratories as a standard test procedure for moisture. The Karl Fisher volumetric method for moisture may be determined by ordinary visual titration or by an electrometric method. The titration endpoint is a color change from yellow to brown. The electrometric method of titration for moisture is more accurate when dark solutions are encountered and enable a lesser experienced technician to correctly identify the endpoint. An automatic titration speeds up the analysis with an

even better degree of accuracy. AOCS Method Ca 2e-84 [1] indicates that the precision for the Karl Fisher moisture determination by the same operator is 0.6% relative.

- skillet moisture—This simple qualitative method determines very rapidly if moisture is present in a sample. It involves pouring a small amount of the sample into a hot skillet previously heated until it began smoking. The absence of moisture is denoted by no action while any popping or spitting indicates the presence of moisture. This qualitative evaluation, usually reported as either wet or dry, can be used to determine if a more quantitative result is required.

2.2 Impurities Analysis

Foreign material in incoming crude edible oils or incorporated during processing must be removed by filtration or another process to produce an acceptable finished product. The following methods and procedures have been effective for identification that a problem existed and confirmation that the impurities were removed before subsequent processing or shipment:

- insoluble impurities—Meal, dirt, seed fragments, and other substances insoluble in kerosene or petroleum ether are the impurities identified by AOCS Method 3a-46 [1] normally identified by trading rules for crude vegetable oils. This method utilizes the residue from the moisture and volatile matter determinations or another sample prepared in the same manner. It is dissolved in 50 ml of kerosene and then vacuum filtered through a Gooch crucible. After washing with warm kerosene and petroleum ether, the crucible is dried and weighed. The gain in weight of the crucible is calculated as percent insoluble impurities.
- filterable impurities (standard disc method)—During processing, small quantities of undissolved impurities may be picked up from polymerized oil deposits, charred materials, salt formed after a reaction, filter aids, bleaching earths, or some other foreign material. These materials should be removed by one or more of the filtering systems in the process, but an evaluation procedure will assure that the systems are operating properly and are properly sized to remove all of the foreign materials before shipment to the consumer. A suggested procedure for filterable impurities evaluation is: Filter a standard quantity (500 grams) of heated (70 to 90°C) oil sample with a vacuum suction funnel through a Whatmann No. 2 filter disc. After filtering, wash the funnel and filter disk with chlorothene. Compare the filter disk to standard

impurities discs previously prepared to identify the degree of acceptability.

* turbidimeter impurities—The amount of undissolved impurities in oil can be rapidly determined with the use of a turbidimeter. Product samples are usually heated to 70°C ± 1.0°C and allowed to stand 5 to 10 minutes in a controlled temperature oven after pouring into the prescribed sample bottles to eliminate air bubbles. The turbidimeter readings can be converted to parts per million impurities with a pre-determined calibrated curve. Experience has shown that meat fats and products containing emulsifiers cannot utilize this method because they contain turbid materials other than those effectively measured by this procedure.

2.3 Trace Metals Analysis

Throughout the processing of edible fats and oils, metals can be encountered, many of which reduce the efficiency of the process or cause deterioration of the product quality. The most notable of these metals are copper, iron, calcium, magnesium, sodium, and nickel. Various procedures have been used for determining trace quantities of the trace metals in edible oil products. Initially, wet chemical analyses were utilized, but improved trace metal determination procedures have been introduced; flame atomic absorption spectroscopy, later improved by replacing the flame with a graphite furnace and, more recently, plasma emission spectroscopy (ICP and DCP), has been introduced. These procedures are normally performed by a technical research laboratory or a centralized quality control laboratory [2]. Several qualitative methods for iron and nickel are available, but the sensitivity for these methods is limited; however, for quality control satisfaction that a product contains less of a trace metal than the established limits of the procedure, these evaluations perform adequately.

2.4 Soap Analysis

In caustic refining, sodium hydroxide is introduced to the oil to react with the free fatty acids and produce soaps. Traces of soap remaining in the oil after refining and bleaching poison hydrogenation catalysts and have a detrimental effect upon the oxidative stability of deodorized oils at levels below 5 ppm. Two AOCS methods are available for determining soap: Cc 15-60, a conductivity procedure, and Cc17-79 [1], a titration procedure. Soap analyses are also an investigative tool for evaluating the cause of early breakdown of frying fats. Trace amounts of caustic cleaning materials left in deep fat fryers will react with the free fatty acids present to produce soaps that shorten the frying life of the shortening, i.e., foaming, off-flavor, and high absorption.

3. MELTING, SOLIDIFICATION, AND CONSISTENCY

The data obtained from melting points, solids fat index methods, and other solidification procedures are all related in predicting the consistency of the finished product. For edible fats and oils products, the liquid/solids levels at the various temperatures in relation to body temperature can give good indications of the mouth feel, gumminess, workability, and overall general behavior at cool, ambient, and elevated temperatures. The melting and solidification procedures are the most frequently and routinely performed evaluations in fats and oils laboratories, both for quality control and product development. They are useful to control production and help identify unknown samples and are critical in new product formulation.

3.1 Melting Point Analysis

Melting point is usually defined as the point at which a material changes from a solid to a liquid. However, natural fats do not have a true melting point. Pure compounds have sharp and well-defined melting points, but fats and oils are complex mixtures of compounds that pass through a gradual softening before becoming completely liquid. This melting procedure is further complicated by the fact that fat crystals can exist in several polymorphic modifications, depending on the specific triglycerides involved and the temperature–time pretreatment (tempering) of the sample. The different crystal forms are often stable enough to exhibit distinctive melting points. Therefore, instead of melting point, melting range or melting interval is more correct. For a melting point, one point within the melting range must be selected with a defined method. Only with ridge and specific definition of the conditions of the fat pretreatment and the test procedure can a melting point be determined. Many methods have been devised to determine melting point or a point close to it, some methods by direct observation and some by an indirect and objective process [3]. The advantages of most melting point methods are the relative simplicity, but the dependence of the melting point on the sample pretreatment and on the method used must be considered disadvantages.

Several fats and oils melting point procedures have been standardized by AOCS and other associations. The melting point methods vary considerably in the endpoint determination, conditioning of the sample, the amount of automation, time requirements, attention required, the degree of melt, and so on.

- capillary melting point—AOCS Method Cc 1-25 [1] is essentially the procedure used by the organic chemist for determining the melting point of pure organic compounds. In this method, capillary tubes, 1 mm inside diameter, are filled to a height of 10 mm with melted fat; the end is sealed and tempered 16 hours at 4 to 10°C. After tempering, the tubes are heated in a bath at 0.5°C per minute, starting 8 to 10°C below the expected melting point, until the fat becomes completely clear. It has been difficult to reproduce results with this method because of the subjective interruption of the completely clear endpoint.
- softening point—This open capillary tube melting point method, AOCS Method Cc 3-25 [1], follows the closed capillary tube procedure, except the tube is not sealed and the endpoint is the physical movement of the fat column under a standardized hydrostatic pressure. This objective determination is a definite advantage for better reproducibility of results.
- slipping point—AOCS Method Cc 4-25 [1] is based on the same principal as the open capillary melting point or the softening point, except only plasticized or solidified samples are filled into metal tubes and heated at 0.5°C per minute while immersed in a brine solution in a 600-ml beaker. The endpoint for this method is the temperature when the fat rises from the cylinder. This method is useful for finished products and the endpoint is determined objectively, but the sample lacks laboratory pretreatment so the results can be influenced by the non-standardized history of the product.
- Wiley melting point—AOCS Method Cc 2-38 [1] was the most popular melting point in North America. A solidified fat disk is solidified and chilled in a metal form for 2 hours or more. The disk is then suspended in an alcohol water bath and slowly heated while being stirred with a rotating thermometer. The Wiley melting point is when the fat disk becomes completely spherical. The subjective interpretation of the endpoint is again a major disadvantage, and slight variation in sample tempering or the heating rate interferes with reproducibility of the results. Another major disadvantage is the constant attention required to determine the endpoint.
- dropping point—AOCS Method Cc 18-80 [1] utilizes a Mettler Instrument Corporation control unit and dropping furnace. A sample cup designed for the furnace is filled with approximately 17 drops of the melted fat sample and tempered in a freezer for 15 minutes. The cold solidified sample is warmed in the temperature programmed furnace

until it becomes fluid enough to flow. When the sample flows, it trips a photoelectric circuit that records the temperature on a digital readout. This procedure has become a standard in most fats and oils laboratories due to the advantages over the other melting point methods, i.e., fully automatic endpoint determination, good correlation of results with Wiley melting point results, less than an hour elapsed time required for complete results, and the ability to analyze products with low melting points.

3.2 Solids/Liquids Contents Analysis

The proportion of solid to liquid fat in a shortening, margarine, or specialty product at a given temperature has an important relationship to the performance of the product at that temperature. This proportion cannot be determined from single point melting analysis or other analysis of the physical properties. Estimates of the solids and liquids contents in a fat at different temperatures over the melting range can be made with calorimetric, dilatometric, or nuclear magnetic resonance procedures. Dilatometric (SFI) has gained wide acceptance for the characterization of solid/liquid contents of fats over the past 40 years. Methods have also been standardized for the use of magnetic resonance (NMR) for this purpose, and differential scanning calorimetry (DSC) techniques have been identified for determination of solid/liquid contents [4].

The usual temperatures at which solids fat index/contents determinations are made are 10.0, 21.1, 26.7, 33.3, and 37.8 or 40.0°C. Generally, the solids fat index may be related to the physical characteristics of the fat and oil products over the temperature range in use [5]. For example, SFI results for margarine base oils are usually determined at three temperatures to indicate properties:

- 10.0°C (50°F): an indication of printability and product spreadability at refrigerator temperatures
- 21.1°C (70°F): an indicator of product resistance to oil—off at room temperature
- 33.3°C (92°F): an indicator of eating or melt-in-the-mouth characteristics

Solids fat index/contents determinations at different temperatures, covering the range where a fat passes from a solid to a liquid, can be plotted to produce a curve that illustrates the changes. The solids curves will show:

- plastic range—Fats and oils products generally have the best consistency for mixing and working within a solids content range of 15 to 25%, which is usually identified as the plastic range.

- melting point—The dilatometric melting point is the temperature at which the solids curve meets the liquid line, that is, the lowest temperature at which the solids content is zero. This melting point is obtained by extrapolating the solids curve to zero either graphically or mathematically.
- flat solids curve—Flatter solids slopes generally have a greater plastic range because the solids remain within the plastic range for a wider temperature range.
- steep solids curve—Sharp melting properties are usually accompanied by a high oxidative stability and characterized by a steep solids slope. Frying, cookie fillers, and nondairy applications are examples of products benefiting from a steep solids slope. These products require a sharp melting profile with good flavor stability while workability and creaming properties are not required.

Since its beginnings in the early 1930s, dilatometry, now referred to as solids fat index, has become the preeminent analysis in fats and oils laboratories. It has become the single most important criterion for basing specifications on the melting behavior and crystalline structure. However, despite its importance, the SFI method is time-consuming and labor-intensive and requires more maintenance than the majority of the other analyses performed. Therefore, the fats and oils industry has long desired a less time-consuming, automated, less expensive procedure that would provide comparable results [2]. However, SFI remains the preferred method in the United States while the European and Asian fats and oils industries have adopted NMR techniques for solids fat contents.

- solids fat index—Dilatometry is the basis of the most widely accepted method for solids measurement in the United States. AOCS Method Cd 10-57 [1] for dilatometric SFI has been used extensively for product development, process control, and quality specifications for fats and oils products. The solids fat index is an empirical measure of the solid fat content. It is calculated from the specific volume at various temperatures utilizing a dilatometric scale graduated in units of milliliters times 1,000. It should be recognized that the SFI results are arbitrary because in the determination assumptions are made and liberties are taken insofar as precise measurement is concerned. The liberties and assumptions are (1) use of volumetric instead of gravimeteric measurements, (2) use of solutions other than mercury as the confining liquid, (3) assumption that the slope of the liquid and the solids lines are parallel, (4) assumption that the slope of the liquid line is the same for all fats, and (5) assumption that the melting dilation is 0.1. Some of the assumptions are of necessity and others are for convenience. However, even with the assumptions and liberties taken, the results have relative significance and are related to other fats and oils properties that are all important with respect to performance and use [6].

- solids fat content (SFC)—AOCS Method Cd 16-81 determines the quantity of solid glycerides in a fat by measuring the pulsed nuclear magnetic resonance (NMR) signal of the liquid fat [1]. The hydrogen atom in a solid crystal does not give a signal, but a hydrogen atom in a liquid environment does. Therefore, NMR measures the actual amount of liquid in a sample. SFI and NMR analysis may be correlated, but the methods do not provide identical results. The most significant differences are that the solids contents at the higher levels, usually at temperatures of 10.0 and 21.1°C (50 and 70°F), are more different than those with the lower solids contents at the higher temperatures [5]. The NMR measurement or solids fat content, using either the pulsed or wideline NMR spectrometers, is a direct measurement while SFI is an empirical measure calculated from the specific volumes at various temperatures using a dilatometer.

- differential scanning calorimetry fat solids content—Measurement of the heat of fusion of fat rather than its thermal expansion appeared to offer promise as a means of determining fat solids content. Therefore, methods were developed for measurement of the solids contents of vegetable and animal fats by differential scanning calorimetry (DSC) in the search for a less time-consuming method for solids fat content determination. Briefly, the methods developed measured the heat of fusion of a frozen, completely solid sample of fat. The area under the melting curve, presented on a strip chart, is measured; then selected partial areas are measured as a percentage of the whole. The temperatures selected for partial area measurement are normally those employed for SFI by dilatometry [7].

    Several advantages were identified for the DSC fat solids methods as compared to SFI: (a) the thermal history determined can provide clues to the tempering and storage conditions for the fat sample; (b) the fat melt curve can help distinguish between two fat products with identical SFI values; (c) a fat solids range from −10°C for soft oils to 70°C for hard fats would be possible; (d) no limit for hardness would be necessary as is the case with SFI by dilatometry; (e) results should be available in approximately 1 hour elapsed time; and (f) correlation with SFI results appeared to be good. The identified disadvantages were (a) accurately weighing the small sample size, 5 to 10 mg, posed the problem of representative sampling; (b) variation for fat solids was not constant and greater at low temperatures [8]; and (c) precision was found to be poorer than with the standard SFI dilatometric technique.

Fat solids analysis by DSC has never been accepted as a quality control tool to replace SFI. Originally, it had appeared to be an ideal choice because it measured the change in heat absorption versus the temperature of the sample programmed from one temperature to another to record a melting profile of a sample as it passes through various crystalline states from a solid to a liquid. The points of dissatisfaction with the DSC solids measurement techniques were primarily that the reanalysis of the same sample was not as reproducible as SFI, that there was a minute sample quantity and weighing accuracy required, and that DSC provided too much information [2]. It appears that DSC can be a very useful research tool but it is not suitable for quality control laboratories.

Direct comparable solids results with SFI, DSC, and NMR techniques on a wide variety of fats and oils have not been attained. Chart 3.1 compares the fat solids contents determinations by SFI, NMR, and DSC for a 72 iodine value (IV) soybean oil basestock and tallow [4]. This illustration indicates good agreement for SFI and DSC for the soybean oil basestock, but the tallow curves show a significant difference in results. The NMR determinations at 10°C (50°F) are much higher than the other two techniques but have somewhat comparable results beginning at 80°F (26.7°C) and the higher temperatures. Chart 3.1 demonstrates that comparable values will not be attained with the three solids measurement techniques. All three techniques have merit and could be utilized for solids fat content determinations; however, the SFI dilatometry has been established as the standard in the United States.

3.3 Solidification Analysis

Solidification of mixed fatty acids is the point where a balance is attained between the heat generated and the heat lost, which will vary according to the rate of crystallization and the degree of super-cooling in the liquid. These evaluations are some of the oldest fat characteristics used for fats and oils control purposes:

- titer—The titer test, AOCS Method Cc 12-59, measures the solidification point of the fatty acids. First, a fat sample must be saponified and dried before determining titer. Then, a titer tube is filled to the 57 mm mark with dried fatty acids and suspended in an air bath, which is surrounded by a water bath at 15 to 20°C below the expected titer result. The sample is stirred until the temperature begins to rise or remains constant for 30 seconds, after which the stirring is stopped and the endpoint determined as the maximum temperature that the

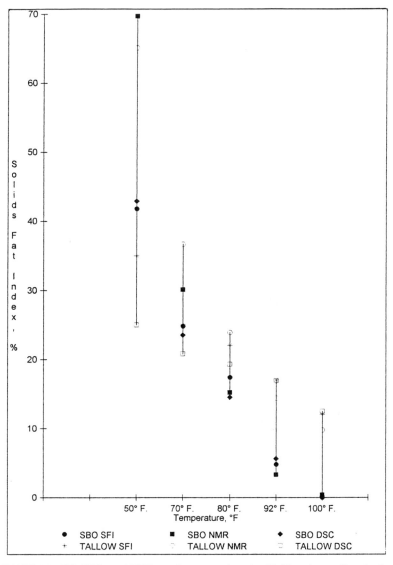

CHART 3.1. SFI, DSC, and NMR results comparison for 72 IV soybean oil and tallow.

fat reaches with the heat of crystallization [1]. Titer is an important characteristic for inedible fats used for soap making or as a raw material for fatty acid manufacture. For edible fats and oils, it is commonly specified for low iodine values hard fats sometimes referred to as titer stocks.

- quick titer—Hydrogenation control of hardstocks or titer stocks is normally done with a quick titer determination since the official titer (AOCS Method Cc 12-59) is too time-consuming for control purposes. A quick titer analysis is performed by dipping the bulb of a glass thermometer into the liquid fat sample and then rotating the thermometer stem between the fingers to cool the fat at room temperature. The endpoint is the temperature reading when the fat on the bulb clouds. Constants for each oil source have been identified for addition to the quick titer results, which approximate the official titer determinations very closely [9]:

Source Fat or Oil	Quick Titer Constant
Cottonseed oil	11.0°C
Soybean oil	13.0°C
Peanut oil	14.0°C
Palm oil	9.0°C
Lard	11.0°C
Tallow	12.0°C

- congeal point—This solidification method, sometimes referred to as setting point, is a measure of the fat itself, rather than the separated fatty acids for titer. Congeal point, AOCS Method Cc 14-59, is determined by cooling a melted sample while stirring until the fat becomes cloudy. The sample is then allowed to remain quietly in the air at 68°F. Under these conditions a temperature rise occurs. The highest temperature attained is the congeal point, which estimates the solidification temperature of the fat [5].
- cloud point—This nonstandard method is a variation of the congeal procedure. An empirical cloud point is obtained by stirring a sample of fat while it is being cooled until the oil has clouded enough to block off a lightbeam of known intensity. The cloud or congeal point values are more closely related to consistency than melting points. A definite relationship exists between SFI at 92°F (33.3°C) and the congeal or cloud point, especially for meat fats.
- cold test—The ability of a salad oil to withstand refrigerator storage is determined by the cold test. It measures the ability of an oil to resist crystallization and is defined as the time in hours for the oil to become

cloudy at 0°C (32°F). For the standardized AOCS Method Cc 11-53, dry filtered oil is placed in a sealed 4-ounce bottle and submerged into an ice bath. The AOCS method stipulates that the oil should be examined after 5.5 hours for clarity to determine if the oil passes or fails. An alternative to the method allows the continuation of the cold test until a cloud develops [1]. The alternative procedure is probably the norm since most salad oil specifications require at least a 10-hour minimum cold test.

The cold test was developed to evaluate cottonseed oil for the production of mayonnaise and salad dressing. An oil that will solidify at the refrigerator temperatures used for the preparation of these products will cause an emulsion break and separation. Currently, the cold test is also utilized to assure that bottled salad oils for retail sales will not develop an unattractive appearance on the grocery shelf.

- chill test—Natural winter oils with soluble waxes can at times successfully pass an extended cold test but develop a cloudy appearance on the grocery store shelf. A nonstandard chill test has successfully predicted this problem when the cold test failed. For the chill test, dry filtered oil is placed in a sealed 4-ounce bottle and held at 70°F (21.1°C) and examined after 24 hours for clarity. Any indication of a cloud or wisp indicates the presence of a wax or hard oil contamination.

3.4 Consistency Analysis

Consistency is generally assumed to be a combination of those effects that tend to give the impression of resistance. Plasticity relates to the capacity of the product for being molded. The factor most directly and obviously influencing the consistency of a fat is the proportion of the material in the solid phase. It is well established that a fat becomes firmer as the solids content increases.

Crystal size also has an effect: the smaller the crystals, the firmer the fat. An example is the fact that grainy lards are softer than smooth lards that have been interestified. The persistence of the crystal nuclei is also a factor. As a fat is exposed to temperature fluctuations, a portion of it undergoes melting and resolidification. The ability of certain fats to retain their original crystal form, regardless of temperature variation, is probably due to their capability to leave behind crystal nuclei that serve as starting points for the development of new crystals when the fat is cooled [10].

Some of the most critical performance factors of fats and oils products are related to the properties commonly referred to as consistency and plasticity. Butter, margarine, and spreads depend upon the consistency of the fat portion and its ability to spread on bread. A wide plastic range and a smooth consistency are mandatory for roll-in shortenings and margarines used for thin layer lamination at refrigerator temperatures. Plasticity is also important for the workability and creaming properties required for shortenings used in the preparation of icings and aerated batters.

- consistency ratings—Undoubtedly, the first method used for evaluating a fat and oil product's consistency was by pressing a finger into the product or squeezing the product in the hand. These rating methods, although very subjective, are still applicable and effective. An experienced evaluator can identify slight differences or imperfections in the finished product more readily than the available instrumentation in regard to body, firmness, softness, and inconsistencies in the feel like sandiness, lumps, or ribby, as well as problems with the appearance like oiling, air pockets, or grainy. Methods for applying a numerical value to the finished product evaluations have been developed independently by most edible fats and oils processors or laboratories for comparison purposes. Suggested methods for shortening and margarine finished product rating are presented in the nonstandardized methods section.
- penetrations—The most widely used method to measure consistency of a plasticized shortening or margarine involves the ASTM grease penetrometer or an adaptation of it. AOCS Method Cc 16-60 for penetration testing identifies the penetrometer as a mechanical device with a support to grip and release the shaft and cone, a support for the sample, adjustment capability to level the device and a gauge graduated in 0.1-mm units, which conforms to ASTM D5, D217, and D937 designations. Most of the method variations used by different laboratories involve the design of the needle or cone. Penetration evaluations measure the depth to which the cone penetrates into the surface of the shortening or margarine after allowing the cone to settle into the product for 5 seconds, starting from a position where the tip of the cone just touches the surface of the sample. The penetration result for each product is the average of four readings performed at each evaluation temperature.

Penetration results are utilized internally for the most part and are not normally a part of customer specifications for fats and oils products. Nevertheless, uniform procedures must be observed to obtain reproducible results. To avoid overheating or overcooling, evaluation samples must be carefully tempered until the desired product temperatures are achieved, usually 24 hours for 1-pound containers. The sample temperature should be stabilized to within 1.0°C before testing for reproducible results. Temperature abuse at any time can change the penetration results, even if the sample temperature is returned to the desired level and especially if the sample has experienced any actual melting.

Consistency of a product must be measured at a number of different temperatures to determine its plastic range, that is, the range of temperatures over which the fat has the capability of being molded or worked. Normally, samples are held at three to five different temperatures ranging from 40 to 100°F (4.4 to 37.8°C) until each sample has equilibrated at the desired temperature. The samples are evaluated to determine the relative softness at low temperature and firmness at high temperature. Products with a wide plastic range are workable at both high and low temperatures. A perfect plastic range, if it could exist, would have the same penetration at all temperatures. AOCS Method Cc 16-60 personal data has indicated that the penetration plastic range falls between 150 and 300 mm/10.

While the elapsed time required for penetration results is too lengthy for process control, it is a valuable tool for finished products control and product development. A penetration curve can confirm that the desired plasticization and tempering conditions have been utilized for an individual product when the SFI and blend composition are controlled within specified limits. Chart 3.2 illustrates the effect of varied chilling unit temperatures upon consistency of soybean oil hydrogenated basestock as measured by penetrations. The results illustrate the softening affect of cooler chilling unit operating temperatures and that the finished product consistency can be predicted from the solids fat content when the plasticization and tempering conditions are controlled.

4. COMPOSITION ANALYSIS

Knowledge of the composition of fats and oils is very important in nearly every phase of fat chemistry and technology although, often, its importance is not fully realized. In fact, progress in the utilization of commercial fats and oils as raw materials in the manufacture of useful products is dependent to a large degree on knowledge of the composition of the starting material and the products derived from it. Methods for determining the composition of fats and oils are important not only because of the fatty acids contents and the pattern of glyceride distribution elaborated by plants and animals, but also because the physical character and end-use performance of fats and oils are directly related to composition.

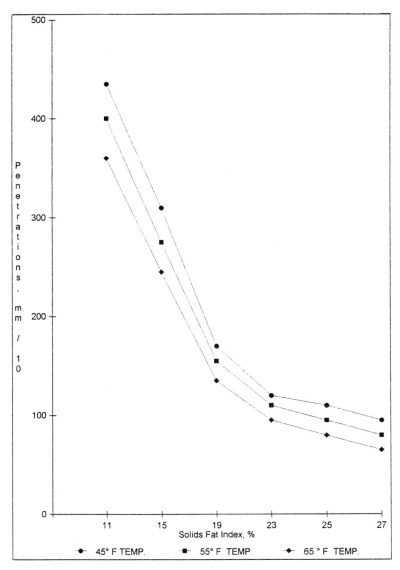

CHART 3.2. Varied chilling unit temperature.

197

nalytical methods for determining compositions of fats and oils have evolved from chemical separations to instrumental procedures. Instrumental methods are attractive to the analytical chemist because of the time savings, better accuracy, and less tedious features. However, most instrumental analysis must rely upon a standardization or calibration procedure, which are usually the original wet chemical analyses.

(1) Saponification value—Saponification value is a measure of the alkali-reactive groups in fats and oils and is useful in predicting the type of glycerides in a sample. Glycerides containing short-chain fatty acids have higher saponification values than those with longer chain fatty acids. The saponification value and the iodine value determination have been useful screening tests both for quality control and for characterizing types of fats and oils. Independently, the results overlap too much to identify individual fats or oils; for example, both domestic vegetable oils and animal fats have saponification values in the 180 to 200 range. Saponification value analyses have been replaced almost exclusively in edible fats and oils processing by fatty acid composition analysis by gas liquid chromatography, except that some purchasers may specify it to prevent lauric oil contamination of domestic oils. Lauric oils have saponification values in the range of 240 to 265, which differs substantially from the 180 to 200 range for domestic vegetable oils.

(2) Iodine value—The iodine value is a simple and rapidly determined chemical constant for a fat or oil. It is a valuable characteristic in fat analysis, which measures unsaturation but does not define the specific fatty acids. Iodine value analyses are very accurate and provide near theoretical values, except in the case of conjugated double bonds or when the double bond is near a carboxyl group. However, unless the previous history of the fat or the type of fat in the product is known, an iodine value may be somewhat meaningless by itself. For example, a product prepared with a meat fat with consistency and performance characteristics similar to a vegetable oil–based product will have a considerably different iodine value. Further, even vegetable oil products with comparable functionality but different source oils will not have like iodine values. Even with these inadequacies, iodine value is a useful tool for process control and product specifications.

Iodine value is defined as the grams of iodine that, added to 100 grams of the fat or oil sample, or in other words, the weight percentage of iodine based on the weight of the sample, adds to the sample. Determination of the iodine value, AOCS Method Cd 1-25, is carried out by adding an excess of the Wijs reagent to the sample, allowing the mixture to react for 30 minutes at 25 ± 5°C, treating the excess reagent with potassium iodide to convert it to equivalent iodine and titrating with thiosulfate reagent and a starch indicator until the blue color disappears [1]. The iodine value procedure must be performed very precisely and timed carefully to be reproducible.

(3) Refractive index—Hydrogenation process control requires a rapid method for the determination of the product endpoint. Refractive index is used by most processors to measure the change in unsaturation as the fat or oil is hydrogenated. By reference to a predetermined curve relating the refractive index to iodine value, a rapid estimation of the iodine value may be made. One source of error in this method is that *trans* acids formed during hydrogenation affect the refractive index but not iodine value.

The index of refraction is the degree of deflection of a beam of light that occurs when it passes from one transparent medium to another. A refractometer with temperature control is used for fats and oils with measurement usually at 25°C. High-melting fats require temperature adjustments to 40 or 60°C depending upon the melting point of the product. Temperature changes affect the results obtained; refractive index decreases as the temperature rises but at the same time increases with the length of the carbon chains also with the number of double bonds present in the fatty acids. AOCS Method Cc 7-25 identifies the procedure used to measure the refractive index for fats and oils [1].

(4) Fatty acid composition—The classical method for identification of fats and oils has been replaced by fatty acid composition analysis determined by gas liquid chromatographic (GLC) patterns. The classic method was based on the identification of a specific fat or oil by means of a combination of its iodine value, relative density, refractive index, and saponification value. The advantages of the GLC procedure are that it permits identification of source oils that cannot be identified by the classical methods, plus the ability to identify the source oils proportions in a blended product. Further, since the fatty acid composition requires only one analysis, it can be made more rapidly and applies equally well to refined and unrefined oils, necessitating only one set of standards.

Gas chromatography includes those chromatograph techniques in which the mobile phase is a moving gas. In general, the procedure involves passing the methyl esters, or transesterified triglycerides, to be analyzed through a heated column by means of a carrier gas such as helium or nitrogen. The components of the mixture are eluted with the gas and detected and measured at the exit end of the column by a suitable means. The retention time is the time required for a given compound to pass through the column. The fatty acid esters exit in the order of saturation. The retention line is indicated on the horizontal axis of Chart 3.2 and is a qualitative index of the substance, and the area under the curve is in each case a quantitative measure of the component.

The fatty acid analysis provides a rapid and accurate means of determining the fatty acid distribution of fat and oil products. This information is beneficial for all aspects of product development, process control, and marketing because the physical, chemical, and nutritional characteristics of fats and oils are influenced by the kinds and proportions of the component fatty acids and their position on the glycerol radical. Fatty acids are classified by their degree of saturation:

- saturated fatty acids—fatty acids in which all carbon atoms in the chain contain two hydrogen atoms and therefore have no double bonds
- monounsaturated fatty acids—fatty acids that have only one double bond in the carbon chain
- polyunsaturated fatty acids—fatty acids that have two or more double bonds in the carbon chain

Each fatty acid has an individual melting point. The melting point of saturated fatty acids increases with chain length and decreases as the fatty acids become more unsaturated; for a given fatty acid chain length, the saturated fatty acid will have a higher melting point than the unsaturated fatty acids. Capric (C-10:0) and longer chain saturated fatty acids are solids at room temperature. The unsaturated fatty acids are chemically more active than the saturates because of the double bonds, and this reactivity increases as the number of double bonds increases. The double bonds are subject to oxidation, polymerization, hydrogenation, and isomerization. The physical characteristics of a fat or oil are dependent upon the degree of unsaturation, the carbon chain length, the isomeric fatty acid forms, and the molecular configuration. Usually, fats are liquid at room temperature when the unsaturates are high and solid when the level of unsaturates are low. However, this generalization can be complicated by *trans* isomers, which have different melting characteristics than the *cis* isomer of the unsaturated fatty acid.

(5) Calculated iodine value—Iodine value measures unsaturates or the average number of double bonds in a fat. Therefore, it is logical that an iodine value can be easily calculated from a fatty acid composition

analysis. The constants for the most common unsaturated fatty acids required for calculation of a triglyceride iodine value are

Fatty Acid		Constant
Palmitoleic	C-16:1	0.950
Oleic	C-18:1	0.860
Linoleic	C-18:2	1.732
Linolenic	C-18:3	2.616

The calculated iodine value is determined simply by multiplying the percentage of each unsaturated fatty acid by its constant and addition of the results. This procedure has not replaced the regular iodine value. It can be utilized as an audit for the chemical iodine value and does provide the capability of obtaining two results from one analysis.

(6) Triglyceride structure—The chemical, physical, and biological properties of fats and oils depend not only on the kind and quantity of participating fatty acids, but also on the positions of these fatty acids in the triglyceride molecule. In general, fats and oils are composed of mixed glycerides rather than mixtures of simple glycerides. When all three fatty acids are identical, the product is a simple triglyceride. A mixed triglyceride has two or three different fatty acids joined to the glycerol. The characteristics of a triglyceride depend upon the position on the glycerin molecule that each fatty acid occupies. As a general rule, fats with uniform triglyceride molecules have beta crystals in the most stable state. Fats that are stable in the beta-prime crystal form contain a mixture of types of triglycerides, which prevents large growth. For example, fully hydrogenated soybean, corn, canola, and peanut oils are essentially all tristearin due to a low level of palmitic fatty acid and form stable beta crystals. Cottonseed and palm oils with high palmitic fatty acid levels have mixtures of stearic and palmitic fatty acids when fully hydrogenated, which form beta-prime crystals. Lard and cocoa butter have high palmitic fatty acid levels but crystallize in the beta form due to the uniform triglyceride structure. For lard, palmitic fatty acid is always found in the 2-position of the glycerin molecule, and cocoa butter always has oleic fatty acid in the 2-position [11].

Analysis of the complex mixtures of triglycerides present in natural fats has been carried out by many methods. Early analytical techniques employing crystallization and countercurrent distribution were not reproducible and required large samples and long analysis times. Extensive evaluations of chromatography techniques has led to high-performance liquid chromatography (HPLC), which partially fulfills the ideal requirements of an accurate, quick, and easy analytical procedure. HPLC procedures have provided rapid methods for the determination of relative amounts of glycerides present in a fat. It can be used to monitor the modification of a fat, as well as to detect adulteration, and as a developmental tool for specialty fats and oils. It appears that triglyceride structure analyses can be very useful research and development tools but are not practical for quality control purposes.

(7) Emulsifier analysis—A rapid, precise, and easily reproducible method for the determination of monoglycerides in fats became an essential requirement from a quality control standpoint when the use of monoglycerides in shortening and margarine blends became widespread. Hydroxyl number determinations, interfacial tension measurement, and alcohol extractions among other procedures were investigated and found not specific and sensitive enough or suitable for quality control purposes. The most satisfactory method for determining monoglycerides and free glycerin identified was based upon their quantitative oxidation with periodic acid. AOCS Method Cd 11-57 for Alpha Monoglycerides presents this titration analysis where the amount of monoglyceride or glycerin is determined by measuring the amount of periodic acid consumed. Free glycerin can also be determined by titration of the formic acid produced [1].

Several other analytical procedures for monoglycerides based on thin layer chromatography (TLC), gas liquid chromatography (GLC), and high-performance liquid chromatography (HPLC) have been developed as substitutes for the periodic acid oxidation method. These procedures provide reproducible results comparable to or better than the periodic acid oxidation method and are fast and easy to perform. However, the total monoglyceride values usually average higher than the titration method. Monoglycerides exist in two isomeric forms, the 2-form and the 1-form. Only the latter responds to periodic oxidation which is the basic reaction in the normally applied chemical reaction of alpha monoglyceride contents [12,13]. The chromatographic procedures are also applicable for determination of other emulsifier types: notably, acetylated monoglyceride and propylene glycol esters.

(8) Antioxidant analysis—Antioxidants are widely used in fats and oils products to delay decomposition processes that result in offensive flavors. Several phenoleic compounds have been identified that inhibit oxidation of fats and oils by interrupting the free radical mechanism of oxidation. The most notable synthetic antioxidants are propyl gallate (PG), butylated hydroxyanisole (BHA), butylated hydroxytoluene (BHT), and

tertiary butylhydroquinone (TBHQ). The US FDA permits the use of the antioxidants at a maximum level of 200 ppm singly or in combination by weight in the vegetable oil portion of the food except when prohibited by a Standard of Identity. The USDA limits the antioxidant use to 100 ppm singly but allows 200 ppm in combination with no single antioxidant exceeding 100 ppm for meat fat products. Both the USDA and the FDA labeling requirements specify that the antioxidant utilized must be listed in the ingredient statement. These requirements plus the beneficial effect of the antioxidants necessitate good analytical control of their additions. Both qualitative and quantitative analytical methods are available for evaluation of the fats and oils products for the presence of the various phenoleic antioxidants:

- qualitative methods—Rapid color endpoint qualitative process control procedures are available for the detection of PG, BHA, BHT, and TBHQ. These evaluations only indicate the absence or presence of the particular antioxidant in the fat and oil product.
- quantitative methods—HPLC and GLC methods are currently utilized to measure the amounts of BHA, BHT, PG, and TBHQ antioxidants in a fat or oil product. These procedures require less than an hour to provide accurate and reproducible results.

(9) Tocopherols analysis—Vegetable oils contain tocopherols that are natural antioxidants that retard oxidative rancidity. The tocopherol content decreases during each step of processing and can be markedly reduced during deodorization since these compounds are volatile under these conditions. Studies have shown that the deodorizer tocopherol loss can vary from 19.8 to 51.2%, depending upon the deodorizer conditions utilized [14]. The amount of tocopherols removed from the oil during deodorization depends on the time, temperature, and stripping steam flow used. It is important that high proportions of the tocopherols survive oil processing to purified optimum oxidation stability. The alternative is the addition of synthetic antioxidants or tocopherols that have been purified from deodorizer distillate of vegetable oils.

Several methods have been developed for tocopherol analysis of the vegetable oils and in deodorizer distillates and soybean oil sludge and residues. The instrumentation and procedures evaluated have involved colorimeters, paper chromatography (PC), thin layer chromatography (TLC), column chromatography (CC), gas liquid chromatography (GLC), gas liquid chromatography/mass spectrometry (GLC/MS), and high-pressure liquid chromatography (HPLC) separation techniques, all of which are also sometimes used in combination. Four different

AOCS Methods have been standardized for tocopherol determinations [1]:

Method	Description
Ce 3-74	Tocopherols in soya sludge and residue by GLC
Ce 7-87	Total tocopherols in deodorizer distillates
Ce 8-89	Tocopherols in vegetable oils and fats by HPLC
Ja 13-91	Tocopherols in lecithin concentrates by HPLC

Oxidative stability or lack of it in the finished fats and oils product may be due to an abnormal reduction of the tocopherols, which act as free radical chain breaking antioxidants. Tocopherol analysis of deodorized vegetable oil products or deodorizer distillate can indicate or prove the reason for a stability problem.

5. FLAVOR, RANCIDITY, AND STABILITY

Consumers use organoleptic evaluations to judge the quality of fats and oils. Therefore, organoleptic evaluation of oil products has long been the most sensitive method of assessing quality, but it is well recognized that it generally lacks precision and reproducibility. Rancidity is considered to be the objectionable flavors that result from the accumulation of decomposition products of either oxidation or hydrolysis reactions. In the development, evaluation, and quality control of edible fats and oils, resistance of the products to oxidative and hydrolytic deterioration is of prime importance. Many chemical methods have been developed to measure oxidative deterioration with the objective of correlating the data with flavor characteristics. Varied acceptability has been experienced in these endeavors, but researchers are still seeking the ideal test. However, experience has shown that when suitable flavor testing methods are employed, the chemical methods become more valuable in assessing the quality as an oil product.

5.1 Flavor Analysis

Flavor in foods is a combined result of the senses of taste and smell, plus those of touch, temperature, and pain. Taste and smell sensations result from contact stimuli, where the stimulating substances must be placed upon the receptive sensory cells. The taste-sensing organs, called taste buds, are grouped together on the tongue and to a lesser extent on the palate, pharnyx, and larnyx [15]. Taste is a four-dimensional phenomenon, consisting of sweet, sour, salt, and bitter; it has been determined that most taste receptors respond to most of them. Obviously, the complex taste stimuli cannot be identified with analyses that measure only one dimension; in addition, flavor identity is further complicated by the presence of other substances that block some sensations [16].

- sensory evaluations—Taste assessment is the most common method of grading finished oil quality. A panel of experienced tasters rate the flavor of the oil according to an established intensity scale. A 10-point scoring system is usually used where panel members assign numbers for the flavor intensity, from which a mean flavor score is determined. A trained flavor panel should agree on flavor intensity scores of plus or minus one unit. Off-flavors are described with descriptive names such as green, grassy, weedy, fruity, beany, watermelony, nutty, raw, painty musty, metallic, oxidized, buttery, reverted, fishy, rancid, tallowy, and so on. Agreement among flavorist regarding off-flavor description is often poor due to individual preferences, age, and background [17].
  Organoleptic, or taste, evaluation will always be necessary and probably will remain the most important technique in flavor evaluation. Taste panels for edible oil evaluations are utilized for several purposes: (a) as a research or developmental tool, (b) to determine consumer acceptance, and (c) as a process quality evaluation. The selection or identification of the panel members for each application are different. Consumer panels are usually a random selection of people who constitute the targeted market. Research and development evaluations require trained panelists with a finite discrimination among different oil flavors and intensities. Process quality evaluators must be able to distinguish between a "bland" or acceptable flavor and an off-flavor. AOCS Method Cg 2-83 provides a standardized technique for the sensory evaluation of edible fats and oils, which encompasses standard sample preparation, presentation of samples, and reporting of sensory responses [1].
- volatile flavor analysis—Even though sensory evaluation is the most important and common way to determine fats and oils quality, the method is time-consuming, tedious, expensive, variable among panel members and not always available. The advancements in food science analytical instruments has stimulated researchers to persue development of methods for the evaluation of the sensory qualities of fats and oils. The AOCS Flavor Nomenclature and Standard Committee studied the objective methods that could complement subjective organoleptic evaluations of oils and found good correlation between actual sensory scores and predicted results by a GLC procedure. The sci-

entific validity of the correlation coefficients and whether GLC analysis performed by one laboratory would correlate with sensory scores obtained at other laboratories was questioned by some scientists. Further evaluations by 94 panel members from eight different laboratories utilizing the same instrumental GLC analysis found results equivalent to the panel members from one specific laboratory [18]. Correlation of Oil Volatiles with Flavor Scores of Edible Oils became an AOCS Recommended Practice, Cg 1-83, and eventually a standardized method [1].

5.2 Rancidity Analysis

Detection of advanced stages of rancidity in a fat or fatty food has never been a problem for people with normal olfactory senses. The sharp pungent odors mixed with stale and musty odors provide the tell-tale evidence of rancidity. The major causes of fats and oils off-flavors are oxidation and hydrolysis. Factors such as temperature, light, moisture, metals, and oxygen contribute to the formation of off-flavors. Preferences for fat and oil products with fresh bland flavors and odors require keeping qualities and incipient rancidity evaluations both during development and as processed.

- peroxide value—Oxidation of lipids is a major cause of their deterioration, and hydroperoxides formed by the reaction between oxygen and the unsaturated fatty acids are the primary products of this reaction. Hydroperoxides have no flavor or odor but break down rapidly to form aldehydes, which have a strong, disagreeable flavor and odor. The peroxide concentration, usually expressed as peroxide value, is a measure of oxidation or rancidity in its early stages. Peroxide value (PV) measures the concentration of substances, in terms of milli-equivalents of peroxide per 1,000 grams of sample, that oxidize potassium iodide to iodine. The iodometric AOCS Method Cd 8-53 [1] is highly empirical, and any variation in procedure may cause a variation of results. Therefore, it is necessary to control temperature; sample weight; and the amount, type, and grade of reagents, as well as the time of contact. It has also been observed that the standardized method has difficulties in the titration endpoint for low PV levels and is inadequate for products like phosphatides, which develop emulsions [19].

   Peroxide value is one of the most widely used chemical tests for the determination of fats and oils quality. PV has shown good correlation with organoleptic flavor scores. However, a peroxide determination does not provide a full and unqualified evaluation of fat quality because of the peroxide's transitory nature and their breakdown to nonperoxide materials. Although a linear relationship has been observed between peroxide values and flavor scores during the initial stages of lipid oxidation, this method alone is not a very good flavor quality indicator because the peroxide value increases to a maximum and then decreases as storage time increases. Therefore, high peroxide values usually mean poor flavor ratings, but a low peroxide value is not always an indication of a good flavor.

- anisidine value—The anisidine value measures the amount of alpha-beta unsaturated aldehydes present in the oil. The method is based on the fact that in the presence of acetic acid, *p*-anisidine reacts with the aldehydic compounds in an oil-producing yellowish reaction product. The color intensity depends not only on the amount of aldehydic compounds present, but also on their structure. Thus, it has been found that the double bond in the carbon chain conjugated to the carbonyl double bond enhances the molar absorbance at 350 nm by a factor of four or five [20]. Anisidine value is a measure of secondary oxidation, or the past history of an oil, and therefore useful in determining the quality of crude oils and the efficiency of processing procedures, but it is not suitable for the detection of fat oxidation. AOCS Method Cd 18-90 has been standardized for anisidine value analysis [1].

- free fatty acid/acid value—Hydrolytic rancidity occurs as a result of a splitting of the glyceride molecule at the ester linkage with the formation of fatty acids that can contribute objectionable odor, flavor, and other characteristics. The flavors resulting from free fatty acid development depend upon the composition of the fat. Release of short-chain fatty acids such as butyric, caproic, and capric cause particularly disagreeable odors and flavors, whereas the long-chain fatty acids (C-12 and above) produce candlelike or, at alkaline pH, soapy flavors [20]. Both acid value and free fatty acid measure the free fatty acid content of fats and oils. Acid value is the amount of potassium hydroxide required for neutralization while free fatty acid utilizes sodium hydroxide for neutralization. Free fatty acid results are normally calculated as free oleic acid on a percentage basis. Free fatty acid results may be expressed in terms of acid value by multiplying the free fatty acid percent by 1.99. The percentage of free fatty acid is calculated as oleic acid for most fats and oils sources, although for coconut and palm kernel oils it is usually calculated as lauric acid and as palmitic acid for palm oil. The standard AOCS Method for Free Fatty Acid is Ca 5a-40 and Cd 3a-63 for Acid Value [1].

Free fatty acid is an important fat quality indicator during each stage of fats and oils processing. It is a measure of deodorizer efficiency and a process control tool for other processes. High free fatty acid results for deodorized oils indicate a poor vacuum of inadequate steam sparging or air leaks if the product color is high with oxidized oil flavor. The deodorized oil free fatty acid level, which has become standard in the United States, is 0.05% maximum, but deodorized oils free fatty acid level is usually 0.03% maximum.

Crude vegetable oils may have abnormally high free fatty acid levels if the seed has been field damaged or improperly stored. Seed and fruit enzyme lipases are activated by moisture, and hydrolysis is initiated, which increases the free fatty acid content. Higher crude oil free fatty acid levels equate to higher refining losses.

Free fatty acid content monitoring during and after all processes, including storage, provides process control results, which identifies potential problems for which corrective actions can be initiated on a timely basis. Free fatty acid is the result of hydrolysis of the fat or oil. Moisture must be present for hydrolysis to develop. This reaction is accelerated with heat and pressure like most reactions.

Free fatty acid titrations identify all acidic materials in the oil. This will include the acid added to chelate metals, acids leached from the bleaching earths, antioxidant acidity, emulsifiers added, and other acidic materials. Deodorization, the final process, must reduce the free fatty acid content to a level that will still meet the specification requirements even when the required additives have high acidity levels.

During deep fat frying, free fatty acid analyses are quality indicators that determine the amount of hydrolysis. Free fatty acid development results from the reaction of water and fats at frying temperature. The rate of hydrolysis development is due to the amount of moisture from foods fried and the frying temperature.

• smoke point—AOCS Method Cc 9a-48 measures the temperature at which smoking is first detected in a laboratory apparatus protected from drafts and equipped with special lighting [1]. The temperature at which smoking will be observed with actual frying or heating situations will be somewhat higher. Smoke point depends primarily on the free fatty acid content because the fatty acids are more volatile than triglycerides. Also, lower molecular weight fatty acids, as well as mono- and diglycerides, have less resistance to smoking. Initially, deodorized shortenings with alpha monoglyceride contents of 0.4% or less and free fatty acid contents of 0.05% or less should have about the same smoke point in the range of 400 to 450°F (204 to 232°C).

5.3 Stability Analysis

Stability of a fat or oil is generally accepted as the storage life of the product until rancidity becomes apparent. Oxidative rancidity is usually the

principal concern although other types of deterioration can occur simultaneously and make the problem more complex. For instance, hydrolytic
rancidity is sometimes mistaken for oxidative flavor degradation, which can
lead to ineffective preventive measures. In general, the methods for measuring fats and oils stability combine the measurement techniques for initial
evaluations with the long-term effects of temperature, light, moisture, oxygen, and other abuses.

Most fats and oils products are tested for flavor stability as a part of
quality control programs to assure that the customer specification limits are
satisfied. The purpose of these evaluations is to confirm that the product
will have a satisfactory shelf life and will not develop an off-flavor prior to
incorporation in a finished product and the shelf life of this product before
use. A number of methods have been developed for evaluating the long-
range stabilitys of fat and oil products, the majority of which are based on
subjecting the sample to conditions that attempt to accelerate the normal
oxidation process. However, a direct correlation between the various evaluations is not possible because oxidation of a fat and oil product is a complex mechanism. Therefore, the required test protocol must be selected
for each product that most adequately simulates the end-use performance.

- AOM stability—The active oxygen method (AOM) is the most commonly used method for estimating the oxidative stability of fats and
  oils products. This method has been used for routine quality control
  and as a research and development tool for new products. AOM employs heat and aeration to accelerate deterioration of the fat and oil
  sample and shorten the time required to reach the endpoint. AOCS
  Method Cd 12-57, AOM for Fat Stability, is performed by placing a
  20-ml sample in a special aeration tube and then heating it in an oil
  bath or a heat block controlled at 97.8 ± 0.2°C (208 ± 0.4°F) with
  an air flow adjusted to 2.33 ml per tube per second. The sample is
  exposed to the elevated temperature and aeration until a predetermined peroxide value is attained, usually 20 PV for meat fats and 70
  or 100 PV for vegetable oil products.

  AOM stability analysis has several major disadvantages. First, the
  procedure involved is highly empirical, requiring close attention to
  detail if reproducible results are expected. The maximum expected
  variation in results between laboratories is ±25% for a 100-hour AOM
  sample. Second, the AOM evaluation is faster than normal aging methods; however, a 100-hour AOM sample will still require 4.2 days to
  reach the expected endpoint.
- accelerated AOM stability—AOM stability evaluations began requiring
  more elapsed time to reach the endpoint after development due to
  product changes that improved the fats and oils product's resistance

to oxidation. A comprehensive study of temperature effects indicated that a satisfactory correlation with existing AOM data could be obtained by increasing the heating temperature to 110°C with a time savings of 60%. The ratio of time required for the standard AOM method to the accelerated procedure is 2.5 [21].

- oxygen bomb—The petroleum industry has used an oxygen bomb method for many years as a means of determining the resistance of gasoline to oxidation or gum formation. This method has been adapted for use in evaluating oxidative stability of edible fats and oils and fat-containing foods. The material to be tested is placed in a glass container, which is inserted into a stainless steel bomb. The bomb is sealed and pressurized with oxygen. The whole bomb is immersed in a bath of boiling water. A pressure recorder is used to plot a continuous curve of oxygen pressure versus time. The length of time in minutes, from the time the pressure reaches an initial plateau at the temperature of the bath until a sharp drop in the pressure occurs, is taken as a measure of the oxidative stability of the sample. Since it was found that fats and oils do not have a defined pressure drop, an arbitrary endpoint based on comparative pressure drops is generally used.

  This method offers several advantages over the standard AOM stability analysis: (1) it is 1.5 times faster because the increased pressure of the reactive gas speeds up the reaction; (2) it is more reproducible; (3) it can be used to evaluate fats and oils as well as fatty foods; and (4) it requires less technician time and attention to perform. A disadvantage for the bomb stability is that only one sample can be tested at a time, which could more than offset the time savings attained by the reduced elapsed time required [22].

- oil stability index—Increased use of two conductivity instruments, Rancimat and The Oxidative Stability Instrument, as an AOM stability alternative necessitated that AOCS initiate a collaborative study to investigate the conductimetric method as an official alternative. As a result of the study, AOCS Method Cd 12b-92 titled Oil Stability Index (OSI) was scheduled to become an official method in 1996 [23].

  These instruments measure the increase in deionized water conductivity resulting from trapped volatile oxidation products produced when the oil product is heated under a stream of air. The conductivity increase is related to the oxidative stability of the oil product. The evaluation procedure is performed by placing 2 grams of the oil sample into the sample tube, which has been preheated to 120°C (248°F), connected on one side to the air source and on the other side to a 50-ml cell of deionized water. The conductance of the water is measured automatically over time with a strip chart recorder and/or data acquisition software. Normally, the oxidation curve indicates the induction period followed by a rapid rising response as the oxidation rate is accelerated [2]. The endpoint for the regular AOM stability is specified as the time in hours required for the sample to reach a peroxide value of 100 meq/kg. The endpoint of this conductivity procedure is the time required for a sudden increase in formic acid production to occur at the end of the induction period [24]. Good correlation has been found to exist between the Rancimat instrument and AOM, and the OSI instrument and AOM at various temperatures.

The conductimetric oxidative stability method can be used in application traditionally utilizing AOM stability, AOCS Method Cd 12-57: trading regulations for finished fats and oils products or quality control during processing. This method automatically delivers rapid oxidation stability data, which requires less technician time and attention. Additionally, this AOM alternative provides a more precise and reproducible method. The conductivity AOM alternate has not solved the unreliability problem for evaluating natural antioxidants. High-temperature oxidative stability tests, including AOM stability, conductimetric methods, oxygen bomb, and oxygen uptake, are unreliable due to the lipid oxidation mechanism at elevated temperatures. Oxidation stability evaluations based on oils oxidized at 100°C have peroxide values in excess of 50 meq/kg while soybean oil flavor is unacceptable at a PV of less than 10 meq/kg [23].

- schaal oven test—The schaal oven test was developed by the biscuit and cracker industry to provide an age stability evaluation for shortenings. Since it is performed at temperatures only moderately warmer than those found in ordinary storage conditions, it provides an index of stability that more nearly rates a product under use conditions. The schaal oven evaluation is normally performed in a forced draft oven at 63°C (145°F). The test sample is stored in beakers covered with a watch glass or in glass jars with loose-fitting caps. The product stability is measured as the number of days before rancidity is detected. Peroxide value can be used as an indicator of rancidity but should only supplement organoleptic evaluations. A distinctive feature of the schaal oven evaluations of both fats and oils, plus fatty foods, is that flavors and odors other than oxidative rancidity can be revealed because it does not rely on very high temperatures to accelerate degradation. Also, a minimum of laboratory equipment is required to perform the evaluation.

The relationship between product shelf life, schaal oven, and AOM stability results varies with each fat and oil product and the processing it has received. Some claims are that one AOM hour or one schaal oven day is equivalent to 15 days of shelf life. However, there is no overall correlation that can be applied in all situations. None of the stability evaluations can be used as an index of shelf stability, except when it is applied to a given type of fat and oil formulation for which a specific relationship has been established [25].

- pastry flavor test—The effects of unsaturation, prooxidants, antioxidants, oxygen, light, moisture, and temperature on lipid products can be measured with the foregoing analytical methods established for the fats and oils products. However, the oxidation and hydrolysis resistance factors can change when the fats and oils products are incorporated into a food product and processed. Shortening and margarine products are major ingredients for bakery products. In this application the fats and oils products must withstand baking temperatures while in contact with moisture and other ingredients without flavor degradation. The pastry flavor test evaluates a shortening or margarine product in a high-fat baked product also containing flour, moisture, and salt for an extended period at baking temperatures. Sensory flavor and odor evaluations of the baked product will identify oxidation, hydrolysis, or other organoleptic problems or changes. This evaluation, outlined in the nonstandardized methods section, can be utilized for product development and to audit the products produced.

6. COLOR AND APPEARANCE

Color and appearance of fats and oils are not monitored solely for aesthetic qualities, although this is an important parameter. There are many reasons why color and appearance are important, but ultimately, it all relates to the cost of processing, the quality of the finished product, and what the product looks like to the end user.

Most oils are yellowish-red or amber liquids. The color is due to the presence of carotenoid pigments and/or chlorophyll pigments, the latter imparting a greenish cast to the oil. Some crude oils can have unexpectedly high pigmentation caused by field damage, improper storage, or faulty handling during crushing, extraction, or rendering. Color measurement will determine the condition of the receipts to identify the necessary blending, processing, and handling procedures; support claims against vendors; and prevent contamination of prime quality oils with problem products. During processing, product appearance may be an indicator of a problem. Improper bleaching, oxidation, ineffective filtration, and other problems can be indicated by color darkening or lack of color change.

Changes in fats and oils finished product appearance are perceived as poor quality product irregardless of the reason or effect upon performance.

Consumers may not consciously notice the color of a bottled oil unless it appears different from the other product on the shelf. Marketing has successfully promoted lighter or whiter oil as better for most salad oils and shortenings. Food processors usually have ingredient specifications that identify the allowed color parameters because the fat and oil product may have the ability to enhance or diminish the appearance of the prepared food product. Product colors must be monitored from receipt to use to ensure real or perceived product quality.

- Wesson color method—AOCS Method Cc13b-45 determines the color of a melted fat or oil product by comparison with red and yellow Lovibond glasses of known characteristics [1]. The oil sample is placed in an optical glass tube with a path length of either 1.0 or 5.25 inches and then viewed by the operator to begin the color match. This is done by superimposing a mixture of red and yellow standards over the reference field, which is adjacent to the oil sample. The 5.25-inch tube is utilized for most samples, except for dark oils, which exceed a 40.0 red at this level. It is standard practice to indicate when the 1.0-inch level has been utilized, but the 5.25-inch level is understood.

    The Wesson method using Lovibond glasses is an abbreviated version of a method originally developed in England for measuring the color of beer. Color is three-dimensional, but the Wesson method ignores the brightness factor and is only interested in the degree of redness. Yellow is necessary to make colors look similar to allow assessment of redness, but the amount of yellow was considered unimportant for this method. Therefore, the use of a fixed yellow value was adopted with only the red viewed as critical. The fixed yellow ratio is 10 yellow to 1.0 red for most oils with red color readings under 3.5; higher yellow settings are specified for the darker oils.

    The Wesson method is the principal color method for the U.S. edible oil industry, which has been utilized for many years primarily because of its simplicity. However, some difficulties are caused by oversimplification: (1) apparent red values are reduced when chlorophyll is present in the oil; (2) brown pigments interfere with red and yellow comparisons; and (3) visual comparisons must be made [26]. Visual color measurement has become less acceptable because the operator must be adept at matching colors and also must have good color vision.

- Lovibond (British Standard)—AOCS Method Cc 13e-92 utilizes a Lovibond Tintometer, which has become the standard in most countries other than the United States. The geometry and color scales for the Wesson and Lovibond methods are different, and consequently, the results are not compatible. The vital parts of the Lovibond Tintometer are the series of red, yellow, and blue permanently colored glass standards. These standards vary from water-white colors to deep reds, yellows, and blues. Each standard color is numbered and subtly different from the one preceding and following it. The addition of the blue color field provides a greater degree of brightness and greenness than for the Wesson method [27].

- spectrophotometric color method for oils—This method represents an attempt to measure the color of fats and oils with an automatic instrument to eliminate the visual judgment requirement for operators. In 1950, AOCS Method Cc 13c-50 was tentatively adopted for the spectrophotometric determination of oil color to replace the manual Lovibond systems. A collaborative study, which included 30,000 color measurements, determined the best wavelengths to measure the color of oil. These wavelengths were then used to develop an equation to relate the spectrophotometric readings to Lovibond values. The calculations were designed to give values identical with Lovibond color values using the Wesson method [28]. In general, the calculated values agree, but wide discrepancies occur with some oils. The photometric color method has not replaced the Wesson method because of the occasional disagreement, economics, and the firm entrenchment of the visual procedure. Nevertheless, the Spectrophotometric Color Method for Oils, AOCS Cc 13c-50, is an official method [1].

- automatic tintometers—Three filter electronic color instruments have been introduced by several equipment manufacturers to replace the manual color procedures. The automatic instruments have been designed to conform to AOCS Method Cc 13b-45, as well as the European procedure or AOCS Method Cc 13e-45. These instruments have performed well with light colored oils but have had poor agreement with darker oils or products containing certain additives like emulsifiers.

- FAC color method—Inedible tallow and grease colors are often too dark or too green to read on the Lovibond tintometer even with the 1.0-inch column. An arbitrary system of color standards was developed for identification of the dark colors. AOCS Method Cc 13a-43 employs standard color tubes for comparison of oils in a similar tube. The standards consist of 26 permanent color tubes numbered from 1 to 45 in odd numbers, with three overlapping series, one normal, one green, and one red. The spacing of the tubes is not uniform on purpose. This method, far from precise, has been used mainly for inedible oils because of its ease of application.

- gardner color—The Gardner scale is a single number scale used mainly to categorize lecithin, paint or drying oils, fatty acids, and some other oil derivatives. These standards are patterned along the same lines as the FAC standard but bear no direct relationship. The Gardner standards consist of glass standards numbered from 1 to 18, lightest to darkest. Gardner color has been standardized with AOCS Method Td 1a-64 [1].

- chlorophyll—The presence of green pigments or chlorophyll is of interest not only because of their impact on the finished product's color, but also because chlorophyll can act as a sensitizer for fats and oils oxidation. These pigments must be removed in the prebleaching process. AOCS Method Cc 13d-55 is used to determine the parts per million chlorophyll content of vegetable oils by spectrophotometric absorption measurements at 630, 670, and 710 nm [1]. The method is not applicable to hydrogenated and deodorized oils because the 670 nm absorption is missing in most processed oils.

- coloring agents determination—Yellow colors are often added to fats and oils products to simulate the appearance of butter. Carotenes, annattos, and apocarotenals are the primary colorants utilized. Yellow color additives are extremely difficult, if not impossible, to control with the two-dimensional Lovibond color measurements. Spectrophotometric measurements of yellow density at wavelengths of 440, 455, or 460 millimicrons, depending upon the product, provide better control for these additives.

6.2 Appearance

The first and most obvious product characteristic evaluated by any consumer of a finished fats and oils product has to be its appearance. Appearance is an important attribute because the initial impression usually influences subsequent judgments, even though most of the appearance properties have only aesthetic value. These characteristics usually have little affect upon performance; however, abnormal appearance characteristics can indicate a handling or storage problem, poor processing techniques, and stability or other quality problems.

PRODUCT APPEARANCE RATINGS

Visual examinations are the best way to get a meaningful evaluation of product appearance as a whole. Appearance ratings are subjective opinions based on experience to assign numerical values for comparison purposes. Product appearance evaluations of solidified products is two-dimensional, i.e., surface and internal appearance characteristics. The product's uniformity of color, hue, and textural qualities are rated both for the product's surface and internally after sectioning, cutting, or scraping. The textural ratings evaluate each product for deficiencies like mottles, streaks, oil separation, and other problems that can be related to specific processing or formulation problems. The apparent color cast and sheen ratings can be related to bleaching deficiencies, color additions, or plasticization problems. Procedures for appearance evaluations and ratings are detailed in the nonstandardized method section. Appearance evaluations are useful measurement tools for product development, as well as process quality control.

7. REFINING AND BLEACHING

Specific refining and bleaching analyses are necessary for crude oil receipts for three distinct purposes: (1) as a basis for settlement of crude contracts under the trading rules of the various oil trading associations, (2) as a yardstick for the efficiency of the refining and bleaching operations, and (3) as indicators for caustic and bleaching earth types and levels for processing. Most crude vegetable oils are traded on the basis of refining loss or neutral oil, refined or bleached color, and flavor. Trading rules for vegetable oils have been established in the United States by three associations: National Cottonseed Producers Association, National Soybean Processors Association, and National Institute of Oilseed Products. All three of these organizations accept the AOCS methods as the basis for trading. For most oils, settlement is based on agreement between the buyer and seller of the results of specified analytical methods within an established variance performed on the same sample. Results beyond the established differences are settled by an Official Referee Chemist's analysis.

- refining loss—About 75 years ago, the cottonseed crushers and oil producers agreed to a series of specifications for crude cottonseed oil. Agreement for price concessions for products less than prime were determined for refining loss, odor, taste, and color. The refining loss penalty assessed was three-quarters of a percent for each percent in excess of 9.0% loss. This agreement led to the development of the "Cup Loss" method for refining a sample of crude oil in the laboratory in a manner that would simulate plant operations to the extent of providing approximate refining losses and refined color assessments. Initially, this method was very subjective, allowing the laboratory technician to rely on his best judgment on how to perform the evaluation for the best loss result. In 1927, it was agreed that a premium would be paid by the buyer for crude cottonseed oils with a loss below 9.0%. This agreement offered the crude mills an incentive to produce better grades of oil but at the same time made loss results suspect because the method relied upon the judgment and expertise of the laboratory technician. Standardization of the laboratory cup refining procedure made the results less variable, and the technicians performing the analysis no longer had to be refining experts [29].

AOCS Method Ca 9a-52 determines the loss of free fatty acids, oil, and impurities when the sample is treated with alkali solutions under the specific conditions of the test [1]. The actual method simulates kettle refining on a laboratory basis with standardized levels and strengths of sodium hydroxide solution, temperatures, and holding periods to produce reproducible results within and between laboratories. The refining loss is calculated by subtracting the refined oil sample weight from the original crude oil sample weight.

- neutral oil and loss—The laboratory cup refining test was acceptable when open kettle refining was common practice, but improvements in refining processes provided losses lower than the laboratory estimates. Various methods for estimating the neutral oil content in a crude oil were explored. These procedures identified the theoretical recoverable oil in the crude oil sample for comparison to actual results to determine refining efficiency. Of the methods techniques evaluated, an International Chemical Union chromatographic procedure appeared most appropriate because it was found to be reasonably accurate, most reproducible, easy, and rapid to carry out and required no special elaborate or expensive equipment [30]. This procedure was the basis for AOCS Method Ca 9f-57, titled Neutral Oil and Loss. The neutral oil and loss method extracts the oil or fatty materials on a column of activated alumina by ether-methanol. The mixture extracted consists primarily of triglycerides and unsaponifiable material. Reproducibility has been found to be good; i.e., two single determinations performed within a laboratory should not differ by more than 0.15 and not more than 0.30 between laboratories [1].

7.2 Bleaching Analysis

Adsorption bleaching is used both in the plant and in the laboratory. The laboratory technique involves the addition of a bleaching earth, carbon, or both to refined oil, heating to 120°C, and held at that temperature for 5 ± 1 minutes while agitating at 250 ± 10 rpm. Afterwards, the earth is removed by filtration and the oil color determined. Bleaching analyses are performed with oils from the cup loss evaluation or specially laboratory refined samples for the bleach evaluations. Two AOCS methods are available for bleaching fats and oils; both utilize the same procedure, but the specified amount of AOCS natural bleaching earth is different and one has a provision for the use of activated bleaching earth for high green soybean and sunflower oil. AOCS Method Cc 8a-52 is applicable to refined cottonseed oils, and AOCS Method Cc 8b-52 is applicable for soybean and sunflower oils [1].

8. PERFORMANCE TESTING

Some essential attributes contributed by fats and oils cannot be directly measured with chemical or physical analytical methods. In these cases, performance testing is the only means for evaluating the fat or oil product's ability to perform the desired functions in a food product. Actual determinations of the performance qualities of an edible fat and oil product are made with small-scale practical tests that evaluate how the product will perform in a finished product. Successful performance tests are designed with standardized conditions and ingredients with critical formulations chosen in regard to the property evaluated to highlight small differences in performance.

Performance testing is essential for the development of new products, especially for fats and oils products designed for a specific food product, formulation, or process. After development, physical or chemical analysis can be related to performance results in most situations; however, continuation of certain performance evaluations is necessary for some products to ensure adequate performance or more timely results in some cases. Initially, most performance testing was designed for bakery products but has now been expanded to every specialty product situation, i.e., baking, frying, candy, coatings, formulated foods, nondairy products, and so forth, whereever tailored shortenings, margarines, oils, and other specialty products are utilized. In many cases the performance tests are developed to evaluate the fat and oil ingredient as it would be used by a specific food processor.

Fats and oils performance testing is most likely performed by all of the processors, but the procedures utilized have never been standardized for the industry. Performance testing is an analytical technique similar to other laboratory evaluations. The procedures must be designed to incorporate good laboratory techniques, which include standardization of equipment, ingredients, procedures, results reporting, and control of the environment.

Performance results must be reproducible to be of value for comparative purposes. Reproducibility is achieved by controlling variables that can only be controlled by the use of standardized methods. The performance test procedures must be written with adequate detail and followed closely. A review of some of the most common performance evaluations follows, and complete performance testing procedures are presented in the nonstandardized methods section of this chapter.

(1) Creaming volume—Cake batter aeration can be affected by the plasticity, consistency, emulsification, basestock formulation, and other fats and oils properties. Creaming volume evaluations measure the ability of a shortening or margarine to incorporate and retain air in a cake batter. In most cases, batter aeration is an indicator of the baked cake volume, grain, and texture and materially affects the handling qualities of the cake batter.

The creaming volume test formula consists of only three ingredients: (1) test shortening or margarine, (2) granulated sugar, and (3) whole eggs. This procedure is the first stage of an old fashion pound cake where all of the cake batter aeration depended upon the creaming properties of the shortening with whole eggs. Batter specific gravities are determined after mixing 15 minutes and again after 20 minutes. Continued aeration, identified by a decrease in batter specific gravity, indicates that the fat or oil product has a stable consistency that has not broken down to allow the release of air from the batter. Specific gravity, grams per cubic centimeter, can be converted to specific volume, cubic centimeters per 100 grams, by multiplying the reciprocal of the specific gravity by 100. Specific volume better illustrates the amount or degree of aeration. This performance test is applicable to emulsified, as well as nonemulsified, products to measure aeration potential in a cake batter.

(2) Pound cake test—In some cases, shortening or margarine creaming value is most accurately measured by preparing a regular pound cake, omitting the chemical leavener, and measuring the volume, grain, and texture of the baked cake. Creaming value, as determined by this method, is affected by the batter mixing temperature. Working range, or creaming range, can be measured by adjusting the finished batter temperature over the desired temperature range. The results obtained in this manner provide a good indication of the creaming range or shortening temperature tolerance. The baked pound cake volume is determined by a seed displacement procedure and the cake appearance rated numerically with a scale [5] similar to that provided below:

Score	Rating	Description
10	Perfect	Fine regular grain; no holes, cracks, or tunnels; very thin cell walls and perfect symmetry
9	Very good	Close regular grain; free of holes, cracks, or tunnels; thin cell walls
8	Good	Grain very slightly open but regular, free of cracks or tunnels, may have occasional hole, good cell wall thickness
7	Satisfactory	Grain slightly open, mostly regular, a few small holes, no tunnels or cracks, slightly thick cell walls
6	Poor	Open or irregular grain, or frequent holes, some cracks or tunnels
5	Unsatisfactory	Very open or irregular grain, or numerous holes, cracks, or tunnels, or thick heavy cell walls; may have solid streaks or gum line
4 and below		Increasing degrees of unsatisfactory performance

(3) Icing volume—Shortening and margarine formulation, consistency, and/or plasticity can affect the product's ability to produce light creme icings. This property can be measured by making a standardized creme icing under controlled conditions. The evaluation procedure is somewhat similar to the creaming volume evaluation where the aeration potential and mixing tolerance of a fats and oils product are measured with specific gravity determinations. However, the results indicate the ability of the product to aerate and retain the incorporated air in a low-moisture icing. The icing ingredients, composed of (1) test shortening, (2) powdered sugar, (3) nonfat milk solids, (4) salt, and (5) slightly less than 12% water, are mixed for 12 minutes at high speed, stopping after 7 and 12 minutes to measure the specific gravity of the icing. These measurements determine the aeration potential and the mixing tolerance of the fat and oil product. A fat and oil product with poor mixing tolerance will either aerate very little or give up a portion of the incorporated air after the first specific gravity determination. This evaluation is applicable to emulsified and nonemulsified fat and oil products to measure the aeration potential in a creme icing.

(4) White layer cakes—Fats and oils perform several important functions in all cake products. These functions include entrapment of air during the creaming process, lubrication of the gluten and starch particles to break the continuity of the gluten/starch structure of the cell walls for a tender crumb, and emulsification or water-holding capabilities that affect moistness and cake keeping qualities. The emulsification value of a fat and oil product may be defined as its ability to make a white

layer cake in which the flour content is low but the sugar and moisture contents are high. The best method for determining emulsion value of a shortening or margarine is to test the product in a high-sugar, white layer cake under standard conditions, evaluating the finished cake for volume, crust characteristics, grain, and texture. Previous results have shown that shortenings with poor emulsification properties produce curdled or separated batters and cakes with low volumes, pale sticky crusts, and a raw flour taste. Properly emulsified products will produce smooth cake batters and cakes with golden brown crusts somewhat dry to the touch without a liquid ring, a fine grain and texture, and good eating characteristics [31]. Overemulsified shortenings produce cakes that dip or have a weak center with a very open grain and a coarse texture.

Three white layer cake performance tests are utilized for the evaluation of general purpose shortenings and margarines: (1) 140% sugar white cake, (2) 105% sugar white cake, and (3) household white layer cake. The cake performance tests are designed for the fat and oil type product to be evaluated, i.e.,

- emulsified product—The high sugar, egg whites, and liquid cake measures the ability of an emulsified shortening or margarine to produce a high-volume, moist cake with a fine grain and texture.
- nonemulsified product—A lower sugar, moisture, and egg cake formulation for the evaluation of the emulsification properties is contributed solely by the product consistency of the all-purpose shortening or margarine product.
- household shortening—Shortenings produced for household use usually have a low emulsifier content for baking purposes but not enough to drastically affect frying performance. Therefore, household recipes are designed for this emulsifier level.

(5) Creme filling test—The ability of a shortening or margarine to take up and hold water while aerating to a low filling specific gravity is an important attribute for creme fillings for snack cakes. Equally important is the filling's resistance to weep or its disappearance into the base cake. Shortening plasticity, heat resistance, and emulsification can be measured by creme filling performance testing. Initially, the ability of the shortening product to aerate the filling is measured during preparation, and subsequently, the weep and heat stability of the fillings are evaluated. This evaluation measures the emulsifier system, as well as the consistency of the shortening as provided by the basestock formulation.

(6) Cake mix evaluation—Originally, cake mix formulations were very similar to bakery cakes and utilized standard "Hi-Ratio" cake shortenings;

however, development of improved cake mixes required rapid aerating shortenings to minimize mixing times for the housewife, while at the same time increasing the product's mixing and baking tolerances. The competitive nature of the cake mix industry has continued the demands for new and improved products, of which shortening has always been a major contributor. A basic white mix cake formulation and make-up procedure can serve to evaluate new or revised emulsifier systems for aeration, eating qualities, and cake shelf life, as well as the shortening carrier for lubrication and consistency.

(7) Puff pastry testing—The characteristic features of a puff pastry short-ening are plasticity and firmness. Plasticity is necessary to effect smooth unbroken layers of fat between dough layers during repeated folding and rolling operations performed at retarder temperatures to achieve over 1,200 layers. Firmness is equally important, because soft or oily fat products can be absorbed by the dough, destroying its role as a barrier between the dough layers. The laminated puff pastry dough rises in the oven when the fat layers melt and expand, also creating steam if the fat contains moisture, which causes the layers to separate and the product to expand and rise without the benefit of a leavening agent. Satisfactory performance is a five- to sixfold increase in the height of a test pastry shell after baking.

(8) Restaurant deep fat frying evaluation—A number of factors are studied when evaluating frying shortenings and oils. During deep fat frying, the fat is exposed continuously to elevated temperatures in the presence of air and moisture. A number of chemical reactions, including oxida-tion and hydrolysis, occur during this time, as well as changes due to thermal decomposition. As these reactions proceed, the functional, sen-sory, and nutritional quality of the frying fat changes and eventually reaches a point where it is no longer possible to prepare quality fried products, and the fat will have to be discarded. The rate of frying fat deterioration varies with the food fried, the frying fat utilized, the fryer design, and the operating conditions.

Good testing procedure limits the variables to the product being evaluated. The method should be designed to incorporate techniques that are practiced in any good laboratory, which include the stan-dardization of equipment, control of the environment, and an ex-plicit evaluation procedure. Frying tests have been attempted in commercial restaurant frying operations; however, cause and effect determinations for the results obtained are difficult to assess due to the changing product mix, handling conditions, and other unknown conditions and contaminants. A laboratory test procedure employing the smallest available commercial fryers to continuously heat the frying shortening and frying of french cut potatoes at prescribed intervals has been found to adequately evaluate frying fats. This procedure allows the frying fat deterioration to be monitored visually, organoleptically, and with chemical and physical analytical testing for assessment of the cause and effect.

The deep fat frying evaluation consists of controlled heating of the test shortening or oils at 360 ± 10°F (176 to 188°C) continuously until the test is terminated. Fresh french cut potatoes (227 grams) fried three times daily for 7 minutes at 3-hour intervals are flavored once daily. Frying observations recorded after each frying include smoking, odor, clarity, gum formation, and a determination of foam development. Foam development described as none, trace, slight, definite, and persistent should also be measured with a foam test daily and each time a change in the observed foam is recorded. Samples are taken after each 24-hour period for analysis of color, free fatty acid, and iodine value for quantitative measurement of darkening, hydrolysis, and polymerization. The frying test is terminated when persistent foam has been observed and substantiated by foam height testing.

(9) Ice cream bar coating evaluation—Ice cream novelties are usually enrobed with a confectionery coating, which utilizes a relatively high fat content. The function of the fats utilized are viscosity control, crystallization, gloss retention, and eating characteristics. These characteristics must be measured to identify the acceptability of the coating fat by actually dipping ice cream bars into prepared coatings. The qualities measured to determine acceptability are drying time, hardening time, brittleness or snap, bar coverage, flavor, eating characteristics, and gloss retention.

9. NONSTANDARDIZED METHODS

A number of nonstandardized methods have been reviewed for the evaluation of fats and oils in the performance testing and other sections of this chapter. Complete test methods for these evaluations are presented on the following pages. All of these evaluations have been successfully employed for product development, quality control, or auditing of current production products and those of the competition. The methods presented are as follows (the numbering corresponds to previous subhead numbering):

Method	Description
3.1	Shortening consistency and plasticity rating
5.3	Pastry flavor test
6.2	Shortening appearance rating
6.3	Margarine appearance rating
8.1	Creaming volume test
8.2	Pound cake test
8.3	Icing volume
8.4	140% sugar white cake
8.5	105% sugar white cake
8.6	Household white layer cake test
8.7	Creme filling test
8.8	White cake mix test
8.9	Puff pastry testing
8.10	Deep fat frying test
8.11	Ice cream bar coating
8.12	Sandwich cookie filler shortening evaluation

Nonstandardized Method 3.1

SHORTENING CONSISTENCY AND PLASTICITY RATING

Description: A systematic, uniform means of rating the consistency and plasticity of crystallized fats and oils products.

Scope: Applicable to all shortenings, margarines, and other plasticized fats and oils products.

Equipment:

(1) Constant 80°F temperature storage room
(2) Illumination, 50 to 100 foot-candles at examination surface from cool white fluorescent lamps
(3) Thermometer, 0 to 120°F

Procedure:

(1) Temper the samples to achieve the required 75 to 80°F examination temperature.
(2) Determine and record the sample temperature.
(3) Examine the consistency and plasticity of the sample by pressing a finger into the product or squeezing it in the hand or both. This procedure is the identification of imperfections in the sample.
(4) Some of the most common consistency problems found with plasticized products are
   • puffy—large air cells or pockets that offer very little resistance when pressed or squeezed
   • sandy—small lumps about the size of grains of sand
   • ribby—alternating thin layers of hard and soft product, which has been described as feeling like a course corduroy cloth surface
   • brittle—a firm resistance that cracks with finger pressure
   • mushy—soft and greasy feeling with very little, if any, resistance
   • lumpy—sizable firm clumps within a softer or plastic consistency
   • chalky—dry putty-like feeling
(5) Plasticity rating evaluates the workability of a crystallized product. A product with perfect plasticity would retain a good body feel without releasing liquid or remaining rigid when worked in the hand. A plastic material will not flow or deform from its own weight, but it should be easily molded with only slight pressure.
(6) Consistency and plasticity should be each rated overall using the following 10-point scale, rating each product without reference to the type of product, age, or any other qualifying factor:

10—Perfect	6—Moderately unsatisfactory
9—Near perfect	5—Unsatisfactory
8—Good or satisfactory	4—Bad
7—Slightly unsatisfactory	3 through 1—Degrees of bad

Nonstandardized Method 5.3

PASTRY FLAVOR TEST

Description: Pastry flavor subjects the test fat or oil product to prolonged high-temperature heat in the presence of moisture, salt, and flour for oxidative stability measurement by sensory flavor and odor evaluations.

Scope: Applicable to all fats and oils products, i.e., shortening, margarine, flakes, chips, liquid shortenings, and oils.

Formula:

Ingredients	Grams
Test fat	100
Pastry flour	150
Salt	3
Water	50

Equipment:

 (1) Mixing bowl
 (2) Pastry cutter
 (3) Pastry cloth
 (4) Dusting flour
 (5) Rolling pin
 (6) Two 1/8-inch diameter steel sizing rods, 8 inches long
 (7) 2-inch diameter cookie or biscuit cutter
 (8) Sheet pan
 (9) Kitchen fork
(10) Scale, graduated in grams, with scale pan and counterbalance
(11) Bakery oven

Procedure:

 (1) Scale the fat, flour, and salt into the mixing bowl.
 (2) Cut the fat and the flour together to obtain small pea-sized lumps.
 (3) Add the water and mix together until the mixture balls together.
 (4) Roll out the dough on a flour-dusted pastry cloth, using the 1/8-inch diameter rods to control thickness.
 (5) Cut four dough circles with the 2-inch cookie cutter.
 (6) Place the four dough circles on the sheet pan and perforate with the kitchen fork.
 (7) Bake at 300°F for 1 hour.
 (8) Cool for 15 minutes and evaluate the flavor and odor using the 10-point scale applicable to fresh oil sensory testing.

Nonstandardized Method 6.2

SHORTENING APPEARANCE RATING

Definition: This system evaluates and rates shortening surface and textural appearance. Each product is rated numerically, with ten (10) indicating perfection and one (1) indicating worst possible, utilizing the shortening appearance grading scale.

Scope: Applicable to all plasticized shortenings, both fresh and aged.

Equipment:

(1) Constant 80°F temperature-controlled area

(2) Illumination, 50 to 100 foot-candles at examination surface from cool white fluorescent lamps

(3) Thermometer, 0 to 120°F

(4) Stainless steel spatula with 6-inch blade

Sample preparation:

(1) The shortening sample must be permitted to achieve a temperature of 75 to 80°F before examination.

(2) A portion of the sample should be removed from one side of the container without disturbing the remaining surface area. Cut a thin slit into the undisturbed portion from the area where the portion was removed.

Procedure:

(1) Determine and record the sample temperature to the nearest °F.

(2) Examine the shortening sample using the special lighting to highlight the sample surface.

(3) Rate the appearance of the shortening using the grading scale on the following page as a guide. All of the items should be considered in determining the overall grade. It is not necessary that the grade on each of the seven basic characteristics be recorded for each sample; however, if a sample is rated a number eight or lower, the major defects observed should be recorded.

(4) All products should be graded objectively without reference to the type of product, age, or history.

Shortening Appearance Grading Scale.

Grade	Color Cast*	Surface Sheen	Visual**	Mottles	Streaks	Vaselation	Oil Separation
				Textural Appearance			
10	None	Excellent	Excellent	None	None	None	None
9	Very slight	Very good	Very good	Very few and small	None	None	None
8	Slight cast	Good	Good	Few and small	None	None	None
7	Moderate cast	Fair	Fair	Numerous and small	Few and small	None	None
6	Definitely off	Slightly dull	Poor	Very few and large	Numerous and small	None	None
5	Definitely off	Dull	Very poor	Few and large	Few and large	None	None
4	Definitely off	Dull	Very poor	Numerous and large	Numerous and large	Very slight	None
3	Definitely off	Very dull	Very poor	Numerous and large	Numerous and large	Slight	None
2	Definitely off	Very dull	Very poor	Numerous and large	Numerous and large	Bad	Slight
1	Definitely off	Very dull	Very poor	Numerous and large	Numerous and large	Very bad	Bad

° Red, yellow, green, or grey cast for the product as a whole or as observed in the thin slit cut into the surface.
°° The usual visual textural characteristics observed are pock marks, pimples, and a spongy appearance.

Nonstandardized Method 6.3

MARGARINE APPEARANCE RATING

Description: A systematic, uniform means of grading the surface and internal appearance and visual texture of margarine products.

Scope: Applicable to all margarine and spread products.

Equipment:

(1) Constant 45 ± 5°F temperature-controlled storage area

(2) Illumination, 50 to 100 foot-candles at the examination surface

(3) Thermometer, 0 to 120°F

(4) Stainless steel spatula with 6-inch blade

(5) Hansen's *Color Comparison Card for Butter*

Procedure:

(1) The margarine samples must have achieved a temperature of 45 ± 5°F before examination.

(2) Carefully open the margarine containers.

(3) Determine and record the temperature of the margarine sample to the nearest °F.

(4) Scrape a portion of the sample surface to expose the interior of the sample.

(5) Examine the margarine samples utilizing the special lighting to highlight the margarine surface.

(6) Rate the appearance of the margarine using the grading scale on the following page as a guide. It is not necessary that the grade for each characteristic be recorded for each sample; however, if a sample is graded an 8 or below, the defects should be recorded.

(7) All products should be graded objectively without reference to product type, age, or history.

Margarine Surface and Internal Appearance Rating Chart.

Grade	Hansen's Color	Surface Sheen	Streaked Color	Surface Oxidation	Dark Specks	Grainy	Water Separation	Pin Holes	Puffy or Porous
10	Target	Excellent	None	None	None	None	None	None	None
9	Target	Very good	None	None	None	None	None	None	None
8	Target less 1	Good	None	None	None	None	None	None	None
7	Target plus 1	Fair	None	None	None	None	None	One or two	None
6	Target less 2	Slightly dull	Very slight	None	None	Very slight	None	Three or four	Very slight
5	Target plus 2	Dull	Slight	None	None	Slight	None	Five or six	Slight
4	Target less 3	Dull	Bad	None	Very slight	Bad	None	Seven or eight	Definite
3	Target plus 3	Very dull	Very bad	Slight	Slight	Very bad	Slight	Nine or ten	Bad
2	Target less 4	Very dull	Very bad	Bad	Bad	Very bad	Definite	Ten plus	Very bad
1	Target plus 4	Very dull	Very bad	Very bad	Very bad	Very bad	Bad	Ten plus	Very bad

Nonstandardized Method 8.1

CREAMING VOLUME TEST

Definition: Creaming volume is a measure of the ability of a plasticized
shortening or margarine to incorporate and retain air in a cake batter.
Batter aeration is an indicator of baked cake volume, grain, and texture.

Scope: Applicable to all plasticized shortenings and margarines for use in
cake production.

Test Formula:

Ingredients	Grams	Use Temperature
Test shortening	227	70 to 80°F
Granulated sugar	454	70 to 80°F
Whole eggs	227	50 to 60°F

Equipment:

(1) C-100 Hobart mixer equipped with 3-quart bowl and paddle
(2) Scale graduated in grams
(3) Scale pan with counterbalance
(4) Thermometer, 0 to 120°F
(5) Specific gravity cup
(6) Stainless steel spatula with 6-inch blade

Procedure:

(1) Scale sugar and shortening into 3-quart bowl and cream at second
speed for 5 minutes. Stop machine; scrape bowl and paddle.
(2) Add half of the whole eggs and continue creaming at second speed
for 2.5 minutes. Stop machine; scrape bowl and paddle.
(3) Add remainder of whole eggs and continue creaming at second speed
for 2.5 minutes. Stop machine; scrape bowl and paddle.
(4) Continue creaming at second speed for 5 minutes. Stop machine, determine and record specific gravity, and then scrape the bowl and paddle.
(5) Continue creaming at second speed for 5 minutes. Stop machine;
record specific gravity and finished batter temperature.

Results Reporting: Creaming volume results can either be reported as specific gravity (g/cc) or as volume (cc/100 g). Volume is calculated by multiplying the reciprocal of the specific gravity by 100. Volume may also
be determined by the "direct method," which involves measurement of
the volume of 100 grams of the creamed batter by displacement of alcohol in a graduated cylinder.

Nonstandardized Method 8.2

POUND CAKE TEST

Definition: Evaluation of the creaming properties of plasticized shortenings and margarines by measuring the volume and rating the grain and texture of a pound cake made without chemical leavening.

Scope: Applicable for all plasticized shortenings and margarines designed for cake production.

Formula:

Ingredients	Grams	Mixing Stage
Test shortening	340	First
Cake flour	340	
Granulated sugar	681	
Salt	7	
Liquid milk	142	
Liquid milk	227	Second
Whole eggs	340	
Cake flour	340	Third

Equipment:

(1) C-100 Hobart mixer equipped with 3-quart bowl and paddle
(2) Scale, graduated in grams with weighing pan and counterbalance
(3) Specific gravity cup, calibrated, and 6-inch stainless steel spatula
(4) Thermometer, 0 to 120°F
(5) Standard paper-lined 1-pound loaf pan
(6) Bakery oven
(7) Cake volumeter, seed displacement type or equivalent

Procedure:

(1) Scale the ingredients for the first stage into the 3-quart mixing bowl and mix 30 seconds at first speed. Stop the mixer and scrape the bowl and paddle.
(2) Cream 2 minutes at second speed. Stop the mixer and scrape the bowl and paddle.
(3) Cream 2 additional minutes at second speed. Stop the mixer and scrape the bowl and paddle.
(4) Add half of the second-stage liquids and mix 1 minute at first speed. Stop the mixer and scrape the bowl and paddle.
(5) Add the flour from stage three and mix 1 minute at first speed.
(6) Add the remainder of stage two liquids and mix 1 minute at first speed. Stop the mixer and scrape the bowl and paddle.
(7) Cream 5 minutes at first speed. Stop the mixer; determine and record the specific gravity and batter temperature.
(8) Scale 510 grams of batter into the loaf pan and bake 60 to 65 minutes at 360°F.
(9) Remove the cake from the pan, cool for 2 hours, determine the weight and volume of the baked cake, and then score the internal characteristics.

Nonstandardized Method 8.3

ICING VOLUME

Definition: Icing volume measures the amount of air that can be incorporated and retained by plasticized shortening or margarine in a low-moisture creme icing.

Scope: Applicable to all plasticized shortenings and margarines designated for icing preparation, both emulsified and nonemulsified.

Formula:

Ingredients	Grams	Use Temperature
Test shortening	227	70 to 80°F
Powdered sugar	908	70 to 80°F
Nonfat milk solids	56	70 to 80°F
Salt	7	70 to 80°F
Water	160	70 to 80°F

Equipment:

(1) C-100 Hobart Mixer equipped with 3-quart bowl and paddle
(2) Scale, graduated in grams, with scale pan and counterbalance
(3) Thermometer, 0 to 120°F
(4) Specific gravity cup, calibrated
(5) Stainless steel spatula with 6-inch blade

Procedure:

(1) Scale all ingredients into the mixing bowl and cream 2 minutes at first speed. Stop the mixer and scrape the bowl and paddle.
(2) Mix at third speed for 5 minutes. Stop the mixer; determine and record the specific gravity. Scrape the bowl and paddle.
(3) Mix at third speed for 5 minutes. Stop the mixer; determine and record the specific gravity and icing temperature.

Results Reporting: Icing volume results can be reported either as specific gravity (g/cc) or as volume (cc/100 g). Volume is calculated by multiplying the reciprocal of the specific gravity by 100. Volume may also be determined by the "direct method," which involves measurement of the volume of 100 grams of the icing by displacement of alcohol in a graduated cylinder.

Nonstandardized Method 8.4

140% SUGAR WHITE CAKE

Definition: Volume, external, and internal measurements of a high-sugar, egg, and moisture cake formulation determine the emulsification value of a shortening or emulsifier system.

Scope: Applicable for evaluation of all emulsified cake shortenings and/or emulsifier systems.

Formula:

Ingredients	Grams	Mixing Stage
Test shortening	250	First
Granulated sugar	638	
Nonfat milk solids	42	
Baking powder	28	
Salt	14	
Cake flour	454	
Water	212	
Egg whites	340	Second
Water	200	

Equipment:

(1) C-100 Hobart mixer equipped with 3-quart bowl and paddle
(2) Scale, graduated in grams, with scale pan and counterbalance
(3) Specific gravity cup and stainless steel spatula with 6-inch blade
(4) Thermometer, 0 to 120°F
(5) Two 8-inch greased layer cake pans lined with paper parchments
(6) Bakery oven
(7) Cake volumeter, seed displacement type or equivalent

Procedure:

(1) Scale the shortening and all the first-stage dry ingredients into the 3-quart mixing bowl and premix for 30 seconds at first speed.
(2) Add the first-stage water and mix for 3 minutes at first speed. Stop the mixer and scrape the bowl and paddle. Mix for 3 minutes at first speed.
(3) Add half of the second-stage liquid and mix at first speed for 1 minute. Stop the mixer and scrape the bowl and paddle.
(4) Add the remaining liquid and mix 1 minute at first speed. Stop the mixer and scrape the bowl and paddle.
(5) Continue mixing for 4 minutes at first speed. Stop the mixer and determine and record the batter specific gravity and temperature.
(6) Scale 400 grams of batter into each of the two 8-inch layer pans.
(7) Bake approximately 22 minutes at 360°F.
(8) Remove the layers from the pans, cool 2 hours, cut each layer in half, score the internal characteristics, and determine and record the weight and volume of the two cake layers.

Nonstandardized Method 8.5

105% SUGAR WHITE LAYER CAKE

Definition: This low-sugar cake formulation evaluates the emulsification value of a nonemulsified shortening.

Scope: Applicable to all nonemulsified cake shortenings.

Formula:

Ingredients	Grams	Mixing Stage
Test shortening	140	First
Granulated sugar	475	
Nonfat milk solids	28	
Salt	21	
Baking powder	28	
Cake flour	454	
Water	227	
Egg whites	255	Second
Water	154	

Equipment:

(1) C-100 Hobart mixer equipped with 3-quart bowl and paddle

(2) Scale, graduated in grams, with scale pan and counterbalance

(3) Specific gravity cup

(4) Stainless steel spatula with 6-inch blade

(5) Thermometer, 0 to 120°F

(6) Two 8-inch layer greased cake pans lined with paper parchment

(7) Bakery oven

(8) Cake volumeter, seed displacement type or equivalent

Procedure:

(1) Scale the shortening and all the first-stage dry ingredients into the 3-quart mixing bowl and premix for 30 seconds at first speed.

(2) Add the first-stage water and mix for 2 minutes at first speed. Stop the mixer and scrape the bowl and paddle. Mix 2.5 minutes at first speed.

(3) Add half of the second-stage liquid and mix 1 minute at first speed. Stop the mixer and scrape the bowl and paddle.

(4) Add the remaining liquid and mix 1 minute at first speed. Stop the mixer and scrape the bowl and paddle.

(5) Continue mixing for 3.5 minutes at first speed. Stop the mixer, determine and record the batter specific gravity and temperature.

(6) Scale 400 grams of batter into each of the two 8-inch layer pans and bake approximately 22 minutes at 360°F.

(7) Remove the layers from the pans, cool 2 hours, cut each layer in half, score the internal characteristics, and determine and record the weight and volume of the two cake layers.

Nonstandardized Method 8.6

HOUSEHOLD WHITE LAYER CAKE TEST

Definition: Volume, internal, and external measurements of a white layer cake prepared with baking ingredients and equipment available in most household kitchens to determine the emulsification value of a household shortening.

Scope: Applicable for the evaluation of all household shortenings.

Formula:

Ingredients	Grams	Mixing Stage
Test shortening	100	First
Granulated sugar	260	
Baking powder	12	
Salt	6	
Cake flour	200	
Liquid milk	160	
Egg whites	132	Second
Liquid milk	96	

Equipment:

(1) Sunbeam household type mixer equipped with beaters and a large bowl, or equivalent
(2) Rubber spatula
(3) Thermometer, 0 to 120°F
(4) Specific gravity cup
(5) Two 8-inch greased layer pans with paper parchments
(6) Scale, graduated in grams, with scale pan and counterbalance
(7) Household oven

Procedure:

(1) Scale the shortening and all of the first-stage dry ingredients into the household mixer bowl.
(2) Cream these ingredients together at No. 1 speed for 1 minute.
(3) Add the first-stage liquid milk and mix for 30 seconds at speed No. 1.
(4) Mix for 2 minutes at No. 2 speed, scraping the sides of the bowl frequently.
(5) Add the second-stage liquids and mix at speed No. 3 for 30 seconds.
(6) Mix at speed No. 4 for 2 minutes, scraping the sides of the bowl frequently.
(7) Stop mixer; determine and record the batter specific gravity and temperature.
(8) Scale 400 grams of the batter into each of the 8-inch greased layer pans and bake for approximately 21 minutes at 360°F.
(9) Remove the layers from the pans, cool 2 hours, cut each layer in half, score the internal characteristics, and determine and record the weight and volume of the two cake layers.

Nonstandardized Method 8.7

CREME FILLING TEST

Definition: This evaluation measures the ability of a shortening and emulsifier system to aerate a creme filling to a specific volume of more than 200 cc/100 g and then maintain a stable weep-free consistency at elevated temperature abuse.

Scope: Applicable to plasticized shortenings designed to produce fillings for snack cakes and other filling requirements.

Formula:

Ingredients	Grams	Use Temperature
Test shortening	567	70 to 75°F
Powdered sugar	908	70 to 75°F
Nonfat milk solids	142	70 to 75°F
Salt	7	70 to 75°F
Water no. 1	71	70 to 75°F
Water no. 2	199	70 to 75°F
Water no. 3	185	70 to 75°F

Equipment:

(1) C-100 Hobart mixer equipped with a 10-quart bowl and paddle
(2) Scale, graduated in grams, with scale pan and counterbalance
(3) Specific gravity cup
(4) Stainless steel spatula with 6-inch blade
(5) Thermometer, 0 to 120°F
(6) 4-inch diameter Pyrex glass funnel
(7) 7.75-inch diameter parchment circle
(8) 50 cc graduated cylinder, 1 cc increments
(9) Holding rack for the glass funnel
(10) Constant temperature cabinet controlled at 100°F
(11) Two plastic bags
(12) A 4″ × 4″ square of devil's food cake

Filling Test Procedure:

(1) Scale the shortening and dry ingredients into the 10-quart mixing bowl.
(2) Add water no. 1 and mix 1 minute at first speed. Stop the mixer and scrape the bowl and paddle.

(3) Mix for 10 minutes at second speed. Stop the mixer; determine and record the specific gravity.

(4) Add water no. 2 and mix 1 minute at first speed. Stop the mixer and scrape the bowl and paddle.

(5) Mix 5 minutes at second speed. Stop the mixer; determine and record the specific gravity.

(6) Add water no. 3 and mix 1 minute at first speed. Stop the mixer and scrape the bowl and paddle.

(7) Mix 5 minutes at second speed. Stop the mixer; determine and record the specific gravity and temperature.

Weep Test Procedure:

(1) Fold the 7.75-inch diameter parchment circle into fourths and cut approximately 0.5 inches off of the point.

(2) Line the glass funnel with the folded and cut parchment.

(3) Weigh 200 grams of the creme filling into the funnel. Carefully smooth the filling above the top of the funnel into a rounded mound.

(4) Place the funnel in the holding rack in the 100°F constant temperature cabinet with the 50 cc graduated cylinder under the funnel stem.

(5) Determine and record the amount of weep in the graduated cylinder after 24 hours at 100°F.

Filling Sandwich Test Procedure:

(1) Slice the 4″ × 4″ square of devil's food cake into two layers.

(2) Scale and spread 50 grams of filling onto one layer and place the mate for the layer on top of the filling.

(3) Cut the cake square into two pieces, each measuring 2″ × 4″ and place each in a separate plastic bag and seal.

(4) Place one of the cake sandwiches in the 100°F constant temperature cabinet and hold the other at room temperature (75°F ± 2°F).

(5) After 6 days, evaluate the cake sandwiches for filling disappearance or soakage into the cake.

(6) Report the filling disappearance findings as none, trace, slight, definite, completely disappeared, or with other more descriptive terms.

Nonstandardized Method 8.8

WHITE CAKE MIX TEST

Definition: A basic white cake mix formulation and make-up procedure for evaluation of the aeration, lubrication, moisture barrier, and structural qualities provided by a cake mix shortening.

Scope: Applicable to all prepared mix cake shortenings.

Formulas:

Dry Cake Mix			Cake Preparation	
Ingredients	Grams		Ingredients	Grams
Cake flour	392		Dry mix	567
Powdered sugar	442		Water	240
Nonfat milk solids	35		Egg whites	66
Salt	10		Total	873
Baking soda	7			
V-90, Baking acid	7			
DCP, Baking acid	7			
Lecithin	1			
Test shortening	99			
Total	1,000			

Equipment:

(1) Sifter with 1/16-inch openings

(2) C-100 Hobart mixer equipped with a 3-quart bowl and paddle

(3) Sunbeam household mixer with beaters and large mixing bowl, or equivalent

(4) Specific gravity cup

(5) Stainless steel spatula with 6-inch blade

(6) Rubber spatula

(7) Thermometer, 0 to 120°F

(8) Two 9-inch greased layer cake pans with parchments

(9) Baking oven

(10) Scale, graduated in grams with a scale pan and counterbalance

Dry Mix Make-up Procedure:

(1) Sift the flour and sugar together two times.

(2) Sift all the minor dry ingredients together five times.

(3) Combine all the dry ingredients and sift together five times.

(4) Place the sifted dry ingredient blend into the 3-quart mixing bowl.

(5) Blend the lecithin into the shortening with the metal spatula.

(6) Add the shortening/lecithin blend to the dry ingredient blend in small pieces while mixing at first speed over a 1-minute period. Scrape the bowl and paddle and mix for 5 minutes at first speed.

(7) Force the mixed cake mix through the sifter with the rubber spatula two times.

(8) Remix in the Hobart 3-quart bowl for 5 minutes at first speed.

(9) Record observations regarding the appearance of the mix and the manner in which the shortening mixed into the dry blend.

Cake Baking Procedure:

(1) Scale 567 grams of the cake mix into the household mixing bowl.

(2) Add the egg whites and water and blend 1 minute at speed no. 1.

(3) Mix 4 minutes at speed no. 5 while continuously scraping the bowl sides with the rubber spatula.

(4) Determine and record the batter specific gravity and temperature.

(5) Scale 425 grams of batter into each of the greased 9-inch cake pans lined with parchment.

(6) Bake at 360°F until done, approximately 25 minutes.

(7) Remove the cakes from the pans and allow to cool for 2 hours.

(8) Cut each layer in half and score and record the internal characteristics.

(9) Determine and record the weight and volume of both layers.

Nonstandardized Method 8.9

PUFF PASTRY TESTING

Definition: This procedure is designed to evaluate the relative performance qualities of shortenings or margarines to produce "rise" or "puff" in the very light, flaky-type pastries used for patty shells, cream horns, turnovers, and the like.

Scope: Applicable to water- or milk-churned margarine or puff pastry fats or to any fat proposed for eventual use in producing puff pastries.

Equipment:

(1) Scale, graduated in grams, with weighing pan and counterbalance

(2) A-200 Hobart mixer equipped with a 12-quart bowl and dough hook

(3) Thick plastic bowl scraper

(4) One 16″ × 24″ sheet pan for each evaluation

(5) Damp towel, capable of covering the sheet pan

(6) Refrigerator capable of maintaining a temperature range of 38 to 42°F

(7) Dusting flour

(8) Rolling pin

(9) Bench brush

(10) Two 1/16-inch diameter metal rods 24 inches long

(11) Two 3/8-inch diameter metal rods 24 inches long

(12) One metal, round, sharp-edged, 3-3/8-inch diameter pastry dough cutter

(13) One metal, round, sharp-edged, 2-3/8-inch diameter pastry dough cutter

(14) Pastry brush

(15) Twelve 4-ounce round, wide-mouth jars for each evaluation

(16) Bakery oven

(17) Tape measure

Formula:

Ingredients	Grams	Stage
Bread flour	1248	Blending
Cake flour	227	
Roll-in test fat	227	
Salt	14	
Pure cream of tartar	14	
Cold water (40 to 55°F)	794	Dough Mixing
Whole eggs	227	
Roll-in test fat	1248	Roll-in

Procedure:

(1) Dough mixing
- Weigh the dry ingredients and dough test fat into the 12-quart mixing bowl.
- Blend 2 to 3 minutes at first speed until the fat is in pea-sized lumps.
- Add the water and whole eggs and mix 45 seconds at second speed. Stop the mixer and remove the bowl from the mixer.

(2) Molding and resting
- Remove the dough from the mixing bowl to a lightly floured sheet pan.
- Mold the dough into an oblong or rectangular shape about 4 inches thick.
- Cover the dough with a damp towel.
- Refrigerate the dough for 30 minutes.
- Remove the dough from the refrigerator.

(3) Sheeting and fat spotting
- Carefully roll out, using even pressure with a rolling pin to form a rectangular sheet about 1/2 inch thick. The sheet should be formed by rolling in two directions at right angles to form a rectangle of dough three times as long as it is wide.
- Spot the test roll-in fat in small blobs or chunks evenly over two-thirds of the dough surface, leaving the right third dough surface bare of fat.
- After all of the fat has been so distributed, fold the bare portion of the dough to the left so that it covers half of the fat spotted dough. Then fold the remaining fat spotted dough surface to the right over the first folded dough. This forms five alternate layers of dough and fat.

(4) First roll-in
- Carefully roll out the laminated dough again, using even pressure with the rolling pin to form a rectangular sheet about 1/2 inch thick and about three times as long as it is wide.
- Refold the dough exactly as described above—right third over the middle third and then the left third over the first fold.
- Place the folded laminated dough, now with 15 layers, back onto the lightly floured sheet pan, cover with the damp cloth again to prevent crusting of the dough and refrigerate for 30 minutes.

(5) Additional roll-ins
  - Take the dough from the refrigerator and repeat the procedure for the first roll-in.
  - Repeat the roll-in procedure every 30 minutes until a total of five roll-ins have been completed to form 1215 alternate layers of dough and fat.
  - After the fifth roll-in, refrigerate overnight on the floured pan with the damp cloth.
(6) Patty shell make-up
  - Remove the chilled dough from the refrigerator, cut off one-third of the dough for make-up and return the remainder to the refrigerator.
  - Cut off one-third of the make-up dough portion and roll out to a thickness of 1/16 inch using the metal bars as guides to insure the correct thickness.
  - Cut 12 round pieces with the 3-3/8-inch cutter. Place the dough round pieces on a sheet pan lined with parchment; dock with a kitchen fork at least six times and brush lightly with water.
  - Roll remaining portion of the make-up dough to a thickness of 3/8 inch using the appropriate metal guides. Cut 12 3-3/8-inch diameter rounds, and then with the 2-3/8-inch smaller cutter remove the center of each dough circle to form a ring.
  - Carefully place a ring on each of the 1/16-inch dough pieces on the sheet pan. Be sure that no excess dusting flour adheres to either side of the rings.
  - Place a 4-ounce straight-sided, wide mouth jar inside each shell just before baking. Place another sheet pan on top of the jars to keep the jars upright during baking.
  - Bake at 380°F for approximately 35 minutes until done.
(7) Evaluation procedure
  - Determine the extent of "puff" or "rise" by measuring the height of five shells stacked carefully one upon the other. The minimum satisfactory performance is a height of 12 inches for five shells.
  - Observe the amount of melted fat on the parchment on the sheet pan surface.
  - Weigh the five shells measured for height.
  - Compare the test roll-in fat results to test results obtained with a commercially accepted product included in the test series when possible. Otherwise, compare the test product results to previously obtained results of satisfactory product tested previously.

Nonstandardized Method 8.10

DEEP FAT FRYING TEST

Definition: This method measures the ability of a frying shortening or oil to resist hydrolysis, oxidation, and polymerization with continuous heating and scheduled food frying, which causes foaming, off-flavors, smoking, darkening, gum development, and other thermal decomposition properties.

Scope: Applicable to all frying shortenings and oils.

Equipment:

  (1) Smallest available commercial fryer and basket
  (2) Thermometer, 0 to 500°F
  (3) Scale, graduated in grams, with scale pan and counterbalance
  (4) Fresh potatoes
  (5) Potato peeler and French cut potato slicer
  (6) Cloud point beaker, etched at 50 grams fat level
  (7) Cork borer, 3/8-inch diameter
  (8) Sharp knife
  (9) Metal ruler with 1/8-inch divisions
 (10) Ring stand with wire gauze
 (11) Exhaust hood
 (12) Refrigerator

Frying Test Procedure:

  (1) Boil out fryers and baskets with a caustic cleaner, rinse with clear water, and neutralize with a vinegar solution and dry with paper toweling.

  (2) Inspect the fryer kettle, heating element, and thermal coupling for wear spots, which would allow brass or copper to contact the frying media. Do not use any equipment that could influence the results of the test.

  (3) Scale the required quantity of the test frying shortening or oil into the fryer.

  (4) Start the actual frying test at the beginning of a day. Preheat and adjust the frying shortening or oil temperature to 360 ± 10°F. The frying media should be under continuous heat until the completion of the test unless the test is interrupted for the weekend. The first frying should be performed as soon as the fryers have been adjusted to frying temperature.

  (5) Peel and french cut only enough fresh potatoes for a single day's frying, 1.5 pounds per fryer. Soak in cold water to remove the surface starch. Before each frying, remove the excess water by draining and blotting between cloths before weighing. Fry 8 ounces of the french

cut potatoes per fryer, three times each 3 hours during one 9-hour period each 24 hours for 7 minutes.

(6) At the completion of each 24-hour period, measure the foam height and determine Lovibond red color, free fatty acid, and iodine value for each frying fat in test.

(7) The french fries should be observed and flavored after each frying to determine and record any off-flavors or appearance problems contributed by the frying fat.

(8) Observe and record the frying fat appearance during each frying. The appearance should be noted after 1.5 minutes frying to allow the moisture bubbles from the french fries to dissipate. The frying fat color, smoking or lack of it, odor, gumming on the kettle, foaming tendencies, and any other changes should all be recorded.

(9) The frying test is terminated when persistent foam is observed and confirmed by a foam test measurement of at least 1 inch. The foaming tendencies descriptions and rating scale are

Observed Foaming Tendencies

Foam Description	Observed Foaming Tendencies	Foam Test (inches)
Fried OK	Normal frying	None
Trace	First indication of foam	1/8
Slight	Foam severity increases	1/4
Definite	Definite pockets of foam	1/2
Persistent	Foam persists until food removed	1

Foam Test Procedure:

(1) Pour 50 grams of test frying fat into the cloud point beaker.

(2) Heat to 392°F with a bunsen burner and then remove heat.

(3) Immediately drop six precut potato pieces, measuring 3/8-inch diameter × 1/2-inch length, into the heated fat.

(4) Record the foam height after 1.5 minutes to allow the moisture bubbles to dissipate. The foam height is determined by measuring the actual height of the foam above the surface of the fat less 1/4 inch to allow for the displacement of the potato pieces.

Frying Stability: Frying life determinations are calculated by summing the number of elapsed hours required to reach the endpoint criteria for each stability measurement and then dividing the total by the number of quality factors. For example:

	Frying hours required to reach				Divided		*Overall*
Persistent foam		1.0%		6.0 Red	by		*Frying*
	Plus	Free Fatty Acid	Plus	on 1 inch Lovibond	three attributes	Equals	*Stability*

Nonstandardized Method 8.11

ICE CREAM BAR COATING TEST

Definition: This procedure measures the performance of ice cream bar coating fats.

Scope: Applicable for all fats designed for use in ice cream bar coatings.

Formula:

Ingredients	Grams	Percent
Test enrobing fat	342	57.0
10× Powdered sugar	120	20.0
Nonfat milk solids	28	4.7
Chocolate liquor	108	18.0
Lecithin	2	0.3

Equipment:

(1) Scale, graduated in grams, with weighing pan and counterbalance

(2) Sifter

(3) Experimental confectionery fat coating machine or household kitchen mixer, beaters, small mixing bowl, and heating device capable of holding temperatures of 125° and 105°F

(4) Stopwatch

(5) Thermometer, 0 to 200°F

(6) Rectangular plastic piece, approximately 0.5 mm thick

(7) Freezer, capable of −10°F

(8) Uncoated ice cream bars

Procedure:

(1) Sift and weigh all of the dry ingredients into the mixing bowl.

(2) Add the lecithin to 125 ± 5°F melted fat and stir to incorporate.

(3) Add approximately one-third of the melted fat to the dry ingredients and mix at a slow speed to make a paste.

(4) Add the remaining melted fat and mix thoroughly.

(5) Attain and hold the coating at 105°F ± 1°F for 30 minutes.

(6) Quickly dip the frozen ice bar into the liquid coating, only allowing it to remain in the coating for a maximum of 3 seconds.

(7) Start the stopwatch as soon as the bar is removed from the liquid coating.

(8) Hold the bar at a 45 degree angle and allow to drain.

(9) As soon as there is no further coating dripping from the bar and the coating has flashed from a high, wet gloss to a dull, dry sheen, record time, to the nearest second, as the *drying time.*

(10) Immediately start scratching the coated surface with a corner of the rectangular plastic piece, using about 2-inch strokes and moderate pressure. When the plastic stops digging through the coating and starts making fine shavings, record the total elapsed time as the *hardening time.*

(11) Bite into the coated bar at regular intervals. When the coating audibly snaps, record the total elapsed time as the *brittleness time.*

(12) Make duplicate determinations of drying, hardening, and brittleness times.

Nonstandardized Method 8.12

SANDWICH COOKIE FILLER SHORTENING EVALUATION

Definition: A uniform test procedure designed to evaluate cookie filler plasticized shortening performance for eating quality, consistency, aeration, potential machinability, and aged stability.

Scope: Applicable to all plasticized sandwich cookie or wafer filler shortenings.

Test Formula:

Ingredients	Grams	Use Temperature
Shortening	375	70 to 78°F
Corn sugar	284	70 to 78°F
6× Powdered sugar	476	70 to 78°F

Equipment:

 (1) C-100 Hobart mixer equipped with 3-quart bowl and paddle

 (2) Scale, graduated in grams

 (3) Scale pan with counterbalance

 (4) Sifter

 (5) Thermometer, 0–120°F

 (6) Specific gravity cup, 200-ml capacity, calibrated

 (7) Stainless steel spatula with 6-inch blade

 (8) Two 8-ounce jars with metal lids

 (9) Forty-eight plain sandwich cookie halves

(10) Two plastic bags with ties

(11) 100°F constant temperature cabinet

Procedure:

 (1) Sift corn sugar and powdered sugar together.

 (2) Scale shortening into 3-quart bowl and mix ½ minute at first speed. Stop mixer; scrape bowl and paddle.

 (3) Add half of presifted sugars to the 3-quart bowl containing the shortening. Mix ½ minute at first speed. Stop mixer, add remaining presifted sugars, and mix ½ minute at first speed. Stop mixer; scrape bowl and paddle.

 (4) Mix 5 minutes at second speed, stop mixer, and record specific gravity and temperature. Scrape bowl and paddle.

(5) Mix 5 minutes at second speed. Stop mixer, record specific gravity and temperature, and evaluate filling appearance and consistency. *Note:* Finished filler temperature should be 90 ± 3°F.

(6) Fill two 8-ounce jars with 100 grams of filler each; store one jar at room temperature (75 ± 2°F) and the other at 100°F.

(7) Prepare 24 sandwich cookies with 4 grams of filler for each cookies.

(8) Place 12 cookies each in two plastic bags with ties; store one bag at room temperature (75 ± 2°F) and the other at 100°F.

(9) Evaluate fillings and cookies three times each week for consistency, flavor, and odor.

10. REFERENCES

1. *The Official Methods and Recommended Practices of the American Oil Chemists' Society,* 1994. 4th ed. Champaign, IL: American Oil Chemists' Society.

2. Sleeter, R. T. 1983. "Instrumental Analytical Methods for Edible Oil Processing: Present and Future," *J. Am. Oil. Chem. Soc.,* 60(2):343–349.

3. Mertens, W. G. 1973. "Fat Melting Point Determinations: A Review," *J. Am. Oil Chem. Soc.,* 50(4):115–118.

4. Walker, R. C. and W. A. Bosin. 1971. "Comparison of SFI, DSC and NMR Methods for determining Solid-Liquid Ratios in Fats," *J. Am. Oil Chem. Soc.,* 48(2):50–53.

5. Erickson, D. R. 1967. "Finished-Product Testing," *J. Am. Oil Chem. Soc.,* 44(11):536A–538A.

6. Mehlenbacher, V. C. 1960. "Newer Analytical Methods for the Fat and Oil Industry," *J. Am. Oil Chem. Soc.,* 37(11):615.

7. Miller, W. J. et al. 1969. "The Measurement of Fatty Solids by Differential Scanning Colorimetry," *J. Am. Oil Chem. Soc.,* 46(7):341–343.

8. Bentz, A. P. and B. G. Breidenbach. 1969. "Evaluation of the Differential Scanning Calorimetric Method for Fat Solids," *J. Am. Oil Chem. Soc.,* 46(2):60–63.

9. O'Brien, R. D. 1987. "Formulation-Single Feedstock Situation," in *Hydrogenation: Proceedings of an AOCS Colloquium,* R. Hastert, ed. Champaign, IL: American Oil Chemists' Society, p. 156.

10. Bailey, A. E. 1950. "Consistency in Plastic Fats," in *Melting and Solidification of Fats.* New York, NY: Interscience Publishers, Inc., pp. 300–308.

11. Weiss, T. J. 1983. "Chemical and Physical Properties of Fats and Oils," in *Food Oils and Their Uses,* Second Edition. Westport, CT: AVI Publishing Company, Inc., pp. 19–21.

12. Rilsom, T. and L. Hoffmeyer. 1978. "High Performance Liquid Chromatography Analysis of Emulsifiers: 1. Quantitative Determinations of Mono- and Diacylglycerols of Saturated Fatty Acids," *J. Am. Oil Chem. Soc.,* 55(9):649–652.

13. Lauridsen, J. B. 1976. "Food Emulsifiers: Surface Activity, Edibility, Manufacture, Composition, and Application," *J. Am. Oil Chem. Soc.,* 53(6):400–407.

14. Sleeter, R. T. 1981. "Effects of Processing on Quality of Soybean Oil," *J. Am. Oil Chem. Soc.,* 58(3):242.

15. Evans, C. D. 1955. "Flavor Evaluation of Fats and Oils," *J. Am. Oil Chem. Soc.*, 32(11):596.

16. Kramer, A. and B. A. Twigg. 1970. "Flavor," in *Quality Control for the Food Industry*. Westport, CT: The AVI Publishing Co., Inc., pp. 108–115.

17. Williams, J. L. and T. H. Applewhite. 1977. "Correlation of the Flavor Scores of Vegetable Oils with Volatile Profile Data," *J. Am. Oil Chem. Soc.*, 54(10):461.

18. Min, D. B. 1983. "Analysis of Flavor Qualities of Vegetable Oils by Gas Chromatography," *J. Am. Oil Chem. Soc.*, 60(3):544–545.

19. Lezerovich, A. 1985. "Determination of Peroxide Value by Conventional Difference and Difference-Derivative Spectrophotometry," *J. Am. Oil Chem. Soc.*, 62(10):1495.

20. Sessa, D. J. and Rakis, J. J. 1977. "Lipid-Derived Flavors of Legume Protein Products," *J. Am. Oil Chem. Soc.*, 54(10):1977.

21. Mehlenbacher, V. C. 1942. "A Study of Methods of Accelerating the Swift Stability Test," *J. Am. Oil Chem. Soc.*, 19(8):137–139.

22. Hassel, R. L. 1976. "Thermal Analysis: An Alternative Method of Measuring Oil Stability," *J. Am. Oil Chem. Soc.*, 53(5):179–181.

23. Hill, S. E. 1994. "Comparisons: Measuring Oxidative Stability," *INFORM*, 5(1):104–109.

24. de Man, J. M. and L. de Man. 1984. "Automated AOM Test for Fat Stability," *J. Am. Oil Chem. Soc.*, 61(3):534–536.

25. Dugan, L. 1955. "Stability and Rancidity," *J. Am. Oil Chem. Soc.*, 32(11):605–608.

26. Chamberlin, G. J. 1968. "Comments on the Use of the Lovibond System," *J. Am. Oil Chem. Soc.*, 45(10):711.

27. Belbin, A. A. 1993. "Color in Oils," *INFORM*, 4(6):648–654.

28. *INFORM* Staff. 1993. "Color Methodology," *INFORM*, 4(6):706–707.

29. James, E. M. 1955. "The Determination of Refining Loss," *J. Am. Oil Chem. Soc.*, 32(11):581–587.

30. Linteris, L. and E. Handschumaker. 1950. "Investigation of Methods for Determining the Refining Efficiency of Crude Oils," *J. Am. Oil Chem. Soc.*, 27(7):260–263.

31. Pyler, E. J. 1952. "Cake Ingredients," in *Baking Science and Technology*. Chicago, IL: Siebel Publishing Company, p. 568.

Chapter 4

Fats and Oils Formulation

1. INTRODUCTION

Fats and oils play a significant role in the formulation and performance of a variety of prepared foods. Most of America's favorite foods could not be prepared without fats and oils. This would include pan and deep fat fried foods, baked products, spoonable and pourable salad dressings, nondairy products, whipped toppings, confectionery products, pastries, peanut butter, spreads, and so on, all of which possess desirable properties attributable to the fats and oils ingredients in the formulation. Fats and oils affect the structure, stability, flavor, storage quality, eating characteristics, and the eye appeal of the foods prepared. To accomplish the desired performance, the developer must recognize that most applications require a fat or oil product with different physical and organoleptic properties.

The development of a fat and oil product for a food application is dependent upon many interlaced factors. These factors may differ from customer to customer, depending upon the equipment used, processing limitations, product preference, customer base, and many other contributors. Tailored fats and oils products are now being designed to satisfy these individual specific requirements, as well as products with a broad general appeal. The design criteria for a general purpose product must be of a broader nature than that for a specific product and process. The important attributes of a formulated fat or oil in a food product vary considerably. In some food items the flavor contribution of the shortening is of minor importance; however, it does contribute a beneficial effect to the eating quality of the finished product. This fact has been evident with the recent introductions of fat-free products. Most of these products lack the eating

251

characteristics contributed by fats and oils. In many products, such as cakes, pie crusts, icings, cookies, and certain pastries, processed fats and oils are major contributors to the characteristic structure and eating character, as well as providing other significant effects upon the finished product's quality.

A thorough understanding of the functions and properties of the various fats and oils products is the basic key to formulation for the desired performance attributes. Fats and oils are very versatile raw materials for which processing methods have been developed to make them even more useful to the food industry, while analytical chemists have devised methods to qualify the product produced. Satisfactory fats and oils product performance is dependent upon several important elements that determine suitability. The fats and oils formulator must identify the important attributes and effectively utilize the different functional properties of the available processed fats and oils to satisfy the prepared foods requirements. Successful production of these products relies on the manipulation of the fat blend to produce suitable physical properties and prevent undesirable changes during and after processing. Important fats and oils performance characteristics that must be considered for any product formulation are

- flavor—Generally, the flavor of a processed fat and oil product should be completely bland to enhance the food product's flavor rather than contribute a flavor. In some specific cases a typical flavor like lard or that for a butter flavored product is desirable for a particular application. However, the reverted or oxidized flavors and odors of most fats and oils are objectionable.
- flavor stability—The bland or typical flavor must be stable throughout the shelf life and the use life of the prepared food product. The fat and oil product must possess the identified degree of resistance to both oxidative and lipolytic flavor degradation.
- physical characteristics—Each fat and oil has a characteristic composition and distribution of fatty acids. The physical, functional, and organoleptic properties of fats and oils are in part a function of the fatty acid composition but also are governed by the fatty acid distribution in the triglyceride comprising the raw materials. Consistency, plasticity, emulsification, creaming properties, spreadability, and other fats and oils properties and the prepared foods produced are affected by the unsaturation and saturation of the fatty acids and their position in the triglyceride, which control melting rate and range.
- crystal habit—Fats and oils are polymorphic, which means that with cooling a series of increasingly organized crystal changes occur until a

final stable crystal form is achieved. The crystal types formed define the textural and functional of most fat-based products. Many prepared foods are mixtures of ingredients held in a matrix of solid fat. The fat is the major functional ingredient, binding the other ingredients together and, at the same time, imparting texture and mouth feel to the product while affecting flavor release and dispersion in the mouth of the solid materials at a controlled rate.

- nutritional concerns—Fats and oils are recognized as important nutrients for both humans and animals because they provide a concentrated source of energy, contain essential fatty acids, and serve as carriers for fat-soluble vitamins. However, research studies have indicated a possible relationship between fats and the incidence of coronary heart disease. Diet modifications, including reductions in fat consumption, saturated fats, cholesterol, and *trans* isomers, have been proposed.
- additives—In addition to emulsifiers, a number of other chemical compounds provide a specific function for edible fats and oils products. The additive categories are antioxidants, antifoamers, metal inactivators, colorants, flavors, crystal inhibitors, preservatives, and emulsifiers.

Current chemical and physical processing techniques provide the processor with the capability of modifying one or more fats and oils properties. It is possible to produce fats and oils products that have little resemblance to the natural fats and oils that provide the ability to formulate "tailor-made" products to suit a particular product or process. In addition, the processing techniques provide the processor with a wider range of alternative raw material sources to improve commercial viability. Some of the objectives for applying the available modification or processing techniques are

(1) Production of a fat and oil product to meet certain performance characteristics not possible with natural source oil and fat products

(2) Potential utilization of a more economic feedstock to duplicate functionality of a more expensive alternative

(3) Improvement in oxidative stability through elimination of the reaction sites

(4) Palatability improvement.

(5) Modification of the crystallization behavior

(6) Providing more nutritionally acceptable products, i.e., probable reduction of saturates and *trans* acids while increasing polyunsaturates

2. SOURCE OILS AND FATS CHARACTERISTICS

Chemically, all fats and oils are triglycerides or esters of glycerol and fatty acids. Since all triglycerides have identical glycerol components, the different properties are contributed by the fatty acids. Three aspects can differentiate the fatty acid components: (1) chain length, (2) the number and position of the double bonds, and (3) the position of the fatty acids with regard to the glycerol. Fats and oils, for all practical purposes, contain fatty acids with carbon chain lengths between 4 and 24 carbon atoms with zero to three double bonds. Saturated fatty acids are chemically the least reactive and have a higher melting point than corresponding fatty acids of the same chain length with one or more double bonds. The most important saturated fatty acids (no double bonds) are butyric (C-4:0), lauric (C-12:0), myristic (C-14:0), palmitic (C-16:0), stearic (C-18:0), arachidic (C-20:0), behenic (C-22:0), and lignoceric (C-24:0). The fatty acids with the most double bonds, or the more polyunsaturated, are the most chemically reactive. The notable polyunsaturated fatty acids or essential fatty acids are linoleic (C-18:2), which is di-unsaturated, and linolenic (C-18:3) which is tri-unsaturated. *Cis, trans,* and *positional* configurations of the unsaturated fatty acids have different physical and physiological properties. *Cis* isomers are the natural configuration while the *trans* and positional isomers develop with processing or heating, but predominately with selective hydrogenation. The effects of chain length, unsaturation, and geometric isomerism on the fatty acids melting point are shown in Table 4.1. The triglyceride structure is affected by which carbon atom of the glycerol the fatty acid is linked to and whether the three fatty acids are the same or different and the position of each. All of these fatty acid and structural variations affect the chemical and physical properties of the resulting triglycerides. Therefore, one of the best characterizations of a fat and oil product is the fatty acid and triglyceride compositions.

2.1 Short-Chain Fatty Acids

The lower saturated fatty acids with only 4 to 10 carbon atoms occur in milk fats and a few vegetable oils. Cow's milk fat contains about 4% butyric fatty acid (C-4:0), which contributes to the characteristic flavor, as well as lesser amounts of C-6, C-8, C-10, and C-12 fatty acids. The laurics are oils characterized by a high level of C-12 saturated fatty acids and relatively low unsaturation. Coconut and palm kernel oils have high lauric fatty acid contents with a relatively low level of unsaturation. In tropical countries of origin, these oils are liquid at ambient temperatures, while in temperate climates the laurics are solid fats. Coconut oil typically contains 8 to 10%

TABLE 4.1. Fatty Acids Characteristics.

Fatty Acid	Carbon Atoms	Double Bonds	Melting Point Degree F	Melting Point Degree C
Butyric	4	0	17.6	−8.0
Caproic	6	0	25.9	−3.4
Caprylic	8	0	62.1	16.7
Capric	10	0	88.9	31.6
Lauric	12	0	111.6	44.2
Myristic	14	0	129.9	54.4
Myristolic	14	1 cis	65.3	18.5
Palmitic	16	0	145.2	62.9
Palmitoleic	16	1 cis	113.0	45.0
Margaric	17	0	142.3	61.3
Margarolic	17	1 cis	135.5	57.5
Stearic	18	0	157.3	69.6
Oleic	18	1 cis	60.8	16.0
	18	1 positional	91.4	33.0
Elaidic	18	1 trans	111.1	44.0
Linoleic	18	2 cis cis	19.4	−7.0
Linolenic	18	3 cis cis cis	8.6	−13.0
Arachidic	20	0	167.5	75.3
Behenic	22	0	175.8	79.9
Erucic	22	1 cis	92.3	33.5
Lignoceric	24	0	183.6	84.2

unsaturates, mostly oleic (C-18:1) fatty acid. Palm kernel oil typically contains 17% unsaturates, also mostly oleic fatty acid. As a result of the fatty acid composition, the laurics exhibit a low melting point combined with an extremely steep solids fat index curve to provide a sharp melting behavior that produces an impression of coolness to the palate. The naturally low unsaturation as oleic fatty acid resists the development of oxidation, but hydrolyzed lauric fatty acids are objectionable at very low free fatty acid levels. Coconut and palm kernel oils also contain the highest level of myristic fatty acid of the vegetable oils typically processed as food fats, approximately 18.5 and 16.2%, respectively.

2.2 Palmitic and Stearic Fatty Acids

Palmitic fatty acid is the most widely occurring saturated fatty acid present in every commercially processed food fat in the United States. Fats and oils sources with higher palmitic (C-16:0) fatty acid contents include palm oil (ca. 44%), lard (ca. 26%), tallow (ca. 24%), and cottonseed oil (ca.

21.5%). A high palmitic fatty acid content usually indicates that the fat and oil product will crystallize in the beta-prime form, which is desirable for plasticity, smooth texture, aeration, and creaming properties. However, the stabilization effect of palmitic fatty acid is also related to its triglyceride position. Lard has a predominantly asymmetric triglyceride structure with palmitic in the 2-position to produce a beta crystal habit. Lard's crystal habit can be changed to beta-prime by interesterification to randomize the fatty acids and improve the ratio of symmetric triglycerides.

Stearic fatty acid is less common than palmitic. It is present in most fats and oils, but at a significant level in only a few natural fats and oils; cocoa butter (ca. 34%) and tallow (ca. 18.6%) have the highest stearic fatty acid content naturally. In the United States, stearic fatty acid is normally the result of hydrogenation of oils high in 18 carbon unsaturates.

2.3 Long-Chain Saturated Fatty Acids

Saturated fatty acids with carbon chains longer than that of stearic fatty acid are major components of only a few uncommon vegetable oils. Arachidic (C-20:0), behenic (C-22:0), and lignoceric (C-24:0) are minor components of peanut oil for a total content of 5 to 8% of C-20 and higher fatty acids. Rapeseed oil contains erucic fatty acid (ca. 41%), which hydrogenates to behenic fatty acid.

2.4 Oleic Fatty Acid

Oleic fatty acid is the most widely distributed natural fatty acid. Olive oil is a very flavor stable oil due to the high oleic fatty acid content (ca. 80%). Peanut oil, lard, tallow, and palm oil also have high oleic fatty acid contents, typically 46.7, 43.9, and 42.5%, respectively. Genetic engineering and tissue culture have commercialized vegetable oil products with high oleic fatty acid contents, i.e., canola oil (ca. 60.9%), high-oleic sunflower oil (ca. 81.3%), and high-oleic safflower (ca. 81.5%). Liquid oils with high oleic fatty acids normally have good flavor and frying stability.

2.5 Linoleic Fatty Acid

Many of the vegetable oils available on a commercial scale are rich in linoleic fatty acid content, which is most likely responsible for the characteristic flavor reversion of these vegetable oils, which necessitates partial hydrogenation to reduce the linoleic fatty acid content. Typically, the vegetable oils with high linoleic fatty acid contents are safflower oil (ca. 77.7%), sunflower oil (ca. 67.5%), corn oil (ca. 59.6%), cottonseed oil (ca. 54.4%), and soybean oil (ca. 53.7%). Corn oil, although high in linoleic fatty acid content, is quite flavor stable due to its relatively high tocopherol content. It also reverts to a flavor that is not as unpleasant as some other vegetable oils. Cottonseed oil also has the advantage of a more pleasant oxidized flavor.

2.6 Linolenic Fatty Acids

Oils with high linolenic (C-18:3) fatty acid contents are good drying oils for industrial uses but relatively poor food oils unless processed to decrease the polyunsaturate substantially. Both canola and soybean oils contain relatively high linolenic fatty acid contents, typically 8.8 and 7.6%, respectively. A classic experiment interesterified 9% linolenic fatty acid into cottonseed oil, which typically contains less than 1.0% of the C-18:3 fatty acid. Subsequent flavor panels identified this modified product as soybean oil. This experiment indicated that linolenic fatty acid was responsible for the reverted flavor in soybean oil [2]. Subsequent hydrogenation practices to reduce the linolenic fatty acid content to less than 2.5% improved the flavor stability of soybean liquid oil products.

The triglyceride structure of canola oil has both linoleic and linolenic fatty acids primarily in the 2-position, similar to high-erucic rapeseed oil. This differs from other oils that usually have a random distribution for linoleic and linolenic fatty acids, and the somewhat lower total unsaturation indicates better residence to oxidation than oils with similar linolenic fatty acids.

2.7 Erucic Fatty Acid

Apart from the high linolenic fatty acid content comparable to soybean oil, traditional rapeseed oil contains a high erucic fatty acid content, typically 41%. Because of the possible physiopathological harmfulness of high erucic fatty acid in the human diet, new mustard seed varieties have been developed and are currently available as canola oil. Canola oil must not exceed a maximum erucic fatty acid content of 2.0% by US FDA regulations. Oils with higher erucic fatty acids are not permitted in U.S. foods by regulation. Rapeseed oil can only be utilized if it has been hydrogenated to a low iodine value to convert the erucic fatty acid (C-22:1) to behenic (C-22:0) fatty acid.

3. PALATABILITY

Palatability consists of the organoleptic and physical properties that make a food product pleasurable or, at the very least, not unpleasant to eat. One of the most important palatability parameters for users of edible fats and oils is flavor. Flavor sensation is a composite of taste, odor, and mouth feel. The elements contributing to taste are generally nonvolatile and limited to one or more of the four sensations described as sweet, acid, salty, or bitter. Volatile substances such as acids, alcohols, aldehydes, amines, esters, and keytones contribute to the odor of fats and oils. The major causes of off-flavors in food oils and fats are oxidation and hydrolysis. Oxidation results from the combining of oxygen with a fat or oil, normally from the presence of air. It is accelerated by exposure to light, heat, certain metals, and sometimes moisture. Hydrolysis is a cleavage reaction of a fatty acid by water. The end product from the hydrolysis of an oil or fat, if complete hydrolysis were to occur, would be glycerine and free fatty acids. The liberated fatty acids impart characteristic flavors that are objectionable even in small concentrations.

Mouth feel is an organoleptic characteristic that must be controlled by the physical properties of the edible fats and oils products. Mouth feel can be controlled by melting properties of the component base oils and the judicious selection of the hydrogenated, fractionated, interesterified, or esterified basestocks in the formulated blends.

3.1 Oxidative Stability

The degree of importance for a fat and oil product's oxidative stability depends upon the intended use of the product, the temperature abuse it will be exposed to, and the shelf life/use life expectancy. Fats differ considerably in the way in which their oxidation and accompanying flavor deterioration proceeds. The more highly saturated fats and oils and hydrogenated products experience relatively little change in flavor during the early phases of oxidation, but off-flavor development is sudden and definite. Unsaturated oils exhibit a gradual flavor deterioration, with a greater tendency to develop unpleasant flavors and odors. The amount of oxygen that must be absorbed to produce offensive flavors corresponds to the fatty acid composition of the oil, the presence of natural or added antioxidants, metal content, and temperature exposure.

The oxidative stability of fat and oil products is determined by the distribution, geometry, and number of double bonds. Oxidation can occur only in the fatty acid portions of the triglyceride molecule because the presence of a double bond is necessary for oxidation to occur under ordinary conditions. It is well established that the more unsaturated a fatty acid is, the less oxidative stability it naturally possesses. Table 4.2 lists the relative oxidation rates of the common 18 carbon fatty acids [3]. Linolenic (C-18:3) fatty acid, which has three double bonds, has a relative oxidation rate about

TABLE 4.2. 18 Carbon Chain Fatty Acids Relative Oxidation Rates.

Fatty Acid		Oxidation Rate
Oleic	C-18:1	1
Linoleic	C-18:2	12
Linolenic	C-18:3	25

double that of linoleic (C-18:2) fatty acid with two double bonds. Oleic (C-18:1) fatty acid, with only one double bond (monounsaturated), is the most stable of the three unsaturated fatty acids. Investigators have demonstrated that the natural or *cis* isomeric fatty acid form has less oxidative stability than the corresponding *trans* isomer. Therefore, a judicious selection of the source oil and/or hydrogenation conditions for the finished product formulation must be made to achieve the required oxidative stability.

Fats and oils oxidative reactions are directly related to the fatty acid composition and more specifically to the type and amount of unsaturation. Iodine value measurements can be used to estimate the oxidative stability of a fat or oil since the products with the highest unsaturation are most likely to experience autoxidation. However, this estimate can be slightly misleading at times, as shown in Table 4.3. The relative oxidation rates

TABLE 4.3. Fats and Oils Oxidative Stability Rating.

Rating	Fat and Oil Source	Inherent Oxidative Stability	Total Double Bonds	Calculated Iodine Value
Worst	Safflower	9.546	168.8	146.1
	Soybean	8.579	153.7	133.1
	Sunflower	8.489	156.3	135.3
	Corn	7.708	148.4	128.4
	Cottonseed	6.895	130.1	112.6
	Canola	5.349	131.3	113.3
	Peanut	4.326	112.6	97.1
	Lard	2.426	68.5	59.3
	High-oleic sunflower	1.894	99.4	85.6
	Olive	1.740	95.6	82.4
	Palm	1.724	60.8	81.8
	High-oleic safflower	1.710	96.8	83.3
	Tallow	1.267	55.6	48.4
	Palm kernel	0.430	20.1	17.2
Best	Coconut	0.360	11.0	9.6

were used to assess an oxidative stability rating for the various fat and oil raw materials. The formula for this calculation was [4]: multiply the decimal fraction of each unsaturated fatty acid present by its relative oxidation rate and then sum these to obtain the expected oxidative stability rating.

The inherent oxidative stability is different for each fat and oil raw material and must be considered when formulating fat and oil ingredients. As the oxidative reaction for fats and oils is related to the type and amount of unsaturation, oxidative stability can be improved by controlling the degree and type of unsaturation. This can be accomplished by source oil selection, blending source oils, and/or processing. Normally, hydrogenation is the process utilized to saturate the unsaturated fatty acids; however, fractionation and/or interesterification can also be used to improve flavor stability.

FLAVOR REVERSION

Flavor reversion is most prevalent with oils high in polyunsaturates. At low levels of oxidation, the degree and type of flavor change is characteristic for each source oil. For example, soybean oil develops a flavor described as beany or grassy. This flavor has been attributed to autoxidation of linolenic (C-18:3) fatty acid. Canola oil, which develops flavors similar to those from soybean oil, also contains a high level of linolenic fatty acid. Sunflower and safflower oils, with high levels of linoleic (C-18:2) fatty acid, reversion flavors, are described as "seedy." Similarly, corn and palm oils develop a distinctive flavor. The reversion flavors are observed long before other objectionable oxidized flavors develop. Additionally, the off-flavors formed during hydrogenation are all oxidative flavor reversions. These flavors have been characterized as straw- or hay-like for soybean oil products.

In the case of most well-processed oils, reversion is not a frequent problem. However, high linolenic (C-18:3) and linoleic (C-18:2) fatty acids are most likely the most important precursors of flavor reversion. Experiments interesterifying linolenic fatty acid into cottonseed oil, which contains very little, if any, of this fatty acid, produced the reverted soybean oil flavor [2]. These results led to the adoption of selective hydrogenation to convert linolenic fatty acid to linoleic fatty acid by saturating one double bond to improve the flavor stability of soybean salad oils. Specifically, the linolenic fatty acid content was lowered from nearly 9% to less than 3%. This was accomplished with selective hydrogenation to reducing the iodine value of the soybean oil to less than 110. Oxidative stability is improved as the hydrogenation level is increased to reduce the unsaturates because it takes away the reaction sites. Chart 4.1 compares the effect of selective and nonselective hydrogenation conditions upon oxidative stability as measured by AOM. Selective hydrogenation conditions increase oxidative stability at a more rapid rate because they favor the reaction sequence of linolenic >

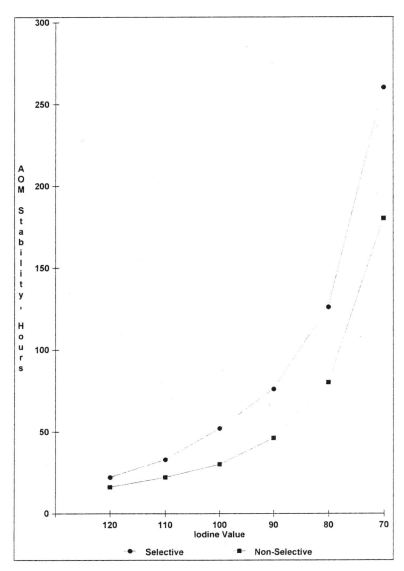

CHART 4.1. Hydro soybean oil oxidative stability.

linoleic > oleic > stearic, rather than reacting uniformly with all the double bonds present.

ANTIOXIDANTS

The oxidation rate for a fat and oil product depends primarily on the number of double bonds and their arrangement. However, oxidative stability is also influenced by the presence of natural or added antioxidants. Antioxidants are chemical compounds that provide greater oxidative stability and longer shelf life/use life for edible fats and oils by delaying the onset of oxidative rancidity. Antioxidants function by inhibiting or interrupting the free radical mechanism of glyceride autoxidation. They function as a free radical acceptor, thereby terminating oxidation at the initial step.

All food products in U.S. interstate commerce are subject to the regulations under the Food, Drug and Cosmetic Act; the Meat Inspection Act; and the Poultry Inspection Act. These regulations establish limitations on the use of antioxidants and other food additives in food products. Antioxidants permitted under these regulations include [5–12]

- tocopherols—Tocopherols are natural antioxidants contained in most vegetable oils. In most applications tocopherol addition levels of 0.02 to 0.06% are sufficient to provide good antioxidant properties. Tocopherol additions have the most application for stabilizing edible fats of animal origin because of the absence of natural antioxidants in these products. Vegetable oils are the source of the tocopherols, and enough of the natural antioxidant survives processing to provide the optimum stability available from tocopherols.
- propyl gallate—Propyl gallate is an effective antioxidant for shelf life or use life improvement of vegetable oils at usage levels of 0.01 to 0.02% or the applicable regulatory agency permitted levels; however, its use is hampered by solubility problems, discoloration, and poor heat stability. Edible oils stabilized with propyl gallate can darken while stored in black iron vessels, packaged in metal containers, or placed in contact with metal processing equipment due to iron-gallate complexing. Additionally, propyl gallate may be inactivated in alkaline systems, particularly at elevated temperatures. These deficiencies have limited the use of this antioxidant in U.S. edible fat and oil products since an alternative more effective antioxidant, tertiary butylhydroquinane (TBHQ), is permitted.
- BHA—Butylated hydroxyanisole (BHA) has a very good carry-through effect; i.e., it can substantially withstand food processing temperatures like those experienced in baking and frying. However, BHA has a strong phenolic odor that is particularly noticeable with the initial heat-

ing of frying fat and oil products described as a chemical odor. Another concern is the development of a pink color with exposure to high concentrations of alkaline metal ions, such as sodium or potassium.

- BHT—Butylated hydroxytoluene (BHT) has a molecular structure and performance similar to BHA. Both BHA and BHT are extremely soluble in edible fats and oils and have practically no water solubility. BHT can experience some darkening in the presence of iron, but the degree is not serious.

- TBHQ—Tertiary butylhydroquinone is the most recent addition to the approved antioxidant list in the United States and has not yet gained approval in many other countries. TBHQ has been found to be the most effective antioxidant for the unsaturated vegetable oils along with several other advantages: (1) no discoloration when used in the presence of iron; (2) no discernible odor or flavor to fats and oils; (3) good solubility in fats and oils; (4) effective in poultry and animal fats, as well as the vegetable oils; (5) some carry-through protection in baked and fried foods; and (6) a stabilizing effect upon tocopherols. One concern with TBHQ utilization is that a pink color can develop with an alkaline pH, certain proteins, or with sodium salts.

Much of the success of antioxidants depends upon their being in chemical contact with the product they are protecting. Therefore, antioxidant formulations containing various combinations of different antioxidant and chelating agents are generally used, rather than individual antioxidant compounds, in most food applications. Not only does the use of such formulations provide a convenience, in that it is easier to handle the diluted antioxidants, but it also permits the formulator to take advantage of the synergistic properties of the different antioxidant compounds. For example, BHA and BHT used in combination provide a greater antioxidant effect than when either is used alone. Propylene glycol and vegetable oils usually serve as a solvent for the antioxidant mixtures with lecithin, citrate, monoglyceride citrate, or mono- and diglycerides included in the formulations as emulsifiers.

Synergism is a characteristic common to many antioxidant mixtures. Synergistic systems take advantage of the greater potency produced by the mixture without increasing the total antioxidant content. A synergist, such as citric acid, has two important functions in antioxidant formulations: (1) it increases the antioxidant effectiveness of the combination, and (2) it ties up or sequesters the trace metals, which are fat oxidizing catalysts, by forming complex, stable compounds (chelates). Other compounds that function as synergists and chelating agents include isopropyl citrate, stearyl citrate, orthophosphoric acid, sodium monohydrogen phosphate, pyrophosphoric acid and its salts, metaphosphoric acid and its salts, calcium disodium EDTA, and disodium EDTA.

3.2 Hydrolysis

Edible fats and oils hydrolysis results in the formation of free fatty acids, di- and monoglycerides, and glycerol, provided the moisture content of the oil exceeds a certain level. Fatty acids have a distinct flavor and odor when their chain length is shorter than 14 carbons. For this reason, vegetable oils containing mostly C-16 and C-18 fatty acids, which are not too unsaturated, do not become unpalatable when only slightly hydrolyzed, i.e., 1.0 to 3.0% free fatty acid. However, with palm kernel and coconut oils, the same free fatty acid level gives a very distinct off-flavor since these oils contain high levels of C-6 to C-12 fatty acids. Hydrolysis may also be caused by enzymes. This occurs normally with oils produced from fruitcoats with a high moisture content, like palm or olive oils. Since free fatty acids and the accompanying substances are largely removed by processing, the concern for hydrolytic cleavage lies with the prepared food process or composition, i.e., fried foods or high-moisture products.

3.3 Mouth Feel

Mouth feel is dependent upon three factors: (1) temperature, (2) taste sensation, and (3) texture. The texture of a food fat is influenced by the liquidity of the product. It imparts tenderness, and richness and improves the eating quality of the prepared food. Mouth feel can apply to both the ability of liquid oils to form an oily film, which is viscosity related, and how well a solid fat melts in the mouth to give a pleasant cooling effect instead of a pasty, waxy feeling that can mask desirable flavors.

The melting point of common triglyceride mixtures can be separated into four liquidity zones to indicate the physical state at refrigerator, room, body, and preparation temperatures as indicated in Table 4.4. Triglycerides in the refrigerator temperature zone will remain liquid under cool conditions. Room temperature zone triglycerides will remain liquid only if consumed at ambient temperatures or higher. The third zone of triglycerides melt near body temperature to give a cooling effect in the mouth. The high-melting triglycerides help maintain plasticity until baking or cooking temperatures are reached. It should be noted that fats and oils are not composed of only one of these triglycerides or even a liquidity zone and that the ratio of each particular triglyceride will determine the melting behavior. In most cases, a liquid oil fraction suspends a solid fat fraction in a seemingly solid product while an apparently liquid oil probably contains dis-

TABLE 4.4. Triglyceride Liquidity and Functionality.

Liquidity Zone	Triglyceride Fatty Acid Pattern	Melting Point		Double Bonds	Functionality
		Degree F	Degree C		
1 Refrigerator temperature	linoleic-linoleic-linoleic	8	−13.3	6	Nutrition
	oleic-linoleic-linoleic	20	−6.7	6	Clarity
	palmitic-linoleic-linoleic	22	−5.6	4	Lubricity
	palmitic-linoleic-oleic	27	−2.8	3	
	oleic-oleic-linoleic	30	−1.1	4	
	stearic-linoleic-linoleic	34	−1.1	4	
2 Room temperature	oleic-oleic-oleic	42	5.6	3	Lubricity
	stearic-oleic-linoleic	43	6.1	3	Clarity
	oleic-oleic-palmitic	60	15.6	2	
	stearic-oleic-oleic	73	22.8	2	
3 Body temperature	palmitic-palmitic-linoleic	81	27.2	2	Structure
	stearic-palmitic-linoleic	86	30.0	2	Aeration
	stearic-stearic-linoleic	91	32.8	2	Moisture barrier
	palmitic-palmitic-oleic	95	35.0	1	Lubricity
	stearic-palmitic-oleic	100	37.8	1	
	stearic-stearic-oleic	107	41.7	1	
4 Preparation temperature	palmitic-palmitic-palmitic	133	56.1	0	Structure
	stearic-palmitic-palmitic	140	60.0	0	Lubricity
	Stearic-stearic-palmitic	142	61.1	0	Moisture barrier
	stearic-stearic-stearic	149	65.0	0	

solved solid fractions. Therefore, it is the predominance of triglycerides from one of the zones that determines liquidity characteristics [13,14].

4. PHYSICAL CHARACTERISTICS

The characteristics of the fats and oils selected for a particular application are of primary importance in the design of a product for a specific or general purpose or use. Formulation of margarines, shortenings, and other fat-based products must be based primarily on an understanding of the relationships between specific physical measurements and the composition of the oil blends and their components along with an appreciation of processing effects. Desirable solids-to-liquid ratios are achieved by blending and processing. The crystalline structure of fats is important in formulation of shortening, margarine, and other fat products because each crystal form has its own physical property for plasticization, hardness, softness, texture, solubility, mouth feel, aeration, and other properties, depending upon the food into which this ingredient is incorporated. Oils are chosen for their peculiar crystal habit resulting from nature or processing conditions. Each fat or oil component has an inherent crystallization tendency that can be modified with various processes to help produce the desired properties. Hydrogenation has been the primary process used to change the physical properties, but interesterification and fractionation can also be used to modify the melting rate and range properties. Fats and oils emulsification properties are accomplished with adjustments of the fat structure and/or addition of surface-active agents. The typical food emulsifiers supplement and improve the functionality of a properly developed fat and oil product to act as lubricants, build structure, aerate, improve eating characteristics, extend shelf life, act as a crystal modifier, provide antisticking properties, act as a dispersant, improve moisture retention, stabilize emulsions, and other beneficial functions. Obviously, no single source oil or fat, process variable, or additive can perform all of these different functions for every application. In the development of the optimum fat and oil system for a particular application, the formulator must consider the usage application, the preparation method, effects of the other ingredients, emulsion characteristics, economics, and any other applicable criteria for the prepared food product.

4.1 Fat Crystal Habit

By ordinary visual observation shortenings, margarines, and other solidified fat products appear to be homogenous solids. However, microscopic examination reveals them to be mixtures of liquid oil and small separate, discrete crystals. The crystal types formed define the textural and functional properties of most fat-based products. Fats and oils are polymorphic and transform systematically through a series of successive crystalline forms without a change in chemical structure. The order of transition has been established with heating and cooling behavior studies as

$$Alpha > Beta\text{-}Prime > Mixed > Beta$$

The rate of transformation is dependent upon the purity of the triglyceride. The more homogeneous fats, which consist of relatively few closely related triglycerides, transform rapidly to the stable beta form. The more heterogeneous fats, which contain a diverse assortment of different triglycerides, transforms relatively slowly. Some fats consisting of a completely randomized assortment of triglycerides can exist almost indefinitely in the beta-prime form and transform further only under unusual conditions.

Each crystal form has its own physical properties, i.e., plasticity, hardness, softness, texture, solubility, mouth feel, aeration, and other properties important to the food product into which the fat is formulated. The differences among the four crystal forms are [15,16]

- Alpha crystal forms are the lowest melting and the most loosely packed arrangement of the molecules. The molecules are as far apart as possible and still remain in the solid state. Crystals in the alpha form are fragile, transparent rosettes about 5 microns in size, which transform readily to higher melting modifications and therefore are seldom encountered in solidified fats and oils products.
- Beta-Prime crystal forms are tiny, delicate, needle-like shapes that seldom grow to more than 1 micron in length and can pack together into dense fine-grained, rigid structures. The beta-prime crystal form tends to structure as a fine three-dimensional network capable of immobilizing a large amount of liquid oil. Beta-prime is the crystal form desired for most solidified products because it promotes plasticity.
- Beta crystal forms are the highest melting, self-occluding, course, large, stable platelets with the closest possible arrangement of the molecules. Beta crystals average 25 to 50 microns in length and can grow to more than 100 microns during extended aging periods. Clumps of beta crystals can be a millimeter or more in diameter and are usually responsible for the appearance of visibly grainy fats and oils products, which can lead to separation of the oil portion.
- In intermediate or mixed crystal forms, depending upon a variety of factors, a fat may exist in one crystalline form or it may be a mixture of several different crystal modifications. The mixed forms are usually

somewhat course crystals that tend to grow to about 3 to 5 microns in length and that aggregate in sizable clumps that gradually grow larger with the passage of time. Flaked fats are usually in the mixed crystal form, especially those containing soybean oil.

The rate at which fats and oils transform from one crystal form to another is governed by the ease with which their component molecules can pack together in the crystal lattice. Similar molecules can pack closely, and their crystals transform rapidly to the large, stable, high-melting form. Molecules that differ in structure cannot pack together closely, which impedes or prevents crystal transformation. Thus, fats that consist of a heterogeneous assortment of triglycerides tend to remain indefinitely in the smaller, loosely packed, lower melting crystal form; source oil examples are partially hydrogenated cottonseed oil and rearranged lard. The opposite is true of the more homogeneous fats, which contain only a few closely related fatty acids types. The rate of transformation of these materials to the larger, high-melting crystals is usually quite rapid [17]. Hydrogenated soybean and canola oils are examples of this type fat, with triglycerides consisting predominately of 18 carbon fatty acids.

Table 4.5 identifies the crystal habit of the most commonly used edible fats and oils [18]. The crystal habit of each common fat or oil is determined by one or more of four characteristics: (1) palmitic fatty acid content, (2) distribution and position of palmitic and stearic fatty acids on the triglyceride molecule, (3) degree of hydrogenation, and (4) the degree of randomization. The stabilizing effect of palmitic fatty acid is related to its triglyceride position and its level in the solid fat portion, which can be increased with hydrogenation. In general, the more diverse the triglyceride structure of the highest melting portion of the fat, the lower is the beta

TABLE 4.5. Fats and Oils Crystal Structure.

Beta	Beta-Prime
Canola oil	Cottonseed oil
Cocoa butter	Butter oil
Coconut oil	Herring
Corn oil	Menhanden
Lard	Modified lard
Olive oil	Palm oil
Palm kernel	Rapeseed oil
Peanut oil	Tallow
Safflower oil	Whale
Soybean oil	
Sunflower oil	

forming tendency. The importance of position and distribution of palmitic fatty acid is evident when comparing cottonseed oil to lard. Even though each of these fats contains approximately 23% palmitic, beta-prime is the crystal behavior of cottonseed oil, while beta is characteristic of lard. This is because palmitic fatty acid predominates at the 1- and 3-positions in cottonseed oil, but lard has a predominately asymmetric triglyceride structure with palmitic fatty acid in the 2-position. Lard's crystal habit can be changed to beta-prime by interesterification to randomize the fatty acids and improve the ratio of symmetric triglycerides. Additionally, the risk of graininess and posthardening due to high concentrations of 18 carbon fatty acids in oils like sunflower can be prevented with either co-randomization or the addition of a hydrogenated fat with a substantial palmitic fatty acid content [19].

Many edible fats and oils products contain various combinations of beta and beta-prime tending components. The ratio of beta to beta-prime crystal helps to determine the dominant crystal habit, but the higher melting triglyceride portions of a solidified fat product usually force the fat to assume that crystal form. The crystal form of the solidified fat product has a major influence upon the textural properties. In normal practice, some vegetable oils are blended with a minimum of 5% beta-prime hard fats, or 20% of a hydrogenated basestock that forms beta-prime crystals. This fat must have a higher melting point than the other component for the entire product to crystallize in the stable beta-prime form. Fats exhibiting a stable beta-prime form appear smooth, provide good aeration, and have excellent creaming properties for the production of cakes, icings, and other bakery products. Conversely, the beta polymorphic form tends to produce large granular crystals and products that are waxy and grainy, with poor aeration potential. The beta formers perform well in applications such as pie crusts where a grainy texture is desirable, frying shortenings where the crystal is destroyed by heating, and liquid shortening where the large granular crystals are preferred for stability and maintenance of fluidity.

4.2 Fat Plasticity

Although plastic fat and oil products are usually thought of as a solid material, they are predominantly fluids. A plastic fat consists of approximately one-third crystalline solid triglycerides, and other components like emulsifiers or other additives are suspended in two-thirds liquid triglycerides. Therefore, plastic fats like shortenings, margarines, and specialty products are composed of a solid phase of fat crystals intimately mixed with a liquid phase of fluid oil. Body is contributed to the fat by the tendency of the solid particles to interlock. Technically, a fat is plastic so long as this interlocking effect is great enough for it to completely resist small deforming stresses. When the stress is increased, a point is reached where the product structure will yield to allow plastic flow. Therefore, the relative hardness and softness, or consistency, of a fat and oil product is a measure of the stress or load required to cause plastic flow or movement.

The internal strength of the material, or its capacity for resisting stress, is determined by the number of contact points between the crystal particles. Very small crystals provide more contact points to produce a firmer fat product than relatively large crystals. With crystals of a uniform size, consistency will be determined by the number of crystals or simply the proportion of liquid and solid fractions. These proportions are in turn determined primarily, but not entirely, by the product temperature. Fat products tend to become soft as the crystal content is decreased by partial melting and become firm as the crystal content is increased when the temperature is lowered. Simply, the three conditions essential for plasticity in a fat and oil product are [20]

(1) A solid and a liquid phase
(2) A fine enough dispersion of the solid phase to hold the mass together by internal cohesive force
(3) Proper proportions of the two phases

Therefore, plasticity or consistency of an edible oil product depends upon the amount of solid material; the size, shape, and distribution of the crystalline material; and the development of the crystal nuclei capable of surviving high-temperature abuse, which serves as a starting point for new desirable crystal growth. In addition, fats and oils product firmness is increased with the smaller beta-prime crystal sizes due to an increased opportunity for the solids particles to touch and resist flow and the interlacing of long needle-like crystals than with more compact beta crystals of the same size. Another factor most directly and obviously influencing the consistency of a plastic fat and oil product is the proportion of the solid phase; as the solids contents increase, the edible fat and oil product becomes firmer. The proportion of the solid phase is determined by the extent of the saturation of the fat and oil product, either from the normal fatty acid distribution or due to processing, i.e., hydrogenation, fractionation, or interesterification.

4.3 Solid/Liquid Relationships

For a fat or fat blend to be plastic, it must consist of both a solid and a liquid phase. The ratio of these two phases determines its consistency, i.e., hardness or firmness characteristics. The most widely adopted method for characterizing this property of a fat product is dilatometry, or the measured change in increased specific volume with increasing temperature owing to melting dilation. The dilatometer is a good tool for measuring relative percent solids but not a perfect one. It is really a means of measuring density empirically adapted to measuring percent solids in terms of what has been identified as solids fat index, or SFI. There are other methods used to measure and report melting dilations; one other scale more popular outside the United States is nuclear magnetic resonance (NMR), which determines absolute solids (SFC) in fats and fat blends. Either of these methods for measuring the solids contents or some other procedure represents the best practical means for identifying and controlling fats and oils formulations.

In the United States, most fats and oils products are formulated using hydrogenation as the principal means for modifying the solids/liquid ratio of the various fats and oils. During hydrogenation, the consistency of a fat and oil and its characteristics related to consistency, such as melting point, softening point, and the amount of solid fat at various temperatures, depends upon the source oil and the hydrogenation conditions utilized. The hydrogenated product's melting point can vary substantially at the same iodine value due to differences in the hydrogenation conditions used. Melting points varying from 96.6 to 127.6°F (35.9 to 53.1°C) have been recorded at an iodine value (IV) of 67 with changes in hydrogenation temperatures from 350 to 265°F (180 to 130°C) [21]. The hydrogenation conditions comparison in Table 4.6 does not reflect the drastic change in melting point at the same iodine value, but the effect of nonselective and selective hydrogenation conditions are illustrated. Both conditions show an increased saturation with elevated levels of stearic and oleic fatty acids with corresponding decreases in linoleic fatty acid but at different rates and amounts of *trans* acids development. The changes in the saturation/unsaturation levels are reflected in the solids/liquid ratios determined by SFI and Mettler dropping point evaluations:

- Selective hydrogenation conditions reduced the polyunsaturated linoleic fatty acids instead of the monounsaturated oleic at a more selective rate than the nonselective conditions. If this preference is high, the hydrogenation proceeds stepwise, and formation of stearic fatty acid is limited until linoleic fatty acid has been almost eliminated.
- A more rapid stearic fatty acid development with the nonselective conditions affects higher solids at the higher temperatures accompanied by a higher melting point.
- The selective reaction caused more isomerization to *trans* isomers and less stearic development to effect high solids at the lower temperatures. Since the higher melting stearic fatty acid content is lower, less

TABLE 4.6. *Hydrogenation Conditions Comparison: Selective versus Non-selective Hydrogenated Cottonseed Oil.*

Hydrogenation Conditions	None	Selective				Nonselective			
Iodine Value	110	92	80	70	60	92	80	70	60
Fatty acid composition, %									
C-14:0 Myristic	0.7	0.7	0.7	0.7	0.7	0.7	0.7	0.7	0.7
C-16:0 Palmitic	22.5	22.6	22.7	22.7	22.7	22.5	22.6	22.6	22.7
C-16:1 Palmitoleic	0.7	0.6	0.5	0.5	0.5	0.7	0.6	0.6	0.5
C-18:0 Stearic	2.5	2.7	3.2	3.9	8.9	4.0	5.7	7.8	12.5
C-18:1 Oleic	18.2	39.0	52.1	61.4	64.4	36.1	48.2	53.8	56.0
C-18:2 Linoleic	54.1	33.2	19.9	10.0	2.1	35.2	21.5	13.8	6.9
C-18:3 Linolenic	0.8	0.7	0.4	0.3	0.2	0.3	0.2	0.2	0.2
C-20:0 Arachidic	0.3	0.3	0.3	0.3	0.3	0.3	0.3	0.3	0.3
C-22:0 Behenic	0.2	0.2	0.2	0.2	0.2	0.2	0.2	0.2	0.2
trans Acids, %	Nil	8.9	17.6	23.4	29.3	4.7	10.5	13.9	19.3
Solids fat index, %									
at 50°F		8.5	19.5	31.0	51.0	7.5	17.0	25.5	42.0
at 70°F		4.0	10.0	17.5	37.0	3.0	8.0	14.0	28.0
at 80°F		2.0	6.0	12.0	32.0	2.0	5.0	10.0	23.0
at 92°F			0.8	3.5	16.5	0.1	1.5	4.0	12.5
at 100°F					1.5	0.1	0.1	0.5	5.5
at 104°F					1.5				2.5
Mettler dropping point, °C		27.3	31.5	33.0	40.0		30.0	35.5	40.0

Laboratory Hydrogenation Conditions	Selective	Nonselective
Pressure, psig	2.5	4.00
Temperature, °F	385	300
Nickel, %	0.02	0.04
Agitation	Fixed	Fixed

solids are present at the higher temperatures to provide a lower melting point.

In the latter stages of hydrogenation, the conditions used have little effect upon the product produced. Isomerization of oleic fatty acid is not particularly affected under different conditions so that the melting point and SFI results are quite predictable from the amount of saturation or iodine value results. Hydrogenation to produce stearin or low iodine value hard fats is the least critical of all hydrogenation operations since any selectivity and isomerization are not factors; unsaturation is very minor since the oil or fat is almost totally saturated. Some oils can be hardened to higher melting points or titres than others because of their characteristic fatty acid composition. Soybean, canola, sunflower, corn, and safflower oils may be hardened to a titer as high as 65°C. However, it is not a good practice to use these high-titer hard fats for plastic shortening formulation because they solidify into undesirable polymorphic forms. The beta tendency, as well as the high titer is a result of the predominance of the 18 carbon fatty acids, which cause the almost completely hydrogenated product to consist of high levels of tristearin. Cottonseed oil, palm oil, and tallow contain a sufficient proportion of fatty acids with carbon chains higher and lower than C-18 to ensure that tristearin is not predominate, and the solidified oil product will assume the more desirable beta-prime crystal form.

5. HYDROGENATED BASESTOCK SYSTEM

Fats and oils processors could formulate products so that a special base fat or series of base fats is required for each different product. This practice, with the ever-increasing number of finished products, would result in a scheduling nightmare with a large number of product heels tying up tank space and inventory. A basestock system utilizing a limited number of hydrogenated stock products for blending to meet the finished product requirements is practiced by many U.S. fats and oils processors. The advantages provided by a basestock system are twofold [22–24]:

(1) Control
   - blending of hydrogenated oil batches to average minor variations
   - improved uniformity by producing the same hydrogenated product more often
   - cross-product contamination reduction by scheduling more of the same product consecutively

- reduction of product deviations from attempts to utilize finished product heels
- elimination of rework created by heel deterioration before use

(2) Efficiency
  - The hydrogenation process is scheduled to maintain basestock inventories, rather than reacting to customer orders.
  - Hydrogenation of full batches instead of some smaller batches meets demands without creating excessive heels.
  - Better order reaction time improves customer service.

The basestock requirements will vary, depending upon the customer base served, which obviously dictates the finished product mix. A wide variety of source oils are available that could be used for basestocks; however, the choices are narrowed first by product performance and then by other factors such as customer specifications, costs, religious prohibition, traditional preferences, crop economics, legislation, transportation, and others. These factors have favored soybean oil in the United States for several decades. Therefore, most U.S. fats and oils processors have basestock systems dominated by soybean oil with somewhat minor representation by the other source oils. The other source oils in most basestock systems serve as beta-prime promoters for plasticity, like palm oil, cottonseed oil, and tallow or those source oils required for a specific or specialty product preparation.

Table 4.7 outlines a soybean and cottonseed oils basestock system with nine hydrogenated basestocks, ranging from a brush hydrogenated 109 IV soybean oil basestock to saturated soybean and cottonseed oil hard fats with an 8 IV maximum. A tenth base oil is refined and bleached soybean oil, the same oil that serves as the feedstock for the soybean oil hydrogenated basestocks. Utilization of a similar basestock system should enable the processor to meet most fats and oils product requirements by blending two or more basestocks, except for some specialty products that can only be made with special hydrogenation conditions and/or catalyst.

The common practice for the production of many of the shortenings, margarine oils, and specialty products in the United States is the blending of hydrogenated and/or natural oil components. However, co-randomization, or interesterification, and the use of fractionated fat components to blend with natural fats and oils can also be utilized in basestock systems. The use of the basestock blending techniques provides the means for developing fats and oils products with an extremely wide range of potential physical properties. Chart 4.2 graphically depicts the diverse solids fat index results for five products to illustrate the variation of percent solid material due to composition and temperature for different applications.

TABLE 4.7. Hydrogenated Soybean and Cottonseed Oils Basestock System.

	Source Oils								
	Soybean Oil							Cottonseed Oil	
Basestock Type	Brush	Flat			Steep		Hard Fat	Steep	Hard Fat
Iodine value	109	85	80	74	66	60	>8	70	>8
Solids fat index									
at 50°F/10°C	4.0 max.	15–21	22–28	38–44	59–65	65–71	*	44–50	*
at 70°F/21.1°C	2.0 max.	6–10	9–15	21–27	47–53	56–62	*	27–33	*
at 80°F/26.7°C		2–4	4–6	13–19	42–48	51–57	*		*
at 92°F/33.3°C				3.5 max.	23–29	37–43	*	8–13	*
at 104°F/40°C					3–9	14–18	*		*
Mettler dropping point, °C	**						50–54		48–50
Quick titer, °C		28–32	31–35	34–37	41–45	45–48		37–40	
Hydrogenation conditions	†	†			†		†	†	†
Gassing temp, °F	300	300			300		300	300	300
Hydrogenation temp, °F	325	350			440		450	440	450
Pressure, bar	3.5	2.0–3.0			0.07–1.0		5.0	0.7–1.0	5.0
Catalyst, % nickel	0.01–0.02	0.02			0.04–0.08		0.04–0.08	0.04–0.08	0.04–0.08
Agitation	Fixed	Fixed			Fixed		Fixed	Fixed	Fixed

* Too hard to analyze.
** Too soft to analyze.
† Optimum conditions will vary considerably, depending upon the converter, agitation, hydrogen gas purity, and so on.

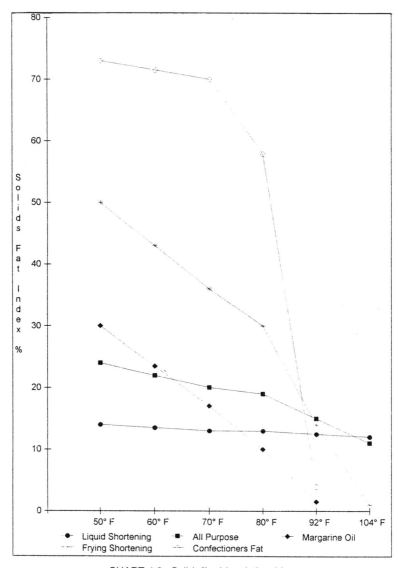

CHART 4.2. Solids/liquids relationships.

276

6. SOLIDS/LIQUIDS CHARACTERISTICS

The different triglycerides present in fats and oils are the main building blocks for application. The primary functions of fats and oils in a food product are lubricity and structure. Lubricity describes imparted tenderness and additions to the richness and improved eating properties of foods; lubricity also provides the feeling of satiety after eating. Fats and oils structural properties affect the consistency of foods through aeration, creaming properties, melting behavior, spreadability, moisture retention, moisture barrier, and other functionalities. The functional properties for fats and oils products can be directly related to the types of triglycerides in the system. As reviewed earlier, the triglyceride types are determined by the fatty acid composition and distribution on the individual triglycerides. Table 4.4 identified the liquidity zones and functionality characteristics for the triglycerides in fats or fat blends made up of palmitic, stearic, oleic, and linoleic fatty acids. As indicated, the trisaturated triglycerides in liquidity zone 4 provide structure and limited lubricity. Zone 3 triglycerides provide structure and improved lubricity because they are solid at room temperature but melt at body temperature. Zone 1 and 2 triglycerides only provide lubricity because of the liquid state at room and refrigerated temperatures.

Vegetable oils of the oleic-linoleic and linolenic fatty acid groups have no inherent structural characteristics because they do not contain any significant quantities of triglycerides represented by zones 3 and 4. However, these structural aspects can be built into the oils with hydrogenation. The flexibility of the hydrogenation techniques provide a wide variety of structures, which can be further modified through blending to provide the solids to liquid ratios required for the varied applications.

Since the main building blocks of fats and oils products are the different triglycerides present, the most exact means to develop a functional product would be to control the triglyceride composition. This approach to formulation requires qualitative knowledge of the triglyceride composition of each individual fat and oil product available for use. This method also assumes that sources for the identified triglyceride structures are available or can be obtained from the source fats and oils available or with processing. A further assumption is that the desired triglycerides can be isolated in sufficient purity to effect the desired functionality. Finally, these components must be available at costs that are competitive with other products available for the same application. The by-products created during processing to obtain the desired components may either improve the economics or contribute a prohibitive cost. Approaches to formulations of fat systems based on replication of triglyceride compositions have achieved some success but are primarily limited to cocoa butter equivalents where the natural product commands a premium price [25].

Most fats and oils product formulation development is based on an understanding of the relationships between specific physical requirements and the compositions of the fats and oils blends and their components and an appreciation of the processing effects. This formulation approach requires an identification of the key functional attributes the fat system is expected to provide and the use of historical knowledge to identify the properties that are most likely to produce the identified performances. It is important to identify the primary and secondary functional attributes and to disregard properties that are unimportant to the end-use product. Efficient utilization of this formulation approach for new fats and oils products is the systematic evaluation of the modified fat system's ability to meet the key primary and secondary functionality requirements. Although triglyceride composition is important to both formulation approaches, the performance evaluation technique usually offers more than one potential solution. This provides the processor with the latitude to choose the most practical product composition and processing sequence and the ability to interchange source fats and oils when necessary.

Therefore, most fats and oils products are identified and formulated according to usage and performance. The graphic SFI curves in Chart 4.2 indicate the differences in plastic range formulated to perform the desired function for several different food products. The product with the very flat SFI slope is a fluid opaque or pumpable liquid shortening that has become popular due to the convenience offered, handling, cost savings in some situations, and lower saturated fatty acid levels. These products are normally formulated with beta crystal-forming base oils and hardstock to produce and maintain fluidity. Solidified products with the flattest SFI curves have the widest plastic range for workability at cool temperatures, as well as elevated temperatures. The all-purpose shortening has the widest plastic range of the solidified products. The confectioners fat and the frying shortening presented in Chart 4.2 have relatively steep SFI curves, which will provide a firm, brittle consistency at room temperature but is fluid at body temperature or only slightly elevated temperatures for good mouth feel.

6.1 Wide Plastic Range Fats and Oils Products

The basic all-purpose formulation has been the building block for shortenings and margarines where creaming properties, a wide working range, and heat tolerance are important. The functionality of an all-purpose product at any temperature is largely a function of the solids content at that temperature. The all-purpose products are formulated to be not too firm at 50 to 60°F (10 to 15.6°C) and not too soft at 90 to 100°F (32.2 to 37.8°C). Initially, a liquid was blended with a hard fat to make a compound shortening that had a very flat SFI curve, which provided an excellent plastic range similar to, but slightly firmer than, the liquid shortening SFI curve illustrated in Chart 4.2. However, the low oxidative stability of these shortenings precluded their use for most products requiring a long shelf life. Currently, most of these products are formulated with a nonselectively hydrogenated soybean oil basestock and a low iodine value (IV) cottonseed or palm oil hardstock. Beta-prime hard fats are added to shortenings both to extend the plastic range, which improves the tolerance to high temperatures, and for crystal type and stability. The beta-prime crystal forming cottonseed oil hardstock functions as a plasticizer for improved creaming properties, texture, and consistency.

Hydrogenation of a fat and oil basestock increases the oxidative stability. As a rule, the lower the base IV, the better is the oxidative stability. However, as base hardness is increased, the level of hardstock required to reach a desired consistency decreases. Hardstock reduction reduces the plastic range and heat tolerance. Therefore, oxidative stability improvements are achieved at the expense of plasticity, and a wide plastic range can be at the expense of oxidative stability. The extent that one attribute can be compromised to improve another must be determined by the requirements of the intended food product. It should also be remembered that oxidative stability is directly related to the level and type of unsaturated fatty acids present; therefore, oxidative stability results do not average. For example, a 50:50 blend of a 40-hour AOM basestock and a 100-hour AOM basestock will not have an AOM stability of 70 hours, but rather will be closer to the fat product with the lower AOM stability [24].

Plastic range is important for bakery use fats and oils products intended for roll-in and creaming applications alike because of the consistency changes with temperature. Fats and oils products become brittle above the plastic range and soft below the range; both conditions adversely affect creaming and workability alike. Shortenings and margarines are normally plastic and workable at SFI values between 15 and 25. Therefore, products with flatter SFI slopes fall within the plasticity window for a much greater temperature range than those with steep SFI slopes. The all-purpose shortening in Chart 4.2 has a 42°F (23°C) plastic range from 50 to 92°F (10 to 33.3°C), while the frying shortening should have equivalent workability if used within the 7°F (4°C) range from 85 to 92°F (29 to 33°C). The frying shortening's use as a roll-in would require very strict controlled temperature use, probably not available in bakeries, and at a temperature detrimental to the laminated baked product. The 42°F working range for the all-purpose product is decidedly more practical.

Two of the soybean oil basestocks outlined previously in Table 4.7 are designed for products requiring a wide plastic range, e.g., the nonselectively hydrogenated 80 and 85 IV basestocks. Even these two basestocks with only a 5 IV difference provide measurable differences in plastic range and stability when cottonseed oil hard fat is added to produce equivalent consistencies at 80°F (26.7°C). The softer 85 IV basestock required 2.5% more hard fat to achieve the targeted 20% SFI at 80°F (26.7°C). The higher hard fat requirement indicates a better heat resistance and a wider plastic range but a lower AOM stability as determined by the higher IV of the basestock. The firmer 80 IV basestock required 2.5% less hard fat to provide the 20% SFI at 80°F (26.7°C), reduced the plastic range by 8°F (4.5°C), but increased the AOM stability to 100 hours versus the 65 hours for the shortening with the 85 IV basestock. Chart 4.3 graphically illustrates the differences in plastic range due to the slight change in basestocks.

The effect of hardstock upon the SFI slope and plastic range is illustrated by Chart 4.4. Cottonseed oil hard fat was added to 85 IV hydrogenated soybean oil basestock to demonstrate the plasticizer's effect; as the hard fat level is increased, the shortening becomes firmer with a flatter slope. The highest hard fat levels are used to formulate roll-in, puff pastry fats, and other products where a plastic, but firm, consistency is required for performance. The use of a partially hydrogenated base plus hard fat to produce a wide plastic range with good creaming properties has been expanded into a whole family of specialized shortening, margarine, and special purpose products. The development of these products has involved the selection of the most suitable hydrogenated basestock and hard fat to produce the desired plastic range and AOM stability. These developments have taken two directions: (1) the addition of an emulsifier or an emulsifier system to an all-purpose fat and oil base or (2) formulating nonemulsified products for a specific functionality. Table 4.8 outlines some of the current applications for these two categories.

6.2 Steep Solids Fat Index Slopes

Plasticity is of minor importance and can be a detriment for products requiring sharp melting characteristics and/or a high oxidative stability. Fat and oil products specially designed for specific frying situations, tablegrade margarines and spreads, nondairy systems, cookie fillers, and confectionery fats require an eating character and flavor stability not possible with blends of nonselectively hydrogenated oils with hard fats. These products require as low an iodine value as possible for oxidative stability with a steep SFI slope to provide a melting point lower than body temperature for good eating characteristics. Two alternatives to obtain the steep SFI curves are

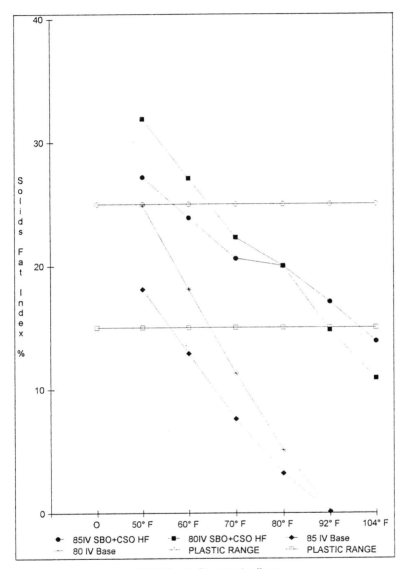

CHART 4.3. Basestock effect.

281

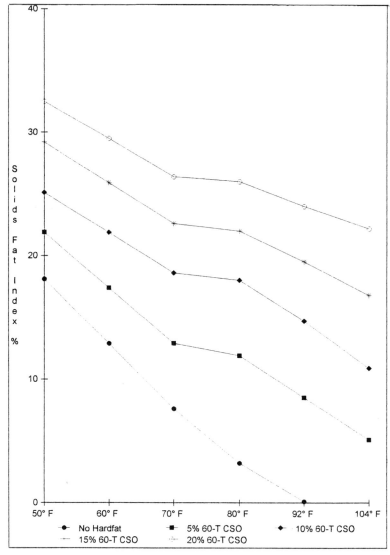

CHART 4.4. Hard fat effect.

282

TABLE 4.8. Wide Plastic Range Applications.

Nonemulsified	Emulsified
All-purpose	Household
Danish roll-in	Cake and icing
Puff pastry	Icing and filling
Cookie	Cake mixes
Pie dough	Specialty cakes
Donut frying	Yeast raised

(1) nature has provided the lauric oils with a steep SFI and a sharp melting point or (2) most liquid oils can be selectively hydrogenated to provide the desired solids melting relationships. Selective hydrogenation is a progressive diminution of the most unsaturated fatty acid groups. When the overall hydrogenation effect is that the fatty acids with three double bonds are nearly all reduced to two double bonds, before those with two are nearly all reduced to one, before the monounsaturates are saturated, then good or high selectivity exists [26]. The progression of a selectively hydrogenated soybean oil can be observed in Charts 4.5a and 4.5b. The fatty acid changes described above can be tracked in Chart 4.5a. The effect of the progressive saturation and isomerization to change the SFI is plotted in Chart 4.5b. Three of the fats and oils products shown in Chart 4.2 all rely on steep SFI slopes for the preferred performance characteristics, i.e., stick margarine oil, frying shortening, and confectioners' fat.

A tablegrade margarine oil must provide somewhat unique physical requirements; it must have a plastic consistency with a relatively sharp melting point. Consumers expect a margarine to melt rapidly in the mouth for a full flavor release, be immediately spreadable out of the refrigerator, and maintain a solid consistency for prolonged periods on the dinner table. These characteristics are provided with a blend of liquid oils and/or a soft basestock with selectively hydrogenated basestocks. Formulation of margarine oils with the multiple components has provided improved product consistencies and, along with good refrigerated storage conditions, allowed the production of margarines and spreads with 100% soybean oil basestocks, which usually convert rapidly to the beta crystal form. The combined multiple components and consistent cool temperature environment has slowed the crystallization rate enough to maintain the beta-prime crystal form for the life of the product. The stick margarine oil SFI curve in Chart 4.2 was formulated with the multiple component concept, i.e., 60% un-

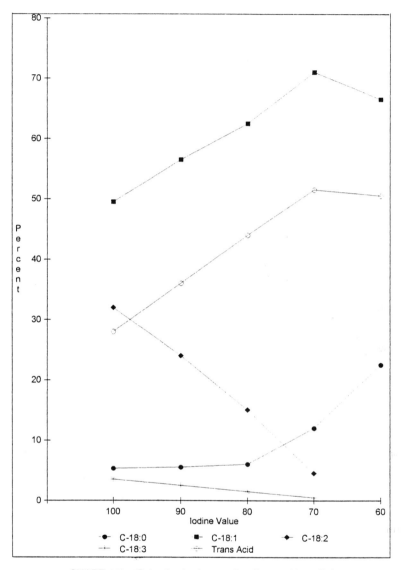

CHART 4.5a. Selective hydrogenation (fatty acid profile).

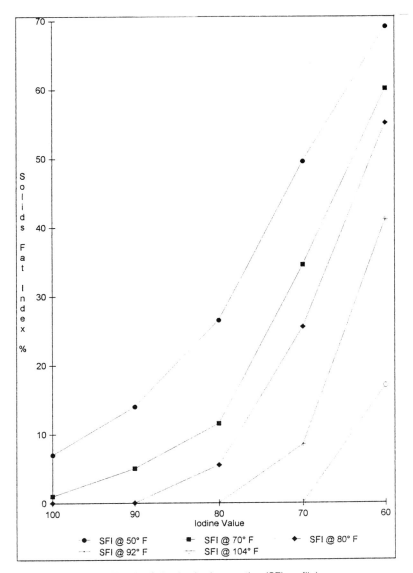

CHART 4.5b. Selective hydrogenation (SFI profile).

285

hardened liquid soybean oil, 25% 74 IV H-SBO basestock, and 15% 66 IV H-SBO basestock. The basestocks utilized represent the peak *trans* isomer development for steep SFI slopes, as represented in Chart 4.5a, which provides the product with a solid appearance at less than body temperature while facilitating rapid melting in the mouth [27].

The frying shortening illustrated meets the restaurant industry requirements for a stable heat transfer media that becomes a part of the food to supply texture and mouth feel and to enhance the food flavor. Solid frying shortenings are usually composed of a single hydrogenated base or possibly two selectively hydrogenated bases for a slightly different slope than available with a single base both with high solids at the lower temperature readings but fall off rapidly to a low melting point for the desired eating characteristics or mouth feel with an excellent oxidative and frying stability. For example, the frying shortening in Chart 4.2 is composed of single hydrogenated soybean oil, a 68 iodine value selectively hydrogenated product. However, it can also be produced by blending the 66 and 74 iodine value soybean oil basestocks detailed in Table 4.7.

Confectionery fats require a very steep SFI curve, which makes them brittle, with a short melting range that ensures a quick melt-down and a pleasant mouth feel. These fats, formulated to resemble the characteristics of cocoa butter, are usually referred to as "hard butter." Quality hard butters have a relatively high solids fat index of above 50%, at room temperature, as lower levels can provide a greasy or tacky feel. The SFI then falls off quickly for a complete melt for most products at 95 to 102°F (35 to 39.2°C). Processes used to prepare hard butters include hydrogenation, interesterification, solvent and dry fractionation, and blending. The hard butters can utilize both lauric and nonlauric source oils. The principal types of hard butters are

(1) Cocoa butter equivalent (CBE)—CBEs are nonhydrogenated specialty fats containing the same fatty acids and symmetrical unsaturated triglycerides as cocoa butter. Careful blending of selected fractions from selected source oils is required to duplicate the triglyceride structure of cocoa butter.

(2) Lauric cocoa butter substitutes—The most widely used lauric oils are palm kernel oil and coconut oil, which may be processed several different ways, i.e., hydrogenated, fractionated, hydrogenated and interesterified, hydrogenated and fractionated, or hydrogenated-interesterified-fractionated.

(3) Nonlauric cocoa butter substitutes—Domestic hard butters are usually blends of soybean and cottonseed oils but have also been produced

with palm, peanut, corn, safflower, and sunflower oils. The nonlauric confectionery fats may be divided into two types:

- selectively hydrogenated—Oils with high levels of unsaturation may be selectively hydrogenated utilizing a sulfur-treated catalyst to produce a high *trans* isomer level and minimizing stearic fatty acid development to optimize solids development at room temperature while maintaining a melting point only slightly higher than body temperature.
- fractionated selectively hydrogenated—The nonlauric fractions selected for hard butters have high SFI contents at room temperature with a better flavor release than the nonfractionated products.

The confectionery fat example plotted in Chart 4.2 is the stearin fraction of partially hydrogenated palm kernel oil. It was designed for confectioners' chocolate and pastel coatings, confectioners' drops, and candy centers but has also found application in nondairy, bakery, and other food products.

6.3 Liquid Opaque or Pumpable Products

Fluid shortening and margarine products are usually suspensions of hard fats, emulsifiers, other additives, or combinations in liquid oils formulated to be pourable or pumpable at room temperature and stable in storage over a range of temperatures. Plastic fat systems have had a dominant role in the preparation of baked, fried, and other food products because of the performance supplied by the plastic-like consistency. Emulsifiers, other additives, and/or hard fats must assume these functions with the fluid fats and oils products. The technologies involved in the development of a fluid fats and oils system are

(1) Base oil technology—Hard fats are a crucial functional property necessary for the development of the crystal structure for suspension stability and crumb strength in bakery and other prepared food products, but too high a solids content limits the fluidity of the system. The liquid oil employed may be only refined and bleached or brush hydrogenated, depending upon the oxidative stability requirements of the finished product.

(2) Additive technology—Liquid or flowable fats and oils products functional additives include emulsifiers, antifoamers, antioxidants, flavors, colors, antisticking agents, and so on. The required additives may have unique crystalline properties, which must be compatible with the base oils to prevent separation or gelation.

(3) Processing technology—The processing techniques utilized to crystallize the fluid fats and oils products may need to be adjusted for the different product compositions or vice versa.

Basically, all liquid opaque, squeezable, or pumpable fats and oils products are flowable suspensions of solid fat in liquid oil with or without additives. The liquid phase may or may not be hydrogenated, depending on the finished product's consistency and oxidative stability requirements. Normally, a low-iodine value, beta forming hard fat is used to seed crystallization for shortening products. It can vary from as low as 1.0% to higher levels, as required to produce the desired finished product viscosity, usually no more than 10%. The liquid opaque shortening plotted in Chart 4.2 is typical of a fluid bread shortening formulated with 10% soybean hard fat crystallized in 90% unhardened soybean oil.

The ease in which beta hard fats convert to the stable beta crystal form makes them ideal for liquid shortening products; the beta crystals do not intertwine to form a matrix that can enmesh the liquid phase to effect a thicker product. However, liquid margarine oil products have been described in the patent literature using either beta or beta-prime type hard fats. Unlike liquid shortenings, beta-prime crystallizing hard fats are suitable for postcrystallization of a fluid margarine product after packaging. Liquid margarine products formulated with beta hard fats are more fluid but must be crystallized before packaging [18].

7. EMULSIFICATION

The emulsifying properties of fats and oils products can be enhanced with adjustments of the fat structure and the addition of surface-active agents. Food emulsifiers supplement, broaden, and intensify the functionality of shortening, margarine, or specialty fat products. Emulsifiers are amphiphilic substances that possess both hydrophilic and lipophilic properties. This dual affinity results in the alignment of the emulsifiers at the interface between two normally immiscible substances to form a bond or an emulsion between the two. In addition to forming emulsions, the surface-active agents or surfactants perform a range of functions that, in some products, are unrelated to emulsification. The functions contributed by emulsifiers in prepared foods include the following:

- emulsification—promote the combination of oil and water in a compatible dispersion for both water-in-oil and oil-in-water
- emulsion stabilization—extend the emulsion stability in dressings, sauces, frozen desserts, and nondairy products

- anti-staling agent—emulsifiers extend the shelf life of baked products by forming complexes with the starch molecule. A soft crumb provides a soft feel, which is the traditional U.S. consumers' test for bread freshness.
- aeration—entrainment of air cells in batters, icings, toppings, and other food products
- texturizer—modify the extensibility of protein to influence the texture of bread and cake products
- lubrication—surfactants provide a higher degree of lubricity to impart "slip" or lubrication properties
- defoamer—stabilize food products or help break undesirable emulsions
- dispersant—solids, liquids, and gas dispersions depend on the reduction of interfacial tension provided by a surfactant
- moisture barrier—emulsify moisture to protect against loss during storage
- oxygen barrier—protect less stable ingredients from oxidation
- crystal modifier—modification of polymorphic form, size, and growth rate of fat crystals
- carrier—solubilizer or dispersant for flavors, colors, and other fat-insoluble additives
- stickiness or tackiness reduction—function as a lubricant to reduce the tendency of a food product to stick to teeth, wrappers, or cutting knives
- wetting—reduction of the interfaction tension between liquid and solid surfaces to cause the liquid to spread more evenly over surfaces
- agglomeration—controlled coagulation of fat particles in a liquid
- palatability improvement—emulsification of a fat system to enhance eating quality

Obviously, no single emulsifier or emulsifier system can perform all of these different functions. A surfactant that will act as an emulsifier for one product or formula may cause instability in others. Propensities of the various types of emulsifiers to exhibit the previously listed properties vary with their chemical structures. Many of the food emulsifiers produced are either derivatives of the first synthetic emulsifier used commercially or mono- and diglycerides or utilize an alcohol other than glycerine to produce the surface active agent with a somewhat different functionality. The emulsifier types most often used for food applications are

- mono- and diglycerides
- propylene glycol esters
- sorbitan esters

- polyoxyethylene sorbitan esters
- polyglycerol esters
- lactated esters
- lecithin

For the selection of the proper emulsifier system, the developer must consider the food product's usage application, preparation method, emulsion type, effects of the other ingredients, economics, and any other applicable criteria for the prepared food product. It is important to clearly identify the properties and the required processing of the food product the fats and oils product's emulsifier system is being designed to help function as desired. The following is a checklist to help identify the influencing factors in the selection of an emulsifier system [28]:

(1) Functional requirements of the food product
(2) Method of processing, i.e., homogenizing, pasteurizing, whipping, cooking, baking, fermentation, pumping, extruding, dry blending, and so on
(3) Finished product form, e.g., liquid, powder, solid, gel, dispersion, emulsion and type, foam, and so on
(4) Consumer preparation technique
(5) Storage requirements
(6) Flavor and mouth feel
(7) Formulation—effect of the other ingredients
(8) Economics
(9) Legal requirements

The roles and functional contributions of the emulsifiers in each different application are not always instantly identifiable; however, the emulsifier system selection is simplified after the functionality requirements are recognized. The next step for selection of the best emulsifier system is to review the functionality of the available surfactants.

7.1 Mono- and Diglycerides

Mono- and diglycerides were the first fatty emulsifiers to be added to foods. These emulsifiers were first used in margarine for Danish pastry and puff pastry shortening. The first U.S. patents for mono- and diglycerides granted in 1938 disclose the surfactants' usefulness in emulsions and margarine [29]. Monoglycerides with only one fatty acid attached to a glycerol molecule and two free hydroxyl groups on the glycerol take on the properties of both fats and water. The fatty acid portion of the molecule acts like any other fat and readily mixes with the fatty materials, while the two hydroxyl groups mix or dissolve in water; thus, monoglycerides tend to hold fats and water together. Cake shortenings with increased aerating and creaming properties, as well as better moisture retention properties contributed by the addition of mono- and diglycerides, were introduced in 1933 also. Soon after, fats and oils products with shelf life extension properties for yeast-raised products provided by mono- and diglycerides additions were introduced to the bread bakers [29].

Mono- and diglycerides are still the most commonly used emulsifiers in the food industry and account for about 70% of emulsifier usage. These surfactants are required in almost every type of processed food product. The major usage is bakery products, prepared mixes, margarines, convenience foods, and frozen desserts. Normally, they are used as a part of a fat product and frequently in conjunction with other emulsifiers. Their lipophilic character causes them to be excellent water-in-oil emulsifiers, as required for margarines. At room temperature, mono- and diglycerides are insoluble in water and have only very limited solubility in oil, except at elevated temperatures.

Mono- and diglycerides are produced in three concentration levels: 40 to 46% alpha monoglyceride, 52% minimum alpha monoglyceride, and distilled or 90% monoglyceride content. They are available in a wide range of physical forms: liquids, semi-liquid, soft plastic, flakes, beads, or powders. The three general physical forms of hard, plastic, and soft have an effect upon the functionality of the mono- and diglycerides emulsifiers. As detailed in Table 4.9, the soft to liquid form of the mono- and diglycerides produces good aeration properties for low-moisture applications, while the hard form exhibits good baked product structure and shelf life extension properties. The intermediate hardness or plastic emulsifiers are the easiest to handle separately but provide compromise functionalities of the hard and soft mono- and diglycerides.

Mono- and diglycerides are on the generally recognized as safe (GRAS) list of *The Code of Federal Regulations* (21 CFR 182.4505), and the levels of use are limited to applications requirements and to standards of identity where applicable. The three different concentration levels (43%, 52%, and 90%) of emulsifiers can normally be used interchangeably and added on the basis of the alpha monoglyceride contribution to the shortening, margarine, or specialty product. Performance evaluations have given some indications that the diglyceride content in the 43% and 52% alpha monoglyceride products may contribute some functionality, but no significant improvements have been determined. The optimum usage range of any emulsifier must be determined for each finished product formulation. Excessively high or low levels of the emulsifier will usually lead to unsatisfactory results.

TABLE 4.9. Mono- and Diglycerides Functionality.

Emulsifier Form	Functionality	Food Product
	Moisture retention	All baked products
	Crumb softener	All baked products
	Anti-staling	All baked products
	Volume improver	All baked products
	Tenderness improver	All baked products
	Grain and texture improver	Cakes
	Batter aeration	Cakes
	Palatability improver	Breads and rolls
Hard	Stickiness retardant	Chewing gum and candy
>5 Iodine value	Antisticking	Pasta
	Oil stabilization	Peanut butter
	Rehydration	Dehydrated potatoes
	Tight emulsion	Margarine
	Freeze/thaw stability	Frozen desserts
	Dispersant	Coffee whiteners
Plastic-like 60–80 Iodine value	Compromise of Hard and Soft	All Products
	Aeration	Icings and fillings
Soft	Water absorption	Icings and fillings
90 Iodine value	Texture improvement	Gravies and sauces
or higher	Loose emulsion	Margarine
	Fat dispersant	Pet foods
	Fat dispersant	Cake donuts

Several mono- and diglyceride deviations are produced by using them as raw materials in the production of esters of edible, water-soluble compounds such as lactic acid, citric acid, acetic acid, diacetyl tartic acid anhydride, and succinic anhydride [29–33]:

- Diacetyl tartaric acid esters of monoglycerides, or DATEM, is a hydrophilic emulsifying agent with excellent dough conditioning properties. By strengthening the gluten in the dough, DATEM enhances gas retention resulting in increased volume in baked products, better dough tolerance to mechanical handling and better proofing stability. Gluten strengthening also improves crumb structure and crust characteristics. Even though DATEM is primarily utilized as a dough strengthener, it also has application in extruded products, icings, and margarines.
- Lactic acid esters of monoglycerides or lactated esters are a group of food emulsifiers used primarily in applications where aeration is re-

quired, such as in toppings, cakes, and icings. The lactated esters have more surface activity and are slightly more hydrophilic than regular mono- and diglycerides. Most lactated esters are processed from saturated or low-iodine value mono- and diglycerides to give optimum aerating properties.

- Citric acid esters of monoglycerides or monoglyceride citrate is a hydrophilic emulsifier prepared for use as a margarine anti-spattering agent, metal chelating, and for use with other surfactants as an emulsion stabilizing agent.
- Acetylated monoglycerides are alpha tending emulsifiers that can be used to promote aeration, volume, foam texture, and stability against syneresis for imitation creams and toppings. Also, acetylated monoglycerides form protective films that are effective against oxidation and moisture loss to keep pie crusts and pizza shells crisp and fresh. Additionally, the liquid acetylated monoglycerides are effective lubricants and release agents for use as trough greases or bakery products depanning agents.
- Succinylated monoglycerides perform as a dough conditioner and strengthener tolerant to mechanical handling and flour variables for uniform good volume, texture, flavor, and crumb softness.

7.2 Propylene Glycol Esters

Propylene glycol mono fatty acid esters are a group of moderately surface active compounds that are effective emulsifiers for cakes and nondairy whipped products such as imitation creams, whipped toppings, and spray-dried topping products. The alpha-tending, propylene glycol monoester (PGME) surfactants are not added for emulsion stability but, rather, to influence the whipping rate, stiffness, volume, and foam stability. These emulsifiers are able to promote an agglomeration of fat globules for the formation of a good and stable foam while forming a protective alpha-crystalline membrane on the liquid droplets, which results in a high degree of aeration.

Cake batters aerate rapidly with PGME and produce baked cakes with high volume but fleecy, elongated, even cell structures. Initially, cakes utilizing propylene glycol monostearate (PGMS) had to be low-fat formulations with high PGMS levels; then, it was determined that a combination of PGMS and mono- and diglycerides were more functional for cake baking. The most effective PGME emulsifiers for cakes are produced with a low-iodine value fat base. For consumer cake mixes, PGME shortenings provide the needed tolerance to mixing conditions at PGME levels from 8.0 to 16.0%, with total monoglyceride levels of 2.0 to 6.0%. Bakery cakes require less functionality from the emulsifier system since more of the aeration is achieved by more efficient mixing procedures and equipment. These shortenings typically range from 2.8 to 8.5% PGME, with 4.0 to 5.5% total monoglyceride.

Whippable emulsions normally contain proteins that provide a great deal of the stability needed by these products in a liquid form. In these emulsions, propylene glycol monoesters function to provide volume, foam texture, and stability against synereses. Structure building consists of agglomerated fat globules to form a network within the foam. The hard form of the 90% monoesters is typically used at 0.2 to 1.5% levels for these products.

Saturated fatty based emulsifiers function more as defoamers than aerators in low-moisture, prepared foods like icings and fillings. However, the softer based PGME emulsifiers can function in these products. An intermediate hardness PGME surfactant made with a 50 IV hydrogenated lard or a 70 IV hydrogenated soybean oil base added to an all-purpose shortening base to obtain 2.5 to 3.5% PGME and 4.0 to 5.0% total monoglyceride emulsifier analysis will perform in icings, fillings, and low and high sugar bakery cakes. The intermediate hardness PGME emulsifier has also been utilized for lean cakes, cookies, and other fat-reduced foods.

7.3 Sorbitan Esters

Sorbitan fatty acid esters are sorbitol-derived analogs of mono- and diglycerides, which are slightly more water-soluble. Sorbitan monostearate, the most common of these surfactants, is a hard solid or beaded material that requires dispersion in water or fat before use.

Sorbitan fatty acid esters are lipophilic emulsifiers that can be used in emulsions where less water-binding properties and enhanced aeration properties are desired. Sorbitan monostearate can be used for icing, whipped topping, and coffee whitener applications for aeration, gloss, and stability characteristics. However, the best known application is probably as a crystal modifier or promoter for confectionery coatings. Sorbitan esters are especially effective additives for the improvement and retention of gloss and a desirable effect upon the solids fat content to improve mouth feel.

7.4 Polysorbate Fatty Acid Esters

Polyoxyethylene sorbitan esters, commonly called polysorbates, are formed from the reaction of sorbitan esters with ethylene oxide. Three of the polysorbate types permitted for food use at limited amounts are polysorbate 60, polysorbate 65, and polysorbate 80. The usage level of these emulsifiers is restricted by food additive regulations; however, relatively small quantities of the polysorbates provide the desired effect in most applications, and higher levels impart a bitter sweet taste. The polysorbates are hydrophilic emulsifiers and act as powerful surface active agents to drastically reduce the interfacial tension between water, oil, and other ingredients; enhance interaction between ingredients; and promote emulsion stability. Polysorbate 60 is particularly effective in formulating oil-in-water emulsions.

These derivatives have found more acceptance than the sorbitan esters. Polysorbate 60 can be used in applications similar to those cited for sorbitan esters or cakes, whipped toppings, coffee whiteners, icings, confectioners' coatings, and the nonstandarized salad dressings. Polysorbate 65 is permitted in ice cream and frozen desserts, mellorine or ice milk, and fruit sherbet, as well as the bakery applications. Polysorbate 80 can be used in a wide range of applications, including special dietary foods and gelatin desserts, as a solubilizing and dispersing agent in pickles and pickled products, and in baking applications.

7.5 Polyglycerol Esters

These compounds are formed by a reaction of fatty acids with polymerized glycerol consisting of two to ten molecules. Careful control of the polymerization and esterification processes produce emulsifiers that can be used to emulsify, thicken, stabilize, defoam, release, plasticize, modify crystallization, enhance gloss, and also prevent sticking, weeping, lumping, and clouding. Some of the typical applications are

- peanut butter—stabilization and plasticization aid to prevent oil separation and weeping while improving spreadability and lubrication to prevent the peanut butter from sticking to the roof of the mouth.
- confectionery coatings—incorporation of 0.5 to 5.0% of a specific polyglycerol ester to enhance and maintain gloss properties while minimizing tempering and handling problems
- margarine—excellent anti-spattering and emulsifying agents; used at 0.1 to 0.5% for margarines and for low-fat spreads at 0.1 to 2.0%
- crystal inhibitor—retards crystal formation in salad oils to increase the length of time before cloud development at use levels of 0.02 to 0.04%

- icings and fillings—rapid effective aeration and weep stabilization for low-moisture icing and filling products at levels of 1.0 to 2.0% of the shortening
- cakes—layer and loaf cakes with good volume, grain, and texture equivalent to mono- and diglyceride emulsified hi-ratio shortenings produced with lower polyglycerol ester use levels of 1.0 to 3.0% of the shortening

7.6 Lactated Esters

Several forms of lactated esters are used in foods. Lactic acid esters of monoglycerides, reviewed previously with mono- and diglyceride, are reaction products of glycerine, lactic acid, and fatty acids. The two most prominent products are glycerol lacto palmitate (GLP) and glycerol lacto stearate. These alpha-tending emulsifiers are used primarily in applications where aeration is required, such as cakes, cake mixes, and whipped toppings to improve volume and texture. Most of these products have almost fully saturated fat bases and exhibit some oil solubility but no water solubility.

Stearoyl-2-lactylates are reaction products of stearic fatty acid and lactic acid converted to the calcium or sodium salts. Sodium stearoyl-2-lactylates, one of the most hydrophilic emulsifiers in the food industry, is used principally as a dough conditioner in baked products, pancakes, and waffles but is also effective in frozen whipped dairy cream and toppings to improve aeration and stabilization of the foam and to improve fat distribution to enhance the whitening effect in coffee whiteners. Calcium stearoyl-2-lactylate, also a dough conditioner, is used in both fresh and prepared mixes for yeast-raised baked products, as a whipping agent for egg whites, and as a conditioning agent for dehydrated potatoes. Low levels, typically 0.25 to 0.5% of the flour, of the stearoyl lactylates complex with starch to delay the staling process in bread. The interaction with the flour gluten results in a finer crumb structure, increased volume, and better crust in the baked bread loaf.

7.7 Lecithin

The processed food industry in the United States utilizes over 350 million pounds of surfactants a year, 90% of which are chemically derived and 10% natural. The most prominent, functional, and commonly used natural surfactant is lecithin. This natural emulsifier is found in eggs, dairy products, vegetable oils, and many of the other food products from nature. Lecithin is a phosphorus-containing lipid, phosphatide being found in all plant and animal organisms. Commercially, the chief source for lecithin is the gums extracted from soybean oil. Lecithin may also be obtained from other plant seeds such as corn, cotton, peanut, and sunflower, but little is available for commercial use.

Lecithin has wide use in food and nonfood products, which makes it a very important by-product of the soybean oil processing industry. The National Soybean Processor Association established criteria in the early 1970s for defining plastic and fluid lecithins, but these lecithin products are only about 15 to 20% of the current products used. Many new types of lecithins have been introduced commercially over the past few years. Some of these new products include oil-free or de-oiled, fractionated, and enzyme-modified lecithins among others. Phospholipids isolated from soybean lecithin functionality in foods are [34,35]

- protein complexing—The ability to complex with the protein in flour (gluten) provides the basis for a good dough conditioner. Lecithin can function as a natural bread dough conditioner at addition levels generally between 0.25 and 0.6% based on the weight of the flour.
- starch complexing—An ability to complex with starch is the basis of some anti-staling agents. Certain lecithin phospholipids like phosphatidyl choline and phosphatidyl ethanolamine form vesicles or liposomes with excellent anti-staling properties.
- fat reducer—Hydrated lecithin can assume a spherical form that rolls on the tongue to give a mouth feel similar to that of fats. Reductions of 20 to 60% of the fat content have been accomplished with cocoa and chocolate products. In chocolate, about 0.25 to 0.35% lecithin is used to reduce viscosity by coating the sugar, increase the "snap" by allowing a reduction in cocoa butter content, and inhibit bloom by allowing a wider range of processing temperatures.
- egg yolk replacer—Commercial lecithin can duplicate the emulsification characteristics of egg yolk in cakes, sweet goods, and other food products. Lecithin or egg yolks provide cake batter fluidity during baking, which prevents a collapse of the cake or a "dip."
- antisticking properties—Lecithin is the antisticking agent used in most pan and grill shortening, pan sprays, baking pan release products, and in the candy industry for hard candies and confectionery. Lecithin mixtures in oil or other solvents form a barrier by coating the equipment surfaces.

- antispattering agent—Lecithin is used in margarines and spreads at 0.1 to 0.5% to act as an antispattering agent, as well as emulsifying to improve texture, absorb moisture, prevent bleeding, improve spreadibility, and act as a browning agent. Spattering is caused by the rapid explosive evaporation of the moisture present in margarine and develops when the finely divided water coalesces to form large drops. Lecithin envelops each water particle with a protective membrane to stop the coalescence of the water droplets.
- wetting—Cocoa powders possess poor wetting properties, irrespective of the fat content, without added lecithin. Cocoa and other instant drink powders are instantly wetted and drawn into cold water additions with lecithin levels of 1.0 to 1.5%.
- antioxidant—Lecithin functions as an antioxidant in some oils at levels of 0.01 to 0.25%. Low-level additions of lecithin can also effectively retard off-flavors caused by hydrolysis by the same action that prevents spattering.

Fats and oils appearance and flavor requirements can limit the use of lecithin, especially for products that must be heated. Lecithin is quite heat-sensitive and can darken, smoke, and develop an offensive fishy odor and flavor at temperatures above 120°F (49°C). Additionally, the level of lecithin added to a fat and oil product may affect both the color of the ingredient and the finished product. Regular fluid food grade lecithin has a brown or amber color and a typical flavor. At the normal low usage levels, the color, flavor, and odor imparted by lecithin does not materially affect the product's appearance and flavor. However, care must be exercised with the use of lecithin product and when formulating to avoid problems.

7.8 Emulsifier Selection Methods

The process of selecting an emulsifier system has been viewed as more of an art than a science requiring experience and performance testing. The synergistic effects, observed with combinations of the surfactants, provide desirable functional properties not possible with single emulsifiers and complicates the selection process. Therefore, development of emulsified shortenings can require a complex evaluation program with numerous surfactant blends of different ratios and use levels. Researchers have concluded that emulsifier selection is not a simple task, and a number of schemes have been developed to assist in the identification of the optimum system for the individual applications. The most well-known selection scheme is probably the hydrophilic lipophilic balance (HLB).

HLB SYSTEM

Emulsifiers can be classified by their affinity to water and oil. Surfactants with an affinity to oil or the lipid phase are classified as "fat loving" or lipophilic emulsifiers. Surfactants that have an affinity to water or the aqueous phase are classified as "water loving" or hydrophilic emulsifiers. Oil-soluble or lipophilic emulsifiers will absorb water or water-soluble materials into the oil or lipid phase. Hydrophilic emulsifiers will absorb the lipid or oil phase into the water or aqueous phase. The measure of an emulsifier's affinity toward oil or water may be identified by its HLB [36,37]. HLB is an expression of the relative attraction of an emulsifier for water and for oil or for the two phases of the emulsifier system being considered. In the HLB system, each emulsifier is assigned a numerical value. An emulsifier that is lipophilic in character is assigned a low HLB value below 9.0, and one that is hydrophilic is assigned a high HLB value above 11.0. Those in the 9.0 to 11.0 range are intermediate. In general, lipophilic emulsifiers such as mono- and diglycerides, have an HLB in the 2 to 4 range and tend to form water in oil emulsions. Hydrophilic emulsifiers, such as polysorbate esters, have an HLB in the 13 to 17 range and tend to form oil in water emulsions. The following indicates the relationship of HLB to water solubility at 70°F (21.1°C):

Emulsifier Behavior When Added to Water	HLB Range
Insoluble	1 to 4
Poor solubility	3 to 6
Opaque with vigorous agitation	6 to 8
Stable opaque dispersion	8 to 10
Translucent	10 to 13
Clear solution	13 and above

The best results are usually obtained with emulsifier systems or by blending two or more emulsifiers to obtain a desired HLB value rather than using only one. A blended emulsifier system, usually described as a synergistic emulsifier blend, provides not only the functional properties of each emulsifier, but also additional functional properties due to interaction effects. The HLB of the blend is essentially a straight-line relationship, which can be determined by multiplying the HLB value of each component in the blend and then adding together these numerical values. The HLB values of various surfactants are listed below:

Emulsifier	HLB Value
Mono- and diglycerides	
40% min. alpha mono	2.8
52% min. alpha mono	3.5
90% min. alpha mono	4.3
Propylene glycol esters	
Propylene glycol mono-diesters	3.4
Propylene glycol mono- and diester	3.5
and mono- and diglycerides	
Sorbitan esters:	
Sorbitan monostearate	4.7
Sorbitan tristearate	2.1
Polyoxyethylene sorbitan esters	
Polysorbate 60	14.9
Polysorbate 65	10.5
Polysorbate 80	15.0
Polyglycerol esters	
Triglycerol mono shortening	6.0
Triglycerol monostearate	6.2
Hexaglycerol distearate	8.5
Lactated Esters	
Lactylated mono- and diglycerides	2.6
Lecithin	
Standard fluid	3.5
De-oiled, 22% phosphatidylcholine	4.5
De-oiled, 45% phosphatidylcholine	6.5

The HLB selection system consists of three steps: (1) determining the optimum HLB value for the intended product, (2) identification of the best emulsifier types, and (3) final HLB adjustments. In this procedure, emulsifiers and blends with HLB values outside the identified HLB range may be eliminated to reduce the trial and error evaluations. The detailed steps to determine the best HLB value are

(1) Select a matched pair of emulsifiers, one lipophilic and one hydrophilic with known HLB values; for example, mono- and diglycerides with a 2.8 HLB are a lipophilic surfactants, and polysorbate 60 with a 14.9 HLB are a hydrophilic surfactant.

(2) Prepare a series of test emulsions with the selected emulsifiers blended to give different HLB values ranging from all lipophilic to all hydrophilic surfactants. This range would be from 2.8 to 14.9 for the two

surfactants suggested in the first step. An excess of the emulsifier blend should be used, or approximately 10 to 12% of the fat and oil level in the final product.

(3) Evaluate the emulsifier's series using appropriate performance methods based on the product requirements. One or more of the emulsifier blends should provide a better emulsion than the others, but if all levels appear good, repeat the series with a lower use level. Conversely, if all are poor, the use level should be increased and the series repeated.

(4) The final test results should indicate within approximately 2 units the HLB range that will function the best for the application. A more accurate HLB value may be determined by another series bracketed around this range if necessary.

The right surfactant chemical type is as important as the HLB value. After the HLB value has been established, it must be determined if some other emulsifier blend would perform better, be more efficient, or be more cost-effective to produce that HLB value. The evaluation object is to select several pairs of emulsifiers covering a suitably wide area of chemical types. Performance evaluations of these blends should indicate the ideal emulsifier blend for the specific application.

The HLB method is an incomplete system for selecting surface active agents. Surfactant selection for emulsions using the HLB system functions well in a pure system when a formula is composed of oil, water, and emulsifiers. However, its use is suspect in many food systems because of their complexity. The effect of surfactants on starch complexing is an important function for some products, but HLB does not measure this attribute. Emulsifiers with the same HLB value may vary from excellent to poor in their ability to produce an acceptable cake; emulsifier selection becomes more complex with the addition of flour, starch, sugar, milk, salt, eggs, or other similar ingredients, some of which contain natural emulsifiers that react and interact with each other. Therefore, the HLB approach can assist in the surfactant selection process, but identification of the optimum emulsifier system for the application still requires experience and detailed experimental work.

EMULSIFIER FUNCTIONALITY TRAITS

Mono- and diglycerides are probably the most multifunctional emulsifiers. Other emulsifiers generally fit a more specific or limited function. The specificity of the surfactant's functions provides a selection guide for emulsifier systems development. Table 4.10 evaluates the functionality degree

TABLE 4.10. Emulsifier Functions.

Emulsifier	Emulsion Stability	Starch Complexing	Dough Conditioner	Crystal Modification	Aeration
Mono- and diglycerides:					
Hard or saturated	2	1	4	3	1
Soft or unsaturated	2	3	4	3	2
Propylene glycol esters:					
Propylene glycol mono- and diesters	5	2	5	1	1
Propylene glycol mono- and diesters and mono- and diglycerides	3	1	4	1	1
Sorbitan esters:					
Sorbitan monostearate	3	5	5	1	3
Sorbitan tristearate	3	5	5	1	5
Polyoxyethylene sorbitan esters:					
Polysorbate 60	3	2	1	3	1
Polysorbate 65	3	3	3	3	2
Polysorbate 80	2	3	3	2	2
Polyglycerol esters:					
Triglycerol mono shortening	2	3	3	3	1
Triglycerol monostearate	3	3	3	3	1
Hexaglycerol distearate	1	3	3	1	2
Lactated esters:					
Lactylated mono- and diglycerides	2	4	5	1	1
Stearoyl-2-lactylates	1	2	1	5	2
Lecithin					
Standard fluid grade	3	3	3	1	4

Emulsifier functionality evaluation: 1 = Excellent, 2 = Good, 3 = Slight, 4 = Poor, 5 = None.

for the emulsifier types most often used for food applications in five different areas [38–40]:

- emulsion stability—Emulsifiers create mixtures of two liquids that are normally immiscible. An emulsion can be defined as a dispersion of droplets of one immiscible liquid with another. Emulsifiers promote formation of dispersed systems by lowering the surface tension at the interface of two immiscible substances. The ability of an emulsifier to act at the interface is due to molecular structure. The polar group of an emulsifier has an affinity for water in which it dissolves. The fatty acid part of the emulsifier has an affinity for oil and dissolves in this phase.
- starch complexing—Starch consists of two types of carbohydrates: amylose and amylopectin. A gel is formed when these carbohydrates are mixed with water and heated. The starch components recrystallize from the gel or retrograde, which is responsible for the staling process of baked products, stickiness in pasta and potato products, and hardening in starch jelly confections. Emulsifiers that can form complexes with starch retard the retrogradation process.
- dough conditioning—Gluten, the protein in wheat flour, forms a network that strengthens doughs. Certain emulsifiers can strengthen the gluten, thereby preventing rupture of the cell walls that have entrapped the fermentation gases. This allows the cells to form uniformly, providing even grain in the finished loaf and doughs that are not sticky or bucky for better machineability.
- crystal modification—Triglycerides are polymorphic, which means that they can exist in more than one crystalline form. Crystal reversion can be retarded by the incorporation of emulsifiers with crystal modification capabilities. The alpha tending surfactants' propylene glycol monoesters, acetylated monoglycerides, and lactylated monoglycerides can stabilize mono- and diglycerides in the more active alpha form for improved functionality. Sorbitan esters prevent the formation of bloom on chocolate products by retarding the beta crystal development. Polyglycerol esters also prevent chocolate bloom and are effective crystal inhibitors for salad oils. Lecithin was the first crystal inhibitor used in salad oil processing.
- aeration—Many food products are prepared in emulsion form and are aerated to a foam. Aerating emulsifiers facilitate aeration and provide faster whipping rates, even cell structures, and higher volumes in baked products and other aerated products such as icings, fillings, and whipped toppings.

EMULSIFIER SELECTION PROCEDURE

Each product's emulsifier system must be tailored to the specific application, since its behavior will be influenced by the processing conditions and the presence of other ingredients. Furthermore, a blend of two or more emulsifiers imparts synergistic effects greater than that possible with a single emulsifier. Development of the optimum system requires analyzing the available surfactants in blends of different ratios. The following guidelines summarize a successful technique for the selection of an emulsifier system for specific applications [30]:

(1) Consult the applicable regulations for restriction on the use of emulsifiers or for specific requirements for standardized products. Only approved emulsifiers should be considered.

(2) Identify the functional requirements of the food system, the method of processing, and the form of the finished product.

(3) Identify the surfactants that contribute the functional properties required by the food system. Table 4.10 summarizes the functions of the emulsifier types most often used for food applications. Also, consult the surfactant suppliers for suggested usage levels, incorporation methods, product functionality, and other available information.

(4) Screen the approved emulsifiers with functional characteristics applicable to the food product using the HLB system guidelines. Identify the most promising blends and evaluate further to identify the optimum emulsifier system.

(5) Optimize the emulsifier system usage level. Emulsifiers have an optimum usage range or plateau that provides the maximum functionality; levels over and under this plateau level provide lesser degrees of the preferred functionality.

8. PRODUCT DEVELOPMENT

Fats and oils are key functional ingredients in a large variety of food products. They have particular physical properties of importance in the processing and final use of both natural and prepared foods. The glyceride structure and composition of the fats present in such foods also differ widely. These fatty acid and triglyceride compositions influence the functional properties of foods and food ingredients that contain even modest fat levels. The relationships between triglyceride composition and the end-use characteristics are evident when the physical and chemical properties are evaluated. Therefore, fats and oils with similar physical and chemical properties, even with different triglyceride compositions, can provide the desired functional characteristics for the food product. Fat systems for foods and food ingredients may be developed in several ways. Three product development methods used to formulate fats and oils products to perform the desired applications are

(1) Applications development—utilization of the qualitative knowledge of the effects of source oils, basestocks, fat fractions, interesterified products, emulsifiers, and other additives contributions to product consistency and behavior in the finished product

(2) Analytical development—determination of the components by their contributions to satisfy predetermined analytical limits

(3) Triglyceride replication—careful blending of selected fractions from various source oils to duplicate the triglyceride structure of another product

8.1 Applications Product Development

Applications fats and oils product development begins with an identification of the key functional attributes that the product is expected to provide to the end-use product and the use of historical knowledge to identify and evaluate the physical properties most likely to produce the intended functionality. Recognition of the primary and secondary functional attributes and disregarding the insignificant properties are important for the product development. The results of these studies provide the information necessary to develop a product specification that identifies the physical and chemical requirements that must be maintained to assure compliance with the requirements of the application. In most cases, the product is required to meet the performance standards of an existing product; however, the product developed may perform better than the original, or the product currently used may be overqualified for the application.

Performance requirements can usually be translated into analytical measurements for the development product's specification limits. For instance, the primary performance requirements of a salad oil can be evaluated with the AOCS Cold Test Method Cc 11-53 [41], which determines an oil's resistance to crystallization while submerged in an ice bath. This analytical method was developed to evaluate the acceptability of salad oil products for the production of mayonnaise and salad dressing. An oil that clouds or solidifies at the cold temperatures used for salad dressing preparation will cause the emulsion to break, which results in product separation. Likewise, the SFI or SFC curves are of principal importance in the specification for the identification of product consistency and performance for shortening, margarine, and other fats and oils products. The usual SFI measurement used to identify the acceptability of a margarine product differs from those used for shortening acceptability due to the primary uses of the two fats and oils products:

(1) Typical margarine SFI measurement points
   • SFI at 50°F (10°C) indicates the printability and spreadabilty of a margarine product at refrigerator temperatures.
   • SFI at 70°F (21.1°C) indicates the product's resistance to oil-off at room temperature.
   • SFI at 92°F (33.3°C) indicates the mouth feel or melt-in-the-mouth characteristics.
(2) Typical shortening SFI measurement points
   • SFI at 50°F (10°C) indicates the consistency at retarder temperatures.
   • SFI at 80°F (26.7°C) indicates the consistency at usual batter or dough mixing temperatures.
   • SFI at 104°F (40°C) indicates the heat resistance of the shortening to high-temperature storage.

These are typical SFI measurement points for standard products and may not be applicable for the product development application; therefore, the measurement points should be established for each product individually. However, in each case where SFI control is exercised, a minimum of three measurement points should be established to identify and help prevent what has been identified as "slope unawareness" [42]. The desired SFI slope of a fat and oil product is the difference between the lowest and highest temperature measurements. Depending upon the allowed tolerances, the slope can vary considerably, sometimes enough to drastically change the performance of the product. For instance, if the desired slope of a particular product from 50°F to 104°F SFI was 18 units, with an allowed tolerance of ±3 at the 50°F SFI and 12 at the 104°F SFI, the actual slope could vary from 13 to 23 units, or ±5 units. Additionally, the slope could vary more than this range if only one or two SFI temperatures were measured without the processor or the formulator being aware that these changes were occurring. On this basis, it is important to identify the SFI measurement temperatures, slope, and the allowable tolerances that do not adversely affect the product performance.

8.2 Analytical Development

Formulation of a fats and oils product to satisfy an existing specification requires a knowledge of the individual component's contribution to each required physical and chemical analysis. Many of the analytical requirements of fat blends, like iodine value and fatty acid composition, are linear, and the blend analysis can be determined by the sum of the weighted averages of the blend components. However, the SFI profile of a fat and oil blend is a nonlinear function of the behavior of the individual constituents, which requires skill and experience to identify the fractional components of the blend to meet the specified limits.

SOLIDS FAT INDEX CALCULATIONS

The contribution of each basestock to a blend SFI can be calculated algebraically; however, it must be recognized that for the higher SFI values the linear calculation may be somewhat more or less than the actual analytical results and that liquid oil has a solubility effect that lowers the SFI values. Experience has shown that this varies with each formulation and can be adjusted by varying the basestock's ratios. For these calculations, the required individual solid fat index data for the component fat and their blends must be determined first by analysis of selected proportions. From these data, factors can be established to help calculate the expected contribution of the individual fat components to the final fat blend:

- liquid oils effect—Liquid oils give a negative contribution or have a dilution effect upon the SFI contribution of hydrogenated basestocks and other fats that are solid at an evaluation temperature. These negative effects change based on the percentage of the liquid oil in the blend and the hardness of the basestock. Table 4.11 shows the calculation of the liquid dilution factor for a blend of 80% refined, bleached (RB) soybean oil with 20% of a 60 IV hydrogenated soybean oil basestock.

  Liquid dilution factors change with each basestock, as well as the ratio of basestocks. The SFI results for blends of RB soybean oil, corn oil, and cottonseed oil with hydrogenated soybean oil basestocks presented in Table 4.12a show the effects of liquid oil dilutions. The dilution factors, summarized in Table 4.12b, also show the negative

TABLE 4.11. Calculation of Liquid Dilution Factor.

SFI Temp.	60 IV H-SBO Basestock SFI	Calculated Basestock SFI Contribution	Actual Blend SFI	SFI Difference	Liquid Dilution Factor
50°F	64.2	×20% = 12.8	8.0	−4.8	÷80% = −6.0
70°F	56.2	×20% = 11.2	4.6	−6.6	÷80% = −8.3
80°F	52.7	×20% = 10.5	3.3	−7.2	÷80% = −9.1
92°F	35.8	×20% = 7.5	0.5	−6.6	÷80% = −8.3
104°F	13.2	×20% = 2.6	0.0	−2.6	÷80% = −3.3

TABLE 4.12a. Liquid Oils and Hydrogenated Basestock Blend Solids Fat Index Results.

RB Soybean Oil, %: Basestock, %:	None 100	10 90	20 80	30 70	40 60	50 50	60 40	70 30	80 20	90 10
60 IV H-SBO										
SFI at 50°F	64.2	55.5	47.2	39.5	32.3	25.8	20.0	14.5	8.0	3.5
SFI at 70°F	56.2	47.2	36.5	31.7	24.2	19.1	14.0	9.2	4.6	
SFI at 80°F	52.7	43.2	34.2	27.5	21.5	16.1	11.7	7.5	3.3	
SFI at 92°F	35.8	28.5	22.0	17.0	12.0	8.5	5.7	3.2	0.5	
SFI at 104°F	13.2	9.9	7.2	5.0	3.0	1.5				
66 IV H-SBO										
SFI at 50°F	61.5	53.0	45.0	37.5	30.7	24.6	19.0	13.3	8.0	2.6
SFI at 70°F	48.5	41.0	34.0	27.5	21.5	16.6	12.5	8.7	5.2	1.7
SFI at 80°F	41.5	34.9	28.5	22.6	17.5	13.0	9.0	5.3	2.0	
SFI at 92°F	23.8	19.0	14.5	10.7	8.0	5.3	3.2	1.4		
SFI at 104°F	4.3	3.0	1.7	0.7						
74 IV H-SBO										
SFI at 50°F	39.1	33.1	28.0	23.2	18.7	14.7	10.5	6.8	3.6	1.2
SFI at 70°F	21.0	17.2	14.0	11.2	8.5	6.0	4.0	1.8		
SFI at 80°F	12.7	10.2	7.9	5.8	4.0	2.2				
SFI at 92°F	1.9									
80 IV H-SBO										
SFI at 50°F	25.4	20.5	16.8	13.8	12.2	8.9	6.8	4.7	3.0	1.0
SFI at 70°F	10.4	8.2	6.4	4.8	3.6	2.6	1.7	0.8		
SFI at 80°F	4.7	3.4	2.3	1.5	1.0	0.5				

308

TABLE 4.12a. (continued).

RB Corn Oil, %: Basestock, %:	None 100	10 90	20 80	30 70	40 60	50 50	60 40	70 30	80 20	90 10
66 IV H-Corn Oil										
SFI at 50°F	62.6	54.5	46.3	38.9	31.1	24.5	19.1	14.0	9.3	4.7
SFI at 70°F	49.8	42.2	35.0	28.2	22.2	17.0	12.7	8.9	5.3	1.8
SFI at 80°F	43.3	36.2	29.1	22.7	17.2	12.7	9.0	5.7	2.9	0.2
SFI at 92°F	24.3	19.4	14.8	10.7	7.4	4.5	2.5	1.0		
SFI at 104°F	4.7	2.7								

RB Cottonseed Oil, %: Basestock, %:	None 100	10 90	20 80	30 70	40 60	50 50	60 40	70 30	80 20	90 10
60 IV H-SBO										
SFI at 50°F	67.1	57.7	49.6	42.3	35.5	29.2	22.5	16.3	10.6	5.5
SFI at 70°F	59.3	49.7	41.7	35.0	26.5	22.3	16.5	12.4	6.7	2.9
SFI at 80°F	55.7	47.8	40.0	32.7	25.7	19.4	13.7	8.9	5.0	1.8
SFI at 92°F	40.0	32.3	25.5	19.8	15.4	11.3	8.0	4.6	3.2	
SFI at 104°F	15.0	11.6	8.5	5.7	3.5	1.4				
66 IV H-SBO										
SFI at 50°F	60.5	53.3	46.0	39.0	32.3	25.7	19.5	13.7	8.5	4.0
SFI at 70°F	49.0	41.5	35.0	28.8	23.2	18.0	13.0	8.7	5.0	2.0
SFI at 80°F	44.3	36.5	30.3	24.7	19.4	14.5	10.0	6.2	5.0	
SFI at 92°F	26.5	20.9	16.5	12.7	9.8	6.6	4.5	2.2	2.8	
SFI at 104°F	5.8	4.0	2.4	0.8						

TABLE 4.12b. Solids Fat Index Calculation Factors.

RB Soybean Oil, %: Basestock, %:	RB Soybean Oil Dilution Factors									Basestock SFI
	10 90	20 80	30 70	40 60	50 50	60 40	70 30	80 20	90 10	
60 IV H-SBO										
SFI at 50°F	−22.8	−20.8	−18.1	−15.6	−12.6	−9.5	−6.8	−6.0	−3.2	64.2
SFI at 70°F	−33.8	−32.3	−25.5	−23.8	−18.0	−14.1	−10.9	−8.3	−6.2	56.2
SFI at 80°F	−42.3	−39.8	−31.3	−25.3	−20.5	−15.6	−11.9	−9.1	−5.9	52.7
SFI at 92°F	−37.2	−33.2	−26.9	−23.7	−18.8	−14.4	−10.8	−8.3	−4.0	35.8
SFI at 104°F	−19.8	−16.8	−14.1	−12.3	−10.2	−8.8	−5.7	−3.3	−1.5	13.2
66 IV H-SBO										
SFI at 50°F	−23.5	−21.0	−18.5	−15.5	−12.3	−9.3	−7.4	−5.4	−3.9	61.5
SFI at 70°F	−26.5	−24.0	−21.5	−19.0	−15.3	−11.5	−8.4	−5.6	−3.5	48.5
SFI at 80°F	−24.5	−23.5	−21.5	−18.5	−15.5	−12.7	−10.2	−7.9	−4.6	41.5
SFI at 92°F	−24.2	−22.5	−19.9	−15.8	−13.2	−10.5	−8.2	−6.0	−2.6	23.8
SFI at 104°F	−8.7	−8.7	−7.7	−6.5	−4.3	−2.9	−1.8	−1.0	−0.5	4.3
74 IV H-SBO										
SFI at 50°F	−20.9	−16.5	−13.9	−12.0	−9.7	−8.5	−7.0	−5.3	−3.0	39.1
SFI at 70°F	−17.0	−14.0	−11.7	−10.3	−9.0	−7.3	−6.4	−5.3	−2.3	21.0
SFI at 80°F	−12.3	−11.3	−10.3	−9.1	−8.3	−8.5	−5.4	−3.2	−1.4	12.7
SFI at 92°F	−17.1	−7.6	−4.4	−2.9	−1.9	−1.3	−0.8	−0.5	−0.2	1.9
80 IV H-SBO										
SFI at 50°F	−23.6	−17.6	−13.3	−10.1	−7.8	−5.8	−4.2	−2.6	−1.7	25.4
SFI at 70°F	−12.6	−10.1	−8.3	−6.6	−5.2	−4.1	−3.3	−2.6	−1.1	10.4
SFI at 80°F	−8.3	−16.4	−6.0	−4.6	−3.7	−3.1	−2.0	−1.2	−0.5	4.7

TABLE 4.12b. (continued).

| RB Corn Oil, %: | 10 | 20 | 30 | 40 | 50 | 60 | 70 | 80 | 90 | Basestock |
Basestock, %:	90	80	70	60	50	40	30	20	10	SFI
RB Corn Oil Dilution Factors										
66 IV H-Corn Oil										
SFI at 50°F	−18.4	−18.9	−16.4	−16.2	−13.6	−9.9	−6.8	−4.0	−1.7	62.6
SFI at 70°F	−26.2	−24.2	−22.2	−19.2	−15.8	−12.0	−8.6	−5.8	−3.5	49.8
SFI at 80°F	−27.7	−27.7	−25.4	−22.0	−17.9	−13.9	−10.4	−7.2	−4.6	43.3
SFI at 92°F	−24.7	−23.2	−21.0	−18.0	−15.3	−12.0	−9.0	−6.1	−2.7	24.3
SFI at 104°F	−15.3	−18.8	−11.0	−7.1	−4.7	−3.1	−2.0	−1.2	−0.5	4.7

| RB CSO, %: | 10 | 20 | 30 | 40 | 50 | 60 | 70 | 80 | 90 | Basestock |
Basestock, %:	90	80	70	60	50	40	30	20	10	SFI
RB Cottonseed Oil Dilution Factors										
60 IV H-SBO										
SFI at 50°F	−26.9	−20.4	−15.6	−11.9	−8.7	−7.2	−5.5	−3.5	−1.3	67.1
SFI at 70°F	−36.7	−28.7	−21.7	−22.7	−14.7	−12.0	−7.7	−6.5	−3.4	59.3
SFI at 80°F	−23.3	−22.8	−21.0	−19.3	−16.9	−14.3	−11.2	−7.7	−4.2	55.7
SFI at 92°F	−37.0	−32.5	−27.3	−21.5	−17.4	−13.3	−10.6	−6.0	−4.4	40.0
SFI at 104°F	−1.7	−3.8	−6.4	−10.0	−12.2	−13.8	−16.0	−17.5	−19.0	15.0
66 IV H-SBO										
SFI at 50°F	−11.5	−12.0	−11.2	−10.0	−9.1	−7.8	−6.4	−4.5	−2.3	60.5
SFI at 70°F	−26.0	−21.0	−18.3	−15.5	−13.0	−11.0	−8.6	−6.0	−3.2	49.0
SFI at 80°F	−33.7	−25.7	−21.0	−18.0	−15.3	−12.9	−10.1	−7.6	−4.9	44.3
SFI at 92°F	−29.5	−23.5	−19.5	−15.3	−13.3	−10.2	−8.2	−6.6	−2.9	26.5
SFI at 104°F	−12.2	−11.2	−10.9	−8.7	−5.8	−3.9	−2.5	−1.5	−0.6	5.8

SFI effects. It is also important to note that the liquid dilution factors are accurate for the first blend only. The effect of a second blend will be different than that of the initial blend because of a change in the triglyceride composition.

- basestock blend effect—Both contribution and dilution constants must be identified for basestock blends when more than one component has SFI values at the evaluation temperatures. Dilution factors are determined by the same method shown for liquid oils for the SFI temperature evaluations where one of the basestocks does not have a solids content. Contribution factors are the portion of the basestock's weighted average, which analyzes as SFI. Contribution factors are determined as shown in Table 4.13 for a 60:40 blend of 60 IV H-SBO with 74 IV H-SBO.

  The blend results at evaluation temperatures where both basestocks have SFI values are almost linear for calculation purposes, but the evaluation temperatures with zero values have a dilutive effect like the liquid oils. These indications are illustrated in Table 4.14a with analyzed blend SFI results of basestock blends. Table 4.14b summarizes the calculated contribution and dilution factors for the same hydrogenated soybean oil basestock blends.

- hard fat effect—SFI analysis of low-iodine value hard fats is very difficult and seldom performed. Therefore, SFI values are not available for the hard fats, but the effect can be determined by evaluating hard fat blends with the desired basestocks to identify the "bump" or increase at each evaluation temperature contributed by a particular hard fat. Hard fat "bump" varies with the blend basestock, evaluation temperature, use level, and the hard fat source or fatty acid composition. Calculation of hard fat bump factors is demonstrated by Table 4.15 for 10% 4 IV H-CSO (hydrogenated cottonseed oil) hard fat blend with 90% of a 91.5 IV hydrogenated soybean oil basestock.

  Table 4.16a illustrates the SFI changes with the addition of hard fats. The "bump" factors for palm and cottonseed oils low-iodine value hard fat blends are calculated in Table 4.16b.

8.3 Triglyceride Replication

Natural fat systems are mixtures of mixed triglycerides. Their functional properties and quality characteristics are directly related to the type of triglycerides in the fat system. Formulation by the triglyceride replication method requires qualitative and quantitative knowledge of triglyceride composition of the target fat system. Characterization of triglyceride compositions are possible but require a well-equipped laboratory and are time-

TABLE 4.13. Hydrogenated Basestock Blend Calculations

SFI Temp.	60 IV H-SBO		74 IV H-SBO		Actual Blend Analysis	60 IV Base SFI Difference	74 IV Base Contribution [+] Dilution [−]
	Actual SFI	60% × SFI	Actual SFI	40% × SFI			
50°F	65.1	39.06	39.8	15.92	54.3	15.24	Diff ÷ 74 IV = 95.7
70°F	54.2	32.52	22.6	9.04	41.0	8.48	Diff ÷ 74 IV = 93.8
80°F	50.5	30.30	15.3	6.12	35.8	5.50	Diff ÷ 74 IV = 89.9
92°F	32.3	19.38	3.3	1.32	19.7	0.32	Diff ÷ 74 IV = 24.2
104°F	11.5	6.90	None	None	4.5	−2.40	Diff ÷ 40% = −6.0

TABLE 4.14a. Hydrogenated Basestock Blends Solids Fat Index Results.

| 109 IV H-SBO, %: | None | 10 | 20 | 30 | 40 | 50 | 60 | 70 | 80 | 90 | 100 |
Basestock, %:	100	90	80	70	60	50	40	30	20	10	None
60 IV H-SBO											
SFI at 50°F	66.1	57.7	50.5	43.7	37.5	31.1	25.3	19.4	13.7	8.9	3.3
SFI at 70°F	58.8	50.3	42.8	36.2	30.0	24.0	18.7	13.5	8.7	4.5	0.7
SFI at 80°F	55.7	46.9	39.7	33.1	27.0	21.1	15.7	10.8	6.5	3.0	0.3
SFI at 92°F	40.5	33.0	27.0	21.9	17.0	12.5	8.8	5.6	3.0	1.1	
SFI at 104°F	20.6	16.7	13.0	9.7	7.0	5.0	3.0	1.5			
66 IV H-SBO											
SFI at 50°F	65.5	56.7	49.5	43.0	36.7	30.5	25.2	19.5	13.8	8.3	3.0
SFI at 70°F	55.5	48.2	41.0	34.0	27.5	22.2	16.6	12.2	8.1	4.5	0.8
SFI at 80°F	50.7	42.7	35.8	29.3	23.2	18.0	13.2	9.0	5.2	2.5	0.4
SFI at 92°F	31.6	26.2	20.7	15.6	12.3	7.6	4.8	2.9	1.3	0.1	
SFI at 104°F	9.0	6.7	4.5	2.8	1.5	0.5					

| 88 IV H-SBO, %: | None | 10 | 20 | 30 | 40 | 50 | 60 | 70 | 80 | 90 | 100 |
Basestock, %:	100	90	80	70	60	50	40	30	20	10	None
66 IV H-SBO											
SFI at 50°F	60.5	56.1	51.5	47.0	42.6	38.2	34.0	29.8	25.7	21.6	17.6
SFI at 70°F	48.3	42.8	38.0	33.5	29.2	24.7	20.6	16.6	12.7	8.9	5.3
SFI at 80°F	43.2	37.8	32.8	28.1	23.5	19.3	15.4	11.5	8.0	4.5	1.5
SFI at 92°F	25.3	21.1	17.1	13.5	10.4	7.3	5.2	3.0	1.3	0.1	
SFI at 104°F	5.7	4.0	2.4	0.8							

TABLE 4.14a. (continued).

74 IV H-SBO, %:	None	10	20	30	40	50	60	70	80	90	100
Basestock, %:	100	90	80	70	60	50	40	30	20	10	None
60 IV H-SBO											
SFI at 50°F	65.1	62.5	60.0	57.0	54.3	51.7	49.3	46.8	44.5	42.1	39.8
SFI at 70°F	54.2	51.0	47.5	44.3	41.0	38.0	34.7	31.7	28.6	25.5	22.6
SFI at 80°F	50.5	46.5	43.0	39.4	35.8	32.3	28.9	25.5	22.1	18.7	15.3
SFI at 92°F	32.3	29.3	26.0	22.8	19.7	16.6	13.7	10.8	8.2	5.7	3.3
SFI at 104°F	11.5	9.5	7.7	6.0	4.5	3.0	1.6	0.3			
66 IV H-SBO											
SFI at 50°F	59.3	57.3	55.7	54.0	52.6	50.5	48.8	47.1	45.5	43.7	41.7
SFI at 70°F	46.7	44.3	42.2	40.0	32.7	35.4	33.0	30.7	28.5	26.6	23.8
SFI at 80°F	42.1	39.5	37.0	34.5	32.2	29.3	26.6	24.1	21.5	18.9	16.3
SFI at 92°F	25.4	22.9	20.6	18.4	16.1	14.0	11.8	9.6	7.5	5.3	3.0
SFI at 104°F	7.5	5.6	5.0	3.5	2.6	1.5	0.7				
80 IV H-SBO											
SFI at 50°F	22.5	24.5	26.4	28.4	30.2	32.0	34.0	36.0	36.9	39.7	41.6
SFI at 70°F	9.7	11.3	12.8	14.2	15.7	17.3	18.8	20.3	21.7	23.2	24.5
SFI at 80°F	5.3	6.3	7.4	8.5	9.6	10.8	12.0	13.3	14.5	15.9	17.0
SFI at 92°F	0.5	0.7	1.0	1.2	1.6	2.0	2.5	2.9	3.4	3.7	4.0

TABLE 4.14b. Solids Fat Index Calculation Factors.

109 IV H-SBO Contribution [+] and Dilution [−] Factors

109 IV H-SBO, %: / Basestock, %:	10 / 90	20 / 80	30 / 70	40 / 60	50 / 50	60 / 40	70 / 30	80 / 20	90 / 10	Basestock SFI
60 IV H-SBO										**60 IV–107 IV**
SFI at 50°F [+]								18.2	75.8	66.1–3.3
SFI at 50°F [−]	−17.9	−11.9	−8.6	−5.4	−3.9	−1.9	−0.6			66.1–3.3
SFI at 70°F [−]	−26.2	−21.2	−16.5	−13.2	−10.8	−8.0	−5.9	−3.8	−1.5	58.8–0.7
SFI at 80°F [−]	−32.3	−24.3	−19.6	−16.1	−13.5	−11.5	−8.4	−5.8	−2.9	55.7–0.3
SFI at 92°F [−]	−34.5	−27.0	−21.5	−18.3	−15.5	−12.3	−9.4	−6.4	−3.2	40.5
SFI at 104°F [−]	−18.4	−17.4	−15.7	−13.4	−10.6	−8.7	−7.1	−5.2	−2.3	20.6
66 IV H-SBO										**66 IV–107 IV**
SFI at 50°F [+]								30.0	63.3	65.5–3.0
SFI at 50°F [−]	−22.5	−14.5	−9.5	−6.5	−4.5	−0.2	−0.2			65.5–3.0
SFI at 70°F [−]	−17.5	−17.0	−16.1	−14.5	−11.1	−9.3	−6.4	−3.8	−1.2	55.5–0.8
SFI at 80°F [−]	−29.3	−23.8	−20.6	−18.1	−14.7	−11.8	−8.9	−6.2	−2.9	50.7–0.4
SFI at 92°F [−]	−22.4	−22.9	−21.7	−16.7	−16.4	−13.1	−9.4	−6.3	−3.5	31.6
SFI at 104°F [−]	−14.0	−13.5	−11.7	−9.8	−8.0	−6.0	−3.9	−2.3	−1.0	9.0

88 IV H-SBO Contribution [+] and Dilution [−] Factors

88 IV H-SBO, %: / Basestock, %:	10 / 90	20 / 80	30 / 70	40 / 60	50 / 50	60 / 40	70 / 30	80 / 20	90 / 10	Basestock SFI
66 IV H-SBO										**66 IV–88 IV**
SFI at 50°F [+]	93.8	88.0	88.0	89.8	90.3	92.6	94.3	96.6	98.3	60.5–17.6
SFI at 70°F [+]				11.3	20.8	39.6	56.6	71.7	84.9	48.3– 5.3
SFI at 70°F [−]	−6.7	−0.8	−0.4							48.3– 5.3
SFI at 80°F [−]	−10.8	−8.8	−7.1	−6.1	−4.6	−3.1	−2.1	0.8	0.2	43.2– 1.5
SFI at 92°F [−]	−16.7	−15.7	−14.0	−12.0	−10.1	−8.2	−6.6	−4.7	−2.7	25.3
SFI at 104°F [−]	−11.3	−10.8	−10.6	−8.6	−5.7	−3.8	−2.4	−1.4	−0.6	5.7

TABLE 4.14b. (continued).

74 IV H-SBO, %: Basestock, %:	10 90	20 80	30 70	40 60	50 50	60 40	70 30	80 20	90 10	Basestock SFI
60 IV H-SBO										60 IV–74 IV
SFI at 50°F [+]	98.2	99.5	95.7	95.7	96.2	97.4	98.0	98.9	99.4	65.1–39.8
SFI at 70°F [+]	98.2	91.6	93.8	93.8	96.5	96.0	97.6	98.2	98.7	54.2–22.6
SFI at 80°F [+]	68.6	85.0	88.2	89.9	92.2	94.8	96.7	98.0	99.1	50.5–15.3
SFI at 92°F [+]	69.7	24.2	19.2	24.2	27.3	39.4	48.1	65.9	83.2	32.3– 3.3
SFI at 104°F [−]	−8.5	−7.5	−6.8	−6.0	−5.5	−5.0	−4.5	−2.9	−1.3	11.5– 0.0
66 IV H-SBO										66 IV–74 IV
SFI at 50°F [+]	94.2	99.0	99.8	102.0	100.0	100.2	100.4	100.8	100.6	59.3–41.7
SFI at 70°F [+]	95.4	101.7	102.4	101.7	101.3	100.3	100.2	100.6	100.5	46.7–23.8
SFI at 80°F [+]	98.8	101.8	102.9	106.4	101.2	99.8	100.5	100.3	100.1	42.1–16.3
SFI at 92°F [+]	13.3	46.7	68.9	71.7	86.7	91.1	94.3	100.8	102.2	25.4– 3.0
SFI at 104°F [−]	−11.5	−5.0	−5.8	−4.8	−4.5	−3.8	−3.2	−1.9	−0.8	7.5
80 IV H-SBO										80 IV–41.6 IV
SFI at 50°F [+]	102.2	101.0	101.3	100.4	99.8	100.2	100.4	99.4	100.0	22.5–41.6
SFI at 70°F [+]	104.9	102.9	100.8	100.8	101.6	101.5	101.4	100.8	100.8	9.7–24.5
SFI at 80°F [+]	90.0	92.9	93.9	94.4	95.9	96.9	98.4	98.8	100.5	5.3–17.0
SFI at 92°F [+]	62.5	75.0	79.2	81.3	87.5	95.8	98.2	103.1	101.4	0.5– 4.0

TABLE 4.15.

SFI Temp.	91.5 IV H-SBO Basestock		Blend SFI	SFI Difference	4 IV CSO Hard Fat Bump
	SFI	Contribution			
50°F	18.1	×90% = 16.29	25.1	8.81	÷10 = 0.88
70°F	7.6	×90% = 6.84	18.6	11.76	÷10 = 1.18
80°F	3.2	×90% = 2.88	18.0	15.12	÷10 = 1.51
92°F	0.1	×90% = 1.09	14.7	14.61	÷10 = 1.46
104°F	NONE	NONE	10.9	10.90	÷10 = 1.09

consuming, tedious analysis. Insight into the molecular composition of fats can be obtained from the fatty acid composition and further refined to differentiate positional and geometric isomers of both fatty acids and triglycerides.

After the molecular structures for the target fat system are identified, individual components from specific source oils and product fractions obtained by hydrogenation, fractionation, interesterification, or a combination of these processes are blended to replicate the desired composition. The source oils and processing normally contain and/or produce unwanted triglyceride structures, which must be separated from the mixture. This process creates one or more by-products, which necessitates additional product development studies to identify cost-effective utilization of these surplus fractions.

This formulation technique has achieved commercial success but is primarily limited in scope and application to cocoa butter equivalents. Cocoa butter is composed of predominantly symmetrical triglycerides, approximately 75%, with oleic fatty acid in the 2-position. It also contains approximately 20% triglycerides that are liquid at room temperature, has a melting range of 32 to 35°C, and softens around 30 to 32°C. The unique triglyceride composition, together with the extremely low levels of diglycerides, contribute to the desirable physical properties of cocoa butter and its ability to recrystallize in a stable crystal form. Therefore, cocoa butter is a simple three-component system consisting of palmitic-oleic-palmitic (POP), palmitic-oleic-stearic (POS), and stearic-oleic-stearic (SOS) triglycerides, and if components with these three triglycerides are blended in the right proportions, the resultant triglyceride replication product will behave like cocoa butter [43]. Commercial products composed of a palm oil middle-melting fraction rich in POP, and fats such as Shea or Sal, are fractionated to obtain triglyceride fractions rich in POS and SOS. Careful preparation

TABLE 4.16a. Low-Iodine Value Hard Fat Blends Solids Fat Index Results.

56-Titer Palm Oil, %: Basestock, %:	None 100	2 98	4 96	6 94	8 92	10 90	12 88	14 86	15 85
109 IV H-SBO									
SFI at 50°F	2.0	4.0	6.0	8.0	10.0	12.0	14.0	16.0	17.0
SFI at 70°F	0.8	3.0	5.2	7.7	9.7	11.5	13.8	15.8	16.8
SFI at 80°F		2.5	4.9	7.2	9.7	11.5	13.8	15.8	16.8
SFI at 92°F		1.8	3.8	5.7	8.0	10.2	12.3	14.5	15.5
SFI at 104°F			1.0	3.0	5.5	7.3	9.3	11.5	12.5
74 IV H-SBO									
SFI at 50°F	41.0	41.5	42.0	43.0	44.5	45.0	46.3	47.5	48.0
SFI at 70°F	23.5	25.5	26.8	29.0	31.0	32.5	34.0	35.0	36.7
SFI at 80°F	16.0	19.5	22.5	25.0	28.0	30.0	32.0	33.7	34.5
SFI at 92°F	4.0	8.0	11.5	15.0	18.0	21.5	25.0	25.7	26.0
SFI at 104°F		1.0	3.0	5.5	9.5	12.0	14.5	16.5	17.5
66 IV H-SBO									
SFI at 50°F	64.0	65.0	66.0	66.5	67.0	67.3	67.8	68.0	68.0
SFI at 70°F	55.0	55.5	57.0	58.5	59.2	60.5	61.0	61.5	61.5
SFI at 80°F	50.5	52.0	53.5	55.0	56.5	58.0	59.0	59.5	59.5
SFI at 92°F	32.0	34.5	37.5	40.3	42.5	44.8	46.5	47.5	47.5
SFI at 104°F	15.0	14.3	17.5	20.8	23.5	26.0	28.3	30.0	31.0

(continued)

TABLE 4.16a. (continued).

4 IV H-CSO Hard Fat, %: Basestock, %:	None 100	5 95	10 90	15 85	20 80
91.5 IV H-SBO					
SFI at 50°F	18.1	21.9	25.1	29.2	33.5
SFI at 70°F	7.6	12.9	18.6	22.6	26.4
SFI at 80°F	3.2	11.9	18.0	22.0	26.0
SFI at 92°F	0.1	8.5	14.7	19.5	24.0
SFI at 100°F		6.1	12.2	17.7	22.6
SFI at 104°F		5.1	10.9	16.8	22.2
28 IV H-CSO Hard Fat, %: Basestock, %:	None 100	5 95	10 90	15 85	20 80
91.5 IV H-SBO					
SFI at 50°F	18.1	23.7	26.8	29.9	31.8
SFI at 70°F	7.6	12.6	17.0	21.1	25.1
SFI at 80°F	3.2	9.6	14.8	19.6	23.7
SFI at 92°F	0.1	4.6	8.9	13.3	17.5
SFI at 100°F		1.8	5.2	8.7	12.4
SFI at 104°F		1	3.5	6.9	10.4

TABLE 4.16b. Solids Fat Index Calculation Factor.

56-Titer Palm Oil, %: Basestock, %:	56-Titer Palm Oil Hard Fat Bump								Basestock SFI
	2 98	4 96	6 94	8 92	10 90	12 88	14 86	15 85	
109 IV H-SBO									
SFI at 50°F	1.02	1.02	1.02	1.02	1.02	1.02	1.02	1.02	2.0
SFI at 70°F	1.11	1.11	1.16	1.12	1.08	1.09	1.08	1.07	0.8
SFI at 80°F	1.25	1.23	1.20	1.21	1.15	1.15	1.13	1.12	
SFI at 92°F	0.90	0.95	0.95	1.00	1.02	1.03	1.04	1.03	
SFI at 104°F	0.00	0.25	0.50	0.69	0.73	0.78	0.82	0.83	
74 IV H-SBO									
SFI at 50°F	0.66	0.66	0.74	0.85	0.81	0.85	0.87	0.88	41.0
SFI at 70°F	1.24	1.19	1.15	1.17	1.14	1.11	1.06	1.12	23.5
SFI at 80°F	1.91	1.79	1.66	1.66	1.56	1.49	1.42	1.39	16.0
SFI at 92°F	2.04	1.92	1.87	1.79	1.79	1.79	1.82	1.51	4.0
SFI at 104°F	0.50	0.75	0.92	1.19	1.20	1.21	1.18	1.17	
66 IV H-SBO									
SFI at 50°F	1.14	1.14	1.06	1.02	0.97	0.96	0.93	0.96	64.0
SFI at 70°F	0.80	1.05	1.13	1.08	1.10	1.05	1.01	0.96	55.0
SFI at 80°F	1.26	1.26	1.26	1.26	1.26	1.21	1.15	1.11	50.5
SFI at 92°F	1.57	1.70	1.70	1.63	1.60	1.53	1.43	1.35	32.0
SFI at 104°F	0.20	0.78	1.12	1.21	1.25	1.26	1.22	1.22	15.0

(continued)

TABLE 4.16b. (continued).

CSO Hard Fat, %:	4 IV CSO Hard Fat Bump				28 IV CSO Hard Fat Bump				Basestock SFI
Basestock, %:	5.0 95.0	10.0 90.0	15.0 85.0	20.0 80.0	5.0 95.0	10.0 90.0	15.0 85.0	20.0 80.0	
91.5 IV H-SBO									
50°F SFI bump	0.94	0.88	0.98	0.95	1.30	1.05	0.96	0.78	18.1
70°F SFI bump	1.14	1.18	1.07	1.02	1.08	1.02	0.98	0.95	7.6
80°F SFI bump	1.77	1.51	1.28	1.17	1.31	1.19	1.13	1.06	3.2
92°F SFI bump	1.68	1.46	1.29	1.20	0.90	0.88	0.88	0.87	0.1
100°F SFI bump	1.22	1.22	1.12	1.11	0.36	0.52	0.58	0.62	
104°F SFI bump	1.02	1.09	1.12	1.11	0.20	0.35	0.46	0.52	

and blending of these components result in a vegetable fat equivalent to cocoa butter in performance [44].

9. SOURCE OILS AND FATS INTERCHANGEABILITY

Substitution of one source oil for another may be desirable due to raw material availability, cost reduction, religious prohibitions, consumer preference, legislation, and so on. The degree of interchangeability possible is determined by (1) the characteristics of the initial product, (2) the substituting source oil or oils, (3) the requirements of the application, and (4) the product knowledge and information available to the formulator. The raw materials of interchangeability are naturally the source oils and fats available to the formulator, which vary in chemical composition and physical properties and may require different processing techniques to become adequate substitutes. The effect of emulsifiers, antioxidants, crystal inhibitors, antifoamers, and other additives must also be considered for successful substitute or alternative formulations.

Source oil interchangeability assumes that a product with a degree of acceptable performance exists. All alternative formulation developments start by identifying the requirements of the product. For example, margarine was developed as a substitute for butter to provide an affordable food source with a more reliable supply. The performance requirements for this substitute product were the flavor characteristics, consistency, and eating characteristics, or mouth feel, of butter. In this case, technology was used to develop a substitute, but not an exact duplicate, for the existing product.

The tools available for source oil interchangeability formulation are (1) analytical methods to identify the characteristics of the existing product, (2) processes that alter the physical and chemical nature of the source oils, (3) performance evaluations to determine if the substitute product has the desired application characteristics, and (4) a data base to screen the potential product components [45].

The degree of interchangeability possible is determined by the characteristics of the initial product, the substituting source oils, and the requirements of the application. In the simplest cases source oil substitution involves (1) direct substitution of one oil for another, (2) blending source oils, or (3) inclusion of a hydrogenated basestock or hard fat to satisfy an SFI or SFC profile.

Alternate formulation development becomes more complex when the source oils have dissimilar characteristics such as the substitution of a semisolid like palm oil for a liquid oil like soybean oil. Likewise, it is necessary to consider the crystal habit of the oil or fat being introduced and its effect upon the product's crystalline form. Beta polymorph is more stable than beta-prime, but the rate of transition from one crystal form to another differs for each oil type. The beta-prime form is preferred for most shortening and margarines because it promotes plasticity, smooth texture, and improved creaming properties. The beta crystal form tends to produce large granular crystals and products that are waxy and grainy with poor aeration potential. The beta formers perform well in applications such as pie crusts where a grainy texture is desirable, frying shortenings where the crystal is destroyed by heating, and liquid shortenings where the large granular crystals are preferred for stability and fluidity maintenance. The ratio of beta to beta-prime crystal formers helps to determine the dominant crystal habit, but the higher melting triglyceride portions of a solidified fat product usually forces the fat to assume that crystal form. In practice, the use of beta-prime hard fats at levels as low as 5% can stabilize a product in the beta-prime polymeric form. Other beta-prime stabilization methods are interesterification, low-temperature storage, and the use of additives such as sorbitan tristearate and diglycerides.

10. REFERENCES

1. Markley, K. S. 1960. "Nomenclature, Classification and Description of Individual Acids," in *Fatty Acids, Vol. 1*, Second Edition, K. S. Markley, ed. New York, NY: Interscience Publishers, Inc., pp. 34, 35, 50–53.

2. Dutton, H. J., C. D. Evans and J. C. Cowan. 1953. "Status of Research on the Flavor Problem of Soybean Oil at the Northern Research Laboratory," *Trans. Am. Assoc. Cereal Chemists*, 11:116–135.

3. Gunstone, F. D. and T. P. Hilditch. 1946. *J. Chem. Soc.* p. 1022.

4. Erickson, D. R. and G. R. List. 1985. "Fat Degradation Reactions" in *Bailey's Industrial Oil and Fat Products, Vol. 3*, Fourth Edition, T. H. Applewhite, ed., New York, NY: John Wiley & Sons, pp. 275–277.

5. Buck, D. F. 1981. "Antioxidants in Soya Oil," *J. Am. Oil Chem. Soc.*, 58(3):275.

6. Sherwin, E. R. 1976. "Antioxidants for Vegetable Oils," *J. Am. Oil Chem. Soc.*, 53(6):430–436.

7. Eastman Chemical Company. 1990. Tenox Natural Tocoperols, Technical Data Bulletin 2G–263.

8. Chahine, N. H. and R. F. MacNeill. 1974. "Effect of Stabilization of Crude Whale Oil with Tertiary Butylhydroquinone and Other Antioxidants upon Keeping Quality of Resultant Deodorized Oil," *J. Am. Oil Chem. Soc.*, 51(3):37–41.

9. Sherwin, E. R. 1992. "Antioxidants for Food Fats and Oils," *J. Am. Oil Chem. Soc.*, 49(8):468–472.

10. Luckadoo, B. M. and E. R. Sherwin. 1972. "Tertiary Butylhydroquinone as Antioxidant for Crude Sunflower Seed Oil," *J. Am. Oil Chem. Soc.*, 49(2):95–97.

11. Sherwin, E. R. and B. M. Luckadoo. 1970. "Studies on Antioxidant Treatment of Crude Vegetable Oils," *J. Am. Oil Chem. Soc.*, 47(1):19–23.

12. Kraybill, H. R. et al. 1949. "Butylated Hydroxyanisole as an Antioxidant for Animal Fats," *J. Am. Oil Chem. Soc.*, 26(8):449–453.

13. Bessler, T. R. and F. T. Ortheoefer. 1983. "Providing Lubricity in Food Fat Systems," *J. Am. Oil Chem. Soc.*, 60(10):1765–1768.

14. Broady, H. and M. Cochran. 1978. "Shortening for Bakery Cream Icings and Cream Fillers," *The Baker's Digest*, 52(12): 23.

15. Hoerr, C. W. 1960. "Morphology of Fats, Oils, and Shortenings," *J. Am. Oil Chem. Soc.*, 37(10):539–546.

16. Hoerr, C. W., J. Moncrieff, and F. R. Paulicka, 1966. "Crystallography of Shortenings," *The Bakers Digest*, 40(4):38–40.

17. Hoerr, C. W. and J. V. Ziemba. 1965. "Fat Crystallization Points Way to Quality," *Food Engineering*, 37(5):90–93.

18. Wiederman, L. H. 1988. "Margarine and Margarine Oil, Formulation and Control," *J. Am. Oil Chem. Soc.*, 55(11):825–828.

19. Posmore, J. 1987. "Application of Modification Techniques," in *Recent Advances in Chemistry and Technology of Fats and Oils*, R. J. Hamilton and A. Bhati, ed. London, England: Elsevier Applied Science Publishing Co., Inc., pp. 172–175.

20. Mattil, K. F. 1964. "Plastic Shortening Agents," in *Bailey's Industrial Oil and Fats Products*, Third Edition, D. Swern, ed. New York, NY: Interscience Publishers, a Div. of John Wiley & Sons, pp. 273–275.

21. Williams, K. A. 1927. *J. Soc. Chem. Ind.*, 46:448–449.

22. Latondress, E. G. 1981. "Formulation of Products from Soybean Oil," *J. Am. Oil Chem. Soc.*, 53(3):185–187.

23. O'Brien, R. D. 1987. "Formulation—Single Feedstock Situation," in *Hydrogenation: Proceedings of an AOCS Colloquium*, R. Hastert, ed. Champaign, IL: American Oil Chemists Society, pp. 155–169.

24. Latondress, E. G. 1980. "Shortenings and Margarines: Base Stock Preparation and Formulation," in *Handbook of Soy Oil Processing and Utilization*, D. R. Erickson et al., ed., Champaign, IL, American Soybean Association and Am. Oil Chem. Soc., pp. 146–154.

25. Thomas, A. E., III. 1981. "Importance of Glyceride Structure to Product Formulation," *J. Am. Oil Chem. Soc.*, 58(3):238–239.

26. Patterson, H. B. W. 1973. "Hydrogenation of Vegetable Oils," *Oleagineux*, 28(12):585.

27. Hastert, R. C. 1990. "Cost/Quality/Health: The Three Pillars of Hydrogenation," in *Proceedings of World Conference, Edible Fats and Oils Processing: Basic Principals and Modern Practices*, D. R. Erickson, ed. Champaign, IL, Am. Oil Chem. Soc. pp. 148–149.

28. Nash, N. H. and L. M. Brickman. 1972. "Food Emulsifiers Science and Art," *J. Am. Oil Chem. Soc.*, 49(8):457–461.

29. Birnbaum, H. 1981. "The Monoglycerides: Manufacture, Concentration, Derivatives, and Applications," *Bakers Digest*, 55(12):6–16.

30. Dziezak, J. D. 1988. "Emulsifiers: The Interfacial Key to Emulsion Stability," *Food Technology*, 42(10):172–186.

31. Krog, N. 1977. "Functions of Emulsifiers in Food Systems," *J. Am. Oil Chem. Soc.*, 54(3): 124–131.

32. Neu, G. D. and W. J. Simcox. 1975. "Dough Conditioning and Crumb Softening in Yeast-Raised Bakery Products with Succinylated Monoglycerides," *Cereal Foods World,* 20(4): 203–208.

33. Henry, C. 1995. "Monoglycerides: The Universal Emulsifier," *Cereal Foods World,* 40(10): 734–738.

34. Silva, R. 1990. "Phospholipids as Natural Surfactants for the Cereal Industry," *Cereal Foods World,* 35(10):1008–1012.

35. Sinram, R. D. 1991. "The Added Value of Specialty Lecithins," *Oil Mill Gazetteer,* September, pp. 22–26.

36. Anonymous 1963. *The Atlas HLB System,* 3rd Edition, Atlas Chemical Industries, Inc., pp. 3–20.

37. Griffin, W. C. and M. J. Lynch. 1972. "Surface Active Agents," in *CRC Handbook of Food Additives, Vol. 1,* T. E. Furia, ed. Boca Raton: CRC Press, pp. 404–410.

38. Anonymous, *Grindsted Emulsifiers,* Grindsted Products, Inc., North Kansas City, MO, pp. 3–5.

39. Anonymous, *Durkee Food Emulsifiers,* Durkee Industrial Foods Group, Cleveland, Ohio, pp. 6–8.

40. Anonymous, *Those Wonderful Disappearing Emulsifiers from Humko Chemical,* Humko Chemical Division, Witco Corp., Memphis, TN, pp. 1–3.

41. *The Official Methods and Recommended Practices of the American Oil Chemists' Society,* 1994. 4th ed. Champaign, IL: American Oil Chemists' Society.

42. Wiederman, L. H. 1978. "Margarine and Margarine Oil: Formulation and Control," *J. Am. Oil Chem. Soc.,* 55(11):826–827.

43. Shukla, V. K. S. 1990. "Confectionery Fats," in *Edible Fats and Oils Processing: Basic Principles and Modern Practices: World Conference Proceedings,* D. R. Erickson, ed. Champaign, IL: American Oil Chemists Society, p. 228.

44. Haumann, B. F. 1984. "Confectionery Fats—For Special Uses," *J. Am. Oil Chem. Soc.,* 61(3):468–472.

45. Young, F. V. K. 1985. "Interchangeability of Fats and Oils," *J. Am. Oil Chem. Soc.,* 62(2): 372–375.

CHAPTER 5

Shortening Types

1. INTRODUCTION

Originally, *shortening* was the term used to describe the function performed by naturally occurring solid fats like lard and butter in baked products. These fats contributed a "short," or tenderizing, quality to baked products by preventing the cohesion of the flour gluten during mixing and baking. Shortening later became the product identification used by all-vegetable oils processors to abandon the lard substitute concept. As the shortening product category developed, the limited application also expanded to include all baked products. Today, shortening has become virtually synonymous with fat to include many other types of edible fats designed for purposes other than baking. In most cases, products identified as shortening will be 100% fat; however, there are exceptions, such as puff pastry and roll-in shortenings, which may contain moisture. Many fats and oils products are now called shortening to distinguish them from margarine. Generally, if the fat product contains at least 80% fat and has the required vitamin A content, it is a margarine. Products that do not meet this criteria have been identified as shortening since they do not have a U.S. Standard of Identity. Currently, a description for shortening would be: processed fats and oils products that affect the stability, flavor, storage quality, eating characteristics, and the eye appeal of prepared foods by providing emulsification, lubricity, structure, aeration, a moisture barrier, a flavor medium, or heat transfer.

1.1 Historical

When and how man learned to use fats and oils is unknown, but it is known that primitive people in all climates used them for food, medicine,

327

cosmetics, lighting, preservatives, lubricants, and other purposes. The use of fats as foods was probably instinctive, while the other applications most likely resulted from observations of their properties and behavior under various environmental conditions. Probably the first fats used by man were of animal origin, which were separated from the tissue by heating or boiling. Recovery of oil from small seeds or nuts required the development of more advanced methods of processing, i.e., cooking, grinding, and pressing processes [1].

The first animal fats used by man were more than likely rendered from wild animal carcasses. As animals were domesticated, their body fat became an important food source. Lard or hog fat became the preferred meat fat for edible purposes, while the other animal fats were utilized for nonedible applications. The more pleasing flavor of lard may have been one reason for its choice for edible purposes. However, the principle reason undoubtedly was the plastic consistency of this fat. At room temperature, lard had a good consistency for incorporation into breads, cookies, cakes, and pastries, as well as most other baked products. Beef fat and mutton tallow were too firm for this use, and the available marine oils were too fluid [2].

Vegetable shortening was an American invention. The expansion of cotton acreage following the American Civil War resulted in large quantities of cottonseed oil. An initial outlet as an inexpensive whale oil substitute disappeared with the development of the U.S. petroleum industry. The European market utilized cottonseed oil in animal feeds and to dilute the more expensive olive oil. This adulteration practice led to tariffs on imported U.S. cottonseed oil, which discontinued all exports to Italy. High lard prices offered a domestic opportunity for cottonseed oil utilization. Initially, meat packers secretly added cottonseed oil to lard. This practice was uncovered when Armour and Company discovered that they had received deliveries of more lard than the existing hog population could have produced. Public disclosure of this adulteration led to a Congressional investigation, which made identification with the descriptive name "lard compound" mandatory [3,4].

Major improvements in oil processing had to take place before substantial quantities of cottonseed oil could replace lard as the preferred fat for baking and frying. Alkali refining procedures were developed in Europe around 1840 but were not employed in the United States until the 1880s. Caustic refining removes free fatty acids and a large proportion of the color pigments, but refined cottonseed oil had limited acceptance for packing sardines and for mixing with lard. Consumers expected lard to be white in color; therefore, straw-colored cottonseed oil had to be bleached to be acceptable for lard compounds. The first cottonseed oil bleaching was accomplished by exposing it to sunlight in large shallow tanks for up to 18 months. Carbon, the first absorbent bleaching agent used was replaced later by Fuller's earth with use patents issued in 1880 [5]. Initially, liquid vegetable oils were converted to solid or plastic fats by blending a high proportion of the vegetable oil with a small quantity of oleostearin or other hard animal fats; however, the unpleasant flavor contributed by the cottonseed oil portion was so strong that the product had very limited acceptance. Attempts to chemically remove the offensive flavor or mask them with spices or flavors were unsuccessful. Deodorization by blowing live steam through the oil at elevated temperatures was introduced around 1891 and was quickly adopted by most American processors [6]. David Wesson later perfected the deodorization process by exposing the oil to superheated steam in a vacuum [7]. The hydrogenation process enabled the vegetable oil processors to become independent of the meat packing industry. Utilizing a British hydrogenation patented process, Procter & Gamble produced the first all-vegetable shortening with cottonseed oil. "Crisco," short for crystallized cotton oil, was introduced at the start of World War I [8]. Either lard or shortening must be quick-chilled to produce the desirable smooth consistency. Internally refrigerated chill rolls were the first improved apparatus developed for crystallizing shortening products. In the early 1930s development of improved heat-transfer equipment for freezing ice cream led to the perfection of a closed continuous internal chilling unit, which replaced chill rolls as shortening plasticization units. Versions of these closed internal chilling systems are still employed to plasticize shortenings [9].

The all-vegetable oil processors had the foresight to abandon the lard substitute product concept and description to offer their hydrogenated products as a new food ingredient that has become known as shortening. Processing improvements had effected the development of shortening products that had definite advantages over lard. Pure vegetable shortenings quickly assumed a position of preeminence and was accepted as a premium product by both the housewife and the commercial user or baker. The bland flavor, uniform color, and smooth texture were undoubtedly prime factors that influenced acceptance by the consumer. These factors also influenced the baker's decision, but the deciding factors had to be the increased stability along with improved creaming properties [10].

Fats and oils chemistry was virtually an unexplored field until a rapid expansion of interest began to be evident in the late 1920s. The opportunities presented by the advancements with hydrogenation of liquid oils attracted research chemists to the lipid chemistry field. This increased research activity led to the development of new commercially important products. These developments helped dispel the previous academic theory that fats and oils were simplistic. It was found that fats and oils were capable of undergoing many of the classic organic chemistry reactions, i.e., isomerization, polymerization, oxidation, esterification, and interesterification [1].

In 1933, the introduction of superglycerinated, High-Ratio [11] shortening brought about significant changes for the baker and the shortening industry. These shortenings contained mono- and diglycerides, which contributed to a finer dispersion of fat particles in cake batters, causing a greater number of smaller sized fat globules that strengthened the batters. Emulsified shortenings allowed bakers to produce cakes with additional liquids, which permitted higher sugar levels. Additionally, the surface active agents improved aerating or creaming properties, which allowed less reliance upon specific crystalline and solids properties for functionality. This improvement also allowed a reduction of shortening levels in some bakery cake formulations without sacrificing aeration and tenderness qualities. Altogether, the superglycerinated shortenings produced more moist, higher volume cakes with a fine grain and an even texture and extended shelf life. As a bonus, it was identified that lighter icings and fillings with higher moisture levels could be produced with emulsified shortenings and that a yeast-raised product's shelf life was extended [12]. Performance improvements were also found in many other food products where aeration, moisture retention, starch complexing, and the other benefits of emulsification helped to improve the functionality of shortenings.

Development of emulsified shortenings added a new dimension to the fats and oils industry; it ushered in the era of tailor-made shortenings. New shortenings specifically designed for a special application such as layer cake, pound cake, cake mixes, breads, sweet doughs, icings, creme fillings, whipped toppings, laminated pastries, and other bakery products were developed and introduced rapidly after World War II [13]. Specialty shortening development fostered further improvements in all aspects of the fats and oils industry and expanded shortenings beyond the baking industry. New food products and concepts developed as dairy analogs, confectionery, foodservice, and other areas were successful in many cases due to the development of functional specialty shortening products.

Increased automation promoted by inflation and labor costs caused food processors to investigate new handling methods for all ingredients in the late 1950s. Processors using large quantities of packaged shortenings began investigating and converting to bulk handling of these products. These systems in many cases required the food processor to install plasticization equipment in addition to storage and unloading facilities for handling heated shortenings. These bulk handling requirements encouraged the development of liquid shortenings. Through unique processing procedures, these products, consisting of suspended hard fats and emulsifiers or other additives in a liquid oil, remained fluid or pumpable at room temperature. It is possible to pump, meter, and bulk store these shortenings without heating. Liquid shortenings are dependent upon emulsifiers or other additives to provide the functionality required for specific product applications, typically cakes, frying, bread, dairy analogs, and others. Liquid shortenings did not eliminate plasticized packaged or bulk handled shortenings; they created another shortening type.

Shortening products are now developed for a specific food product and in many cases for a particular process to prepare that product. Shortening chips and stabilizers were both developed to accommodate particular processes while providing a specific functionality. Both of these products were a departure from the traditional solid plastic consistency of the more recent liquid shortening product forms, and both products types are flake form. Shortening chips provide an alternate process to dough lamination to develop a flaky consistency for new products produced with unique processes while the stabilizers improved functionality and eye appeal of packaged products for improved marketing appeal.

Technology advances have increased the storehouse of fats and oils knowledge, allowing the introduction of more advanced products for all aspects of the food industry. Specialty shortenings have helped create entirely new food products and improved product extensions for the retail consumer and the foodservice industry. The word *shortening* no longer identifies the function performed by a fat or oil product, nor does it indicate the type or consistency of the fats and oils product. Shortening products are now produced in solid plastic, liquid, flake, and powdered forms for a diverse application range: bakery products, dairy analogs, snack foods, nutritional supplements, confections, and other prepared foods.

1.2 Source Oils

Shortenings are a unique food ingredient in that a high degree of interchangeability among the raw materials is possible for many products and uses. However, in order for a particular oil to substitute for another in a given product, it may be necessary for it to undergo additional processing steps, which may increase its cost to become an adequate replacement. After this additional processing, if the replacement fat or oil is substitutable in terms of physical and analytical properties in the end product, then the price becomes a major consideration for employing the raw material replacement. Experience has shown that small cost differences in competing source oils can markedly change the proportion of the oils used in a shortening.

Shifts in the utilization of various fats and oils in the composition of an individual shortening were more common in the past than at present.

Source oil labeling requirements have made alternate formulations for the same product difficult. However, shortening customers may still substitute alternative source oil–produced shortenings that have comparable performance characteristics. Table 5.1 tracks the changes in the source oil utilization for U.S. shortenings from 1940 through the 1993-1994 crop year.

Lard utilization as a shortening agent had decreased to a low point in 1940 because cottonseed oil had become the dominant source oil for shortening production. Interestingly, soybean oil had the second highest shortening usage at this early date and accounted for almost 18% of the shortening total. The next decade, which included World War II, was a time of change for shortening's source oil utilization, as illustrated by a comparison of the 1940 and 1950 usage: soybean oil replaced cottonseed oil as the highest volume vegetable oil at just short of half the total requirement (48.7%) and the introduction of interesterified or crystal modified lard in 1950 helped return this raw material to shortening production. Interesterification modified lard's triglyceride structure to provide a consistency, appearance, and creaming properties comparable to the all-vegetable shortenings.

Soybean oil rose from a minor, little known, problem-related oil before 1940 due to the unavailability of other source oils during World War II. After the war, these volume gains were in jeopardy unless adequate technology could be developed to improve the flavor stability of soybean oil products. At the close of World War II, it was learned that the German oil seed industry had developed a recipe to cure soybean oil reversion, which

TABLE 5.1. U.S. Shortening Source Oils Usage [14–16] (millions of pounds).

Source Oil	1940	1950	1960	1970	1980	1990	1995
Coconut oil	18	20	10	45	103	34*	67*
Corn oil	1	1	4	12	W	304	100
Cottonseed oil	823	549	365	276	189	252	212
Palm oil	NR	NR	NR	90	88	98**	87**
Peanut oil	23	12	2	16	NR	NR	NR
Soybean oil	212	841	1169	2182	2651	4004	4673
Lard	17	177	480	430	378	264	325
Tallow	58	31	268	522	684	637	374
Unidentified	44	27	3	7	86		121
Total	1196	1727	2301	3580	4224	5793	5959
Per capita (lb)		11.0	12.6	17.3	18.2	22.3	23.0

* Estimated.
**Total edible.
W = withheld; NR = none recorded.

amounted to the treatment of deodorized oil with citric acid to chelate prooxidant metals [18]. Citric acid treatment improved the flavor stability of soybean oil along with the other technologies developed, i.e., improved hydrogenation catalyst, antioxidants, surface active agents, and nitrogen blanketing. The changes, along with other processing improvements and controls, helped soybean oil reach and maintain a dominate position as a source oil for shortenings in 1960, with just over 50% of the total oil requirements. Soybean oil's share of the shortening requirements has risen to over 78% for the 1995 crop year.

Palm oil threatened to become a major source oil for U.S. shortenings in the mid-1970s. It grew from a level too small to report in 1960 to over 16% of the total shortening raw material requirement in 1975. Attractive palm oil costs had encouraged most fats and oils processors to investigate its use wherever possible. Palm oil was found to be an excellent plasticization agent to force shortening's crystal habit to beta-prime, and the potato processors used large quantities for frying. Therefore, the growth of palm oil use in shortenings was primarily at the expense of cottonseed oil and tallow. Palm oil usage leveled off at about 5% of the shortening requirement before a drastic decline fueled by unfavorable publicity highlighting nutritional concerns with saturated fatty acid's effect upon atherosclerosis. Palm oil usage in shortening dropped to less than 1.5% for the 1995 reporting period.

Coconut oil and the other lauric oils are not among the more desirable oils for most shortening products because of their short plastic range, tendency to foam in deep fat frying when mixed with other fats, and the soapy flavor that develops with hydrolysis. However, the short plastic range contributed by a sharp melting point is advantageous for fillers for cookies and candies; it provides excellent product "get away" in the mouth. Coconut oil has an excellent frying stability, when isolated from other oils, because of its high level of saturates and was a popular frying media for Mexican foods, which probably contributed to the high level used for shortening production in 1980–1981. This use also suffered from unfavorable publicity that convinced the foodservice industry to discontinue using coconut oil for frying and change to a shortening with a more healthy image. This change undoubtedly accounted for most of the corn oil utilization for shortening in 1990, which decreased considerably after the 1991–1992 crop year, indicating that other vegetable oils, probably soybean oil, have replaced it for foodservice frying.

Both lard and tallow became important shortening raw materials after meat fats regained popularity in the mid-1950s. Two of the reasons for the meat fats' improved status were (1) major fast-food marketing promoting the beefy flavor imparted to french fries by frying in tallow shortenings and

(2) interesterification of lard to give it equivalent performance in baking shortenings to the hydrogenated vegetable oils. Later technology was developed to replace interesterified lard with tallow and vegetable oil blends for more attractive product costs. Meat fat usage continued to grow until cholesterol concerns brought pressure on the major end users to provide food products that had better nutritional images. Meat fats' usage in shortenings accounted for less than 11% of the total 1995 raw materials after controlling 25 to 30% of the requirements for over 30 years.

The total usage of U.S. shortenings more than tripled in the 40 years between 1950 and 1990, while the per person usage doubled. Increased use of shortenings partly reflects the changing eating habits of Americans—in time, place, and frequency of eating. Convenience and snack foods have risen sharply in popularity, and similar growth has occurred in the foodservice industry. It has been estimated that currently about 45% of the U.S. food dollar is consumed away from home. These changes have affected the shortening types and quantity requirements to meet the foodservice and food processor industries' demands for specialty fats for product and performance improvements while experiencing a decline in household shortening demand.

Source oil changes are caused by economics in many cases, but many of the recent changes have been due to required improvements of the nutritional image. The nutritional challenges ahead may require that technology efforts be directed toward changes in the source oil's composition. Plant breeding technology and genetic engineering have made it possible to produce oils that are enhanced in a single component or to essentially eliminate an undesirable component. Table 5.2 compares the fatty acid compositions of three oil seed modifications that are available commercially. Sunflower and safflower hybrids are available with high oleic fatty acid con-

TABLE 5.2. Hybrid Oil Seeds—Fatty Acid Composition Changes [18–20].

Fatty Acid Composition, %	Safflower Oil Regular	Safflower Oil High-Oleic	Sunflower Oil Regular	Sunflower Oil High-Oleic	Rapeseed Oil Regular	Rapeseed Oil Canola
C-16:0 Palmitic	7	5	6	3	4	4
C-18:0 Stearic	2	2	7	5	2	2
C-18:1 Oleic	13	81	19	84	34	55
C-18:2 Linoleic	78	12	67	7	17	26
C-18:3 Linolenic			<1	<1	7	10
C-20:1 Gadoleic					9	2
C-22:1 Erucic					26	trace

tents and substantially reduced linoleic contents. Canola oil is the result of rapeseed or mustard seed modifications to essentially eliminate the erucic fatty acid (C-22:1). Some other changes that have been achieved with oil seed plants are low-linolenic soybean and canola oils, high-lauric canola oil, and high-stearic soybean oil [22,23]. The modified oils currently available have not been designed or used for shortenings, except for canola oil, which still requires hydrogenation when utilized. However, theoretically, plant biotechnology should be able to provide almost any desired oil composition. This theory is reinforced by the high-lauric, fatty acid canola oil development. Thus, sometime in the future farmers may start growing performance and the desired nutritional aspects into oils rather than relying on processing to provide the desired shortening performance characteristics.

1.3 Shortening Product Forms

Three shortening types or forms have emerged to satisfy the requirements of the food industry, i.e., plasticized semi-solids, liquid or pumpable, and flakes, beads, or powders. Additionally, almost all shortening packaged products can be handled in bulk quantities.

- plasticized semi-solid shortenings—Most plasticized semi-solid shortenings are identified and formulated for optimum performance in the end product. General purpose shortenings are identified as all-purpose, unemulsified, emulsified, animal vegetable blends, or the like, while the trend is to classify most other shortenings by the intended usage, i.e., cake and icing, mellorine, bread, frying, and so on.
- liquid shortenings—The liquid or pumpable shortening designation covers all fluid suspensions that consist of a hard fat, usually beta tending, and/or a high-melting emulsifier dispersed in a liquid oil. This shortening type was developed for bulk handling and metering at room temperature and easier handling of packaged product that can be volumetrically measured and poured.
- flakes, beads, or powdered shortenings—Shortening flakes, beads, or powders describe the higher melting edible oil products solidified into these forms for ease in handling, quicker remelting, or for a specific function in a food product.
- bulk handling—Food processors that use a shortening product in sufficiently large quantities most frequently purchase these products in liquid bulk form. Shipments are made in 40M- to 50M-pound capacity tank trucks or 150M- to 180M-pound capacity railcars. Bulk handling systems for shortenings that are normally plasticized for package shipment require the food processor to have processing equipment in ad-

dition to storage facilities, specifically systems for chilling and plasticizing the shortening products. Freshly plasticized shortenings may be added directly to some products without tempering where crystallization properties are assisted by emulsifiers like cake mixes; however, other products require tempering to perform properly.

2. PLASTICIZED SHORTENING APPLICATIONS

Development of a shortening product for a food application is dependent upon many factors. These factors may differ from customer to customer, depending upon the equipment, processing limitations, product preference, customer base, and many other requirements. Shortening products are now being designed to satisfy individual specific requirements, as well as offering products with a broad general appeal. The design criteria for a general appeal product must be of a broader nature than that for a specific product or process. Categorically, three steps enter into the development of a functional shortening product:

(1) Product requirements—identification of the shortening functionality requirements for end-use product application, shelf life, and the process requirements of the industry or a particular food processor's operation

(2) Composition—shortening composition development, which includes the fats and oils components and the supplemental additives, with the tolerance ranges for each identified to provide both optimum and uniform functionality

(3) Processing conditions—chemical and physical processing conditions specific to the shortening, which are critical to application performance. The processes involved may begin with raw material sourcing and continue through tempering and delivery to the customer.

The important attributes of a shortening in a food product vary considerably. In some food items the flavor contribution of the shortening is of minor importance; however, it does contribute a beneficial effect to the eating quality of the finished products. This fact has been evident with the recent introductions of fat-free products. Most of these products lack the eating characteristics contributed by fats and oils. In most applications shortenings are multifunctional. Shortenings provide a heat transfer medium, lubricity, and flavor to fried foods; aeration, lubricity, and structure to cakes, icings, fillings, creme fillers, and whipped toppings; lubricity, and structure to baked pastries like crusts, Danish, and puff pastry; and lubricity with increased shelf life and softness to yeast-raised products like bread and sweet rolls. Therefore, shortening is a major contributor to the characteristic structure of most prepared food products and has a significant effect upon the finished product quality.

The physical characteristics of the fats and oils utilized for a shortening are of primary importance in the design of a product for specific use. Oils can be modified through various processes to produce the desired properties. Hydrogenation has been the primary process used to change the physical characteristics of shortenings. The melting point and solids profile of an oil can be completely altered with this process and the changes controlled by the conditions used for the hydrogenation process. Control of the hydrogenation conditions and the endpoint enables the processor to better meet the desired physical characteristics of the shortening products.

Although formulation strongly influences consistency, crystal structure, and performance, the method of plasticization and tempering is also critical to application performance. Shortenings are plasticized before filling to make them uniform throughout, to effect a more attractive appearance, and for performance improvement. The manner in which a shortening is solidified has a pronounced effect upon formulation, size, and the rate of crystal transformation. Solidification and texturizing of shortening is usually achieved by a shock-chilling process where heated fat is rapidly cooled in scraped surface tubular heat exchangers followed by working or crystallization units, filling and quiescence tempering. The chiller units are designed to remove the heat of crystallization, transform crystals, and perform mechanical work, while the crystallizers improve the plasticity and texture of the solidified shortening. After filling, most baking shortenings are tempered for 24 to 72 hours at 85°F (29.4°C) to attain optimum consistency and creaming properties. The tempering process raises the temperature of all the individual crystals uniformly to the desired point without agitation during the later stages of crystal transformation.

3. LIQUID SHORTENING APPLICATIONS

Prior to the use of emulsifiers, antioxidants, and antifoamers, fats and oils products relied solely on fatty acid composition, crystal habit, plasticization, and tempering of the various products for performance. The aerating function of plastic shortenings was correlated with the polymorphic form of the triglyceride while shelf life and frying stability were affected by the level of unsaturation or saturation to resist oxidation. Fats and oils that exist in the small beta-prime crystal form aerate batters much more thoroughly than those in the large beta form, and saturation of the unsaturated fatty acids eliminates a reaction site for oxygen to extend oxidative stability for shelf life and frying stability purposes.

With the addition of emulsifiers, antioxidants, and/or antifoamers, the crystal form and saturation levels of the base oils are still important, but these additives supplement the functionality previously performed solely by fats and oils composition. These additives have significantly reduced the dependence of solid fats for their performance, which led to the development and introduction of fluid or liquid shortenings for use in both yeast-raised and chemically leavened baked foods, foodservice deep fat and pan frying, and for dairy analog products. Liquid or pumpable shortenings combine the highly functional characteristics exhibited by plastic shortenings with the bulk handling characteristics of a liquid oil. A liquid shortening is a stable dispersion of solids with the proper polymorphic form in a continuous oil phase that is both flowable and pumpable over a temperature range of 60 to 90°F (15.6 to 32.2°C). The solids are derived from either hard fats or emulsifiers, sometimes both. The choice of liquid oil is governed by the level of oxidative stability required, and the solids choice is dependent upon the specific end-use application.

The type and level of solids are important considerations in producing a stable fluid suspension. In contrast to plastic shortenings, it is desirable to formulate beta-stable shortenings whose large crystals tend to form a stable dispersion. Aeration properties for cakes, normally associated with a smooth, plastic consistency developed by beta-prime source oils, rapid chilling, and tempering, are achieved for liquid shortenings by the appropriate emulsifier system. Oxidative stability for products requiring lubrication with a bland flavor is improved by the addition of antioxidants. Dimethylpolysiloxane is the highly functional antifoaming agent that inhibits oxidation to allow a liquid opaque shortening to perform like a higher melting plasticized frying shortening.

The crystallization process for liquid shortenings is equally important to the base oil composition to provide a beta crystal matrix in a concentration that will maintain a viscosity low enough for pumping or pouring but high enough for a prolonged stable suspension. Liquid shortening processing involves proper ingredient selection, proper dissolution of additives and hard fats in the liquid oil, and controlled crystallization of the product. Liquid shortening crystallization procedures for low and high hard fat level products are presented and discussed in Chapter 2.

Liquid shortenings do not require tempering after crystallization and can be shipped to customers both in package and as bulk product immediately after processing. In both cases, ease of use and handling over plasticized shortenings are advantages. Packaged liquid shortenings may be poured from the container and measured volumetrically. Liquid shortenings handled in bulk quantities do not require heated storage facilities and can be pumped and metered at room temperature.

4. FLAKES, CHIPS, AND POWDERED SHORTENING APPLICATIONS

Shortening flake describes the high-melting edible oil products solidified into a thin flake form for ease in handling, quicker remelting, or for a specific function in a food product. The traditional flaked products have been the saturated oil products known as stearins, titers, or low-iodine value hard fats. The hard fat flakes have been joined by products formulated for special uses where the flake form is desirable. These specialty flaked products are shortening chips and stabilizers for various food applications. A third form of the low-iodine value, high melting point fats and oils products is powdered shortenings.

Hydrogenation selectivity is not important for the hard fats or stearins because the reactions are carried almost to complete saturation. However, selectivity is very important for the proper functionality of both the shortening chips and some of the stabilizers. Shortening chips are selectively hydrogenated to attain a steep SFI slope with a melting point as low as possible while still allowing the product to maintain the chip form after packaging, during shipment to the food processor, while processing into the finished product, and until that product is prepared by the consumer. Most chip products can be formulated from the basestock system with a blend of selectively hydrogenated basestock or with specially hydrogenated bases with sulfur proportioned catalysts for steeper SFI slopes, depending upon the desired melting point and SFI requirements. Most domestic oil shortening chips usually have melting points in the range of 110 to 118°F (43 to 48°C). Lower melting point domestic oil–based shortening chips many times require refrigeration before use to maintain the chip form without fusing together in the package. Shortening chips and flakes can also be made with hydrogenated lauric oils, either coconut or palm kernel. Melting points for the lauric oil chips normally range from 101 to 108°F (38 to 42°C). Flavors, colors, spices, and other materials can be encapsulated in the shortening chips for a protected incorporation in food products.

The bakery products stabilizers are more nonselectively hydrogenated, with melting points normally ranging from 110 to 125°F (45 to 52°C). The flattest SFI slopes for these products help to maintain softness while stabilizing the coatings, glazes, icings, and other bakery-type products. The icing stabilizers are usually prepared from domestic oils to obtain the desired analytical and physical characteristics.

Two types of powdered shortenings are produced: (1) spray-dried fat emulsions with a carrier and (2) spray-chilled or -beaded hard fat blends. The spray-dried powdered shortenings are partially hydrogenated shortenings encapsulated in a water-soluble material. Shortenings can be homogenized in solution with a variety of carriers, i.e., skim milk, corn syrup solids, sodium caseinate, soy isolate, and others. Emulsifiers may be included with the shortening for finished product functionality. Fat contents usually range from 50 to 80%, depending upon the original emulsion composition before spray drying [24]. The spray-dried powdered shortenings are used in some prepared mixes for their ease in blending with the other dry ingredients.

Hard fats can be powdered or beaded without the aid of a carrier. Three principal methods of forming powder or beaded fats are practiced in the United States: spray chilling, grinding flaked product, and spray flaking and grinding. Spray chilled hard fats have a disadvantage for feeding or blending accuracy. The spherical shape of the spray-chilled powders may act as roller bearings to give erratic feeding rates with vibratory feeding systems or stratify in blends of dry materials. Beaded products produced by grinding flakes or spray flaking for immediate grinding have granular shapes that can be metered at uniform rates with vibratory or screw feeders and resist stratification or separation in mixes with other granular materials.

5. REFERENCES

1. Markey, K. S. 1960. "Historical and General," in *Fatty Acids: Their Chemistry, Properties, Production, and Uses*, Second Edition, Part 1, Klare S. Markley, ed. New York, NY: Interscience Publishers, Inc., p. 13.
2. Mattil, K. F. 1964. "Plastic Shortening Agents," in *Bailey's Industrial Oil and Fat Products*, Third Edition, Daniel Swern, ed. New York, NY: Interscience Publishers, a Division of John Wiley & Sons. pp. 265–271.
3. Jones, L. A. and C. C. King, ed. 1990. *Cottonseed Oil*. Memphis, TN: National Cottonseed Products Association, Inc. and the Cotton Foundation, pp. 5–7.
4. Meyer, L. H. 1975. "Fats and Other Lipids," in *Food Chemistry*, Lillian Hoagland Meyer, ed. Westport, CT: The AVI Publishing Company, Inc., p. 55.
5. Black, H. C. 1948. "Edible Cottonseed Oil Products," in *Cottonseed and Cottonseed Products*, Alton E. Bailey, ed. New York, NY: Interscience Publishers, Inc., pp. 732–733.
6. Morris, C. E. 1949. "Mechanics of Deodorization," *J. Am. Oil Chem. Soc.*, 26(10): 607.
7. Wrenn, L. B. 1995. "Pioneer Oil Chemists: Allbright, Wesson," *INFORM*, 6(1):98.
8. Wrenn, L. B. 1993. "Eli Whitney and His Revolutionary Gin," *INFORM*, 4(1):12–15.
9. Rini, S. J. 1960. "Refining, Bleaching, Stabilization, Deodorization, and Plasticization of Fats, Oils, and Shortenings," *J. Am. Oil Chem. Soc.*, 37(10):519.
10. Slaughter, Jr. J. E. 1948. "Plasticizing and Packaging," in *Proceedings of a Six Day Short Course in Vegetable Oils*, August 16–21, American Oil Chemists Society, Champaign, IL, p. 120.

11. The term *High Ratio* has been copyrighted by the Procter & Gamble Co.

12. Hartnett, D. I. 1977. "Cake Shortenings," *J. Am. Oil Chem. Soc.*, 54(12):557.

13. Robinson, H. E. and K. F. Mattel. 1959. "Fifty Years of Progress in the Technology of Edible Fats and Oils," *J. Am. Oil Chem. Soc.*, 36(9):434–436.

14. Mattil, K. F. 1964. "Plastic Shortening Agents," in *Bailey's Industrial Oil and Fat Products*, Third Edition, Daniel Swern, ed. New York, NY: Interscience Publishers, a Division of John Wiley & Sons, p. 296.

15. *Fats and Oils Situation.* 1977. Economic Research Service, USDA, July.

16. *Oil Crops Outlook and Situation Yearbook*, 1985. Economic Research Service, USDA, August.

17. *Oil Crops Yearbook*, October,1996. Economic Research Service, USDA, pp. 45 and 57.

18. Dutton, H. J. 1981. "History of the Development of Soy Oil for Edible Uses," *J. Am. Oil Chem. Soc.*, 58(3):234–236.

19. Vaisey-Genser, M. and N. A. M. Eskin. 1982. *Canola Oil Properties and Performance*, Publication 60, Canola Council of Canada, Winnipeg, Canada, p. 13.

20. Purdy, R. H. 1985. "High Oleic Sunflower: Physical and Chemical Characteristics," Paper 127, 76th AOCS Annual Meeting, Philadelphia, PA.

21. Filler, G., M. J. Diamond, and T. H. Applewhite, 1967. "High Oleic Safflower Oil. Stability and Chemical Modification," *J. Am. Oil Chem. Soc.*, 44(4):264.

22. Haumann, B. F. 1996. "The Goal: Tastier and 'Healthier' Fried Foods," *INFORM*, 7(4): 320–334.

23. Del Vecchio, A. J. 1996. "High-laurate Canola," *INFORM*, 7(3):230–242.

24. Weiss, T. J. 1983. "Shortening—Introduction," in *Food Oils and Their Uses*, Second Edition, Westport, CT: AVI Publishing Co. Inc., pp. 129–130.

CHAPTER 6

Baking Shortenings

1. INTRODUCTION

The term *baking* applies not only to the production of bread, but to all food products in which flour is the basic material and to which heat is applied directly by radiation from the walls and/or top and bottom of an oven or heating appliance. More particularly, baking includes the production of items such as bread, cake, pastry, biscuits, crackers, cookies, and pies where flour is the essential and principle ingredient for the base product; baking also includes the toppings, frostings, fillings, and so on that finish the baked product. Baking as practiced by the homemaker is an art with an occasional failure, whereas baking as practiced by the professional baker is an engineering science, in which the volumes involved make uniformity an absolute necessity and failures are disastrous. Survival of the commercial baker requires quality production at the lowest costs, which requires uniform, consistent performance from all of the ingredients employed.

Shortenings are a very important ingredient for the baking industry due to the fact that they comprise from 10 to 50% of most baked products. Shortenings contribute to baked products in a number of ways: (1) imparting "shortness," richness, and tenderness to improve flavor and eating characteristics; (2) enhancing aeration for leavening and volume; (3) promoting desirable grain and texture qualities; (4) providing flakiness in pie crusts, danish, and puff pastry; (5) providing lubrication to prevent the wheat gluten particles from adhering together to retard staling; (6) affecting moisture retention for shelf life improvement; and (7) providing structure for cakes, icings, and fillings [1].

All-purpose shortening with a wide plastic range, made by blending a nonselectively hydrogenated basestock with a low-iodine value hard fat, has been the building block for many different general and specialty baking shortenings. Chart 6.1 traces the products developed, using the all-purpose shortening type formulation to produce the required functionality for many different bakery products. The technology that made production of the all-purpose shortening possible has led to the development of a number of specialized shortenings. Hydrogenation and blending of different base-stocks and source oils to control SFI, melting characteristics, and the crystal habit has affected consistency control to allow the fats and oils processor to design shortenings with specific functional properties. The addition of emulsifiers has provided emulsification for higher moisture levels, aeration, moisture retention, starch complexing, antistaling, palatability improvements, and other functionalities to further extend the performance of bakery shortenings. The two techniques for wide plasticity shortenings, emul-

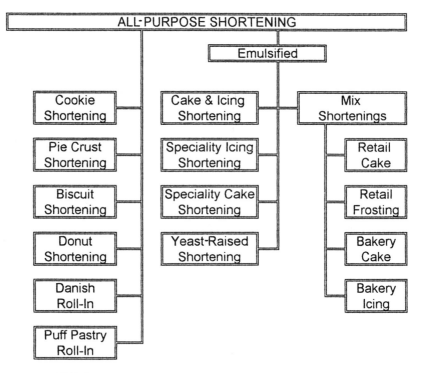

CHART 6.1. Wide plastic range bakery shortening development.

sified and nonemulsified, have resulted in the development of the specialized products outlined in Chart 6.1.

In the years following the introduction of mono- and diglycerides, a number of new emulsifiers were developed that provided shortening processors and bakers with better tools. Properly used, these emulsifiers reduced the dependence on plasticized or solid shortening crystal forms for aerating and structural properties. The more highly emulsified shortenings were first used by the retail cake mix industry to shorten mixing times, increase cake volumes, and improve mixing tolerances. The more sophisticated emulsifier systems developed were able to assume the entire aerating function of crystalline fat and made the commercialization of fluid shortenings possible. Likewise, development of surfactants that were able to complex starch or interact with flour proteins was partially responsible for the acceptance of liquid shortenings for yeast-raised products. The attributes of the emulsifiers and the development of several unique processing techniques allowed the introduction of fluid shortenings consisting of suspended solid hard fats and emulsifiers in a liquid oil base for cakes and another for yeast-raised products. These liquid shortening products met some of the baking industries' ideal shortening desires, i.e., a functional shortening that could be stored unheated in a tank and that could be metered and pumped in exact proportions to the desired mixer. The areas where liquid shortenings have fallen short of the ideal expectations are that they usually require agitation to maintain uniformity and are very specific as to functionality; i.e., liquid cake shortenings can only be utilized for cake production, which may also be limited to the type cake, and are ineffectual for icings, fillings and other batter or yeast-raised products. Fluid bakery shortenings have had the most success with large cake and continuous bread producers that have dedicated automated product lines.

Shortening flake products have quietly evolved as specialty products for specific bakery products. In almost all cases the product functionality is dependent upon the solids to liquid ratios or SFI and the melting point. These characteristics, along with the flake or beaded form, provide convenient handling, quick melting, or a controlled melt for a special function in a baked product. The traditional flaked product, low-iodine value hard fat, is produced for various uses that are usually concerned with adjustments in the melting characteristics of a product. Shortening chips and stabilizers are blends of oils or a single source oil specially hydrogenated to attain specific functional characteristics for use in biscuits, pizza crusts, breads, yeast-raised rolls, icings, donut production, various prepared mixes, and other bakery products.

2. ALL-PURPOSE SHORTENINGS

Bakery all-purpose shortenings are compromise products designed for performance in baked products, icings, and fillings, as well as deep fat frying. Typically, all-purpose shortenings have been formulated with 3 to 15% essentially fully saturated beta-prime hard fat added to a partially hydrogenated basestock, a blend of tallow with vegetable oil, or interesterified lard. The functionality of an all-purpose shortening at any temperature is largely dependent upon the solids content at that temperature. All-purpose shortenings are formulated to be not too firm at 50 to 60°F (10 to 15.6°C) and not too soft at 90 to 100°F (32.2 to 37.8°C). Beta-prime hard fats are added to extend the plastic range, to improve the tolerance to high temperatures, and to establish the shortening's crystal habit. Almost fully saturated cottonseed oil, palm oil, or tallow are the usual sources for the beta-prime crystal-forming hardstocks, which function as plasticizers for improving creaming properties, texture, and consistency. Some typical vegetable and animal–vegetable blended all-purpose shortening formulations are presented in Table 6.1.

Hydrogenation of a fat and oil basestock increases the oxidative and frying stability. Selective hydrogenation conditions and lower basestock iodine values provide the best oxidative and frying stability. However, nonselective hydrogenation conditions produce flatter SFI curves and, consequently, a wide plastic range, and as base hardness is increased, the level of hardstock required to reach a desired consistency decreases. Hardstock reduction reduces the shortening's plastic range and heat tolerance. Therefore, oxidative stability improvements are achieved at the expense of plasticity, and

TABLE 6.1. All-Purpose Shortening Formulations.

Composition %	All-Vegetable		Animal–Vegetable		
80 IV H-Cottonseed oil	90.0				91.0
88 IV H-Soybean oil		88.0		89.0	
109 IV H-Soybean oil			14.5		
Unhardened tallow			82.5		
60-Titer cottonseed oil	10.0				
56-Titer palm oil		12.0			
59-Titer tallow			3.0	11.0	9.0
Total, %	100.0	100.0	100.0	100.0	100.0
Solids fat index, %					
at 50°F	26.0±		32.0±		
at 80°F	20.0±		21.0±		
at 104°F	9.0 min.		9.0 min.		

a wide plastic range can be at the expense of oxidative and frying stability. Additionally, oxidative stability results do not average; oxidative stability is directly related to the level and type of unsaturated fatty acids present in a blend. Therefore, a 50:50 blend of a basestock with a 40-hour AOM with one that has a 100-hour AOM will not have a 70-hour AOM stability; it will be closer to the 40-hour AOM of the most unsaturated basestock [2]. The extent that one attribute can be compromised to improve another must be determined by the requirements of the intended food product.

Plastic range is important for bakery all-purpose shortenings because of the consistency changes with temperature. As is evident, products with the best plastic range are those that are softest at low temperatures while also firmest at high temperatures. A perfect plastic range, if it could be produced, would have the same desirable consistency at both high and low temperatures. Nevertheless, shortenings become brittle above the plastic range and soft below this range; both conditions adversely affect creaming and workability properties. Shortenings are normally plastic and workable at SFI values between 15 and 25. Therefore, all-purpose shortenings that depend upon consistency for performance in baked products are formulated with flat SFI slopes that remain plastic for a wider temperature range than those shortenings with steeper SFI slopes.

Creaming properties and plastic range in a shortening are dependent first upon the proper basestock and hard fat selection and then upon the chilling and tempering conditions employed. All-purpose and other wide plastic range shortenings are normally quick-chilled to 60 to 70°F (15.6 to 21.1°C), followed by a crystallization or working stage, packaging, and then tempering at 85°F (26.7°C) for a 24- to 72-hour period, depending upon the package size. The effect of tempering is illustrated in Chart 6.2, which compares the effect of 60-titer cottonseed oil hard fat and 56-titer palm oil hard fat stabilized all-purpose shortenings tempered at 80 and 85°F for 48 hours. Consistency was measured by penetration evaluations at 80°F (26.7°C) for shortenings with SFI values at 80°F ranging from 15 to 21% with each hard fat. The plastic range for this penetration method was defined as too soft above 300 mm/10 and too firm below 150 mm/10. This evaluation indicated that acceptable all-purpose shortenings could be produced with either hard fat but

(1) Higher levels of palm oil hard fat were required to produce comparable SFI values to the product with cottonseed oil hard fat.

(2) Equivalent shortening SFI values with palm oil and cottonseed oil hard fats did not provide equivalent shortening consistency as measured by penetration. The palm oil–hardened shortening was consistently softer than cottonseed oil–hardened shortening at the same SFI value. About

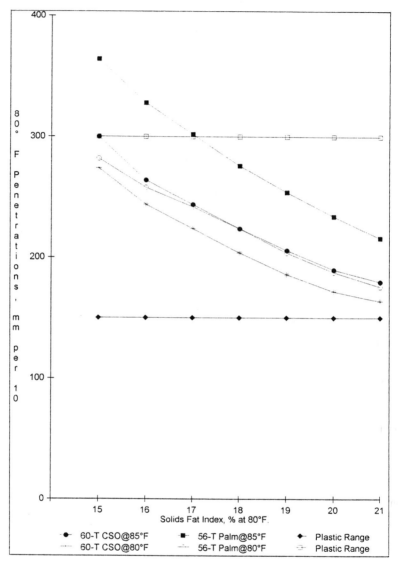

CHART 6.2. All-purpose shortening (80° and 85°F tempering comparison).

2% higher solids were required with palm oil hard fat to achieve a consistency equivalent to the cottonseed oil hard fat stabilized shortening.

(3) Cottonseed oil–hardened shortenings tempered at 85°F (29.4°C) had penetration results comparable to the palm oil–hardened shortenings tempered at 80°F (26.7°C). Palm oil–hardened shortenings may perform better when tempered at 80°F rather than 85°F, especially with SFI values at 80°F below 17%.

Penetrations performed at only one temperature do not indicate the working range for a shortening. Working range can be measured by penetrations over a range of temperatures, especially those that the shortening may be exposed to in end-use applications. Most bakeries and foodservice kitchens have limited temperature control, and a wide range of temperatures are likely to prevail. Therefore, all-purpose shortenings must be formulated and processed to be tolerant to broad variation in working temperatures and resistant to breakdown during mixing.

All-purpose shortenings are probably most used by retail and foodservice bakers where creaming properties, emulsification qualities, icing value, wide working range, water absorption, and stability in baked and fried foods are very important. In an all-purpose shortening, an attempt is made to obtain a balance of these properties so that it may be used with good results for any purpose. By varying composition and processing, it is possible to make a shortening that will be superior in any one respect, but it is impossible to produce one that will be excellent in all respects because some of the qualities are antagonistic to one another. For example, additives beneficial for a certain performance characteristic are usually detrimental for another:

- Dimethylpolysiloxane performs as an antifoamer to extend frying stability, but it is detrimental to the baking characteristics of certain types of cakes and cookies.
- Some antioxidants that retard oxidation to extend flavor stability have reactions with baking acids to give pink or purple colors.
- Emulsifiers that provide many beneficial contributions for baked products are especially detrimental to frying stability.

Shortening end users have had to decide which type of compromise to make for their production: use an all-purpose shortening that requires an inventory of only one shortening or utilize a series of specialty shortenings requiring more inventory. Both of these paths are taken.

3. EMULSIFIED ALL-PURPOSE SHORTENING

A major advancement in shortening and baking technology occurred in the 1930s. Development and introduction of superglycerinated shortening brought about significant bakery product improvements. The addition of mono- and diglycerides to all-purpose shortenings allowed the production of high-sugar and -moisture cakes with improved volume and eating characteristics, a finer grain and texture, and an extended shelf life [3]. The higher ratio of sugar and moisture to flour and other cake ingredients led to the application of the term *High Ratio* for these shortenings [4]. This shortening also provided the means for improving icings and fillings by providing additional aeration capabilities with higher moisture levels for increased volumes with better emulsion stability. A dough conditioning and antistaling effect was also realized in yeast-raised products to improve dough handling properties and to extend the shelf life of these bakery products while softening the crumb, which improved the eating characteristics [5]. Performance improvements were also found with many other bakery products where creaming properties, starch complexing, and emulsion stability properties were important. The addition of mono- and diglycerides to the high ratio shortenings allowed these significant bakery products improvements. Functionality improvements over standard unemulsified, all-purpose shortenings were attained in most areas, with the exception of applications where moisture retention, aeration, or protein modification are detrimental. For example, the use of this shortening type for frying results in excessive smoking, darkening, and foaming.

Formulation changes were necessary for icings, yeast-raised products, and cakes, most of all to realize the full advantage of the emulsified, all-purpose shortenings. The required changes were identified by the shortening developers to gain acceptance for the new shortening product. The most drastic changes were for cakes; bakers had to change their rules to balance cake formulas to accommodate the more efficient performance. The rules for balancing cake formulas with unemulsified, all-purpose shortenings were compared to the revised rules developed for emulsified, all-purpose shortenings in Table 6.2.

Emulsified, all-purpose shortenings also made cake mixing more efficient. The creaming method was almost universally used with unemulsified, all-purpose shortenings. This method required multistage mixing, starting with blending the shortening and sugar together to incorporate air into the shortening to form a cream. Second and third steps were then used to incorporate the liquids, flour, and other dry ingredients. The blending method, introduced with emulsified, all-purpose shortenings, was able to eliminate the creaming stage because of the efficiency of the emulsifier to

TABLE 6.2. Cake Formula Balance.

Cake Ingredient	All-Purpose Shortenings	
	Unemulsified	Emulsified
Sugar	No more than flour	Exceed flour to 140% max
Shortening	No more than eggs	No more than eggs
Liquids	Equal to flour	Equal or exceed sugar slightly

trap air in the batter. All of the shortening and dry ingredients were blended with a portion of the liquid milk or water, with the eggs added during the last mixing stage. Batter mixing times were reduced from 15 to 20 minutes for the creaming method and 9 to 12 minutes for the blending method.

Formulation of the emulsified, all-purpose shortenings involved a determination of the amount and type of emulsifier to add to the existing all-purpose shortening formulations. The intermediate hardness mono- and diglyceride chosen extended the multifunctional capabilities of the shortening. Optimum cake volume, moisture retention, structure, grain, and texture, plus the yeast-raised products improvements, are provided by emulsifiers prepared from fully saturated fats. However, icings and fillings are smoothest and aerate more with monoglyceride products prepared from unhydrogenated oils. A compromise to facilitate performance for both cakes and icings is attained by using an emulsifier prepared from an intermediate hardness hydrogenated oil or blend of meat fats, i.e., 72 to 78 iodine value range for most vegetable oils or a 55 to 60 iodine value range for meat fat bases. The alpha monoglyceride levels used, 2.5 to 3.2%, for emulsified, all-purpose shortenings were also chosen carefully to complement both cake and icing production. Icings and filling products require lower emulsification levels than most cake products for optimum results. Therefore, the alpha monoglyceride levels for emulsified, all-purpose shortenings will produce satisfactory results for most bakery products but cannot provide superior results for any product.

Shortening producers continually evaluate new, different, and improved emulsifiers to identify a superior cake and icing or all-purpose, emulsified shortening. To date, most emulsifiers follow the mono- and diglycerides pattern, which surfactant variations, which improve cake functionality, have a negative effect upon icings and vice versa. However, some emulsifier systems have been developed that improve the all-purpose functionality over the performance of the intermediate mono- and diglyceride emulsifiers. These emulsifier systems incorporate the performance of a high-aerating surfactant like polysorbate 60, polyglycerol ester, or ethoxylated monogly-cerides with the intermediate hardness mono- and diglyceride used for the standard all-purpose, emulsified shortening and possibly a propylene glycol monoester (PGME), also of intermediate hardness. The aeration-producing surfactants are balanced with the structure and texture-building capabili-ties of the intermediate hardness PGME and mono- and diglycerides for cake and icing type products. These shortenings have been formulated to replace the standard emulsified, all-purpose shortenings pound for pound in most bakery formulations to improve volume, texture, and emulsion stability.

Incorporation of mono- and diglycerides reduced the dependence of shortenings upon the crystalline properties, solids to liquid ratios, and mix-ing procedures to develop creaming properties. Where nonemulsified, all-purpose shortenings relied upon the base fats to contribute all functionality, the added emulsifiers enhanced these properties for greater functionality. Nevertheless, a wide plastic range provided by beta-prime crystal formu-lation, adequate solids to liquid ratios for a wide working range, and atten-tion to plasticization procedures is still important for any general-purpose shortening. Temperature control during shipping, storage, and use has also been improved considerably, but all-purpose shortenings must still have a wide plastic range to resist breakdown during mixing and make-up into the various bakery products. Shortenings that are brittle, lumpy, sandy, or with other textural problems that hinder proper incorporation into the baked products provide substandard nonuniform baked items.

4. RETAIL CAKE MIX SHORTENINGS

The prepared cake mix industry recognized that market expansion re-quired mixes that provided ultra-convenient, foolproof performance to pro-duce good tasting moist cakes for consumption over an extended period after baking. These requirements have differentiated cake mix shortenings from bakery cake shortenings to create specialty or tailor-made products. Cake mix shortening requirements include a plasticity to meet the demands of the mix manufacturer's compounding equipment, emulsification capable of batter aeration with a high moisture content, and oxidative stability for an extended package shelf life before preparation, as well as tenderizing the cake crumb.

Historically, the introduction of emulsified shortenings helped initiate the cake mix industry. Originally, and for an extended time, cake mixes utilized the same all-purpose shortenings developed for commercial bakery cakes. Although the sugar/flour ratio in prepared cakes ranges from 90 to 125%, it was recognized that these mixes not only required monoglycerides, but performed better with higher levels than those used in bakery shortenings. It was also recognized that the conventional mono- and diglyceride used in the bakery all-purpose, emulsified shortenings, even when used at elevated levels, did not produce the required effect under the unique conditions imposed by the normal use of packaged retail mixes. It also soon became obvious that some mono- and diglycerides types performed better than others. It was also observed that a shortening containing a particular type of monoglyceride might function satisfactorily with one flour in one manufacturer's formulation but not as efficiently in another situation. For a period of about 10 years, 1947 through 1957, a great number of specialty cake mix shortenings using variations in mono- and diglycerides both for base oil or fat source and for hardness or degree of hydrogenation of the base oil or fat before glycerination were used by various manufacturers. Each of these factors had some effect on the performance, and generally more desirable cake mix results were obtained with the use of mono- and diglycerides made with base fats of 30 iodine value or less or by blends of these with softer mono- and diglycerides at alpha monoglyceride levels of 3.0 to 4.0% of the shortening.

During this same 10-year period, there was a tremendous growth in prepared mix sales, and a healthy competition between companies developed with the goal to produce bigger and better cakes. Improvements in all ingredients and improved technology on the part of the manufacturers in compounding mixes helped in achieving the product image of a big, relatively fine-grained cake with good eating quality and containing as much moisture as possible. Dispersion of the shortening onto the dry mix ingredients and tempering to maintain the optimum crystalline structure were compounding concerns solved by improved methods for extruding and plating shortenings onto the dry mixes. Shortening's contribution toward the improvement goals was the judicious selection of available emulsifiers, usually monoglycerides and sometimes the sorbitan monostearates and/or polyoxyethylene derivatives. Efficient emulsifier systems permitted better batter aeration in stable emulsions where the increased air content carried over to improve the volume of baked cakes. Cake mix manufacturers also used higher package weights with correspondingly higher liquid requirements to achieve greater cake volumes. These product and processing improvements also allowed the introduction of a modified homemaker preparation requiring only a two-stage mixing procedure.

In the late 1950s and early 1960s another major improvement for cake mixes occurred. Shortenings containing alpha tending emulsifiers were developed and introduced to the cake mix industry. The first of these emulsifiers to be used was the glyceryl lactyl palmitates (GLP) and later propylene glycol monoesters (PGME). These emulsified shortenings increased the ability of the mixes to improve fat dispersion and air incorporation. Although these emulsifiers are considered to be especially effective for aeration of cake batters, soluble protein has been found to be the key aerating ingredient, but stable emulsions are obtained only in those systems containing the alpha tending emulsifiers [6]. Introduction of PGME emulsified shortenings allowed the development of a more convenient one-stage cake preparation process. The use of PGME shortenings in a one-stage cake mix produced layer cakes with a large volume, a very fine grain, moist eating qualities, and a flat crown. The use of emulsifiers to improve the performance of cake mixes did not stop with the introduction of PGME cake mix shortenings, but this emulsifier has continued to be a part of the emulsifier systems used in almost all cases. Synergistic emulsifier systems, developed to further increase the moistness and improve the eating properties of the prepared mix cakes while still retaining large cake volumes with fine grain and texture characteristics, have been the target of some of the most recent product improvements. The synergistic systems employed have been composed of polyglycerol esters or polysorbate 60, mono- and diglycerides, PGME, and lecithin. These emulsifier systems, along with other formulation improvements, have allowed the preparation instructions to increase the added moisture and to add liquid oil to improve the polyunsaturated level of the baked cake.

The aerating and emulsifying potential of a mix shortening is determined not only by the emulsifier system, but it is also governed by dispersibility. Dry mix preparation techniques replace the cake batter development bakers realize with more involved mixing procedures than the single-stage cake mix preparation. Therefore, shortenings with high-aerating and emulsifying potential can fail because of plasticity or dispersibility problems. The prepared mix industry has prompted the development of tailor-made shortenings in respect to plasticity because of the different demands upon the shortening made by the mix manufacturer's compounding equipment. A typical dry mix process begins with the addition of plasticized shortening to a ribbon blender containing preblended dry ingredients and mixed thoroughly. After blending, the mix is moved to holding bins and then to the finisher. The finisher usually consists of a turbolizer, which is basically a hollow cylinder fitted with a shaft with high-speed mixing blades. Its function is the complete contact and plating of all the dry ingredients with the shortening to produce a uniform dry powder. After finishing, the dry mix may be stored in holding bins again before packaging [7,8].

Too hard a shortening is difficult to distribute throughout a mix unless special methods are used. Too soft a shortening will tend to soak into the flour and will cause filling and packaging problems. This condition permits packing and lumping of the mix and tends to allow "bridging" and "tunneling" of the mix in the surge bins with resultant conveying and packaging problems. Ultimately, shelf stability problems caused by possible soakage of the fat into the packaging materials may even result. Performance-wise, grease-soaked flour will not absorb moisture readily, and soft shortenings tend to be dispersed in the aqueous phase in spherical, rather than in irregular, form when the mix is made into batters. Cakes from such batters will exhibit a spherical and thickened cell wall structure atypical of usual grain and textures.

All things considered, the hardest shortening that provides the best performance and can be handled and distributed effectively by the mix manufacturer's equipment is recommended. Normally, the special requirements in plasticity for tailor-made or specialty shortenings may be met by changing the proportions of hard to soft components of the basestock and hardstock blends. However, too much of a hard fat component can permit sudden crystallization and pelletizing when heated fats are dispersed in cake mixes. Therefore, the improved functional qualities of the high-aerating cake mix shortenings are influenced by (1) the emulsifier system, (2) the specific fat formulation, (3) the amount of fat used in the mix, and (4) the method of incorporating the shortening into the prepared mix.

Consumer desires for even more convenient preparation methods for cake mixes created a need for another cake mix shortening development. Cake mixes designed for baking in conventional ovens do not perform satisfactorily with microwave preparation. Experimentation revealed that microwave cakes require higher shortening and moisture contents. It was also determined that increased emulsification levels were necessary to retain moistness with microwave baking and that a different level was required for each cake variety, i.e., white, yellow, and devil's food [8].

5. SPECIALTY BAKERY CAKE SHORTENINGS

Attempts to use the same high-aerating plasticized shortenings developed for the cake mix industry in bakery cakes has met with failure. The critical nature of the high-aerating emulsifier system does not provide the tolerance necessary for the variations in formulas and conditions encountered in the many bakeries across the country. In addition, formula variations within a single bakery are of such latitude that the materials will not function properly in existing formulations and with mixing procedures used with standard emulsified shortenings. Bakery cakes require extensive formula revisions and changes in mixing procedures to take advantage of the high-aerating emulsifiers, and most bakers have found that these shortenings do not have the tolerance required even with changes in ingredients, formulations, and mixing procedures. It was necessary to broaden the tolerance of the shortening and still take advantage of the best characteristics of the individual emulsifiers with synergistic combinations of the emulsifiers developed especially for the baking industry. The levels of the different types of emulsifiers must be carefully balanced to give optimum volume, symmetry, grain, eating characteristics, and moisture retention with tolerance levels that complement the requirements of the cake baker.

Most bakery cakes prepared from "scratch" are baked with a plasticized all-purpose, emulsified shortening. Most bakers have probably evaluated the more specialized cake shortenings but opted to continue with all-purpose shortenings to help reduce the number of shortening products inventoried. However, many retail and small wholesale cake bakeries may be using specialized cake shortenings incorporated in bakery cake mixes. The use of a prepared mix provides the small baker with the advantages of specialty ingredients while still maintaining a manageable inventory. The specialized bakery cake mix shortenings require slightly different emulsifier systems than their retail counterparts. The major factors contributing to the different requirements are the mixing procedures employed and the desired finished cake characteristics. Bakery cake production requires lower emulsifier levels and less complex, slightly modified blends due to the mixing equipment and procedures employed. Mechanical emulsification provided by bakery cake mixers is much more efficient than the household mixers, and a less fragile cake is required to withstand the required handling while still maintaining desirable eating characteristics.

6. LIQUID CAKE SHORTENING

Economic conditions and automation presented the need and emulsifier development provided the functionality to develop liquid cake shortening technology. Liquid cake shortenings combine the highly functional characteristics exhibited by plastic shortenings with the bulk handling characteristics of liquid oils. Liquid opaque shortenings are characterized by the presence of low levels of highly functional solids suspended in a liquid oil matrix. Liquid cake shortenings are dependent upon the emulsifiers contained to provide most of the functionality necessary for cakes. A variety of emulsifiers including glyceryl lacto esters, propylene glycol monoesters, and mono- and diglycerides provide aeration, tenderizing, and moisture retention properties to liquid cake shortenings.

Liquid shortening's viscosity may be adjusted with low-iodine value hard fat. The type and level of solids are important contributors to a stable fluid dispersion. Unlike plastic solid shortenings, liquid shortenings must be formulated with beta tending hard fats. Large beta crystals provide a stable dispersion that resists a viscosity increase or gel when crystallized properly. Beta-prime crystalline liquid shortenings can experience a viscosity rise, which eventually thickens the product to a solid consistency.

Equally important to proper formulation for fluidity are the processing conditions used to crystallize the liquid shortenings in the stable beta crystalline form. It is important to promote crystallization that is stable and in the proper concentration range for a low enough viscosity for pumping but high enough for adequate suspension stability. Process "B" for liquid shortening crystallization described in Chapter 2 should provide acceptable results, but many other procedures have also been patented and used by different liquid shortening producers. The main objective is to produce a stable fluid shortening that will not increase in viscosity.

Processing techniques developed to suspend emulsifiers and possibly hard fats in a liquid oil base have produced pumpable liquid shortenings at room temperature, which accommodated bakery automation. Overall, the liquid shortenings were easier to handle and were more readily dispersed in cake batters, which enabled them to function more effectively as aerating agents. Liquid cake shortening's ability to entrap and maintain air produced cakes with greater volume, more even grain, and texture with good eating quality, but the cakes were too tender to handle without formula modifications. The changes in the rules for cake formula balance for liquid shortenings are compared to the two all-purpose plastic shortening's requirements shown in Table 6.3.

The cake formulation changes indicated for liquid cake shortenings were necessary primarily because of the lower solids fat index at use temperatures and the increased aeration provided by the more efficient emulsifier systems. These shortening characteristics produced extreme cake tenderness, which was counteracted by the indicated cake formula modifications.

TABLE 6.3. Liquid Cake Shortening's Cake Formula Balance Changes.

Cake Ingredient	All-Purpose Shortenings		Liquid Shortening
	Nonemulsified	Emulsified	
Shortening	Base	Increase	Decrease
Sugar	Base	Increase	No change
Liquids	Base	Increase	Increase
Leavening	Base	Increase	Decrease
Mixing time	Base	Decrease	Decrease

Mixing time reductions were due to the fluidity of the shortening, which rapidly mixes with both the dry and wet ingredients in cake batters, and more rapid aeration effect by the emulsifier systems with the increased moisture levels.

Another prime consideration for liquid cake shortenings is the base oil and its oxidative stability. Base oils with high polyunsaturated fatty acids levels generally have a low oxidative stability as measured by AOM or OSI. Oxidation resistance can be improved by the addition of antioxidants, usually TBHQ, or by the selection of a lightly hydrogenated oil product that is still fluid at room temperature. If a high stability is required and antioxidants are not permitted, high-stability base oils produced with fractionation process technology are another alternative for liquid shortenings. Most liquid cake shortenings utilize a partially hydrogenated vegetable basestock, usually soybean oil, with an iodine value of 100 to 109. The emulsifier systems utilized usually contain mono- and diglycerides with either glycerol lacto esters and/or propylene glycol monoesters. In many cases, the emulsifier systems are designed for specific customers since liquid cake shortenings remain very specialized products, both from a shortening processor's and the baker's perspective.

7. ICING AND FILLING SHORTENINGS

Buttercreme icings are sugar-based, relatively low-moisture (10 to 15%) confections used to provide protective coverings for cakes. Creme fillings are shelf-stable, aerated confections used inside of cakes. Creme fillings are also sugar-based but more highly aerated than buttercreme icings and contain higher quantities of moisture (20 to 25%). These two bakery confectionery products are related to the extent that what is true functionally for creme fillings is true to a lesser extent for buttercreme icings. The major desirable properties for creme fillings and icings are

(1) Yield—Aeration contributes to the eating qualities, lightness, and cost of the finished product. Buttercreme icings are usually aerated to specific gravities of 0.75 to 0.80 g/cc or a volume of 125 to 135 cc/100 g, while creme fillings are usually aerated to lower specific gravities of 0.45 to 0.50 g/cc or 200 to 250 cc/100 g volume. The aeration potential of both products is a primary function of the shortening emulsifier system and its plasticity, which is controlled by the basestock formulation for liquid to solid ratios, crystalline structure, and plasticization conditions.

(2) Syneresis resistance—Whenever two materials with different water activities are placed in intimate contact with each other, moisture tends to migrate to the material with the lower water activity. Cakes, either injected or filled with creme fillings or covered with buttercreme icings, become excellent candidates for syneresis. Shortenings with emulsification functionalities capable of holding the required moisture contents resist syneresis. Textural evaluations of the icing or filling can indicate emulsion stability; smoothness indicates a tight emulsion with a probable low water activity, while a curdled appearance indicates a loose emulsion with a probable high water activity.

(3) Eating characteristics—Acceptable eating characteristics for icings and fillings are dependent upon three factors:
   - smoothness—Confection sugars with large particle sizes or insufficient moisture levels may contribute a graininess or a gritty sensation from undissolved sugars.
   - getaway—Rapid melt in the mouth is the desired physical sensation while a slow getaway will "mask" some of the desirable flavors. Getaway can be affected by (a) moisture level; (b) emulsion character, either a tight or loose emulsion; (c) SFI slope and melting point of the shortening; and (d) degree of aeration.
   - flavor—Offensive or off-flavors can be contributed by the shortening, either its base fat composition or emulsifier components. Different emulsifiers are prepared with glycerin, propylene glycol, or sorbitol components, which have bittersweet flavors that can be detected at levels usually above use levels. Any off shortening flavors would probably assume the flavor of the oil product before processing or a reverted flavor.

(4) Heat stability—Icings and fillings must be resistant to temperature abuses that the finished product will be subjected to in normal circumstances. Structure for an icing or filling is provided by the shortening and affected by both the base oil selection and processing. The base fat melting point must be high enough to resist melting at the temperature conditions expected before the finished product is consumed. Heat stability may be built into the shortening by the choice of basestocks and hard fats for the base shortening and the emulsifier system.

(5) Body—A somewhat firm but spreadable consistency without undue "drag" is necessary for a smooth icing application to cakes. Likewise, creme fillers need a somewhat firm, but not stiff, body to stand up between cake layers or when injected into cakes. Filling and/or icing structure is impacted by the shortening's consistency and emulsifier system.

A wide range of formulations is available and in use for creme fillings and buttercreme icings. There has also been considerable variation in the shortenings chosen to prepare these products. Butter, margarine, and all-purpose shortenings are used for icing production, but the largest usage has been all-purpose, emulsified shortenings since the early 1930s and, more recently, the shortenings developed specially for icing and creme filling production. The cake and icing or "high ratio" type shortenings utilize plastic or shortening-like mono- and diglycerides at a compromised level to produce satisfactory cakes, icings, fillings, yeast-raised products, and other products. It has also been recognized that mono- and diglycerides alone do not provide all the performance desired in regard to icing and filling volume and stability. Specialty shortenings combining proper fat structure with emulsifier systems composed of hydrophilic and lipophilic components have been proven advantageous. Icings and fillings made with these shortenings aerate rapidly for high volumes, with superior eating qualities contributed by a rapid getaway in the mouth to eliminate the lingering greasy eating characteristics formerly associated with bakery icings and fillings. The effect of the individual components of the specialty icing and filling shortenings are [9–11]

(1) Shortening base composition—The shortening base selected for most specialty icing and filling shortenings are characteristic of basestock and hard fat selections for all-purpose shortenings. Hard fat helps effect a flat SFI slope for a broad plastic range and firmness for icing and filling structure. However, excessive hard fat additions result in a loss of shortening plasticity for more difficult incorporation along with less "getaway" in icing and filling results. Softer hard fats like 25 to 30 iodine value hydrogenated cottonseed oil may be selected to allow slightly higher addition levels than possible with the low iodine value hard fats of 4.0 or less. The beta-prime hard fat levels utilized to produce the most heat-stable fillings and icings range from 15 to 20%.

(2) Emulsifier system—Extensive testing has verified that mono- and diglyceride emulsifiers alone do not provide all of the icing and filling performance desired for aeration, structure, and stability. A synergistic system composed of lipophilic (fat loving) and hydrophilic (water loving) emulsifiers has affected icings and fillings with greater specific volumes, superior eating qualities, and stable tight emulsions. The emulsifier system components are
  • lipophilic emulsifiers—Many icing and filling shortenings utilize two water-in-oil surfactants in synergistic emulsifier systems: mono- and diglyceride and lecithin. The natural emulsifier, lecithin, affects fluidity and emulsion stability. Low levels of soft mono- and diglycerides

produced from unhydrogenated or relatively high-IV (90 or less) base-stock provides the best performance in icings and fillings; levels higher than 2.5% alpha monoglyceride decrease icing and filling stability [12]. Shortenings containing low-IV mono- and diglycerides produce very tight emulsions that restrict aeration at all addition levels.

- hydrophilic emulsifiers—Polyglycerol esters of oleic acid or 80 to 85 IV shortening basestock, as well as polysorbate 60 or 80, are particularly good aeration promoters in icings and fillings. These hydrophilic emulsifiers also provide emulsion stability to prevent oil migration from the icing or filling to the cake.

Effective icing and filling shortenings are the result of combining a firm wide plastic range base with a synergistic emulsifier system. Even still, specialty icing and filling shortenings vary with each producer, and variations are provided for individual customer requirements; an average all-vegetable formulation consists of 15 to 20% beta-prime hard fat, a soft mono- and diglyceride of 1.0 to 1.5%, and fluid grade lecithin at approximately 0.25% with either 1.5 to 2.0% of a soft polyglycerol ester or 0.4 to 0.75% polysorbate 60. The balance of the formulation is composed of an 80 to 88 IV basestock. Meat fat formulations utilize the same emulsifier systems with equivalent hard fat levels with interesterified lard but lower hard fat levels (10 to 12%) when a blend of tallow and soft hydrogenated soybean or cottonseed oil are utilized as the base fats.

8. ICING STABILIZERS

Flat icings, frostings, and glazes constitute a large group of bakery accessories for the baked or fried products. This is the simplest type of icing, which consists essentially of powdered sugar with enough water added to give a consistency suited to the application. These icings may be pourable or plastic and are commonly used on donuts, sweet rolls, danish, breads, petit four cakes, other specialty cakes, and so on. Due to moisture loss, water icings can become dull and hard to flake off of the product in a very short time. Also, when packaged for distribution, the icings or glazes can stick to the wrapping material and fingerprint easily. Low-level additions of a properly processed icing stabilizer provides elasticity and antisticking properties and cause more rapid setting properties to refuse fingerprints.

Icing stabilizers are hydrogenated with melting points centered at 113, 117, and 125°F (45, 47, and 52°C) and are formulated both with and without lecithin, depending upon the desired degree of fluidity in the finished

TABLE 6.4. Icing Stabilizers: Typical SFI and Melting Point Analysis.

Mettler dropping point:			
°F	125.0	117.0	125.0
°C	45.0	47.0	52.0
Solids fat index, %:			
at 10.0°C/50°F	67.0	65.0	77.0*
at 21.1°C/70°F	58.0	60.0	74.0*
at 26.7°C/80°F	53.0	58.0	72.0*
at 33.3°C/92°F	38.0	45.0	71.0*
at 37.8°C/100°F	21.0	30.0	58.0*

*Estimate only—Product too hard to perform SFI analysis.

icing or glaze. Icing fluidity is increased with lecithin additions up to 0.5%, but this functionality reverses at higher levels. Icing stabilizers are hydrogenated to meet the SFI and melting point requirements designed to provide the intended performance. The slope of the SFI curve is important to provide the desired icing or glaze set but still retain the elasticity so that product does not become brittle and flake off the baked or fried product. Typical target solids fat index limits corresponding to the melting points for the three icing stabilizers usually employed are shown in Table 6.4.

Icing stabilizers are flaked to permit ease in handling, scaling, and rapid melting during the icing and glaze make-up. Typically, 1 to 6% of the icing stabilizer is melted and slowly added to the icing or glaze while mixing to prevent solidification of the stabilizer before it is incorporated into the mix. Flaked icing stabilizers are produced on chill rolls that solidify the product into the thin flake form to provide a large surface area that promotes rapid melting when heated for use in the finished icings or glazes.

9. BREAD SHORTENING

Based on the level of usage, shortening is one of the minor ingredients in the production of white bread. Functionally, however, it is one of the more important ingredients because it influences dough mixing, handling, proofing, and bread volume, as well as imparting desirable eating and keeping qualities to the baked bread. All of these effects are due to the lubricating properties of the shortening. An adequate shortening level must be used to produce bread with a good appearance and eating qualities. The optimum shortening level may vary slightly from one bakery to another due to variations in mixing, proofing conditions, and flour quality; however, most bread formulations will contain about 3.0% (100% flour basis) shortening for a good quality loaf of bread.

For many years prime steam lard was considered the ultimate in bread shortening. Then, the addition of mono- and diglycerides changed the concept of bread from a firm chewy type to a soft compressible product. Emulsified shortenings or separate additions of the mono- and diglycerides enabled the production of a soft bread that remained soft for long periods of time both on the grocery store shelf and in the home. The homemaker's "squeeze test" became a quality control measure for freshness observed by bakers and ingredient suppliers. The successes of mono- and diglyceride prompted the development of other surfactants. Emulsifier concentrates emerged that permitted a somewhat greater flexibility in the production of bread of varying degrees of softness, improved handling properties, and greater tolerance to flour variations.

Concerns that some surfactants could present a health hazard and that the increased bread softness created might be a deceptive practice led to limitations of the allowed emulsifiers and use levels permitted by the standard of identity for bread and rolls enacted in 1952. The U.S. FDA standards limited the amount of the approved surfactants to about 0.5% based on the flour weight. In November 1978, the standard of identity for bread, rolls, and buns was amended to remove the use limit for mono- and diglycerides, diacetyl tartaric acid esters (DATEM), propylene glycol monoesters (PGME), and other ingredients that performed a similar function. The U.S. FDA-approved softeners and/or dough strengtheners for yeast-raised products are listed in Table 6.5 with the use level restrictions and functional effects.

TABLE 6.5. *Softeners and Dough Strengtheners for Yeast-Raised Products.*

Surfactant	Softener	Dough Strengtheners	Allowed Use Level, % Flour Basis
Mono- and diglycerides:			
Saturated type	Good+	Slight	GMP
Unsaturated type	Slight	Slight	GMP
Lecithin	Good	Slight	GMP
Diacetyl tartaric acid esters	Good	Good	GMP
Propylene glycol monoesters	Good+	Slight	GMP
Sodium stearoyl-2 lactylate	Slight	Good	0.5 max.
Calcium stearoyl-2 lactylate	Slight	Good	0.5 max.
Succinylated monoglycerides	Good	Good	0.5 max.
Ethoxylated monoglycerides	Slight	Good	0.5 max.
Polysorbate 60	Slight	Good	0.5 max.
Sodium stearyl fumarate	Good	Good	0.5 max.
Lactylic stearate	Slight	Good	0.5 max.

GMP = good manufacturing practices.

For conventional bread production, most plastic shortenings of either animal or vegetable origin supplemented with softeners and conditioners may be used effectively. However, all vegetable shortenings are probably preferred now for nutritional concerns regarding cholesterol and saturated fatty acids. Most all vegetable conventional bread shortenings are partially hydrogenated and deodorized products, like the 110 IV hydrogenated soybean oil basestock (identified in Table 4.6) shipped in tank quantities with antioxidants added to protect the flavor before use.

In the continuous bread mixing process introduced in the 1950s, shortening and emulsifiers continue to provide the same benefits as the conventional process. Continuous bread shortenings have three basic components that can be produced together or added separately:

(1) Base shortening—Almost any fat can be used successfully for the base shortening component for continuous bread production. Both hydrogenated and unhardened meat fats or vegetable oils have been used successfully. Comparisons between fat types have shown that the quality of bread is unaffected by the type of fat used, except that a salad oil with an absence of solids has a depressing effect that requires a higher flake or hard fat addition level.

(2) Flakes—The higher mixing and proofing temperatures employed for continuous bread processing require higher melting shortenings for gas retention and structure. A minimum amount of the shortening must be present as a solid at the dough proofing temperature. Hard fat or flakes made from all the available U.S. domestic fats and oils are adequate to provide solids in the 104 to 115°F (40 to 48.9°C) proofing temperature range normally used for continuous mixed bread doughs.

(3) Emulsifier systems—Numerous evaluations have confirmed that hard, highly saturated mono- and diglycerides, with a 5.0 maximum IV, are superior to softer mono- and diglycerides for starch complexing for shelf life enhancement, dough strength, and bread texture. The softer mono- and diglycerides have been shown to contribute to a weaker dough, which allows fermentation gases to escape to result in bread with open grain, poor texture, and reduced volume. In addition, one or more of the softeners and/or dough conditioners listed in Table 6.5 may be utilized.

Most of the fat-soluble surfactants that are most effective have high melting points, and the hard fats must be melted into the shortening before addition to the continuous dough operation. Most continuous operations either use separate components or purchase liquid shortenings processed with the three components for use at room temperature.

10. LIQUID BREAD SHORTENINGS

Continuous dough process requirements for a fat system composed of three components—base fat, hard fat flakes, and emulsifiers—presented uniformity problems that affected bread quality. Hard fat additions increase the melting point of the fat system as desired for dough structure and gas retention; however, merely adding a predetermined amount of flakes does not eliminate the opportunity for uniformity problems. An improper balance of hard fat and surfactants results in an increase in cripples from handling while transferring the proofed bread to the oven. Experimental work has indicated that the solids fat index values should be in the 10 to 15% range at the high temperatures encountered in the development and processing stages for continuous doughs. Solids fat index control of the fat system is a better control primarily due to the variability of the base fats; however, most bread bakers do not have the facilities to perform the SFI analysis. These control requirements and the convenience of handling a shortening without heating made liquid bread shortenings attractive for the bread baker.

Liquid bread shortenings have the same general definition as the other fluid products, i.e., a stable dispersion of crystalline solids in a liquid oil matrix that is fluid and pumpable at room temperature. Liquid shortenings designed for bread production are normally composed of a liquid vegetable oil base, a high melting point hard fat, and a crumb softening agent, either alone or in combination with a dough conditioner. Softener and dough strengthener additives to the liquid shortening provide a complete system for a high degree of uniformity for the bread baker, which also eliminates another ingredient to inventory and scale.

A review of liquid bread shortening's evolution is shown in Table 6.6 [13,14]. The first liquid bread shortening, introduced in the early 1960s, had an opaque or milky appearance and was composed of a liquid soybean oil, partially hydrogenated for oxidative stability with the hard fat requirement for continuous bread processing. Later, mono- and diglycerides made from a low-iodine value hard fat were incorporated to soften the bread crumb. The softening effect results from an interaction of amylose, the linear starch component, with the fatty acid chain of the monoglyceride, which subsequently reduces the starch tendency to retrograde. Polysorbate 60 was probably the first dough conditioner added to liquid bread shortenings. The function of a dough conditioner is to modify the rheological characteristics of the bread dough to strengthen it for better handling characteristics. All of the liquid bread shortenings up to this point of the evolution had high hard fat levels, as indicated by the SFI levels of the first three shortenings in Table 6.3. The high levels of

TABLE 6.6. Liquid Bread Shortening Evolution.

	Surfactant				
	None	Softener	Softener and Conditioner		
Mettler dropping point					
°F	50.0±	53.9±	54.4±	53.3±	40.6±
°C	122.0±	129.0±	130.0±	120.0±	105.0±
Alpha monoglyceride, %	None	3.0±*	2.5±*	5.75±*	4.0±°
					5.0±†
Dough conditioner, %	None	None	2.0±**	5.0±†	4.9±‡
Solids fat index, %					
at 10.0°C/50°F	13.0±	11.0±	11.0±	8.5±	6.5±
at 21.1°C/70°F		10.0±	10.0±	7.5±	3.0±
at 26.7°C/80°F	12.0±	10.0±	10.0±	7.5±	
at 33.3°C/92°F		10.0±	10.0±	7.5±	2.0±
at 40.0°C/104°F	10.0±				1.5±
at 43.3°C/110°F		5.0±	9.0±	3.0±	
AOM stability, hours	30±	25±	20±	30±	20±
Antioxidant	None	None	None	TBHQ	TBHQ

*Mono- and diglycerides, 5 IV maximum.
**Polysorbate 60.
†Ethoxylated mono- and diglycerides.
‡Sodium stearoyl lactylate.

hard fat were required in the continuous mixing process for good bread side wall strength. This function was partially assumed by the conditioner with the addition of ethoxylated mono- and diglycerides, which allowed a hard fat reduction for an improved softness of the bread loaf. The last variation shown for liquid bread shortenings is a conditioner modification to a blend of ethoxylated mono- and diglycerides (EMG) with sodium stearoyl lactylate (SSL), and still retaining mono- and diglycerides as a crumb softener. This conditioner change was made to strengthen the gluten structure and to attain a better mixing tolerance for the dough, and it allowed an increased moisture absorption, which extended crumb softness and shelf life with an increased loaf volume. The synergistic effect of combining the surfactants and hard fats with the liquid oil carrier, which also acted as a lubricant, has permitted some bakers to reduce their usage levels by 15 to 20% over the addition of the same composition added separately. Nevertheless, some bread bakers still prefer to formulate and add their own softeners and conditioners to a liquid shortening similar to the original nonemulsified product.

11. SWEET YEAST-RAISED DOUGH SHORTENINGS

Sweet doughs are yeast leavened but distinguished from bread, buns, and dinner rolls by a sweeter taste and a richer eating sensation. Sweet doughs are much richer in shortening, mildness, and sugar than are bread-type doughs and, in addition, usually contain eggs. The function of the enriching ingredients is to soften the texture and make it flakier, improve the taste and color, and increase the nutritional quality of the product.

Improvements in sweet doughs came about as an outgrowth of the use of mono- and diglycerides in cake batters. It was a natural step to enrich sweet doughs by the use of a high-ratio shortening containing an emulsifier to aid in a greater absorption of the liquids, thus yielding softer, moister, longer keeping products. This occurred in the late 1930s. Eventually, the emulsifier content was raised to higher levels in special sweet dough shortenings to further enhance these properties, making possible the tender, rich sweet rolls, coffee cakes, and yeast-raised donuts, with a relatively long shelf life produced today.

Shortening bases with low SFI contents are preferred to impact tenderness to the yeast-raised shortenings. The best surfactants are the softeners that have the greatest effect upon the amylose component of the starch. Therefore, hard mono- and diglycerides are typically used in sweet dough shortenings at levels to provide between 8.0 and 9.0% alpha monoglyceride contents. These shortenings can be produced with either meat fats or all-vegetable shortening bases, which act as carriers for the hard emulsifiers and provide lubricity. Typically, a meat fat shortening would be composed of lard with 20% of a low-iodine value, tallow-based mono- and diglyceride to produce a plasticized shortening with an $8.5 \pm 0.5\%$ alpha monoglyceride content. An all-vegetable product could be a plasticized blend of 80% of an 80-IV hydrogenated, deodorized soybean oil basestock with 20% of a low-IV soybean oil–based mono- and diglyceride also with an $8.5 \pm 0.5\%$ alpha monoglyceride content.

12. COOKIE SHORTENINGS

The formulas for some types of cookies are very similar to those for layer cakes, except for a lower moisture content, which may have influenced the dictionary definition for cookies, i.e., "a small sweet cake, usually flat." The term *cookie* actually covers a wide variety of bakery products. There are hundreds of types of cookies made by the various branches of the baking industry. Cookies may be classified into hand-cut, bag type, dropped, ice box, wire cut, Dutch machine type, English hard-goods, American stamped type, and many others too numerous to list. However, two broad general classifications can be made: hard and soft.

Cookie varieties contain from 10 to 30% shortening, which is definitely an essential component for most cookie formulations. The type and amount of shortening utilized affects the machine response of the cookie dough and the quality of the baked cookie. Shortening is one of the principal ingredients for increasing tenderness, keeping qualities, grain and texture, and adding a richness quality to cookies. It must provide the proper solids contents for cohesion and spread, yet still be a tenderizer. In soft cookie production, plastic shortening is creamed with sugar to incorporate air bubbles that are trapped in the liquid phase of the shortening. High SFI shortenings do not have enough oil volume for adequate aeration, and low SFI shortenings do not have the facility to hold the air until mixing is complete. Also, cookie shortenings must have controlled plasticity or the ability to work into the blend during mixing to provide a smooth, lump-free cookie dough with a minimum of oiling out. Pockets of shortening in a cookie dough will melt and leave voids and uneven areas in baked cookies. Low heat stability shortenings break down with machining to release liquid oil and cause distorted greasy cookies. However, some oiling is necessary for wire-cut, rotary molded, and other cookie processes for extrusion or release. It is necessary to maintain a balance between the solid and liquid phases of the shortening for machinability, aeration, tenderness, and shelf life functions. All-purpose shortenings with wide plastic ranges, a beta-prime crystal habit, and good heat stability are normally adequate for soft cookie production. Emulsified shortenings produce soft cookies with a gummy mouth feel due to the increased moisture retention after baking. Normally, this effect has been viewed as undesirable, but it was the basis for a heavily marketed soft cookie category recently.

Hard cookies merely require lubrication from the shortening component. The low-moisture content of cookies causes the sugar to solidify to a candy-like mass instead of a liquid or mobile syrup as in cakes with higher moisture levels. The shortening function is to disrupt the continuity of the caramelized sugar mass [15]. Usually, a selectively hydrogenated shortening with a narrow plastic range is preferred for hard cookies for lubricity, shelf life, and enhancement of the eating character.

13. COOKIE FILLER SHORTENINGS

Filler is the name given to the creme or filling between two cookie pieces that form a sandwich cookie or a sugar wafer. All cookie fillers have the same basic ingredients, i.e., in order of prominence, sugar, shortening, salt, flavor, and lecithin. Anhydrous corn sugar and powdered sugar are the sweeteners used that melt rapidly in the mouth. Powdered sugar imparts a warming sensation while corn sugar gives a cooling sensation. The filler consistency and eating character is determined to a large extent by the shortening used. The requisites for a good sandwich cookie or wafer filler shortening have been delineated as

- quick get away in the mouth
- oxidative stability capable of providing 6 to 12 months' filler shelf life
- produce a filler consistency that does not become sandy, hard, or disappear with age or normal heat abuse
- aeration potential of 0.75 to 0.95 g/cc specific gravity in most fillers within approximately 10 to 15 minutes mixing time. *Note:* See Chapter 3 for Cookie Filler Performance Evaluation Method 8.12.
- filling consistency to be "tacky" but still break or shear at depositor operating temperatures, usually 90°F ± (32.2°C ±)

Cookie bakers have essentially five different options for their filler shortening requirements:

(1) Coconut oil—Unhardened coconut oil with a 76°F (24.4°C) melting point imparts a cool, pleasant sensation in the mouth to the filler but is seldom used since it will melt and separate from the filler with only moderate heat. Hydrogenated coconut oil with a 92°F (933.3°C) melting point provides an excellent eating quality with fair aeration, but the heat generated in many filling machines will cause unacceptable filler oil-out. Hydrogenated coconut oil with a domestic oil hard fat added to adjust the melting point to about 110°F (43.3°C) usually eliminates the depositing problems, but the cookie filler eating quality suffers due to the high melting characteristics.

(2) Coconut oil blends—Plasticized blends of coconut oil with other hydrogenated oil to obtain an 82 to 94°F (27.8 to 34.4°C) melting point produce acceptable eating quality fillers. However, with increased production rates for the cookie sandwich machines, these products have oil-out problems. Reformulation of these blends to higher melting points helps to obtain maximum production efficiencies while retaining some of the sharp melting characteristics from the reduced coconut oil levels.

(3) Domestic oil blends—Many sandwich cookie filler shortenings are produced with a blend of hydrogenated domestic vegetable oils, i.e., soybean, cottonseed, and so on. The melting points generally range from 98 to 112°F (36.7 to 44.4°C), depending in part upon the type of filler

TABLE 6.7. Cookie Filler Shortenings.

Probable Composition, %	Coconut Oil			Coconut and Domestic Oil Blend	All Domestic Oil Blend	Animal-Vegetable Blend	Special Hydrogenation
	76°	92°	110°				
Coconut oil	100			69			
5 IV H-CNO		100					
I IV H-CNO			98	25			
66 IV H-SBO					78		
74 IV H-SBO					20	75	
80 IV H-SBO					2		
60-Titer CSO			2	6			
Tallow						25	
75 IV SH-SBO							100
Total	100	100	100	100	100	100	100
Solids fat index:							
at 50°F	59±	57±	63±	53±	39±	27±	58±
at 70°F	29±	33±	41±	24±	24±	15±	43±
at 80°F	0	8±	16±	14±	17±	12±	34±
at 92°F		3±	7±	8±	7±	6±	12±
at 104°F			4±	4±	3±	1±	1±
Mettler dropping point, °C	24.5±	34±	43.5±	45±	40.5±	39±	38.5±
AOM, hours	200+	200+	200+	200+	100+	75+	200+

depositing equipment in use. The oils are selectively hydrogenated to produce as steep a SFI slope as possible to provide good eating qualities and oxidative stability.

(4) Animal vegetable blends—These cookie filler shortenings were attractive to the low-priced cookie producers. The flavor and creaming properties of these blends are acceptable, but the mouth feel and getaway are usually poor in comparison to the other products.

(5) Specially hydrogenated domestic oils shortening—Special hydrogenation utilizing a sulfur proportioned nickel catalyst produces a hardened soybean and cottonseed oil shortening with characteristics like hydrogenated coconut oil, i.e., a steep SFI slope with a sharp melting point. Cookie fillers produced with these specialty products hydrogenated to a 102°F (38.9°C) melting point have a good get away in the mouth, acceptable performance with at least one type of filler/depositor, and exceptional oxidative stability.

Typical compositions, solids fat index profiles, and melting point analysis are listed in Table 6.7 for five cookie filler shortening options.

14. PIE CRUST SHORTENINGS

Pie crusts are low in moisture and high in shortening. The ratio of ingredients, teamed with the preparation method, prevents the formation of a continuous gluten network through the dough and results in a baked product that is flaky. A porous crust structure is not desired because pie crusts must support and retain without leakage, fillings of moderate viscosity, and high moisture contents. Pie crusts are generally divided into three classes based upon flakiness:

(1) Flaky—When broken, flaky crusts exhibit a fracture along different lines at different levels and show a distinct separation in layers parallel to the surface. Long flake crusts require a minimum of mixing with a high shortening content to maintain the desired texture.

(2) Mealy—When broken, mealy crusts break in a straight line and have a surface resembling a cookie. Mealy crusts are usually mixed more thoroughly to allow rougher handling both in make-up and distribution.

(3) Medium flake—Short or medium pie crusts have properties between flaky and mealy. The shortening particles are purposely left about the size of a pea, and the moisture is incorporated with a minimum of mixing to keep the shortening particles discrete.

Nearly all of the shortening functions in pie crust are associated with its capacity to lubricate, tenderize, and provide flakiness. Lard qualifies as an ideal pie crust shortening because its solid components are primarily symmetrical in configuration and do not set up in a tight-knit matrix. Flakiness is imparted by the high melting, coarse, large, beta crystals, which form grains and the liquid oil, which separate the dough into layers. Lard, the standard pie crust shortening, has an SFI profile suitable for flakiness and low-temperature plasticity due to its beta crystalline habit. Most pie lard shortenings are developed around the addition of 2 to 3% low-iodine value lard hard fat for consistency control. The fatty acid composition and resulting consistency of lard receipts can have wide variations related to diet, climate, and position on the animal's body. Hard fat additions increase the SFI values most at the higher temperatures with a lesser effect at the lower temperatures to effect a flatter SFI slope. Lard can also be partially hydrogenated for consistency control. Hydrogenated lard essentially maintains the original SFI slope with gradual increases across the entire SFI profile. Consistency control of pie lard is also achieved through the plasticization

TABLE 6.8. Pie Crust Shortenings.

	Composition, %					
Source Oil	Lard			Soybean Oil		Soybean and Cottonseed Oil
Lard	100		97			
Hydro. lard		100				
59-Titer lard			3			
95 IV H-SBO				95		
88 IV H-SBO					60	
80 IV H-SBO						92
60 IV H-SBO					40	
63-Titer SBO				5		
60-Titer CSO						8
Total	100	100	100	100	100	100
Crystal habit	Beta	Beta	Beta	Beta	Beta	Beta-prime
Solids fat index, %						
at 50°F	29.0±	35.5±	32.0±	25.0±	34.0±	26.0±
at 70°F	21.5±	26.0±	25.0±	15.0±	25.5±	18.0±
at 80°F	15.0±	19.5±	18.0±	13.0±	22.5±	16.0±
at 92°F	4.5±	10.0±	10.0±	10.0±	13.5±	10.0±
at 104°F	2.9±	7.0±	7.0±	7.5±	4.5±	6.0±
Mettler dropping point, °C	32.5±	38.0±	41.0±	45.0±	42.0±	39.0±

and tempering conditions used. Tempering packaged lard at 70 to 80°F (21.1 to 26.7°C) develops a plastic, rubbery consistency, while product tempered at 40°F (4.4°C) becomes loosely structured and brittle [16].

The procedures used in preparing pie crusts have been developed around the peculiar characteristics of lard. Use of identical procedures with the standard all-purpose shortenings formulated for a wide plastic range and a beta-prime crystalline structure results in substandard pie crusts. It is necessary to mix the pie dough at colder temperatures and use them immediately for the standard all-purpose shortenings to perform successfully. However, all vegetable shortenings formulated to the beta crystalline structure provide pie doughs equivalent in performance but with a bland flavor instead of the lardy flavor from the lard products. Table 6.8 compares the composition, SFI, and melting characteristics of the lard products to all-purpose–type shortenings prepared with vegetable oils.

15. BISCUIT SHORTENINGS

Biscuit means one thing in the United States and another in the British Isles and Europe. The word is French and means "twice cooked," applying originally to thin, flat breads used on shipboard. These breads had to be baked twice to expel as much moisture as possible to increase their keeping qualities. This description certainly doesn't fit the U.S. biscuit developed by the American colonists in the late 18th century. American biscuits are better described as "quick breads" made with baking powder, instead of yeast or baking soda, in combination with buttermilk as the acid ingredient for leavening. Biscuits, long a staple of the diet in the southern states, have become a favorite nationwide. The American consumer can prepare biscuits from "scratch" or a dry mix, bake refrigerated biscuits from a can, purchase them as a bakery item, or order them at fast-food or full-service restaurants.

Biscuit doughs cannot withstand the handling that yeast-raised doughs require to develop the flour gluten or protein. Biscuits are mixed by first "cutting" the shortening into the flour, until it is walnut- to pea-sized to assure that the flour is not "greased," and then followed by the moisture addition and minimal mixing only to wet the flour without excessive gluten development. The mixing time is dependent upon the shortening's physical consistency. A soft, plastic shortening will incorporate into the mix rapidly to develop the dough and produce a tough biscuit. A firm, brittle shortening will shatter in small pieces that do not incorporate into the dough easily and produce a flaky light biscuit when baked. Many biscuit bakers refrigerate their shortenings to help assure that the biscuit doughs are not overmixed. Firm hydrogenated vegetable oil blends or an antioxidant-treated tallow with a relatively high-oxidative stability are often tailor-made to the individual food processors or foodservice operation for biscuit manufacture. Since plasticity is a detriment for biscuit shortenings, blends with steep SFI slopes and good oxidative stability are preferred with consistencies only soft enough to handle conveniently. Crystal structure is not critical, except that a beta crystal shortening with a grainy texture probably provides a flakier biscuit.

16. DANISH PASTRY ROLL-IN SHORTENINGS

Danish pastry was first introduced to the United States around the turn of the century. Danish pastry is a rich, yeast-leavened sweet dough that has a high fat content. The distinguishing feature of Danish preparation is the interleaving of dough sheets with layers of shortening that, upon baking, give a separation of strata and an open network of crisp and flaky layers. Initially, it was produced with a hard, imported Danish butter, and the product was closer to puff pastry than the Danish production today. The character of Danish has been gradually changed; first by substitution of domestic butter and then margarine and eventually specialty shortenings. The current Danish pastry still has good flakiness but is also tender or short [17]. The steps for hand make-up production of Danish pastry are

- A sweet yeast-raised dough is prepared in the usual manner, except that mixing ceases as soon as the ingredients have been thoroughly incorporated. Danish doughs are further developed during the rolling in process.
- The dough is immediately divided into strips of an appropriate size and then allowed to rest for 15 to 20 minutes in a retarder at 35 to 40°F (1.7 to 4.4°C) with a relative humidity of not less than 85%. The primary purpose of the retarder is to slow yeast fermentation while the dough is relaxing.
- The dough strips are sheeted to a thickness of one-half to three-quarters of an inch, three times as long as wide, and roll-in shortening is spotted over two-thirds of the length of the dough sheet.
- The unspotted third of the dough is folded over the center third and then the remaining third on top, making three layers of dough and two layers of roll in shortening.
- The laminated dough is again rolled to approximately ½-inch thickness and folded into thirds. *No additional roll-in shortening is used.*
- The folded dough pieces are then allowed to rest 20 to 30 minutes in the retarder.

- The laminated dough is sheeted and again folded in thirds twice more with a 20- to 30-minute rest period in the retarder between each fold.
- After rolling and folding for the last time, most Danish doughs are allowed to rest 4 to 8 hours or overnight before make-up of the individual rolls or coffee cakes.
- The usual recommended proofing temperature for Danish is 90 to 95°F (32.2°C), with only enough humidity to prevent crusting of the pastry surface. Proof time is usually short, 20 to 30 minutes, to allow some expansion of the dough from the yeast, being careful to keep from melting the roll-in shortening and destroying the layering. The rolled in shortening must have an opportunity to melt and expand the pastry in the oven to create a flaky texture.
- After proofing, Danish pastry is baked at 380 to 400°F (193.3 to 204.4°C) for 15 to 20 minutes, depending upon the unit size.

Several mechanized Danish pastry systems have been developed to reduce the work load involved with hand make-up. The development of dough and shortening pumps, dough and shortening extruders, sheeters, dough folding, rotary and guillotine cutter, cooling devices, and many others have permitted automation of Danish production. The automated systems do not have the capacity for adapting to variations in product consistency, softness, and other characteristics as readily as bakers did with hand preparation. Therefore, more tailor-made roll-in shortenings became a necessity for the automated systems to produce quality Danish pastry.

The richness or amount of roll-in shortening in Danish pastry is often varied over a wide range. The most popular "American" type has approximately half as much roll-in shortening as flour. Danish has been made with as low a roll-in quantity as 20% of the flour; however, the finished product represents a poor imitation of traditional Danish pastry. It has been established that the quantity of roll-in determines the number of roll-ins required. Rich Danish doughs or those with high roll-in levels require more folds to prevent leaking and oily crust characteristics. Doughs with 70 to 80% roll-in (flour basis) require a minimum of four folds; doughs with 50% roll-in should be given three folds; and doughs with lesser roll-in quantities should have reduce folds to obtain some flakiness.

The most important features of a roll-in shortening for Danish are plasticity and firmness; the roll-in shortening should be of medium firm consistency, slightly firmer than the average all-purpose shortening. Plasticity is necessary because the shortening must spread or roll out in unison with the dough and then remain as unbroken layers between the dough layers during the repeated sheeting and folding operations. It must remain lump-free after retarding at refrigeration temperatures. At this point, the dough and fat layers have been compressed to a thickness of approximately 0.025 of an inch. Hard, brittle shortenings will rupture the dough layers while sheeting to destroy the lamination, which provides flakiness. Firmness is equally important because a soft or oily shortening can be partly absorbed by the dough during sheeting or melted during proofing, reducing its role as a barrier between the dough layers, and the lamination effect is lost [18]. Furthermore, the baked Danish must have a good mouth feel, so the shortening should have a rather sharp melting point in the vicinity of body temperature. Briefly, the ideal Danish pastry roll in shortening should be functional over a temperature range of 50 to 90°F (10 to 32.2°C) with a low enough melting point to provide a good mouth feel to the finished product, rather than a waxiness or greasy mouth feel associated with high-melting fats.

Formulation of the fat blend, as well as the plasticization and tempering conditions, are essential elements for the preparation of Danish roll in shortenings. Evolution of these roll-in shortenings have been necessitated by changes in bakery production practices, product improvements, and introduction of new baked pastry items. As the melting points and SFI profiles in Table 6.9 indicate, roll-in shortenings have tended to be compromises between either good functionality or good eating characteristics [19,20]. This has been the general rule because the high-melting saturated fat components were excessive for good eating characteristics. On the other hand, if the high-melting fractions were lowered to improve the edibility, then structure and functionality at working temperatures suffered.

TABLE 6.9. Danish Roll-in Product Evolution.

Typical Analytical Characteristics	Domestic Butter	Margarines		Roll-in Shortening Melting Point Variations		
		Imported Danish	Baker's	Low	Mid	High
Solids fat index, %						
at 50°F	34.0	43.5	29.0	39.0	24.0	26.0
at 70°F	12.0	25.8	18.0	27.0	20.0	20.0
at 80°F	4.0	20.6	16.0	22.0	19.0	18.0
at 92°F		11.1	13.0	11.5	16.0	17.0
at 104°F		2.0	5.0	2.5	11.0	14.0
Mettler dropping point, °C	32.0	40.2	45.0	39.0	45.0	52.0

Danish roll-in shortenings are usually compounded with one to two basestocks and a beta-prime hard fat. The basestocks utilized are determined by the SFI slope desired, usually a blend of a selectively hydrogenated basestock with a nonselective softer basestock, and stabilized with the hard fat for the steeper SFI slopes formulated for eating quality [20]. The flatter SFI slope products are formulated like all-purpose shortenings with an 85 to 95 IV nonselective basestock stabilized with a higher level of the hard fat component. The hard fat for these products may be hydrogenated to a less saturated IV endpoint to allow a higher addition level to flatten the SFI without increasing the melting point to intolerable waxy mouth feel levels.

Plasticization conditions are especially important for roll-in shortenings to produce the desired workability at the refrigerator temperatures used for Danish pastry make up. The plasticization process is normally performed as outlined in Chapter 2 in Chart 2.10 using a tubular scraped wall heat exchanger to rapidly chill the molten oil blend from a temperature not in excess of 20°F (11.1°C) above nor more than 10°F (5.6°C) below the product melting point to a temperature within a range of about 60 to 70°F (15.6 to 21.1°C) to initiate development of the beta-prime crystal nuclei. Immediately following the chilling unit, the shortening is worked or kneaded for a time sufficient to further develop the beta-prime crystalline phase, which results in a product temperature rise before filling of 10 to 15°F (5.5 to 8.3°C). Typically, properly plasticized roll-in shortening forms a slight mound in the container as it fills. After filling, the packaged shortening is held at an elevated temperature, usually 85°F (29.4°C) for 48 to 72 hours, until the crystal structure of the hard fraction reaches equilibrium and forms a stable crystal matrix.

17. PUFF PASTE ROLL-IN SHORTENING

Puff pastry is an expanded, flaky, baked product best known for making turnovers, patty shells, and creme horns. It is prepared by layering puff paste shortening into a tough, but pliable, dough in the same manner as with Danish pastry. Layers of fat are interleaved between layers of dough so that, when baked, the dough layers expand. Danish pastry, which is a yeast-leavened dough, has a relatively soft and porous structure in the baked dough layers. Laminated, nonleavened puff pastry doughs produce a very open network of crisp and flaky layers. Puff pastry dough formulations generally fall within certain limits, i.e., equal weights of flour and puff paste shortening with water at half the flour weight, salt to flavor, and a low level of cream of tartar, vinegar, or lemon juice to facilitate the sheeting operation. Two general methods of incorporating puff paste into the pastry dough are in general use:

- French method—The puff paste is applied by hand or extruder onto sheeted dough, folded, resheeted, and refolded several times to form the thin interleaved layers of fat and dough.
- Scotch or blitz method—The puff paste is added to the dough at the end of the mixing period and mixed only briefly to distribute the shortening before sheeting and folding to form layers of fat and dough.

Puff paste shortening has several functions in puff pastry. In the conventional French method of preparation, a portion of the shortening is incorporated into the puff pastry dough during the mixing operation. This portion, usually about 15% of the total shortening requirement, functions in the usual role of any shortening for a dough: lubrication. The major portion of the shortening is spread between the folds of the dough. This portion of the shortening gives puff pastries their lift and flaky texture. The shortening is rolled into thin layers, alternating with the dough during the course of folding to form 1,215 layers with 6 three folds. This provides a natural cleavage point during baking. The fat layers expand upon melting, aided by steam produced by the moisture in many puff paste shortenings to create the internal lift. The properties desired for puff paste roll-in shortening to produce puff pasty with excellent volume when baked with even layers resembling the pages in a book are

- plasticity over a wide temperature range that permits stretching during the sheeting and folding process but remains as unbroken layers, characterized by a flat SFI curve
- firm and waxy but not brittle consistency, equivalent to the dough at retarder temperatures, to produce uniform, thin layers. Puff paste shortening with a too firm consistency can rupture the dough layers during the sheeting operation. Too soft a consistency product will soak into the dough layers to effect a tender flake that does not puff or raise. An evaluation of puff paste consistency is to knead a small sample in your hand. A good product will act like modeling clay. It will change shape without feeling brittle or lumpy and, conversely, will not become sticky or greasy [22].
- distinguished from other shortenings in that it is one of the higher melting edible fat products produced from animal fats and/or vegetable oils with final melting points on the order of 115 to 135°F (46.1 to 57.2°C)
- preparation both as emulsions of oil or fat with water (usually 10%) or as an anhydrous product without the addition of water. Hydrous

puff pastes usually contain approximately 0.5% alpha monoglycerides and/or 0.1 to 0.2% lecithin for emulsion stability. Mono- and diglycerides prepared with low-iodine value basestocks provide optimum emulsion stability. Traditionally, bakers have thought that, for puff paste to be functional, it must contain water, but anhydrous products have proven to be as functional to better than many hydrous roll-in shortenings [21].

- beta-prime crystal habit for a fine three-dimensional network capable of immobilizing a large amount of liquid oil

Animal vegetable blended products were the preferred puff paste shortenings because of more even puff probably due to the somewhat sticky consistency and the oxidized flavor and odor of the all vegetable products. These all vegetable products were almost true compound type shortenings prepared by blending about 25% cottonseed oil hardfat with unhydrogenated vegetable oils, usually cottonseed or soybean oil. The sticky puff paste penetrated into the dough layers to raise unevenly when baked and the polyunsaturates oxidized rapidly which produced an off flavor and odor. All vegetable oil puff pastes were improved by the use of brush and partially hydrogenated basestocks with beta prime hardfats to produce the high melting point and flat SFI slopes required with a better oxidative stability. The typical SFI and melting point ranges for puff paste shortenings are given in Table 6.10.

For decades, puff paste was solidified on the surface of a chill roll, held in troughs to allow formation of the stable crystal form, and then texturized before packaging. The advantage of this process was the long resting time between chilling and working or texturizing, which ensured an almost ideal state of crystallization before the working step. Even slow crystallizing fat blends could be processed to produce firm, plastic puff paste without a tendency toward postcrystallization. The chill roll plasticizing systems are

TABLE 6.10. Puff Paste Typical Solids Fat Index and Melting Point Ranges.

Solids fat index, %	
at 50°F	26.0–40.0
at 70°F	24.0–38.0
at 80°F	22.0–34.0
at 92°F	21.0–28.0
at 104°F	17.0–24.0
Mettler dropping point,	
°C	49.0–53.0
°F	120.0–127.0

labor-intensive and, therefore, expensive, and a high risk of contamination existed. Thus, process development to employ the same tubular scraped wall chilling units as other shortening or margarine products became high priority projects. Processing techniques have been identified that produce acceptable puff paste products, which generally consist of [23,24]

- prolongation of the product cooling and working time by reducing the equipment throughput to approximately half the normal production rate for standard shortenings for crystal development
- postcrystallization being reduced or controlled as much as possible to obtain good plasticity; in package temperature rise after packaging is a measure of postcrystallization
- programmed cooling temperatures during crystallization for shock chilling in the initial stages followed by more moderate temperature reductions in succeeding chilling units
- addition of a cooling stage after chilling or insertion of a working step between cooling steps used to dissipate the heat of crystallization and create a more stable crystal structure to reduce tempering requirements. Normal tempering at 85°F (29.4°C) can require up to 10 days, depending upon the package size, to attain the desired texture and plasticity. These additions speed up puff paste tempering but have not totally eliminated this requirement for optimum performance, especially for slow crystallizing formulations.
- quiescent resting tubes with sieve plates capable of holding 10 to 15% of the hourly capacity incorporated into many puff paste systems, which allows the product sufficient time to exchange the super-cooled condition for crystallization equilibrium. This time lag converts the product from a viscous fluid to a solid suitable for forming into parchment-wrapped 5-pound prints, 2½-pound parchment divided sheets, or 50-pound containers.

18. SHORTENING CHIPS

Shortening chips were developed to provide a flaky baked product similar to Danish pastry, but without the traditional roll-in process. The chips can be incorporated into a dough or batter just before the mixing process is complete to distribute them throughout. Upon baking, shortening chips melt, causing small pockets that simulate the flakiness of laminated doughs. Flavors, colors, and/or spices can be encapsulated in the shortening chips to leave pockets of color and flavor where the chips have melted during baking. Shortening chips have been commercially available in three flavors:

plain, butter, and cinnamon sugar. These products have found application in biscuits, pizza doughs, breads, simulated Danish, croissants, dinner rolls, cookies, pie crusts, and various dry mixes such as biscuit, pancake, muffin, waffles, and others.

Shortening chips are selectively hydrogenated to attain a steep SFI slope with a melting point as low as possible to allow the product to maintain a chip form after packaging, during shipment to the food processor, while processing into the finished product, and until heated for preparation by the foodservice operator, baker, or homemaker. The chips can be formulated from the basestock system with a blend of the 60 and 66 IV bases, or a product can be hydrogenated specially to meet the desired solids fat index and melting point limits. Shortening chips made with soybean oil usually have melting points ranging from 110 to 118°F (43.3 to 47.8°C). Lauric oil–based shortening chips offer a sharper and lower melt than the typical shortening chips made with soybean or another U.S. domestic oil. The melting points for the lauric oil–based chips normally range from 101 to 108°F (38.3 to 42.2°C). Shortening chips have also been prepared with domestic oils hydrogenated with a sulfur proportioned catalyst. The specially hydrogenated vegetable oil products do not have quite as sharp a melt as the lauric oil–based products, but they are not susceptible to soapy off-flavors that develop with hydrolysis of lauric fatty acids. Shortening chips can experience high surface moisture contents during solidification, which can expedite the hydrolysis reaction.

Shortening chips are flaked to produce a product form that can be easily handled and incorporated into prepared foods while still maintaining the chip integrity. Flakes are produced on chill rolls that solidify the hardened oils into chips that are thicker than the usual flaked product. The thicker chip product allows the product to maintain a larger form during distribution and incorporation into the finished product. The chip products are usually packaged in 50-pound cases and may require refrigeration to dissipate the heat of crystallization and to maintain the desired chip integrity.

19. CRACKER SHORTENINGS

Crackers contain little or no sugar with a low moisture content, but moderately high levels of shortening: 10 to 20% based on the flour weight or 6 to 14% total basis. Snack cracker formulas vary more widely than saltines and usually contain higher shortening levels for richness and flavor enhancement. The chief emphasis in the selection of a cracker shortening is a bland or mild flavor with a good oxidative stability. Plastic range is not of significant importance since the major function of the shortening is lu-

TABLE 6.11. High-Stability Cracker Shortenings—Typical Analytical Data.

Analytical Evaluations	Hydrogenated Soybean Oil	Meat Fat and Vegetable Oil Blend
Solid fat index, %		
at 10.0°C/50°F	43.0 ±	39.0 ±
at 21.1°C/70°F	27.0 ±	28.0 ±
at 26.7°C/80°F	21.0 ±	24.0 ±
at 33.3°C/92°F	9.0 ±	17.0 ±
at 40.0°C/104°F	5.0 ±	11.0 ±
Mettler dropping point, °C	43.0 ±	46.0 ±
AOM stability, hours	200+	100+

brication. Most cracker shortenings utilized in the United States are composed of hydrogenated soybean oil with a relatively steep solids fat index to melt at or slightly above body temperature with a long AOM stability, but the shortenings can also be made with hydrogenated lard or blends of tallow with hydrogenated vegetable oils. Typical analytical data for an all-vegetable and a blended animal and vegetable, high-stability shortening utilized for crackers are shown in Table 6.11.

20. REFERENCES

1. Baldwin, R. R., Baldry, R. P. and R. G. Johansen. 1972. "Fat Systems for Baking Producers," J. Am. Oil Chem. Soc., 49(8):473–477.
2. Latondress, E. 1980. "Shortening and Margarines: Base Stock Preparation and Formulation," in Handbook of Soy Oil Processing and Utilization, D. R. Erickson et al. ed., Champaign, IL: American Soybean Association and American Oil Chemists Society, pp. 146–154.
3. Colth, H. S., Richardson, A. S. and V. M. Votaw. 1938. U.S. Patent 2,132,393 and 2,132,398.
4. The term High Ratio has been copyrighted by the Procter & Gamble Co.
5. Epstein, A. K. and B. R. Harris. 1938. U.S. Patent 2,132,046.
6. Howard, N. B. 1972. "The Role of Some Essential Ingredients in the Formation of Layer Cake Structures," The Bakers Digest, 46(10):28–37.
7. Painter, K. A. 1981. "Functions and Requirements of Fats and Emulsifiers in Prepared Cake Mixes," J. Am. Oil Chem. Soc., 58(2):92–95.
8. Paulicka, F. R. 1990. "Shortening Products," in Edible Fats and Oils Processing: Basic Principles and Modern Practices: World Conference Proceedings, D. R. Erickson, ed. Champaign, IL: Am Oil Chem. Soc., pp. 205–206.

9. Brody, H. and W. M. Cochran. 1978. "Shortening for Bakery Cream Icings and Cream Fillers," *The Bakers Digest,* 52(12):22–26.

10. Moncrieff, J. 1970. "Shortenings and Emulsifiers for Cakes and Icings," *The Bakers Digest,* 44(10):62–63.

11. Weiss, T. J. 1983. "Bakery Shortenings," in *Food Oils and Their Uses,* Second Edition, Westport, CT: AVI Publishing Co. Inc., pp. 153–154.

12. Stauffer, C. E. 1996. "Bakery Product Applications," in *Fats & Oils.* St. Paul, MN: Egaan Press, p. 73.

13. Oszanyi, A. G. 1974. "Complete Fluid Shortenings for Bread and Rolls," in *Proceedings of The Fiftieth Annual Meeting of the American Society of Bakery Engineers,* March 4–7, 1974, Chicago, IL., pp. 48–53.

14. Werner, L. E. 1981. "Shortening Systems," in *Proceedings of The Fifty-Seventh Annual Meeting of the American Society of Bakery Engineers,* March 2–4, 1981, pp. 61–70.

15. Mattil, K. F. 1964. "Bakery Products and Confections," in *Bailey's Industrial Oil and Fat Products,* Third Edition, Daniel Swern, ed. New York, NY: Interscience Publishers, a Division of John Wiley & Sons, p. 382.

16. Kincs, F. R. 1985. "Meat Fat Formulations," *J. Am. Oil Chem. Soc.,* 62(4):815–816.

17. Gordon, W. C. 1961. "Rolled-in Coffee Cakes and Danish Pastry," in *Proceedings of The Thirty-Seventh Annual Meeting of The American Society of Bakery Engineers,* March 6–9, 1961. Chicago, IL. pp. 215–220.

18. Colburn, J. T. and G. R. Pankey. 1964. "Margarines, Roll-ins and Puff Pastry Shortenings," *The Bakers Digest,* 38(4):66–72.

19. Anonymous. 1992. *Product Catalog,* Van den Berg Food Ingredients Group, Joliet, IL., p. 9.

20. Vey, J. E. 1986. "Danish," in *Proceedings of The Sixty Second Annual Meeting of The American Society of Bakery Engineers,* March 2–5, 1986. Chicago, IL, pp. 111–120.

21. Kriz, E. R. and A. G. Oszlanyi, October 12, 1976. U.S. Patent 3,985,911.

22. Weiss, T. J. "Margarine," in *Food Oils and Their Uses,* Second Edition, Westport, CT: AVI Publishing Co. Inc., pp. 205–207.

23. Klimes, J. 1990. "Shortening–Europe Formulation and Processing," in *Edible Fats and Oils Processing: Basic Principals and Modern Practices: World Conference Proceedings,* D. R. Erickson, ed. Champaign, IL: Am. Oil Chem. Soc., pp. 207–213.

24. Alexandersen, K. 1986. "Margarine Production," in *Proceedings: World Conference on Emerging Technologies in the Fats and Oils Industry,* A. R. Baldwin, ed. Champaign, IL: Am. Oil Chem. Soc., pp. 66–70.

CHAPTER 7

Frying Shortenings

1. INTRODUCTION

Deep fat frying has been a part of Chinese cooking for centuries. It started and remained an art until it gradually evolved into a science prompted by a need for better and faster food preparation methods. Deep fat frying has become one of the more important methods of food preparation used by the foodservice, snack, and baking industries, as well as the home kitchen. Frying technology development has determined that many of the principals apply equally to all types of commercial deep fat frying, but each type of frying also has distinctive requirements not shared by all. Since frying fats play such a unique role in the food preparation technique, the increase in acceptance of fried foods parallels the advancements made in the processing of improved frying shortenings along with the improvements in frying equipment and the other ingredients.

The deep fat frying process consists most simply of (1) partially or totally immersing the food prepared for frying into (2) a body of heated frying fat, which is (3) contained in a metal vessel, and (4) maintaining the food in the fat at the appropriate frying temperature for (5) the duration required to cook the product. Going into the kettle are (a) frying fat, (b) heat, and (c) the food prepared for frying. Emerging from the kettle are (a) steam and steam-entrained frying fat, (b) volatile by-products of heating and frying, (c) the finished product, and (d) with filtering, the crumbs or foreign solid by-products of the frying operation [1].

The result of deep fat frying is a food with a distinctive structure. The outer zone consists of the surface area, which contributes to the initial visual impact. This surface is generally an even, golden brown color resulting from a browning reaction that occurs when the sugars and proteins

385

in the product react in the presence of heat. The degree of browning depends on the time and temperature of frying in combination with the chemical composition of the food being fried, rather than on the shortening or source of fat or oils in the fryer.

The crust of crisp exterior skin formed on the food by dehydration during frying is the second part of the outer zone. Heat during frying reduces the moisture content of this layer to 3% or less, and the water driven off is responsible for most of the steam released during frying. The void created by this moisture loss is filled by absorbed frying shortening. This absorbed fat exerts a tenderizing effect on the crust, as well as contributing flavor, crispness, and pleasant eating characteristics. The amount of frying fat absorbed by the food, which varies with the product being fried, is governed by the ratio of crust to core; potato chips with a large surface area and very little core absorb approximately 30 to 40% fat, while french fried potatoes, which have a smaller surface and crust area in relation to the core area, absorb only 7 to 10% fat [2].

Usually, the function of a heat exchange fluid is to transfer heat from the source to the material being heated, or the reverse if cooling is involved. In most situations the heat exchange fluid is an inert material that experiences little change during the process, and its sole purpose is heat exchange. Frying shortenings are a notable exception to the heat exchange fluids commonly used for heat transfer. The deep fat frying process utilizes the fact that heat is rapidly and efficiently transferred to the product being processed via direct contact of the heat transfer agent with the food, but instead of acting as an inert heating medium, it becomes a part of the finished product. Then, because the frying fat is a food, it must be easily broken down into its component parts for digestion when eaten. Therefore, it must be considered as a relatively unstable heat exchange fluid that is exposed to its five natural enemies while performing its intended functions:

(1) Heat—Frying temperatures ranging from 300 to 425°F (148.9 to 218.3°C) are necessary to properly prepare the different fried food products. Unfortunately, exposure to high temperatures accelerates all of the breakdown reactions of fats and oils. Heating alone with no frying taking place will eventually cause a frying shortening to break down, rendering it unacceptable for further use.

(2) Air—Oxygen from the air is necessary to sustain human life, but it also reacts with the double bonds in the frying shortenings to oxidize the unsaturated fatty acids, which results in offensive odors and flavors and promotes gum formation or polymerization. Laboratory testing has proven that oxygen presence is necessary for frying fat deterioration

and that the rate of change is directly proportional to the degree of exposure of the fat surface to oxygen [3].

(3) Moisture—All food products contain moisture, which causes hydrolysis of fats and oils, resulting in an increased fat absorption in most foods. Free fatty acids themselves do not produce the increased absorption, lower smoke point, and objectionable flavors, but when split from the triglyceride, they leave mono- and diglycerides and free glycerin, which cause the changes [4].

(4) Contamination—Any material associated with the frying process that causes the frying media to deteriorate or accelerates the process is a contaminant. The food fried is an unavoidable contaminant. For instance, it may leach different fat types into the frying shortening like chicken fat, tallow, fish oils, or other materials that reduce the frying stability, while foods like onions and fish are obvious contributors of strong flavors. Some examples of other frying fat contaminants are

- trace metals—Most metals are prooxidants that exert a marked catalytic effect to accelerate fat breakdown, but some metals are much more active than others. These prooxidants can be picked up during processing or storage, from frying equipment, the food fried, or some other contact with a metal. Two metals that promote more rapid breakdown of frying than others are brass and copper.
- soap or detergent—Residue of these materials from cleaning storage tanks, fryers, or utensils is the addition of an emulsifier to the frying fat, which will catalyze fat breakdown.
- gums or polymerized fats—Addition of polymerized fats or oils to fresh oils act as catalysts to accelerate the formation of more gums, which contribute to foaming and darkening.
- burnt food particles—Food particles allowed to remain in the frying fat impart a bitter, caramelized, and/or burnt taste along with an unappealing appearance to the food fried and accelerate frying shortening breakdown.

(5) Time—The extent of a frying shortening's exposure to the effects of the other enemies determines the degree of product deterioration.

There are two beneficial frying fat quality factors affected during the frying operation. One is the steam released during frying, and the other is the addition of fresh shortening to replace the fat absorbed by the food fried. Steam formed from the moisture released from the food mixes intimately with the fat, and when given off, it carries with it the odor- and flavor-bearing volatile by-products of frying that would otherwise accumulate in the frying fat to adversely affect the flavor and odor of the fried food. This steam continually scrubs or purges the frying fat of the potential off-flavors and odors each time food is fried, even though it is the same moisture that causes hydrolysis.

Fresh shortening must be added to the fryer to compensate for the fat removed by the fried food. This addition helps to overcome the changes to the frying fat brought about by heat and the other frying fat enemies. Obviously, the frying fat will remain in better condition when higher replacement shortening quantities are required. The ratio of the fryer's capacity to the rate at which the fresh shortening is added to replenish the fryer is referred to as turnover rate, or the number of hours required for the addition of fresh frying shortening equal to the amount of fat maintained in the fryer. Because oxidative changes occur continuously in heated fats, turnover must be related to the total period that the fat is heated, rather than only the actual time the product is fried. Obviously, the quality and, especially, the flavor of the frying fat will be maintained at a more desirable level with the highest turnover rate, or stated another way, the shortest turnover time equates to the best frying fat condition. In general, an operation with a turnover of less than a day should never have to discard used frying shortening because of breakdown, except in the case of product abuse or a contaminant. Operations with slower turnover rates need to include this product quality and economic factor in their frying shortening selection criteria. In terms of turnover, food processor operations frying products like salty snacks, donuts, prefried frozen foods, blanched foods, and so on have two definite advantages over the foodservice restaurant operations:

- a complete shortening turnover on less than a daily basis due to the volume of food fried and the absorption rate of the food product
- frying on a continuous basis during the entire time that the fat is at frying temperature

In the foodservice kitchens, the nature of the deep fat frying operation is quite different. The rate of turnover varies considerably between restaurants but averages 20 to 35% per day. The low turnover is caused by:

- a lower volume of food fried, which lowers the absorbed fat removal from the frying kettle
- frying that is not continuous but has peaks and lulls which results in the fat being heated for long periods without any fat removal by absorption

Under restaurant conditions it is close to impossible to achieve a turnover rate high enough to avoid discarding used frying fat, while most snack and bakery frying operations should never have to discard frying shortening.

2. FRYING SHORTENING ADDITIVES

Hydrogenation contributes frying stability to the frying shortenings by saturation or isomerization of the double bonds to take away the reaction or weak point in the frying shortenings. Another significant contributor to frying stability is the antifoamer dimethylpolysiloxane. Addition of this additive to a frying shortening at levels of 0.5 to 2.0 parts per million (ppm) effectively retards oxidation and polymerization, which promotes foaming during frying. Frying stability increases of three to ten times the original frying stability results have been confirmed by controlled laboratory frying evaluations. The degree of the stability increase is dependent upon the stability of the frying fat without the antifoamer. Frying shortenings with a high frying stability before the antifoamer addition benefit more than products with a lesser initial frying stability. This additive has substantially changed the frying parameters for frying operations with poor or slow turnover ratios.

The use of dimethylpolysiloxane as an antifoaming agent for frying shortenings is well known and widely practiced. Foam suppression is an indirect result of the inhibition of oxidation, which means that dimethylpolysiloxane is actually an antioxidant that prevents the buildup of oxidation products that are foam promoting during frying. The use of dimethylpolysiloxane as an antifoamer was first reported by Martin [5], who showed that a concentration of 0.03 ppm was sufficient to inhibit oxidation of the frying oil, as indicated by changes in a variety of physical and chemical properties of the frying shortening over a prolonged heating period. Later, Babayan [6] showed that low concentrations of dimethylpolysiloxane would also raise the smoke point of an oil by as much as 25°F (13.9°C), which indicated that dimethylpolysiloxane is effective as a surface to air barrier for frying shortenings. Freeman et al. [7] also concluded that the protective effect is due to a monolayer of dimethylpolysiloxane on the oil to air surface, which retards oxidative polymerization. Due to the inert nature of the dimethylpolysiloxane molecule, silicone-based antifoamers will not react with the frying shortenings and remain effective until reduced to low levels by removal with the food fried.

Dimethylpolysiloxane, a defoaming agent may be safely used in processed foods at a level not exceeding 10 ppm, according to 21 CFR 173.340 [8]. The effective level for frying fats is less than 0.5 ppm, and evaluations indicate that use levels should not exceed 2 ppm. The concentration of dimethylpolysiloxane in oils extracted from fried foods increases as the concentration of dimethylpolysiloxane in the frying shortening is increased. Food absorption of the antifoamer is probably dependent upon the mechanical pickup of excess dimethylpolysiloxane from the frying fat, which occurs because of the low solubility of dimethylpolysiloxane in fats and oils. Droplets of excess dimethylpolysiloxane actually appear in the frying fat when the concentration exceeds 1 ppm. This excess dimethylpolysiloxane will adhere to any available surface, such as the walls of the fryer, a fryer basket, or the fried food [7].

The extremely low solubility of dimethylpolysiloxane can result in dispersion problems for the fats and oils processor if not handled properly. In addition to the insolubility of dimethylpolysiloxane, it has a much heavier density than the frying shortenings. Therefore, the antifoamer is only dispersible in frying shortenings and must be agitated thoroughly before and preferably during packaging to keep it in suspension. Dispersion problems lead to high concentrations of dimethylpolysiloxane in portions of the frying shortening, which violate the U.S. FDA regulations and create foam, rather than prevent it. Dimethylpolysiloxane levels somewhere between 10 and 50 ppm promote immediate foaming of frying shortenings.

Dimethylpolysiloxane cannot be used indiscriminately because it has an adverse effect upon some food products [9,10]:

- Unintentional incorporation of the antifoamer into shortenings for cake and icing preparation will cause aeration problems that result in poor volumes for the finished products.
- Cookies prepared with shortenings containing as low as 0.5 ppm dimethylpolysiloxane have experienced spread problems.
- The antifoamer has been blamed for defoaming cake donut batters at the surface to yield a deformed donut, increased fat absorption, and poor sugar and glaze adherence to both cake and yeast-raised donuts.
- Potato chips fried in dimethylpolysiloxane-treated frying shortenings lack desirable crispness.

Antioxidants are materials that can retard the development of oxidation in fats and oils. Vegetable oils contain natural antioxidants, tocopherols, that can survive most processing and frying conditions. In addition, several phenolic compounds have been identified that can also increase oxidative stability. It has been postulated and proven that deep fat frying conditions of high temperature and steam distillation rapidly deplete phenolic antioxidants. Tests performed with BHA, BHT, and TBHQ showed that all three phenolic antioxidants were volatilized to depletion at approximately the same rate. However, in spite of the disappearance of the antioxidants, their use does effect an increased AOM stability of the used frying shortening and a longer product shelf life than an identical product fried in antioxidant free shortenings. Evaluations comparing BHA-, BHT-, and TBHQ-stabilized frying shortenings showed that TBHQ provides the maximum protection at frying temperatures, as well as the best "carry-through" effectiveness for the fried foods. An assumption for this result is that TBHQ breakdown products may also be effective antioxidants. BHA and BHT were also found somewhat effective but at a much lesser degree than TBHQ [11,12].

Even if the antioxidants did not provide any protection at frying temperatures, the protection provided to the frying shortenings before the frying operation is beneficial. Fats and oils oxidation has two different phases. During the initial or induction period, oxidation proceeds at a relatively slow, but uniform, rate. Then, after a certain amount of oxidation has occurred, the reaction enters a second phase characterized by a rapidly accelerating rate of oxidation. Antioxidants cause an increase not only in the resistance to oxidation, but also in the amount of oxidation required to produce offensive flavors and odors. The protection afforded the frying shortening to maintain a low oxidation level before entering the frying kettle equates to improved oxidative frying stability.

Oxidation can also be accelerated or catalyzed by trace amounts of metals in frying shortenings. Fats and oils can obtain metal contents from the soils where the plants are grown and later from contact during crushing, processing, storage, and transportation and from the frying equipment or the food fried. Copper has been identified as the most harmful metal with iron, manganese, chromium, and nickel following. Many frying shortenings are hydrogenated with nickel catalysts to increase the oxidative stability; if appreciable levels of the catalyst are left in the hydrogenated oils, the protection afforded by lowering the unsaturation level will be somewhat negated by the catalytic effect of the nickel content. Chelating agents are used after the deodorization process by most processors to complex with the prooxidative metals to avoid this problem. The most widely used chelating agent is citric acid. Phosphoric acid has been used by some fats and oils processors, but problems with overchelating can easily occur to cause a free fatty acid increase, darkening, and flavor reversion to a watermelon flavor during frying. This same problem can also occur with citric acid, but it is a weaker acid that provides a higher margin for error.

3. FRYING SHORTENING SELECTION

Frying shortening selection criteria is the same for foodservice and food processor operations. In both cases the frying shortening must be matched to the performance requirements of the product fried, the limitations of the frying equipment, and the demands of the operation. Ideally, a frying shortening should be selected on the basis of the eating, keeping, and other functional characteristics that are desired in the fried food. It is generally agreed that frying shortenings should possess and maintain the following characteristics during frying: (a) a light color, (b) surface free from foam, (c) a clear appearance free from burnt particles, (d) a bland flavor that enhances the eating quality of the fried food, and (e) no offensive smoke emitted. All fresh frying shortenings possess these characteristics; however, with continued use, the frying shortenings change under the influences of heat, air, moisture, contaminants, time of use, the frying equipment, and the other deteriorating factors associated with the frying process. If the frying fat is permitted to deteriorate excessively, flavor, appearance, and eating characteristics of the fried food will be at risk. Again, it is important that the frying shortening be matched to the performance requirements of the product fried, the limitations of the frying equipment, and the demands of the frying operation. The frying shortening type or composition that can satisfy these requirements can be determined by careful identification of the prerequisites for each operation. Three areas for consideration in the selection of the proper frying shortening are

(1) Product characteristics—Fried food product characteristics must naturally be a major frying shortening selection consideration. Product criteria can be divided into three functional areas to help identify the ideal frying shortening for the intended food application and operation:
   - palatability—Frying shortenings must have a bland flavor or a flavor that complements the food fried. Melting point and SFI profile for hydrogenated frying shortenings depend upon the temperature range over which the food will be eaten and the amount of fat absorption to provide the desired eating sensation or mouth feel. Mouth feel and palatability characteristics are also related to the physical form of the frying shortening at room temperature. Products high in liquid oils or that melt rapidly in the mouth tend to give a quicker flavor release than products with higher melting points. A high melting point product can mask or slow the food flavor release. Further, the relationship of the other ingredients must be considered. An all-purpose shortening with a high melting point gives donuts a pleasant creamy mouth feel; however, the use of the same shortening to fry potato chips will produce a product without flavor and a waxy mouth feel.
   - product appearance—The solids to liquids ratio of the absorbed fat influences the appearance of fried foods. Higher SFI values at the consumption temperature will present a dry surface appearance, and as the level decreases, a shiny appearance develops. The degree of apparent differences and importance varies with food types. Appearance-wise, the frying shortening will provide an oily surface or the

illusion of greater fat absorption if it has a low melting point and a high liquid oil content. Shortenings with higher melting points provide different shades of dullness, depending upon the degree of hydrogenation.

- product feel—Crystallization properties of the frying shortening, which coats the product surface, determines the product feel. A soft or low melting frying shortening contributes a degree of oiliness, and high-melting products impart a more dry surface by solidification at the consumption temperature. Frying shortening consistency changes during frying can significantly change the product appearance and feel.

(2) Frying life—The unavoidable exposure of frying shortening to the adverse conditions in varying degrees dictates that frying stability is a major economic and quality factor that frying operations must consider when choosing a frying shortening. The frying stability of a shortening is dependent upon the fat composition, physical characteristics, and how it is used. Exposure time, as determined by turnover rate, is one of the most important factors for identification of the shortening characteristics required to maintain the required fried product quality for the maximum time. High-turnover situations will allow the use of frying shortenings with a lower stability, which usually equates to higher liquid oil levels. Low-turnover situations demand high-stability shortenings, which normally equates to a higher degree of selective hydrogenation. The factors that adversely affect frying stability or frying shortening life are

- Oxidation is the combining of oxygen with the unsaturated fatty acids causing off-flavor and odor development along with color darkening. Oxidation is accelerated at the higher temperatures necessary for frying foods. The oxidation rate is roughly proportional to the degree of unsaturation; linolenic fatty acid, with three double bonds, is much more susceptible than oleic fatty acid, with only one. Therefore, shortenings with high polyunsaturated fatty acid levels have the least resistance to oxidation, which precedes polymerization. The frying shortening factors that affect the oxidation rate are (1) the degree of hydrogenation or the unsaturation level, (2) the addition of antifoamers, (3) the addition of antioxidants, (4) the tocopherol or natural antioxidant level, and (5) the use of chelating agents.
- Polymerization is the combining of many triglyceride molecules to form three-dimensional polymers. Heat accelerates polymerization, which physically results in an increased frying fat viscosity that reduces its ability to transfer heat. Lower heat transfer results in a higher fat absorption, and polymerization continues to the degree

that it eventually results in foaming. Foam develops when the heated fat will not release the moisture from the food but keeps it trapped while also incorporating air. Unsaturated fatty acids have the least resistance to polymerization.

- Hydrolysis is a breakdown of the fat induced by the presence of moisture, accelerated by heat, and results in free fatty acid, diglycerides, monoglycerides, and glycerin development, which causes off-flavors, smoking, increased fat absorption, and darkening of the frying shortening.

(3) Shelf life—When the fried food is consumed immediately, as is the case for restaurant operations, product shelf life is not a necessary consideration for the frying selection. However, finished products that will experience severe or prolonged storage periods before consumption must have shelf life as an important part of the selection criteria. Oxidative stability must continue after the frying period for products packaged and marketed through a distribution system. The frying shortening oxidative stability requirements will depend upon the product's level of fat absorption, packaging protection, expected shelf and use life, and the product itself. Oxidative stability analysis of the frying shortening is the major indicator of the fried product's shelf life. Hydrogenation, or the degree of unsaturation, is the determining factor for increasing fried product's shelf life.

4. FOODSERVICE DEEP FAT FRYING SHORTENING APPLICATIONS

Deep fat frying is a very important cooking method for the foodservice industry. Deep frying is chosen both for rapid food preparation and for producing foods with a desired texture and flavor. The high temperatures used cause rapid heat penetration and provide short cooking times. Fast-food preparation allows a greater customer turnover during peak periods and better service to customers during the slow periods. Deep fat frying is also economical since the restaurant fries only what is ordered, which eliminates leftovers; frozen or refrigerated foods may be prepared quickly, which eliminates food loss due to spoilage; and the heat is concentrated for little waste of utilities.

In foodservice kitchens, the rate of turnover varies considerably, but the average is in the 20 to 35% range. The low rate of turnover is the result of a relatively low fat absorption and the fact that frying is not continuous. Frying to order or even anticipated demand results in peak and slack periods. Restaurant fryers must be kept at frying temperature to prepare customer orders during the slack periods, as well as the rush periods. Low turnover and the necessity for frying many different types of foods in most restaurant frying operations make it almost impossible to avoid the necessity to discard abused frying shortening. Determination of the point to discard used shortening has a significant economic impact: high shortening cost when the used fat is discarded too early and poor food quality if the discard point is too late. Some of the most used frying shortening indicators that restaurant kitchens use to determine when to discard their used product and replace it with fresh shortening are

- color—Used frying shortening is discarded when the heated oil matches a certain color standard or when visibility through the heated oil is impaired at a definite distance.
- smoke—Used frying shortening is discarded when smoking reaches a predetermined disagreeable point. Smoke is caused by hydrolysis and oxidation, which lowers the smoke point.
- food condition—Used shortening is replaced when the fried food reaches a predetermined flavor, odor, and/or greasiness point. This method evaluates the product from a customer viewpoint.
- foaming—The amount of foaming is used by some operators to determine the discard point; with experience, five different stages of foaming may be identified by observation:

Foam Description	Observed Foaming Tendencies
None	Normal frying
Trace foam	First indication of foam
Slight foam	Foam severity increases
Definite foam	Definite pockets of foam
Persistent foam	Foam persists until food is removed

Any of the indicators can be used successfully to identify the discard point when it is correlated with the frying operation to serve as a reliable indicator of the used fat's condition. Some operators replace their frying shortening after a prescribed time or after a specific quantity of food has been fried. This practice cannot provide a uniform product quality even when the discard point is for good quality frying shortening, which has an extremely negative impact on economics and eliminates the incentives to improve the frying shortening's performance.

Since each foodservice operation is unique regarding its menu, product mix, equipment, and operation, fats and oils processors must produce more than one frying shortening to meet the performance expectations. Three general types of plasticized foodservice frying shortenings are produced in addition to the specialized or tailor-made product for specific customers: all-purpose shortenings; animal–vegetable blends; and heavy-duty, all-vegetable frying shortenings.

4.1 Foodservice All-Purpose Shortening

All-purpose shortenings are compromise products designed for performance in baking and frying. Typically, all-purpose shortenings have been formulated with 3 to 15% essentially fully saturated beta-prime hard fats added to a soft partially hydrogenated basestock, a blend of tallow with vegetable oil, or interesterified lard. Some fats and oils processors add low levels of dimethylpolysiloxane to all-purpose shortenings to improve the frying stability. The antifoamer level must be kept low, probably 0.5 ppm or less, to maintain creaming properties for cake, icing, cookie, and other baked product production.

All-purpose shortenings were the first plastic shortenings used for frying and have a better frying stability than most oil products but less than the other types of plasticized shortenings formulated for frying. A dry fried product appearance is contributed by a melting point of 105 to 120°F (40.6 to 48.9°C). Restaurant kitchens that bake and deep fat fry only occasionally may choose to use an all-purpose shortening to reduce the number of products inventoried.

4.2 Animal–Vegetable Blended Frying Shortenings

Animal fats and blends with vegetable oils have been used as frying shortenings for many years. These products have been attractive to frying operations because of a usually low initial cost and good frying stability, and some claimed a preference for the meaty flavor transferred to the fried foods. A large market did exist for a blend of undeodorized tallow with RBD cottonseed oil for frying french fries. This product was produced with good quality, caustic-refined tallow blended with about 10% cottonseed oil to depress the melting point. The characteristic tallow flavor was complemented by the "nutty" flavor contributed by liquid cottonseed oil to impart a "distinctive" flavor to french fries. Tallow has a natural resistance to oxidation and polymerization contributed by its high oleic fatty acid content.

The use of meat fat frying shortenings in restaurants has decreased dramatically because of nutritional concerns and pressure from special interest groups. Most foodservice operators that utilized meat fats or blends for frying have changed to cholesterol-free, all-vegetable frying shortenings. However, one fats and oils processor has initiated a steam distillation process to strip meat fats of serum cholesterol to return them to use as frying shortenings [13].

4.3 Heavy-Duty Vegetable Frying Shortenings

All vegetable frying shortenings were among the first shortenings specifically formulated for a specific function. These shortenings, often referred to as heavy-duty are processed for maximum frying stability by hydrogenation to render them resistant to oxidative changes during the frying operation. Plasticity or creaming properties is not beneficial for frying shortenings since the product is melted for use. Therefore, specific crystal types are not a prerequisite, and most source oils can be hydrogenated for frying shortening performance. Most heavy-duty frying shortenings are composed of a selectively hydrogenated soybean oil or another source oil that is usually determined by economics and availability. Selective hydrogenation is utilized to provide high solids at the low temperatures that drop off rapidly to provide a good mouth feel for the fried food when eaten. Saturation of linolenic and linoleic fatty acids to predominantly oleic fatty acids also substantially increases the shortening's frying stability by removing the opportunities for oxidation. Dimethylpolysiloxane, the antifoaming agent, is added by all processors, and some also add TBHQ antioxidants to protect the product before addition to the frying kettle, as well as the carry-through protection for the fried foods. Plasticization conditions are controlled to provide a uniform consistency, and the product does not require tempering after filling. Plasticized frying shortenings should be ready to ship to the customer after the product has solidified enough to palletize or stack properly. These shortenings provide economy to the frying operation through good-quality fried foods over an extended frying period.

4.4 Liquid Frying Shortenings

Liquid opaque frying shortening introduced in the late 1950s appeared to be a contradiction to frying technology that correlated frying stability with saturation, which eliminates the reaction sites for oxidation. The liquid product had a milky appearance and could be poured from the container like a salad or cooking oil, and it had a better frying stability than the best heavy-duty or animal–vegetable frying shortenings, even though its composition was high in polyunsaturates. The secret for the liquid shortening's superior frying stability was dimethylpolysiloxane. The addition of this antifoamer effectively retarded oxidation and polymerization to more than double the frying life of the then current best, heavy-duty plasticized frying shortening relying on hydrogenation for stability. The opaque liquid shortening enjoyed this supremacy over all other frying shortenings until the antifoaming agent was added to their competition. This addition returned the heavy-duty frying shortening to the most stable frying shortening status, but liquid frying shortening had gained a following with foodservice operators because of the convenience of a pourable product, the respectable frying stability, and a slightly oilier fried food appearance. Many foodservice operations also prefer the relatively high level of unsaturation, which may be important for nutritional considerations.

Two types of liquid frying shortenings have evolved since their introduction: the original opaque-type product and a competition product that is clear at room temperature. Opaque liquid frying shortenings are usually formulated with 2 to 6% of a beta crystal forming hard fat in 100 to 110 IV partially hydrogenated soybean oil. The opaque liquid shortenings must be crystallized carefully to stabilize the hard fractions to maintain a pourable consistency. The clear liquid frying shortenings are hydrogenated and then fractionated to provide the high-stability translucent product at room temperature. Winterization facilities are normally used to fractionate an 88± IV hydrogenated soybean oil to obtain a liquid fraction with a 92± IV. The clear liquid frying shortenings obviously do not require crystallization, but exposure to temperatures below the chill room temperature of 65°F (18.3°C) will cause crystallization of this product. Frying stabilities of the clear liquid frying shortenings are slightly better than the opaque liquid frying shortening, but the difference is usually indistinguishable by the foodservice operator. A comparison of the analytical physical characteristics of the two liquid shortening types (shown in Table 7.1) illustrates the differences in fatty acid composition and SFI profile, which influences the appearance and frying stability differences.

Liquid frying shortenings have been formulated with source oils other than soybean oil for their distinctive flavors developed during frying.

TABLE 7.1. Foodservice Liquid Frying Shortenings.

	Opaque	Clear
Fatty acid composition, %		
Palmitic C-16:0	10–11	8.5–9.5
Stearic C-18:0	6–8	3.5–4.0
Oleic C-18:1	42–46	61–65
Linoleic C-18:2	31–37	20–24
Linolenic C-18:3	2–3	0.2–1.5
Iodine value	101–107	88–94
Solids fat index, %		
at 10.0°C/50°F	3–7	5–14
at 21.1°C/70°F	3–6	0.2–0.7
at 26.7°C/80°F	3–6	
at 33.3°C/92°F	2–6	
at 40.0°C/104°F	0.5–5	

Peanut and corn oils are typically the alternate source oils used for this purpose. Peanut oil has been a favorite for fish frying since it was reported that the Queen of England preferred her fish and chips to be fried in peanut oil. Peanut oil can be added with hydrogenated soybean oil to prepare a liquid frying shortening with equivalent to better frying stability due to its high oleic fatty acid content. Corn oil is usually partially hydrogenated or added at low levels for a characteristic musty corn flavor. This flavor has been favored for replacement of the meat fat flavor and for other foods where a corn-type flavor complements the fried product.

5. FOODSERVICE PAN AND GRILL SHORTENINGS

Griddle and pan frying are the major foodservice breakfast food preparation methods. It has been estimated that 75 to 80% of all U.S. breakfast menu items are prepared on a grill or in a frying pan. In recent years many fast-food chains have increased sales volume and profits by adding a breakfast menu. This has enabled them to make more efficient use of their restaurants for reduced overall costs.

Any cooking oil, unemulsified shortening, margarine, or butter can be used for pan and grill frying. Butter and margarine have been the choice of many chefs for the butter flavor and color contributions to the fried product. However, both require careful, low-temperature preparation to avoid scorching the milk solids, while the moisture content causes spattering and the food to stick to the hot metal surface. Salad oils would seem to be the logical solution to the moisture and milk solids problems, but they tend to polymerize rapidly on the hot surface and create unsightly gums, and the food still sticks to the pan or grill unless the surfaces are properly seasoned after each cleaning. Seasoning a pan or grill is accomplished by carefully building up thin layers of polymerized oil to fill in the metal pores until a slick, firm surface is created.

Unemulsified shortenings would appear to be a logical solution to the gum problem with liquid oils. It has been hydrogenated for a better resistance to polymerization. All-purpose shortening does resist gumming, but the food still sticks to the hot surface if it isn't seasoned after cleaning. Many chefs solve the sticking, gumming, spattering, scorching, and surface preparation problems by clarifying either butter or margarine and using the oil portion for grilling or pan frying. The separated oil contained a buttery color and flavor, it didn't spatter excessively, and the food didn't stick to the hot surface; the natural emulsifier in butter, which is added as lecithin to margarine, functions as an antisticking agent. Therefore, since lecithin is oil-soluble, the clarified butter or margarine oil contained this antisticking agent.

Clarification of either butter or margarine results in at least 25% waste. In addition, the butter flavor of margarine does not carry through to the fried foods to an appreciable extent. This situation became an ideal specialty product development project for fats and oils processors. The fast-food chains added urgency to this development because of their desires to produce the same product quality in hundreds of units. Obviously, clarification of butter or margarine did not fit into the fast-food quality commitments, especially with their high turnover of kitchen personnel.

Foodservice operators have readily accepted pan and grill shortenings that are butter flavored and colored with lecithin as an antisticking agent. These products have proven to be more than adequate specialty product replacements for butter, margarine, or cooking oils previously used for grilling and pan frying. Table 7.2 reviews the three general types of pan and grill shortenings available: two liquids and a solid. The distinguishing difference between the two liquid types is the salt addition. Salt enhances the butter flavor to give this product a more pleasing standalone flavor, making it more closely duplicate the flavor of clarified butter or margarine. The finely milled salt added to the liquid pan and grill shortening can only be dispersed because it is not soluble in oil. Usually, the salt is dispersed in the pan and grill shortening after it has been crystallized, just prior to packaging, to minimize the abrasive action of undissolved salt on chilling equipment and pumps. Lecithin is added to all the products to provide antisticking properties and prevent spattering during frying. Both the salted and the salt-free liquid products are formulated like opaque frying shortenings with a soft soybean oil basestock and a soybean oil hard fat, usually the same formulation as for liquid frying shortenings. The antifoamer, dimethylpolysiloxane, is added to counteract some of the foaming produced by lecithin in pan frying and as an antioxidant.

Solid pan and grill shortenings usually contain coconut and/or palm kernel oils with a hydrogenated soybean oil and/or cottonseed oil. The function of the lauric oil addition is to provide a sharp melt, promote foaming, and give oxidative stability. The foaming action complements lecithin to provide antisticking and antispattering properties. Stability is more important for the solid pan and grill products because the usual kitchen procedure is to keep a container of melted product on the grill or stove, instead of waiting for the product to melt before each food preparation. Additionally, the solid pan and grill products perform better in bun and toaster applications where a reservoir of melted product is maintained. Liquid products polymerize and require constant attention to prevent service interruptions. The colors and flavors utilized for all pan and grill shortenings

TABLE 7.2. Pan and Grill Shortenings.

Pan and Grill Type	Pourable	Salted/Pourable	Solid
Usual source oil	Soybean oil	Soybean oil	Coconut oil, palm kernel oil, soybean oil, cottonseed oil
Other ingredients	Lecithin, color, butter flavor, dimethylpolysiloxane, antioxidants	Salt, lecithin, color, butter flavor, dimethylpolysiloxane, antioxidants	Lecithin, color, butter flavor, dimethylpolysiloxane, antioxidants
Solids fat index, %			
at 10.0°C/50°F	3 to 7	3 to 7	20 to 35
at 21.1°C/70°F	3 to 6	3 to 6	8 to 20
at 26.7°C/80°F	3 to 6	3 to 6	5 to 15
at 33.3°C/92°F	2 to 6	2 to 6	1 to 5
at 40.0°C/104°F	0.5 to 5	0.5 to 5	0
Packaging	Gallon, 5 quart, 17½ pound, 35 pound	Gallon, 5 quart, 17½ pound, 35 pound	6 pound can, 30 pound pail
Applications	Pan frying, grilling, soups, sauces, gravies, basting, brush-on dressing, corn-on-the-cob	Seasoning, basting, corn-on-the-cob, popcorn, seafood dip, brush-on dressing, soups, sauces, gravies, pan frying, grilling	Bun toaster oil, bun dressing, pan frying grilling

must be heat stable but more so for the solid products because of the longer heating periods.

6. BAKERY FRYING SHORTENING APPLICATIONS

Bakery deep fat fried products consist of several hundred varieties but can be classified into three general classifications: (1) yeast-raised, (2) chemically leavened or cake, and (3) fried pies or pastries. The frying methods for all three classifications may be either surface frying, which requires turning the product approximately halfway through the frying procedure, or submerged frying where the dough pieces are held beneath the surface of the frying fat during most of the frying cycle. Almost all of the fried products may also be subjected to various finishing or coating techniques to apply toppings or fillings. In many cases, the product attributes must be modified to accommodate the toppings and/or application methods. The toppings utilized include (1) fluid topping such as glazes and enrobing icings; (2) dry coatings like powdered, granulated, or cinnamon sugar; and (3) combinations of the fluid and dry coatings, which involve the application of a liquid coating followed by sprinkling or tumbling with a dry coating. The fillings are either injected after frying or incorporated during the product make-up. Fillings include cinnamon sugar, fruits, jellies, cremes, marshmallow, and so on. The three bakery fried product classifications may be produced in various shapes, subjected to various coatings or topping products, and filled with many different products to produce literally hundreds of different varieties.

Bakery frying shortenings have three distinct functions: (1) heat transfer medium, (2) major ingredient in the finished product, and (3) binder for casings such as glazes and sugars. Frying shortening's solids/liquid ratios' effect upon bakery fried products are identified and reviewed on Table 7.3. Either all-vegetable or meat fat–based shortenings will perform adequately for this application. The same general SFI requirements should be applicable for either fat or oil source utilization. However, the crystal habit provided by the source oils utilized may have an effect for a particular bakery operation. Crystallization, or set time, required for the absorbed shortening to solidify to the desired degree will vary with crystal types. Beta-prime crystal tending products set up more quickly than beta crystal shortenings. Therefore, it may be desirable to choose the frying shortening source oils or at least the hard fat portion to develop a crystal type that complies with an operational time requirement for solidification for acceptable sugar or glaze adherence.

TABLE 7.3. *Typical Bakery Frying Shortening Solids Fat Index Profile.*

Solids Fat Index		Characteristic Affected	Solids/Liquids Relationship Effect
Temperature	% Solids		
10.0°C/50°F	33 to 38	Shelf life	The correct solids content will
	21 to 26		create a moisture barrier, keeping moisture inside the donut.
21.1°C/70°F	21 to 26	Appearance	Too high an SFI value leaves a visible layer of fat on the donut, which provides a waxy mouth feel and can promote flaking of the sugar coating. Too low an SFI content will leave the crust oily, which can promote oil soakage of the sugar.
26.7°C/80°F	19 to 24	Sugar pickup	The correct ratio of hard to soft
33.3°C/92°F	12 to 17		fractions in the frying shortening composition will help ensure proper sugar pickup. Too high SFI-decreased sugar adherence. Too low SFI-increased sugar disapearance.
40.0°C/104°F	7 to 12	Eating Quality	Donut frying shortening's SFI content at temperatures above body temperature have a direct effect upon eating quality. SFI values above 12% to 104°F may cause a waxy, unpleasant mouth feel.

Bakery frying shortenings normally do not contain dimethylpolysiloxane because of the adverse effect of the antifoamer upon the performance of the fried products:

- Dimethylpolysiloxane-treated frying shortenings have apparently defoamed cake donut batters at the surface to produce a deformed donut.
- Glaze and sugar retention problems have been attributed to the use of dimethylpolysiloxane in frying shortenings.

- Increased fat absorption with antifoamer treated shortenings has been attributed to the maintenance of a low frying fat viscosity for longer periods.

Conveniently, the high absorption of bakery fried products coupled with the demands that allow continuous frying provides a high turnover rate. In most cases the turnover rate is high enough that properly cared for frying shortening should never have to be discarded. Therefore, the antifoamer additive is not required for bakery shortenings.

Normally, fresh frying shortenings do not produce the desired fried characteristics for bakery fried products until after a "break-in" period. Since bakery frying shortenings normally have a high turnover rate because of the type of frying and the rather high absorption rate of the fried foods, the products are developed to be fried in a frying shortening that has been used and has experienced a degree of change from a fresh shortening. The desired frying shortening condition has been called the "quality period" where the donuts produced have the preferred shape, appearance, absorbed fat, and eating character. This point has been thought to correlate with free fatty acid; however, no particular free fatty acid level has been identified. In fact, different researchers have identified a different ideal free fatty acid level dependent upon the oxidative stability of the shortening product. It is more likely that the ideal frying shortening condition is dependent upon the amount of polymerization. The fresh shortening that is added to the frying kettle to compensate for the fat that is removed by absorption helps to overcome the changes caused by heat and other factors.

6.1 Cake Donut Shortenings

Dictionaries define donut as a small ring-shaped cake fried in deep fat, either (a) made of a rich batter leavened with baking powder called cake donuts or (b) made of yeast-leavened dough called raised donuts. No other fresh bakery product has as many variety possibilities, except perhaps cookies. Donuts are produced commercially by retail bakeries, grocery in-store bakeries, and wholesale bakeries as well as specialty donut shops. It has been estimated that donuts, both cake and yeast-raised combined, are as much as 10% of the total bakery product category. Cake donuts are different from other products prepared in bakeries in three major respects [14]:

(1) No container is used to determine shape. A cake donut is a free flowing block of dough floating in the frying fat. The characteristics of the batter and how fast it sets and seals controls the total symmetry of the cake donut.

(2) The entire preparation is compressed into a very short period of time. The frying time for a typical donut is 30 to 45 seconds on each side at 375°F (190.6°C).

(3) Up to 25% of the final product composition is added during the heating process—the frying shortening.

Most bakeries use cake donut mixes instead of preparing them from "scratch," primarily because they are one of the most sensitive products with respect to formulation and ingredients. Prepared mix suppliers test each ingredient, especially the flour, to determine the formula modifications necessary to produce quality cake donuts. Flour for cake donuts is performance tested to make a product with proper water absorption, shortening absorption, specific volume, height, spread, and several other overall performance characteristics. The amount of shortening and the emulsifiers in the mix are controlling factors for fat absorption along with the milk level for the mix formulation and floor time, mix time, water level, and batter temperature during make-up. Vegetable oils are generally used in the production of cake donut mixes. Plastic or solid shortenings provide a cake donut with a very rough, coarse crust. Oil produces a cake donut with a smooth crust and a lower shortening absorption level. Cake donut mixes normally will contain mono- and diglycerides emulsifiers alone or in combination with an intermediate hardness PGME at a level of no more than 0.5% of the total mix. The purpose of the emulsifiers is to control shortening absorption, extend shelf life, and tenderize to provide a shorter bite. Lecithin is almost always incorporated into cake donut mixes to act as a wetting agent and for batter flow characteristics. Lecithin helps the cake donut batter flow uniformly to the center to assist in making a uniform formation resembling a star.

A typical fat content of a cake donut will be 20 to 25% total extractable fat. Of this range, which is necessary for good eating quality, 16 to 20% of it is absorbed frying shortening. In addition to enhancing eating quality, it has another important function; it is the mechanism or glue to hold donut coating sugar onto the surface of the donut. Typically, donuts to be sugar coated are cooled to a surface temperature of 90 to 95°F (32.2 to 35.0°C) before actual contact with the donut sugar in a tumbler. Donut coating sugar mixes are prepared with dextrose (corn sugar), starch, shortening, and small amounts of other ingredients such as salt and flavor. The shortening utilized in a donut sugar usually has a flat SFI profile, typically ranging from 16 to 20% solids at 50°F (10°C) down to 4 to 6% at 104°F (40°C). The process for preparing sugar coating mixes is based on encapsulation techniques for protecting the sugars from moisture and fat penetration. A typical donut tumbler resembles a 55-gallon steel drum laid on its side with a slight pitch with a perforated interdrum where the donuts and coating sugar mix tumble. The sugar is held on the donut surface by the liquid fraction of the frying shortening. Therefore, the SFI profile and the crystallization rate of the frying shortening are extremely important. Shortenings with slow crystallization rates have excellent sugar pickup but migrate through the coating sugar within 5 to 10 minutes to create brown, unappealing oil spots on the sugared donuts' surface. Low frying shortening SFI values at sugaring temperature and below accelerate fat migration, and the donut will have an oily, greasy mouthfeel. Alternately, excessively high frying shortening SFI values result in poor donut sugar pickup and waxy eating characteristics. The typical bakery frying shortening SFI profile presented in Table 7.3 applies to cake donuts, as well as the other bakery fried products. Fortunately, for bakery production, one frying shortening is applicable to as the fried products produced.

6.2 Yeast-Raised Fried Products

Yeast-raised fried products are basically lean, yeast-raised sweet doughs. Raised donuts are the most recognized variety for this classification of the bakery fried products. Other varieties within this classification include honey buns, fried cinnamon rolls, bismarks, long johns, and so on. All of these bakery fried products are made from a sweet dough fermented with yeast to obtain the leavening action or expansion. After being fermented, the doughs may be sheeted to the desired height and cut into donut or other shapes. Alternate make-up methods include air-pressure cut or extrusion-type procedures. After proofing, the yeast-raised products are fried and finished with various toppings and/or fillings. Glazed, raised donuts are the most widely known and popular yeast-raised fried product.

Shortening functions in a yeast-raised fried product mix as a tenderizer. Generally, plastic sweet yeast dough shortenings are used for the raised donut doughs and the other products in this fried food classification. These shortenings contain emulsifiers for dough conditioning and shelf life extension. The addition of lecithin is purported to condition the crust of yeast-raised donuts for better glaze retention after frying [15]. Bakery frying fats have traditionally been chosen with a fairly high melting point to accommodate glazing and sugaring requirements. Some bakeries even adjust the melting point with hard fat additions in the summer time to compensate for higher operating and distribution temperatures. Bakery fried products' eating characteristics are complemented by a slightly high melting character, which provides a pleasant creamy mouth feel.

6.3 Fried Pie Shortenings

A fried pie is a crispy, fruit-filled, fried pastry usually made in a semi-circle or oblong, rectangular shape and glazed. The fried pie should not be viewed the same as a baked fruit pie. The most obvious difference is that the baked pie has a pan to support it during manufacture, distribution, and until eaten by the consumer. The crust of the fried pie serves as the container; thus, it must have enough strength to hold the product together during production, distribution, and consumption. Still, the crust must be tender enough to provide a satisfactory eating quality. The fried pie crust formula utilized is generally a lean pie crust type with approximately 25 to 35% shortening on a flour basis or 15 to 20% on a total basis. The doughs are mixed at a relatively warm temperature and long enough to uniformly distribute the ingredients and obtain a degree of flour gluten development.

The pie crust is sheeted to approximately one-eighth-inch thickness and cut into the desired shape and the filling deposited in the center. The overlapping half of the dough is laid over the top of the filling, trimmed, and sealed. Improper sealing of the fried pie will cause break-outs during frying, which contaminate the frying shortening. After sealing, the fried pie is deep fat fried using the submerged method for 4 to 6 minutes at 375 to 380°F (190.6 to 193.3°C). Immediately after frying, some fried pies have a glaze applied that is similar to yeast-raised donuts.

Two shortenings are utilized in the production of fried pies: the shortening used to prepare the pie crust and the frying shortening. The shortening utilized in the crust can be the same as that used for baked pie doughs or a standard all-purpose shortening. Fried pies are successfully fried in the same bakery frying shortenings utilized for the yeast-raised and cake donut products. However, the shortenings used to fry the pies may require more maintenance to remove contaminants from the fried pies themselves.

7. SNACK FRYING SHORTENING APPLICATION

Potato chips, corn chips, tortilla chips, pork skins, puffed snacks, and similar products are fried and packaged for consumption within 2 to 3 months after preparation. Frying shortenings are a crucial element in the finished quality and therefore consumer acceptance of the fried snacks, which utilize more than 10% of the fats and oils available in the United States [16].

Deep fat fried snacks are first dehydrated by the heat transfer media, which concentrates the snack's flavors, and then it becomes a major component of the snack. Absorbed frying oils in the spaces vacated by moisture contribute to texture, flavor, eating characteristics, and appearance of the fried snack. Most fried snack products have high fat absorption rates on the order of 30 to 45%, which makes the frying media important ingredients for these products. The frying fat usage is very high in most snack processing plants. Frying fat is absorbed rapidly by the high production rates and is constantly replenished with added fresh fat, which minimizes frying fat deterioration.

Snack frying shortenings are not abused as much as restaurant or bakery frying shortenings due to the extremely high turnover rate, continuous frying, and the controls exercised over frying temperatures, fat replenishment, and so on. In snack frying, deterioration of the frying shortening is reflected in the finished snack. The longer a frying oil is used, the greater the decrease in the oil quality, which results in a lower shelf life. It has been estimated that corn chips fried in peanut oil with a 40-hour AOM stability, with a 4- to 5-hour turnover rate, at 400°F (204.4°C) should have a shelf life of more than 90 days at 70 to 80°F (21.1 to 26.7°C) before becoming rancid [17]. Rancidity is the stage in fat oxidation that is characterized by the development of easily recognized sharp, acrid, and pungent off-flavors and odors.

The choice of snack frying shortening is influenced by flavor, mouth feel, texture, product appearance, and snack type. The fried snack finish or appearance will be oilier with a frying shortening that is a liquid at room temperature and drier or grayer with higher degrees of hydrogenation. Liquid oils tend to provide a quicker flavor release, while high-melting products tend to impart gummy textures that mask or slow the flavor release. All of the frying shortenings used are processed to be neutral or bland in flavor, but their presence in the snack product enhances the flavor of the product, which is sometimes described as a cooked flavor or effect. Snack frying shortenings are limited to vegetable oils, except for the lard used to fry animal products such as pork skins.

The additives used in snack frying shortenings are antioxidants, chelating acids, and dimethylpolysiloxane. BHA, BHT, and TBHQ are the antioxidants most utilized for snack foods. The phenolic antioxidants protect the oils from oxidation during shipment and storage between the processor and the fryer where they are apparently destroyed by the high frying temperatures. Nevertheless, test results have shown a carry-through protection for snack products fried in shortenings containing antioxidants [16], and TBHQ appears to be the most effective [11]. Dimethylpolysiloxane is added to frying shortenings designed for restaurant use to retard foaming, but snack frying oils should never require an antifoamer due to the rapid turnovers experienced. However, it also offers oxidative protection during frying by forming a barrier to prevent penetration of oxygen into the oil [7]. Weiss recommended that dimethylpolysiloxane should not be used for potato chip frying because of an adverse affect upon the texture of the finished chip [18]. This result may have been the result of mechanical pickup of excess dimethylpolysiloxane from the frying shortening. Antifoamer concentrations above 1 ppm will quickly adhere to any available surface [7]. Citric acid is added by the frying shortening processor to chelate metals that accelerate oxidation; this addition is in the cooling tray of the deodorizer and again possibly with an antioxidant mixture after deodorization. Phosphoric acid should not be used as a chelating agent for frying oils; it causes a catalytic reaction to increase fatty acid development and darkening.

The extent to which the melting point of the frying shortening affects food palatability depends upon the temperature range that the food will be eaten and the amount of fat absorption. Snack food processors prefer frying shortenings with a low degree of hydrogenation for liquid or low melting products for their high-fat products that are consumed at room temperature. Many potato chip manufacturers use refined, bleached, and deodorized cottonseed oil for frying their product because of its stability compared to other oils and the characteristic "nutty" flavor developed with oxidation. Some other snack food manufacturers use peanut oil for the same reasons; peanut oil reverts to a peanut flavor with oxidation. Corn and sunflower oils are also used for some specialty snack frying, but generally, a hydrogenated soybean oil is the most attractive frying media from a quality and economic position. Typically, soybean oil snack food frying shortenings are selectively hydrogenated to significantly increase the oxidative stability with little or no increase in saturated fatty acids for an oil that melts below body temperature. Blends of source oils to take advantage of a particular oil's desirable reverted flavor and the economics of a hydrogenated soybean oil have also been popular. A typical snack food frying shortening probably does not exist. Most snack food processors have specific frying shortenings for their products, which they have developed either independently or with their supplier's assistance. These frying shortenings will differ among the snack food processors, depending upon the product produced and the processor's perception of the customer's preferences.

7.1 Nut Meats Oil Roasting

Peanuts, pecans, almonds, brazil nuts, and other nut meats are oil roasted as a snack food, as well as for use in candies, salads, desserts, and so forth. In oil roasting, the nuts are dehydrated, and a browning reaction occurs throughout the nut to change the texture, appearance, and flavors. Nut meats have relatively low moisture levels (typically a 5% average) but high oil contents. As a result, fried nuts absorption levels are very low, only about 3 to 5%. Even at high production rates with continuous frying, the frying fat turnover rate will not match even the low restaurant frying shortening replacement levels. Therefore, the frying stability and the oxidative stability of the frying shortening are very important attributes during and after the frying process. Coconut oil has been the frying medium of choice by most nut fryers. It resists oxidation because of a high saturated fat content and has a sharp melting point below body temperature but is a solid at room temperature for good salt adherence. Some nuts are fried in liquid oils like peanut oil, cottonseed oil, and liquid frying shortenings, but these oils have limited frying and oxidative stability. Selectively hydrogenated domestic oils with melting points below body temperatures and long oxidative stability are the closest substitute or replacement for coconut oils for nut oil roasting.

8. REFERENCES

1. Robertson, C. J. 1966. "The Principals of Deep Fat Frying for the Bakery," *The Bakers Digest,* 40(10):54–57.
2. Stevenson, S. G., M. Vaisey-Genser, and N. A. M. Eskin. 1984. "Quality Control in the Use of Deep Frying Oils," *J. Am. Oil Chem. Soc.,* 61(6):1102–1108.
3. Rock, S. P. and H. Roth. 1964. "Factors Affecting the Rate of Deterioration in the Frying Qualities of Fats. I. Exposure to Air," *J. Am. Oil Chem. Soc.,* 41(3):228–230.
4. Brooker, S. G. 1975. "Frying Fats–Part 2," *Food Technology in New Zealand,* 10(5): 19–21.
5. Martin, J. B. 1953. U.S. Patent 2,634,213.
6. Babayan, V. K. 1961. U.S. Patent 2,998,319.
7. Freeman, I. P., F. B. Padley and W. L. Sheppard. 1973. "Use of Silicones in Frying Oils," *J. Am. Oil Chem. Soc.,* 50(4):101–103.
8. *Code of Federal Regulations,* 1993. Title 21, Section 173.340.
9. Weiss, T. J. 1983. "Chemical Adjuncts," in *Food Oils and Their Uses,* Second Edition, Westport, CT: AVI Publishing Co., Inc., pp. 112–113.
10. Chrysam, M. M. 1985. "Tablespreads and Shortenings," in *Bailey's Industrial Oil and Fat Products, Volume 3,* Thomas H. Applewhite, ed., New York, NY: A Wiley-Interscience Publication, p. 114.
11. Buck, D. F. 1981. "Antioxidants in Soya Oil," *J. Am. Oil Chem. Soc.,* 58(3):277–278.
12. Warner, C. R. et.al. 1986. "Fate of Antioxidants and Antioxidant-Deprived Products in Deep-Fat Frying and Cookie Baking," *J. Ag. and Food Chem.,* 34(1):1–5.
13. Haumann, B. R. "The Goal: Tastier and 'Healthier' Fried Foods," *INFORM,* 7(4): 320–334.

14. Block, Z. 1959. "Uniform Doughnut Production—Cake and Yeast Raised," in *Proceedings of the Thirty-Fifth Annual Meeting of the American Society of Bakery Engineers*, March 2–5, 1959, Chicago, IL, p. 258.

15. Dixon, J. R. 1976. "Ingredients in Doughnut Mixes," in *Proceedings of the Fifty-Second Annual Meeting of the American Society of the Bakery Engineers*, March 1–3, 1976, Chicago, IL, p. 164.

16. Min, D. B. and D. Q. Schweizer. 1983. "Lipid Oxidation in Potato Chips," *J. Am. Oil Chem. Soc.*, 60(9):1662–1665.

17. Vandaveer, R. L. 1985. "Corn Chip Frying Oils," *Chipper/Snacker*, May, pp. 35–38.

18. Weiss, T. J. 1983. "Frying Shortenings and Their Utilization," In *Food Oils and Their Uses*, Second Edition, Westport, CT: AVI Publishing Co. Inc., p. 173.

CHAPTER 8

Dairy Analog Shortenings

1. INTRODUCTION

The early stages of the development and introduction of dairy analogs began during World War II when a shortage of dairy products caused a need for simulated foods. Acceptance of the dairy substitutes can be attributed, to a large extent, to rapid advances in fats and oils, as well as emulsion technology and food product formulations advancement. In all three of these areas, technologies have been developed to enable food processors to produce dairy analogs, which not only closely resemble the natural dairy products, but also to include many improvements. The advantages these products offer the household and institutional users are: (1) ease of handling, (2) extended shelf life, (3) tolerance to temperature abuse and bacterial spoilage, (4) source oil selection to satisfy religious dietary requirements, (5) nutritional values control, and (6) an economic advantage.

Two types of dairy-like products are produced with fats and oils products other than butterfat. Filled dairy products are those dairy analogs that have been compounded or blended with any fat or oil other than milk fat. These dairy analog types include mellorine, filled milk, evaporated milk alternatives, filled cheeses, and margarines. Imitation dairy products do not contain any milk product; however, casein and whey, which are milk protein, have been allowed. This group includes coffee whiteners, whipped and aerated toppings, imitation milk, imitation cheese, and dip bases. Dairy analogs originally resembled the original milk products quite closely, but now many of these products have matured to the point that improvements have been made to change or modify some of the products to provide a

413

better shelf life, function better for the intended performance, or some other desirable characteristic.

Dairy analog products are basically emulsions of specially processed fats and oils products in water with varying quantities of protein, sugar, stabilizer, emulsifiers, flavors, colors, and buffer salts added to give each product the desired physical appearance and eating properties. Dairy analogs can be prepared in a variety of physical forms, ranging from dry mixes to pressurized containers. Regardless of the final physical form, the quality of the finished product is contingent upon the selection and use of proper ingredients:

- dairy analog shortening—Fat is the most important ingredient used in dairy-like products. It establishes the eating properties, the physical appearance, and the stability of the finished product. A shortening that performs well in one dairy analog application may be unsatisfactory for another; therefore, it is necessary to match the finished product requirements to the performance characteristics of the shortening. The various shortening types utilized for nondairy applications are identified in Table 8.1, characterized by melting point, SFI, and fatty acid composition.
- protein—The principal functions of protein in dairy-like products are to contribute to the stability, body, and viscosity of the finished product. Protein may also serve as agents to entrap gases in whipped toppings or as dispersing agents or protective colloids in emulsions. The protein used may come from a number of sources such as liquid skimmed milk, nonfat milk solids, caseinates, gelatin, whey, egg, or soy protein. Whichever is selected, emphasis must be placed on the use of a bland, flavorless protein that will not detract from the flavor of the finished product.
- sugars or carbohydrates—Sugar provides sweetness and body, aids in solubility, and affects the viscosity or density of the finished product. Available sources are corn syrup, corn syrup solids, dextrose, cane, or beet sugars.
- stabilizers—Stabilizers increase the body of the emulsion and help to prevent syneresis or water separation. In many products a combination of stabilizers may be required to achieve the stability required.
- emulsifiers—The surfactant or surfactant system for each product can vary with the individual demands of the finished product and the manufacturer's process.
- buffer salts—Buffers are added to certain dairy-like products to maintain the desired pH to minimize body variations and to improve the colloidal properties of the protein employed.

TABLE 8.1. *Dairy Analog Shortening Potentials.*

Source Oils:	Milk Fat		Coconut Oil		Soybean Oil		Palm Kernel and Cottonseed Oil			Soybean and Cottonseed Oils		Soybean Oil
							Hydrogenated and Interesterified					
Fats and Oils Products:	Butter	RBD	Hydrogenated		Selectively Hydrogenated					Hydrogenated Fractionated	Specially Hydrogenated	Liquid Shortening
Mettler dropping point												
°F	95	76	92	110	95	106	97	102	112	99	102	88
°C	35	24.4	33.3	43.3	35.0	41.1	36.1	38.9	44.4	37.2	38.5	31.1
Solids fat index, %												
at 10°C/50°F	33	59	57	63	41	57	64	68	69	72	58	3.5
at 21.1°C/70°F	14	29	33	41	24	45	55	56	58	63	43	2.5
at 26.7°C/80°F	10		8	16	16	40	38	40	50	55	34	2.5
at 33.3°C/92°F	3		3	7	3	20	8	12	27	25	12	2.0
at 40.0°C/104°F				4		4		4	14	5	1	1.5
Iodine value	31.5	9	1	>1	74	67	3	3	3	59	75	107
Fatty acid composition, %												
C-4:0 Butyric	3.6											
C-6:0 Caproic	2.2	0.5	0.5	0.6								
C-8:0 Caprylic	1.2	7.1	9.5	9.3			2.0	2.0	2.0			
C-10:0 Capric	2.5	6.0	6.5	6.4			3.0	3.0	3.0			
C-12:0 Lauric	2.9	47.1	46.0	46.8			46.0	48.0	40.0	0.5		
C-14:0 Myristic	10.8	18.5	16.9	16.4			17.0	16.0	14.0	0.6	0.3	
C-14:1 Myristoleic	0.8											
C-15:0 Pentadelandic	2.1											
C-16:0 Palmitic	26.9	9.1	8.5	8.5	10.8	10.8	9.0	8.0	12.0	16.4	12.5	10.6
C-16:1 Palmitoleic	2.0									0.4	0.4	
C-17:0 Margaric	0.7									0.3		
C-18:0 Stearic	12.1	2.8	10.4	11.2	8.2	13.2	21.0	20.0	27.0	12.2	10.8	6.5
C-18:1 Oleic	28.5	6.8	1.2	1.0	74.0	74.0	2.0	3.0	2.0	67.4	74.8	44.5
C-18:2 Linoleic	3.2	1.9	0.2		6.0	2.0				1.4	0.3	36.5
C-18:3 Linolenic	0.4	0.1								0.3		2.0
C-20:0 Arachidic		0.1								0.4	0.5	
C-20:1 Gadoleic	0.1											
C-22:0 Behenic										0.3	0.4	

415

- flavors—Numerous natural and artificial flavors are available for use in these dairy-like systems. The flavor experienced with a fresh product is not necessarily the flavor that the product will have after the fats and oils crystal habits have stabilized.

2. NONDAIRY CREAMER SHORTENINGS

Nondairy creamers are not imitation cream. They are formulated systems similar in functionality to the natural dairy product, with the advantages of longer shelf life, convenient product forms, and uniform quality and performance. Generally, a nondairy creamer or base may be defined as a stabilized fat source, a creaming agent, or cream substitute. Nondairy creamers combine five basic ingredients: shortening, protein, carbohydrates, stabilizers, and emulsifiers with water to form a stable product with a delicate flavor that disperses quickly in coffee without feathering or oiling off. A good whitener effectively approximates the appearance, quality, and taste of coffee cream. Nondairy creamers are marketed in three different physical forms:

(1) Liquid coffee whiteners are processed, transported, and marketed in the liquid state. This ready-to-use product form is utilized primarily in homes and restaurants and usually has a limited shelf life, slightly better than dairy creamers. Acceptable liquid coffee whiteners must have an unusually high degree of emulsion stability to remain in a uniform emulsion on standing after preparation and prior to sale to prevent oiling off or feathering when added to coffee. It must also withstand freeze–thaw cycles without separating and maintain a viscosity simulating the natural dairy product. A heavy-bodied product will not disperse in coffee just as a thin or separated product that oils off is not acceptable. Whitening ability, which is controlled by the total amount of solids present and the fineness of the dispersed phase, must be uniform. Coffee whiteners must also maintain a bland flavor and be odor-free for the life of the product. The concentration of fat can vary from 5 to 18% for the liquid coffee whiteners. A shortening or fat with a relatively low melting point and a narrow plastic range indicated by a steep SFI profile is desirable for the liquid creamers. Higher melting points may impart a greasy mouth feel to fluid whiteners. Liquid fats interfere with dispersibility due to absorption into the protein and coalescence. The shortening most often utilized by the liquid whiteners is 76° and 92°F coconut oil, selectively hydrogenated soybean oil with a 95°F melting point, and liquid opaque shortening. Coconut has a steep SFI and sharp

melting point for good mouth feel and eating characteristics but can develop soapy flavors due to hydrolysis of the lauric fatty acid content, especially with the high moisture content of the liquid whiteners. The 74 IV selectively hydrogenated soybean oil product has a melting point and SFI profile very similar to butter fat. Liquid shortenings have found acceptance because of the ease of handling and the high polyunsaturated level in the finished coffee whitener.

(2) Frozen liquid coffee whiteners are processed, frozen, and shipped to the retail markets in a frozen state with directions for defrosting before use. The characteristics and processing techniques for this type of whitener are similar to the liquid-type coffee whitener. But since this product is frozen and maintained in the frozen state prior to use, a shorter shelf life after thawing can be tolerated, but the product must have good freeze–thaw stability to prevent separation when the product is thawed for use. Normally, the same shortenings are utilized for frozen coffee whiteners as the regular liquid product.

(3) Spray dried nondairy creamers are processed as a free flowing dry powder. During processing of the nondairy powders, the fat is coated with emulsifier, allowing it to mix with the protein slurry. Homogenization forms globules, which are tiny, stabilized droplets of fat. An emulsified, homogenized fat exhibits a high degree of whitening in coffee. The carbohydrates and the other ingredients coat the outside of the globules. This slurry is sprayed in a fine mist into the hot dryer chamber. This causes the water to flash off, and the resulting powder falls to the bottom of the dryer. The nondairy powders must have a superior oxidative stability to withstand the processing abuse and to provide the shelf life stability required, which is one of the main advantages for use along with room temperature storage where the other nondairy and natural dairy creams require refrigeration. Additionally, the dry product must exhibit good flow properties; i.e., clumping and caking must be avoided. While the principal use for the nondairy powders is for whitening in coffee, a number of other uses have emerged where the powders can replace dairy products or other new applications have been developed. At least five different types of the spray-dried nondairy creamers are produced:
- coffee whiteners—These retail marketed products which have fat contents between 16 and 40% (dry basis), are specifically processed, low-density powders, which allow for desirable packaging properties, as well as "sink and dispersion" characteristics in coffee. The lower fat levels are utilized with the light or low-fat coffee whiteners for reduced calories.

- aerated whiteners—Specially processed dry powders with encapsulated air within the nondairy creamer to produce a cappuccino effect when added to regular coffee.
- reconstituted creamers—Dry powders designed to be reconstituted, pasteurized, homogenized, and packaged for restaurant or retail marketing as liquid products for use in dairy-type applications.
- vending creamers—Dry powders with fat contents in the mid-30%, with a higher density than the regular coffee whiteners.
- ingredient bases—Dry powders with relatively high densities, typically 50%, are produced as ingredients or bases for liquid beverages, puddings, gravy mixes, whipped toppings, dips, and so on.

The lipid systems for coffee whiteners have been tailored to meet the performance requirements of the product type into which they are formulated, whether liquid, frozen-liquid, or the various spray-dried products. Nondairy powdered creamers usually require higher melting point shortenings. Lower melting point fats incorporated into spray-dried whiteners may cause the powders to lump at high temperatures and may disperse poorly in hot coffee. Generally, the shortening requirements for nondairy coffee creamer powders have been composed of hydrogenated coconut oil with a 110°F (43.3°C) melting point interesterified and hydrogenated palm kernel oil with a 112°F (44.4°C) melting point or a selectively hydrogenated soybean oil with a 106°F (41.1°C) melting point. The spray-dried coffee whiteners usually require higher melting fats for shelf life, anticlumping characteristics, good dispersion in hot liquids, and whitening.

Emulsifiers are employed in all three nondairy creamers to combine immiscible fat and water, help maintain a stable emulsion, and create the right amount of fat agglomeration in order to achieve the major objective—lightening the color of coffee. A number of different emulsifier systems can be formulated for the nondairy powders including mono- and diglycerides, polysorbate 60 and 80, glycerol lactoesters, lecithin, propylene glycol monoesters, sodium stearoyl lactylate, and others, depending upon functionality and the processor's preference. In almost all cases, the emulsifier requirements are added independently of the shortening requirement.

3. WHIPPED TOPPING SHORTENINGS

Whipping cream stability problems brought about the development of nondairy whipped toppings. Whipped toppings have become popular for both commercial and consumer use as toppings on puddings, sodas, cakes, ice cream, fruit, and pastries, as well as extensive use as cream pie bases. Nondairy whipped toppings are more functional than whipping cream because manufacturers can use a more desirable fat characterized by a specific solid fat index profile with complementary emulsifier systems. Whipped toppings are marketed in a variety of forms, all of which have similar formulation characteristics:

- liquid whipped topping—This basic topping is an oil-in-water emulsion containing fat, protein, sugar, stabilizer, and emulsifier and requires mechanical whipping to produce a topping with the desired overrun and dryness. It is usually packaged in pure-pak containers for retail, foodservice, and food processor applications.
- topping concentrate—An oil-in-water emulsion similar to the liquid whipped topping but containing less water. Prior to use, milk, skim milk, cream, juice, or water is added, and the mix is agitated to produce the desired overrun and dryness. This is a popular form for preparation of bakery cream pie fillings and cake toppings.
- aerosol topping—An oil-in-water emulsion similar to a liquid topping, but packaged in a pressurized container. The topping is automatically whipped as it passes through the aerosol spray nozzle. This package and application is popular for the retail and foodservice markets.
- powdered toppings—This oil-in-water emulsion, which is spray-dried to contain a minimum of water, is one of the most difficult toppings to formulate and manufacture. When reconstituted with milk, skim milk, or water, it is mechanically whipped to attain the desired stiffness and overrun. The powdered topping form offers a longer shelf life than either the liquid or aerosol toppings and the end product ranks high in consumer appeal.
- frozen ready-to-use topping—This complete product is marketed in retail grocery stores in plastic recloseable containers in convenient sizes for the household consumer. Marketing this product in the frozen state substantially improves the product's shelf life. The use life in the home refrigerator after thawing is probably 3 to 6 weeks. Normally, the ready-to-use toppings are formulated with sodium caseinate because milk solids do not lend themselves to freezing. The ready-to-use toppings are processed like the other toppings, except after pasteurization and homogenization, the finished topping mix is sent through a continuous whipping machine and brought to its optimum specific gravity. It is then filled into plastic containers and rapidly frozen, usually with blast freezers set below $-20°F$ ($-28.9°C$).

Preparation of a satisfactory whipped topping is more complex than most dairy analogs. Proper balance of the individual ingredients for the finished aerated topping is necessary to produce an appealing and commercially desirable whipped topping. The common ranges of basic ingredients as a percentage of the finished topping are generally in the ranges shown in Table 8.2 [1].

Most of the nondairy toppings are made by combining the ingredients and pasteurizing the mixture. The mix is then homogenized and cooled to 40°F (4.4°C) or lower before packaging. The finished mixes usually require 18 to 24 hours tempering or aging before satisfactory whipping performance can be expected.

The selection of the optimum emulsifier system for a whipped topping is quite important, since overrun, dryness, stiffness, mix stability, topping stability, and, to a degree, body and texture depend on it. The emulsifier concentration may vary from 0.4 to 1.0% of the total weight of the topping; according to the emulsifier system selected. The emulsifier system choice depends upon the ultimate form of whipped topping; liquid, frozen-liquid, or dry powder. Soft mono- and diglycerides and/or glyceryl lactoesters or propylene glycol monoesters are usually used in dried and liquid toppings. In the fluid whipped toppings, whipping time can usually be reduced by adding more soft monoglycerides or polysorbates to the formula. However, higher emulsifier levels usually increase the viscosity of the finished whipped topping. Polysorbates or hard mono- and diglycerides are employed to make toppings that have freeze–thaw stability. Adding a hard mono- and diglyceride usually lowers the specific gravity of the whipped topping and increases the whipping time required.

Experience has indicated that whipped toppings should contain 25 to 35% fat to achieve whipping and body characteristics equivalent to natural cream. Lower fat contents may be used with the addition of high sugar,

TABLE 8.2. Basic Ingredients as a Percentage of Finished Topping.

| | Range, % Total Weight Basis | |
Ingredients	Low	High
Shortening	25.0	35.0
Protein	1.0	6.0
Sucrose	6.0	12.0
Corn syrup solids	2.0	5.0
Stabilizers	0.1	0.8
Emulsifiers	0.4	1.0
Salts	0.025	0.15
Water	46.0	64.0

stabilizer, and emulsifier levels to provide body. However, a topping containing less than 25% fat is generally characterized by a slack body with a poor mouth feel, stability, and texture. Most whipped toppings are formulated with fats that have narrow plastic ranges reflected by a steep SFI slope. The shortening used must have enough fat solids at whipping temperature to give rigidity to the whipped product with a melting point in the 95 to 102°F (35 to 38.9°C) target range for rapid getaway in the mouth. The use of higher melting shortenings provides significantly better body and standup stability but causes a distinct greasy and waxy mouth feel and aftertaste. Typically, the type of shortenings chosen for whipped topping are characteristic of those products outlined in Table 8.1, processed with selective hydrogenation, interesterification, fractionation, or specially hydrogenated with a sulfur proportioned catalyst all with melting points from 95 to 102°F.

4. CHEESE ANALOG SHORTENING

Cheese analogs have replaced cheese in a variety of applications, primarily because of economic and improved performance for certain applications. Natural cheese is basically made up of fat (24%), protein (20%), water (46%), and minerals, and a small amount of carbohydrates, flavored by any one of a number of processes, combine to provide the flavor and texture properties. A nondairy replacement involves the use of a fat source other than butter, a protein other than milk solids, and the compounding of a flavor system that duplicates as near as possible the natural cheese counterpart. Functionally, an imitation or substitute cheese product must duplicate the original cheese product's performance more uniformly. Functional characteristics, such as firmness, slicing properties, melting properties, shredding, and so forth are all controllable through formulation and processing conditions and, once attained, can be reproduced with uniformity. The major uses developed for cheese analogs have been pizza, salads, frozen entrees, sandwiches, frozen appetizers, dips, spreads, sauces, and snacks.

Replacement of butter fat with a shortening is one of the major factors and first steps in the duplication of a dairy product. Cheese analogs require a fat with a melting point close to body temperature, a relatively steep SFI slope, good oxidative stability, and a bland flavor. A desirable SFI slope provides both good eating quality and solid fats at the temperature required for slicing and shredding but allows the product to melt at elevated temperatures without oiling out. The fatty acid composition of the dairy analog shortening utilized is also important. Short-chain fatty acids from lauric oils can interfere with the flavor development in some cheese varieties and hydrolyze in others due to the high moisture content to produce soapy flavors.

Deviation from butter fat properties is necessary for some dairy analog products, but for imitation cheese products a shortening with similar properties has performed most satisfactorily. Selectively hydrogenated domestic oils like the 74 IV product in Table 8.1 have performed more than adequately for cheese analogs. These shortening types have relatively steep SFI slopes and melting points like butter fat, do not contain lauric fatty acids, have good oxidative stability as indicated by the low polyunsaturates level, and have bland flavors.

5. FROZEN DESSERT OR MELLORINE

Frozen desserts or imitation ice cream were probably the first dairy product analogs produced after margarine. The name *mellorine* was adopted by several states as the generic name for frozen desserts made with fats other than butter fat. Unlike margarine, the dairy industry controlled mellorine by producing it in their ice cream plants and distributing it as a line extension. Mellorine is a filled milk product because it is produced with a fat source other than milk fat but still contains milk solids contributed by nonfat milk solids or skim milk. It is generally made by the same process as used for ice cream in two basic types: (1) soft serve and (2) hardened. Soft serve was introduced to the American public through dairy stands dispensed directly from batch freezers in these retail outlets or restaurants. Complete soft serve mixes are frozen in the batch freezer and dispensed as the customer watches in cones, sundaes, or milk shakes. Hardened mellorine products are packaged in the traditional pints, quarts, gallons, and bricks or made up into novelty items such as coated bars, cups, and so on and quick-frozen for distribution through the freezer sections of retail stores. Two factors that contribute heavily to mellorine quality are fat content and overrun. The fat content for hardened mellorine products has been varied from as low as 4% to over 16%. Generally, higher fat levels are rated better quality, with 10% fat usually considered the minimum for good quality hardened product. Hardened mellorine with a lower overrun has also been rated as preferred product. Soft serve products are judged differently. A 6 to 8% fat level with a high overrun is considered a quality soft serve mellorine product.

The mixing and manufacture of frozen desserts is handled in much the same manner as ice cream with regard to pasteurization, homogenizing, freezing, and so on. Nonfat milk solids or condensed skim milk can be used as the protein source, and hydrogenated vegetable oil or an animal fat can be substituted for the butterfat. However, with mellorine production, the dairyman has a choice of shortening products; processed fats have been tailored for frozen dessert applications. Shortenings with a melting point close to body temperature and a steep SFI slope were developed to solve "churn out" problems with soft serve mellorine products. Churn out is freckling, or graining out, of the fat, which results in a "gritty" mouth feel in the frozen soft serve mellorine. Churn out occurs when the fat separates from the mix in lumps that are difficult or impossible to reemulsify into a smooth mixture. A coating of fat over the surfaces of the freezing unit is also an indication of churn out. The selectively hydrogenated soybean oil shortening, with a 95°F melting point identified in Table 8.1, has performed well for soft serve products. It can also be used for hardened products but better results are obtained with the slightly firmer 106°F selectively hydrogenated shortening. For hardened frozen desserts, the melt down, chewiness, dryness, and texture are improved by the higher SFI contents and melting point. However, these same qualities are detriments for soft serve products to promote churn out.

Both the emulsifiers utilized and the shortening composition can affect the stability of the mellorine mix against churn out. Hard mono- and diglycerides and/or polysorbate-type emulsifiers are used in mellorine and ice cream. The hard mono- and diglyceride emulsifiers help produce a fine air cell structure and improve whipping performance. Polysorbate 80 or polyglycerol, 8-1-0, provides optimum dryness and a smooth product with good standup qualities as required for packaged product.

6. SOUR CREAM ANALOG AND DIP BASES

Imitation sour creams are used extensively for party dips, salad dressing, potato toppings, sauce enrichments, cold soups, and many other applications. These dairy analogs can be produced with lower fat levels, are more resistant to wheying off, have longer shelf life, and are usually lower priced than the natural dairy product. Most processors use a direct acidulation process with edible organic acids, rather than the conventional sour cream process of injecting a bacterial culture into the pasteurized product. In most cases, analog sour creams are produced with 14 to 18% fat with nonfat milk solids or sodium caseinate, stabilizers, sugars, emulsifiers, flavoring, and an acid media. Processing includes mixing, pasteurizing, and homogenizing the product before packaging. Sour cream analogs are fluid as filled and require 10 to 12 hours tempering at refrigerated temperatures for the fat to crystallize while the protein and stabilizers thicken the product to use consistency.

A number of different fat compositions have been utilized for imitation sour creams and/or dip bases. Initially, coconut oil, either refined, bleached, and deodorized (RBD) with a 76°F melt or hardened to 92°F melt, was preferred for the fast getaway provided by tropical oils high in lauric fatty acids. Selectively hydrogenated cottonseed or soybean oils with a 95°F melt, coupled with a steep SFI slope, provide good mouth feel and product stability without the possibility of soapy flavor development due to hydrolysis. Specially hydrogenated or the hydrogenated and fractionated domestic oil blends can provide a more stable consistency over selective hydrogenated oil products and retain mouth feel quality even though the melting point is slightly higher.

7. FLUID MILK ANALOGS

Fluid milk analogs can be produced as either filled or imitation products. The formulation of filled milk products is relatively simple. Whole milk is replaced with skim milk or buttermilk that has been homogenized with a fat source other than butter fat. Filled milk contains about 3.5% fat and requires about 3% alpha monoglyceride (fat basis) for emulsification. The fat component should have a melting point below body temperature to avoid greasy mouth feels [2]. Three of the shortenings outlined in Table 8.1 would be likely filled fluid milk fat candidates, i.e., 76°F coconut oil, 95°F melt selectively hydrogenated soybean or cottonseed oil, and the opaque liquid shortening. The liquid shortening product offers a high polyunsaturate to saturate level or P/S ratio compared to the other fat products:

	P/S Ratio
Liquid shortening	2.3 to 1
Coconut oil	0.02 to 1
Selectively hydrogenated soybean oil	0.3 to 1
Butter fat	0.06 to 1

Imitation fluid milk products should not contain any dairy products, except casein and whey proteins, but should maintain the same nutritional level as dairy fluid milk products. Sodium caseinate and whey have typically replaced milk solids in most imitation fluid milk products [3]. The same shortening candidates apply to imitation as with the filled fluid milk products.

8. SWEETENED CONDENSED MILK ANALOGS

Confectioners have used a considerable amount of regular sweetened, condensed milk for caramels, candy centers, fudges, nougats, kisses, toffees, and similar confections. Sweetened condensed milk is produced from pasteurized and homogenized fluid milk, which is first condensed with a vacuum and then sugar is added before condensing it further to a ratio of 3 solids to 1 liquid. Sweetened condensed milk contains approximately 8.5% butter fat, 21.5% nonfat milk solids, 42% sugar, and 28% water. Sweetened condensed filled milk analogs can be produced by substituting condensed skim milk for condensed whole milk and adding a shortening and emulsifier for the fat content. This product is then condensed with a vacuum at 95 to 110°F to develop the cooked or caramelized milk flavor desired for the confectionery products. The suitable shortening products in Table 8.1 for this application encompass all those with melting points centered from 95 to 102°F. The important fat source characteristics are a relatively sharp melting point, a bland flavor, and good oxidative stability to prevent off-flavor development during the milk processing, production of the confection product, and its shelf life.

Production of an imitation sweetened, condensed milk product requires a more extensive composition of ingredients to replace the nonfat milk solids, which consists of protein, lactose, and minerals. Caseinates, whey, starches, and other protein sources have been used to formulate the imitation products. The shortening requirements have been satisfied with essentially the same products as used with the filled sweetened condensed milk analogs.

9. REFERENCES

1. Anonymous. 1968. "Guidelines to the Formulation of Whipped Toppings," Product Information Bulletin, Atlas Chemical Industries, Inc., Wilmington, DE, pp. 1–5.
2. Weiss, T. J. 1983. "Imitation Dairy Products," in *Food Oils and Their Uses*, Second Edition, Westport, CT: AVI Publishing Co. Inc., p. 295.
3. Henderson, J. L. 1971. "Special Milk Products Including Imitations," in *The Fluid-Milk Industry*. Westport, CT: The AVI Publishing Co., Inc., p. 455.

CHAPTER 9

Household Shortenings

1. INTRODUCTION

Shortenings produced for household use have not become tailored products for specific applications like the products produced for the food service and food processor industries. Household shortenings marketed in the United States are still truly all-purpose shortenings. These shortenings must be formulated for cooking, baking, frying, candy making, and any other home food preparation. The two basic types of household shortenings available are differentiated by composition, i.e., all-vegetable or animal–vegetable blends.

1.1 Historical

In the United States, the style of cooking and baking has been based on the use of plastic fats. The early immigrants to the United States were predominately of northern European extraction, who were accustomed to solid fats of animal origin and consequently preferred the solid fats rather than the liquid oils used predominately in southern Europe or Asia [1]. Consumers tend to maintain their old habits; therefore, it is not surprising that less than 100 years ago, lard was the most commonly used fat in most U.S. households. It possesses several unique characteristics compared to the other available fats. It has a plastic solid consistency at room temperature, which provides creaming capabilities that other softer or firmer products could not duplicate. Lard also has a white color and a mild flavor that was much less objectionable than most of the alternative products available at that time.

427

Vegetable shortening with all the physical characteristics of a plastic animal fat was an American invention; it was created by the cotton-raising
industry and perfected for soybean oil utilization. Major improvements in
fats and oils processing had to take place before vegetable oils could be
accepted as lard substitutes by the homemaker. The color, flavor, texture,
and functional properties had to be as much like their lard prototype as
possible. First, all American fats and oils processors learned how to bleach
and then to produce a bland-flavored cottonseed oil that could be blended
with lard as an extender. Initially, vegetable oils were converted to solids
or plastic fats by blending them with a harder fat to give the required body
to the final product. These compounds, which normally contained some
lard, usually had a stiff consistency; a dull, dry appearance; and a characteristic odor and flavor. Deodorization introduced around 1890 and later
perfected by David Wesson removed the strong offensive flavor contributed
by cottonseed oils in the compound shortenings [2]. Then using an English
patent for hydrogenation, Procter & Gamble introduced the first all-
vegetable household shortening with cottonseed oil at the beginning of
World War I. This shortening was called "Crisco," which was short for
"crystallized cotton oil" [3]. In 1933, Procter & Gamble implemented another significant change for shortenings: the addition of mono- and diglycerides, which dramatically improved the performance of baking shortenings. This improvement was tempered somewhat for household shortenings
because of the required all-purpose performance. The alpha monoglyceride
level had to be maintained at a low enough level to retain a high smoke
point for frying but high enough to show an improvement in baking performance. The all-hydrogenated household shortening had evolved into a
product with a creamy white color, bland odor and flavor, smooth texture
and plastic consistency with an extended shelf life.

Advancements in segments of edible oil processing provided distinct
quality and performance differences between all-vegetable and meat fat–
based household shortenings with the advantage moving back and forth.
However, technology progressed to the point where it was difficult and
often impossible for the consumer to distinguish one from the other. Both
products were formulated to be smooth, creamy, white products for general
purpose functionality in all household cooking and baking. In the 1960s
marketing studies indicated that the meat fat/vegetable blended shortenings
were purchased most for frying applications, while all-vegetable shortenings
were more often purchased for baking. These studies led the fats and oils
processors to reduce the emulsifier level in meat fat/vegetable blended
household shortenings to raise the smoke point and increase the frying
stability.

Household shortening reformulation activities were also centered around nutritional concerns beginning in the 1960s. Initially, all-vegetable shortenings were reformulated to increase the level of essential fatty acids, i.e., linolenic (C-18:3) and linoleic (C-18:2). Until this time, household shortenings were hydrogenated to reduce the polyunsaturate levels, which ranged from 5 to 12%, for functionality and oxidative stability. In response to research findings, which advised a higher intake of the essential fatty acids, many all-vegetable household shortening compositions were modified to contain 10 to 30% of the polyunsaturated fatty acids [4].

Since heart attack risk increases with higher serum cholesterol levels that can be affected by diet, some health advisory organizations have recommended diet modifications to achieve lower serum cholesterol levels. These diet modifications include reducing consumption of total fat, saturated fat, and cholesterol [4]. Meat fat household shortenings have suffered from the results of these studies since only animals have the ability to produce cholesterol. However, these concerns have also contributed to reduced sales of all household shortenings because solid or plastic shortenings are probably related to saturated fats by the consumer.

A decrease in household shortening consumption can also be related to other causes, such as convenience foods and a higher frequency of meals away from home. Prepared mixes, frozen foods, microwave foods, fast-food restaurants, home-delivered foods, and other convenient foods have severely reduced home preparation of foods requiring household shortenings.

2. HOUSEHOLD SHORTENING PRODUCT REQUIREMENTS

The product requirements for a household shortening differ somewhat from the products prepared for foodservice and food processor applications. The important attributes for a household shortening, sold through retail outlets to consumers with extremely varied eating and cooking habits, are more difficult to identify than the more specific requirements for the institutional products. The qualities necessary for an ideal household shortening should include the following:

- appearance—The appearance of a household shortening is important from a psychological standpoint; it is associated with quality. Whiteness or freedom from color with a satiny sheen suggests a creamy, smooth product texture. The container fill also affects appearance quality impressions. Many fats and oils processors have had requirements for the amount of shortening on a 3-pound can lid and the appearance of the concentric rings in the surface of the product.

- flavor—Household shortenings flavor should be completely bland, unless intentionally flavored, to enhance a food product's flavor, rather than contribute a flavor. The bland flavor must be sustained throughout the expected life of the shortening, which includes distribution, retail store shelf time, and the time after purchase until use.
- consistency and texture—The shortening should have a plastic, smooth, and creamy consistency and texture to make the product easy to measure and mix. A smooth, plastic shortening distributes easier for measuring and more evenly in a batter or dough than a hard shortening.
- wide plastic range—The original shortening consistency should be maintained over a wide range of temperatures. Neither fluctuating home temperatures or the different temperature requirements of the products being prepared should affect the body or consistency of the shortening.
- creaming properties—Household shortenings should aerate quickly and hold the absorbed air through the mixing cycle and baking. Creaming properties are important for cakes, cookies, icings, fillings, toppings, and other aerated products.
- moisture retention—Moistness and eating quality of the finished products should be enhanced by the household shortening. Moisture retention capacity is directly related to a finished product's ability to stay fresh.
- smoke point—Consumer shortenings should have a high enough smoke point to ensure odorless, smokeless pan and deep frying throughout repeated frying.
- stability—Three areas are concerned with stability: (1) package stability—the flavor of the shortening must remain fresh and bland at room temperature for the use life of the product; (2) frying stability—flavor of the shortening during frying must be retained at its original purity with no breakdown as revealed by smoke and odor; and (3) baking stability—the baking performance must be retained at the original quality with no breakdown as revealed by off-flavors in the baked product.
- uniformity—Each shortening purchase should perform just like the previous product. Product uniformity provides expected baking, frying, and cookery performance with each shortening purchase.
- nutritional—Household shortening formulations must adjust to the latest research findings to provide the most healthful product to the homemaker with uniformity of performance.

3. HOUSEHOLD SHORTENING FORMULATION

Household shortenings must be designed for multipurpose use to provide creaming properties, a wide plastic range, heat tolerance, frying stability, baking performance, oxidative stability, light color, bland flavor, moisture retention, high smoke point, and other properties identified as important for a successful product. All-purpose shortenings are prepared by blending a partially hydrogenated basestock or basestock blend with a low-IV hard fat. The same beta-prime hard fats used in bakery-wide plastic range shortenings are applicable to household shortening formulation. Animal–vegetable household shortenings normally contain a tallow hard fat while the all-vegetable products are usually stabilized with either palm or cottonseed oil hard fats. These beta-prime hard fats function as plasticizers for improving creaming properties, texture, and consistency. Most all-vegetable household shortenings currently utilize cottonseed hard fats due to special interest pressures to discontinue tropical oil use, which included palm oil.

Household shortenings are formulated with a moderate level of mono- and diglyceride emulsifier for cake baking. Alpha monoglyceride content is maintained at a low level to prevent excessive smoking during frying. A shortening's smoke point decrease is related to the glycerine content of the mono- and diglyceride added. Smoke point drops sharply with slight free glycerine increases up to 0.12%, where it remains constant at about 200°F (93.3°C). Thus, most household shortening's alpha monoglyceride contents are maintained at 2.0 ± 0.4% to provide emulsification for baking but retain a fairly high smoke point for frying performance. Many fats and oils processors added the mono- and diglycerides to the shortening at the end of the batch deodorization cycle or in the cooling tray of a semicontinuous deodorizer to reduce the free glycerine content to improve the shortening's smoke point [5].

The basestock utilized for all-vegetable household shortenings evolved from 100% cottonseed oil to blends with soybean oil or, for some processors, all soybean oil. For quite some time, 80 to 85 IV hydrogenated basestocks like those used for bakery all-purpose shortenings were utilized until higher polyunsaturate levels were encouraged for nutritional purposes. The shortenings with firmer basestocks contained 8 to 15% linoleic fatty acid. The higher polyunsaturated household shortenings had linoleic fatty acid contents increased to 28 ± 2% with a linolenic fatty acid level of 1.5 to 3.0%. These formulation changes increased the all-vegetable household shortening's polyunsaturated to saturated fatty acids ratio (P/S ratio) to equivalent levels of 1 to 1 or 3 grams of each for a 12-gram serving. The basestock for the shortening high in polyunsaturated fatty acids has to be

TABLE 9.1. All-Vegetable Household Shortening.

Analytical Characteristics	1957	1995
Solids fat index, %		
at 10.0°C/50°F	29.0	18.0
at 21.1°C/70°F	22.0	16.0
at 26.7°C/80°F	18.5	15.0
at 33.3°C/92°F	13.5	12.5
at 40.0°C/104°F	8.5	8.0
AOM stability, hours	135	35
Iodine value	74	93

formulated with a basestock or blended basestock with a 100 to 110 iodine value. These basestock changes reverted the product back to almost a classic compound shortening of a liquid base blended with a hard fat. The softer basestock effected a flatter SFI slope and a poorer oxidative stability, as shown by a comparison of a household shortening's SFI and AOM stability results produced in 1957 and 1995 (Table 9.1).

Meat fat shortenings were prepared with interesterified lard as the basestock until it was acknowledged that the animal fat shortenings were used primarily for frying. The compositions were changed to a blend of tallow with a lightly hydrogenated soybean oil to soften the shortening to allow the addition of tallow hard fat for heat stability and plasticity. BHA and BHT antioxidants are added to the meat fat shortenings for oxidative stability to help assure that the shortening has at least a 6-month shelf life. Additionally, regular lard is still available in selected retail markets. However, it is usually deodorized and stabilized with lard hard fat and antioxidants, BHA and BHT. Most of the lard is used for pie crust and ethnic foods preparation and frying.

Butter-flavored and -colored household shortenings introduced in the 1950s have had limited acceptance but enough to remain an active product. The beneficial effect is probably more psychological than functional. A yellow or butter-like color may be retained in some baked products; however, the color and flavor tends to disappear with most frying and baking.

4. HOUSEHOLD SHORTENING PLASTICIZATION

Household shortening functionality is influenced by two other processes in addition to composition: (1) chilling, which initiates the crystallization process, and (2) tempering, where the desirable crystal nuclei are developed and stabilized. The chilling unit outlet temperature control limits are determined by fill tests to identify the conditions necessary to produce the desired workable consistency and plasticity. However, household shortenings are normally quick-chilled from a heated liquid to 60 to 65°F (15.6 to 18.3°C) for all-vegetable product and 70 to 75°F (21.1 to 23.9°C) for an animal–vegetable product followed by a crystallization or working stage, filling into containers, and then tempering at 85°F (26.7°C), usually for 48 hours.

During the plasticization process where the shortening is transformed from a liquid to a plastic solid by chilling and working, creaming gas is incorporated into the product. The function of the creaming gas, preferably an inert gas like nitrogen, for household shortenings is

- white creamy appearance
- satiny surface sheen
- texture improvement
- homogeneity
- volume increase
- reduced serving weight
- reduced calories per serving
- reduced saturated fat grams per serving
- less dense product for easier handling

All-vegetable shortenings normally have 13 ± 1% nitrogen creaming gas incorporated into the product. Some meat fat shortenings are also packaged with 13 ± 1% nitrogen added, but most contain 18 to 25% and are designed as precreamed. Additional nitrogen assists the animal–vegetable shortenings formulated with tallow basestocks to maintain a more workable consistency and increases the product volume. Normally, the precreamed meat fat shortenings are packaged with 2 pounds, 10 ounces of product in the same container that is marketed with 3 pound of the all-vegetable shortening with 13 ± 1% nitrogen incorporated.

5. HOUSEHOLD SHORTENING PACKAGING

Consumer product packaging requirements must be considered carefully. Product quality, point of purchase appeal, and the cost of the finished product are all affected by the package selection. Packaging cost is the second most significant portion of the finished product cost after raw materials [6]. The product protection afforded the product by the container directly affects the shelf life and customer satisfaction. And the package appearance and graphics must appeal to the grocery shopper.

The 3-pound metal can, measuring 5-1/8-inch diameter by 5-3/4-inch height, with a key opening tear strip and a captive lid, was the standard for household shortenings for many years. This container was carefully filled so that the shortening surface was smooth with a satiny sheen and a curlicue on the top just barely touching the center of the lid. Shortening on the lid in excess of the size of a dime was considered a serious defect. The shortening can was filled and allowed to solidify on a slowly moving time delay conveyor before seaming on the lid and applying the label to attain the desired product appearance.

The standard 3-pound can has gone through several economic changes to reduce the cost of the finished product. First, the expensive tear strip and key was eliminated and replaced with a plastic overcap for the consumer to reclose the can after removing the lid with a standard can opener. Next, the tin plate in the body of the can was replaced by either a fiber-wound composite material or plastic. The lid and bottom of the composite can and the lid of the plastic can remained tin plate, requiring the consumer to open the product with a can opener. Now, the metal lid has been replaced with a thin foil tear off top. Throughout all of these packaging innovations, the container has retained the customary appearance of a 3-pound can of household shortening.

Household shortenings have also been packaged in other containers, including glass jars, cellophane bags, parchment lined cartons, metal pails, and plastic pails or tubs. Most of these packages, with the exception of the glass jar, are still used for shortening products, at least on a regional basis. The most recent packaging introduction for consumer shortening is the shortening stick. Packaging for this product has changed from a parchment wrapped quarter-pound stick similar to the margarine quarters formed on Morpak equipment, to a parchment-wrapped stick in a foil container. The margarine forming and packaging equipment necessitated a product composition change away from the wide plastic range shortening product to a firm product that was not representative of the product concept known as household shortening. The parchment-wrapped stick in a foil container allows a return to the wide plastic range type shortening. The focus of this introduction is that customers have changed to include more consumers with full-time, away-from-the-home employment; singles; and smaller households, which indicates that lessor quantities of shortening are being used in the home.

6. REFERENCES

1. Black, H. C. 1948. "Edible Cottonseed Oil Products," in *Cottonseed and Cottonseed Products,* Alton E. Bailey, ed. New York, NY: Interscience Publishers, Inc., p. 737.
2. Wrenn, L. B. 1995. "Pioneer Oil Chemists: Albright, Wesson," *INFORM,* 6(1): 98.

3. Wrenn, L. B. "Eli Whitney and His Revolutionary Gin," *INFORM,* 4(1): 12–15.

4. Anonymous, 1994. *Food Fats and Oils,* Washington, DC: Institute of Shortening and Edible Oils, Inc., pp. 6–7 and 22.

5. Gupta, M. 1996. "Manufacturing Process for Emulsifiers," in *Bailey's Industrial Oil and Fat Products, Volume 4,* Fifth Edition, Y. H. Hui, ed., New York, NY: A Wiley-Interscience Publication, pp. 594–595.

6. Leo, D. A. 1985. "Packaging of Fats and Oils," in *Bailey's Industrial Oil and Fat Products, Volume 3,* Fourth Edition, T. H. Applewhite, ed., New York, NY: A Wiley-Interscience Publication, p. 311.

CHAPTER 10

Margarine

1. INTRODUCTION

Margarine is a flavored food product containing 80% fat, made by blending selected fats and oils with other ingredients and fortified with vitamin A, to produce a table, cooking, or baking fat product that serves the purpose of dairy butter but is different in composition and can be varied for different applications [1]. Margarine was developed to fill both an economic and a nutritional need when it was first made as a butter substitute. Its growth has resulted because it could be physically altered to perform in many varied applications. There are over 10 different types of margarine produced today, including regular, whipped, soft-tub, liquid, diet, spreads, no-fat, restaurant, bakers, and specialty types, which are packaged in as many different packages. These margarines are made from a variety of fats and oils, including soybean, cottonseed, palm, corn, canola, safflower, sunflower, lard, and tallow. Margarine products cater to the requirements of all the different consumers: retail, foodservice, and food processor.

1.1 Historical

Margarine was developed in 1869 after a prize was offered by Emperor Louis Napoleon III of France for an inexpensive butter substitute. Butter production was lagging far behind demand because of a short supply of milk in all of Western Europe. Large population shifts from farms to factories during the Industrial Revolution had created a demand for butter that the milk supply could not meet, which caused butter prices to escalate. The situation was particularly serious in France because it was experiencing

437

a depression and a war with Prussia was imminent [1]. Attempts had been made to create a butter-like food for years, but a French chemist won the prize the first year it was offered. Hippolyte Megè-Mouriès obtained French Patent Number 86480 for his development, which he named "oleomargarine," a combination of the Greek word for "pearl-like" because it had a pearly luster when crystallizing and the fat source, oleo oil, derived from beef fat [2].

The beef fat for the first margarine was rendered at a low temperature, and then the separated oil was drawn off into trays and left to crystallize. The crystallized beef fat was then wrapped in filter cloths and cold pressed. The soft fraction from this pressing, called oleo oil, was the fat used for the original margarine. The oleo oil was mixed with milk and salt, which was then chilled and churned to solidify the emulsion. The excess water was drained off and the margarine worked into a plastic mass and packed into barrels for distribution and sale [3]. The first margarine was primitive by today's standards. It was very firm and brittle at cold temperatures but nutritionally equivalent to butter and cost only about half as much. It was an extraordinary achievement for the time and increasingly gained favor throughout Europe during the next decade, and accounts show that it was produced in the United States as early as 1874 still principally from oleo oil [4].

Margarine was not economically feasible in the United States initially because of the refining techniques available. However, as these techniques were improved and margarine sales increased, it attracted political attention. Concern from dairy farmers led the 1886 Congress to pass a series of antimargarine laws, which were to last for 64 years. The Oleomargarine Act of 1886, together with subsequent amendments in 1902 and 1930, imposed taxes on white and then on yellow margarine. The act also required that the new table spread be labeled "oleomargarine." Despite the restrictive legislation by the United States Congress and many individual states, there was a slow, but steady, increase in margarine acceptance as manufacturers continually improved the product and economic conditions dictated.

We still occasionally hear or see oleo used, which is actually a misnomer today because oleo oil is used very infrequently as a margarine oil. However, oleo oil continued to be the fat source for margarines until the world population grew to the point that the animal fat supply could not meet the consumer demand for this table spread. This demand was in spite of the growing cattle herds on the Great American Plains, which made the United States the major oleo oil supplier to the world for a period until the discovery of the hydrogenation process. This process, along with the improvements in vegetable oil refining and the use of steam distillation to deodorize, made it possible to transform a liquid oil into a hardened fat suitable for use in margarine. This development gave the margarine manufacturers a wider range of raw materials, which resulted in improvements in the texture and plasticity to make margarine products more acceptable to the consumer.

During the 1920s, margarine quality was considerably improved. Hydrogenated vegetable oils were used to a greater extent but initially as blends with animal fats. Then coconut oil came into wide use and accounted for about half of the oil used for margarine. Coconut oils offered several advantages over oleo oil and the blends that had been used. It could be hydrogenated and processed into firm margarine with the solid and stable shape associated with table spreads but still melt sharply in the mouth more like butter. The oleo oil margarines had high melting points with poor getaway in the mouth. Coconut oil became the preferred margarine source oil to account for about 45% of the total margarine fat requirements in 1920 [5]. During the 1930s depression, farm prices collapsed and farmers received high tariff protection against imported products, which included coconut oil. These excise taxes made the use of coconut oil as the base fat for margarine prohibitive.

The restrictive costs for coconut oil promoted the development of margarines with the available domestic oils: cottonseed and soybean oils. These oils were selectively hydrogenated to produce a relatively steep SFI with a melting point close to body temperature. The products developed did not have the sharp melt and quick getaway as experienced with coconut oil, but the eating character was more like butter and the product spread better than either coconut oil–based margarine or butter at cold temperatures [6].

Nutrition became an issue as early as 1923 when Nucoa brand margarine was the first to be fortified with vitamin A. This practice became universal among margarine producers about 1937. Margarine national advertising claims of nutritional suitability for children led to Federal Trade Commission hearings that lasted for over 4 years. A notable scientist, Dr. Anton J. Carlson, testified that the fats in Nucoa margarine and butter were equally digestible but that the margarine had more of the polyunsaturated fatty acid linolenic acid, which had been established as one of the fatty acids essential for normal growth and skin maintenance [2].

An important milestone for margarine was the promulgation of the Definition and Standard of Identity for Oleomargarine by the Food and Drug Administration in 1941. Further recognition of margarine's food value was the U.S. Department of Agriculture's classification of margarine as one of the items in its Basic Seven food groups. This recognition gave margarine an official identity of its own and removed the "imitation butter" stigma from the product. Then on July 1, 1950, after 28 months of debate, the

Margarine Act of 1950 was passed, which ended the federal margarine tax system that had been in force since 1886. Restrictive state laws also began to be repealed, but it was 1967 before Wisconsin became the last state to repeal a law prohibiting colored margarine, and the final state margarine tax was not repealed in Minnesota until 1975 [6].

The depression and later World War II hardly constituted a proper environment for new product development. However, the war did increase the consumption of margarine as a result of a butter shortage. With the relaxation of government regulations in 1950, the margarine industry was ready for new product developments. The postwar era saw an increase in personal income for a substantial rise in the standard of living. Margarine, which still had the image of an economic, inferior substitute for butter, could have suffered a substantial market share loss. However, the margarine industry reacted to the challenge by offering the consumer a wide variety of products with better quality, flavor, improved packaging, and higher prices with the beginning of the emphasis on nutrition. Some of the notable post–World War II era margarine developments were [2,3,6]

- 1947: Coal-tar oil-soluble dyes were replaced with carotene, a form of vitamin A, for coloring margarines. Later, the use of coal-tar dyes were prohibited for food use in favor of natural coloring materials, which for margarine consist primarily of carotene extracts, red palm oil, and annatto.
- 1950: Aluminum foil inter-wrap for margarine quarters offered more product protection from oxidation than parchment and provided a quality image.
- 1952: A softer, cold spreadable stick margarine was introduced.
- 1952: Soft whipped margarine in a tub was introduced and withdrawn due to poor reception.
- 1956: A premium margarine with a lower melting margarine base oil with butterfat as a flavoring agent, which required refrigeration, was successfully introduced.
- 1957: A unique new process for a whipped margarine in stick form, packaged six sticks to a pound, was developed.
- 1958: Corn oil margarine was introduced, which successfully capitalized on the corn oil nutritional image.
- 1959: Additional corn oil margarines introduced had high polyunsaturated fatty acid levels.
- 1962: The first soft margarine in a table service plastic container made from safflower oil with a high P/S ratio was an immediate success upon introduction.

- 1963: Liquid margarine in a squeezable plastic container was introduced.
- 1964: Diet margarine containing half the calories of regular margarine was introduced. The U.S. FDA questioned the legality of the product but lost an ensuing court case, which confirmed diet margarine as a table spread product.
- 1965: Soft soybean oil margarine was introduced in an aluminum cup.
- 1968: Soft margarine was introduced in a decorated plastic container.
- 1975: There was a 60% fat content product introduced that could not be identified as margarine, which must contain 80% fat, or diet margarine, which must contain 40% fat or half the calories of regular margarine.
- 1978: Whipped spread in a 2-pound container was introduced.
- 1981: Margarine butter blends containing as much as 40% to as low as 5% butter were introduced by several margarine and butter processors. These premium products, priced between butter and margarine, have a more definable butter flavor, with the improved spreadability and health benefits of margarine.
- 1989: Lower fat content spreads of less than 20% were introduced.

During the 1950s, a series of developments in nutritional and medical science opened a new aspect of the margarine business. Studies of the diets of other countries, animal research, and dietary experiments began to make the biochemical word *cholesterol* familiar to the public. These reports concluded that fats high in polyunsaturated fatty acids would lower serum cholesterol levels [8–10]. The margarine industry reacted to the nutritional issues by introducing products containing source oils identified as high in polyunsaturates and formulation changes to provide the maximum level of the essential fatty acids. Table 10.1 tracks the changes in the source oil utilization for U.S. margarines from 1950 through 1994 [11–13]. Soybean oil has been the dominate source oil since 1951, probably due to economics and a high polyunsaturated fatty acid content. The appearance of corn oil during the 1950s, with a real increase during the 1960s, indicates the acceptance of the premium nutritional margarines. The other vegetable oils listed indicate premium entrees that had some success but were later changed to another source oil or discontinued. Safflower oil was an example of the discontinued usage in favor of another source oil, in this case, soybean oil. Animal fats have dropped from the only fat source to a very limited usage in low-cost products. This trend has continued with the increased consumer awareness of cholesterol and saturated fats.

Source oil changes were somewhat minor in comparison to the formulation changes in reaction to the nutritional concerns. Blends of two or

TABLE 10.1. U.S. Margarine and Spread Source Oil Usage
(millions of pounds).

Source Oil	1950	1960	1970	1980	1990	1995
Corn oil	1	68	185	223	208	NA
Cottonseed oil	513	168	68	25	NA	NA
Safflower oil		13	22			
Soybean oil	382	1370	1410	1653	1749	1684
Animal fats	16	76	99	104	35	41
Unidentified			10	11	110	122
Total	937	1695	1794	2016	2102	1847

NA = not available.

more hydrogenated basestocks, one of which was a very selectively hydrogenated basestock capable of creating a "hump" at the 70°F (21.1°F) SFI determination, were the standards for stick margarines before the high polyunsaturated fatty acid levels necessity became important. Formulation for high polyunsaturates levels, but still meeting the other stick-type margarine requirements, was accomplished by blending selectively hydrogenated firm basestocks with liquid oils that had not been hydrogenated.

The soft-tub margarines introduced in 1962 were formulated with higher levels of polyunsaturates than stick margarines initially and even higher levels were possible. High-liquid oil levels can be used because of the package, and a softer product consistency is expected. The soft-tub product does not require the firm print consistency necessary for the stick margarines. The tub package also affords the product more protection than a parchment or foil wrap to allow more liquid oil of a lesser resistance to oxidation. The immediate success of the soft-tub margarine is illustrated in Table 10.2. This margarine type captured over 25% of the consumer market in less than 10 years after introduction.

The per capita tablespread consumption history plotted in Chart 10.1 indicates that margarine volume surpassed that of butter for the first time in 1957. However, margarine has not achieved the high personal usage butter enjoyed prior to 1940. The overall trend for table spread consumption is decreasing. Current nutritional concerns regarding fat consumption have certainly played a major role in reducing the per capita usage of the tablespread products. This concern was a major factor for the development, introduction, and acceptance of reduced fat spread products.

Table 10.2 identifies the shift in consumer table spread popularity to low-fat spreads [1,14]. The spreads market share increased from less than 5% in 1976 to more than 74% in 1995. The reasons for this growth were

TABLE 10.2. Consumer Margarine/Spread Market Share.

Year	Margarine, % Stick Forms	Soft-Tub	Spread, % (All Forms)	Total Pounds, Millions
1945	100			614
1950	100			937
1955	100			1334
1960	100			1695
1965	100	NR		1904
1970	79	21		1975
1975	75	25		2133
1980	63	24	13	2226
1985	58	16	26	2155
1990	50	9	41	2213
1995	19	7	74	1587

NR = not recorded.

low price, availability in sizes greater than 1 pound, and fewer calories from a lower fat content than margarine, combined with an acceptable flavor, mouth feel, melt characteristics, and other functional attributes. The first spreads were introduced at 60% fat, but the majority are now produced with 52% fat or less but still utilize the regular margarine base oils. Packaging for spreads has been in all the familiar margarine forms: stick, soft-tub, soft whipped, and so on. The spread products' popularity continues to grow, as shown by the market share results in Table 10.2, even though the high moisture content produces some serious defects in comparison to margarine or butter, i.e., poor baking performance and a less creamy eating sensation. Obviously, the benefit of a low-fat content for a reduced calorie content and a perceived healthful composition or a lower unit cost are attractive to the consumers. Some spread producers have carried this concept even further to produce consumer table spread products with less than 20% fat.

U.S. consumer margarine and spread sales began to decline after 1982; demand for these products has decreased almost 22% over the 1982–1995 period. Table 10.3 market share evaluations [14] verify the reduced consumer product sales but indicates that the retail consumer sales were countered by foodservice demand improvements through 1985, which indicated that consumers were eating away from home more often. However, even with continued foodservice demands, the overall demand for margarine and spread products has declined due to substantial consumer sales decreases.

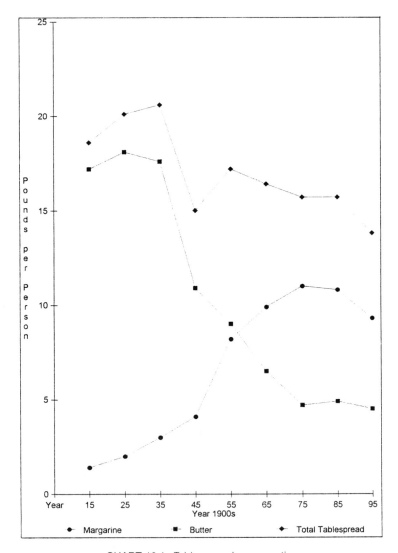

CHART 10.1. Table spread consumption.

TABLE 10.3. Margarine/Spread Markets.

	Year					
	1970	1975	1980	1985	1990	1995
Market share, %						
Consumer	88.7	89.1	86.0	82.8	80.1	72.4
Foodservice	3.4	4.0	6.5	9.0	14.6	19.5
Food processor	7.9	6.9	7.5	8.2	5.3	8.1
Market share, MM-lb						
Consumer	1,975	2,133	2,226	2,155	2,213	1,587
Foodservice	34	101	138	193	402	426
Food processor	173	165	194	213	146	177
Total	2,182	2,399	2,558	2,561	2,761	2,190

The food processor margarine and spreads appear to be declining somewhat but at a much slower rate than consumer products. These trends indicate that the U.S. consumer is aware of the nutritional message to reduce fat intake. Consumer margarine and spread products represent visible fats that are controlled by the homemaker.

2. MARGARINE FORMULATION

Most countries that allow the sale of margarine have laws to regulate its composition. The United States has two standards of identity for margarine, which are similar but not identical. Vegetable oil margarines are regulated by the U.S. Food and Drug Administration (FDA), and margarine made with meat fats are regulated by the United States Department of Agriculture (USDA). Margarine is defined as a plastic or liquid emulsion food product containing not less than 80% fat and 15,000 international units per pound of vitamin A. All products containing less than 80% fat are required to be labeled as spreads, except that product with 40% or less fat should be identified as light, or lite, margarine [15]. It may contain one or more aqueous-phase ingredients and one or more optional ingredients with specific functions.

2.1 Milk Products or Protein

The optional ingredients of the aqueous phase include water, milk, or milk products. Traditionally, this was added as a cow's milk product but now can be water with or without an edible protein component. The suitable edible protein components include whey, albumin, casein, caseinate, or soy protein isolate in amounts not greater than reasonably required to accomplish the desired effect. The factors controlling the protein choice are primarily flavor considerations and, to a lesser extent, performance in frying and cooking similar to butter. Initially, the margarine standards promulgated in 1941 required that the aqueous system contain 10% milk solids; however, that requirement has been removed, and most margarines produced with milk or another protein contain considerably lower levels, usually less than 1.5%. A high level of the margarine products produced currently are milk-free, which presents less microbiological problems and helps to conform to the requirements of several religious diets.

2.2 Emulsifiers

Several different food grade emulsifiers are allowed as optional ingredients for margarine. Emulsifier systems are used to hold the fat and the water phase together and impart specific performance characteristics to the finished product. Originally, only lecithin and mono- and diglycerides were allowed at limited use levels, but the standards were opened to any surfactant with a generally recognized as safe (GRAS) designation at a level required to provide the desired effect or within the restrictions for the specific surfactant. Nevertheless, consumer margarines normally still rely on a two-component emulsifier system of lecithin and mono- and diglycerides. Lecithin is usually added at levels of 0.1 to 0.2% for its antispattering properties and natural emulsifying properties, and it helps effect a quicker salt release in the mouth. Mono- and diglycerides are added to most margarines for emulsion stability or protection against weeping, usually at levels below 0.5% alpha monoglyceride. Hard mono- and diglyceride products provide a tight emulsion that can effect a delayed flavor release, while the soft mono- and diglycerides provide a looser emulsion for a quicker flavor release. Foodservice and food processor margarines formulated for specific applications may contain different emulsifier systems and levels suited to the functionality requirements of the finished product.

2.3 Flavoring Materials

Salt, which is sodium chloride, or potassium chloride for sodium-free dietary margarines, is added for flavor and also acts as a preservative. Flavoring substances approved for food use and nutritive carbohydrate sweeteners or sugars are considered optional ingredients by the margarine standards. Many synthetic butter flavors are available for use in margarine. These are usually based on mixtures of compounds that have been identified as contributing to the flavor of butter, such as lactones, butyric fatty acid esters, ketones, and aldehydes. Diacetyl, a primary constituent of many butter flavors, is formed in butter during the culturing process at concentrations of 1 to 4 ppm.

Flavor perception is influenced by mouth feel, which is determined by the rate at which a margarine product melts in the mouth. Mouth feel is controlled by the melting characteristics of the fat portion and the tightness of the emulsion, which is a function of the emulsifier, protein, and stabilizer selections, as well as the processing techniques. Ideally, these characteristics should be balanced to allow the margarine to melt in the mouth and release the flavors to provide a pleasant eating sensation, rather than a pasty, waxy feeling that masks the flavor system or an immediate release that is overpowering and of short duration.

Bakery margarine's flavor and color are the primary differences between margarine and shortening for baking applications. The flavor level for baker's margarine is usually stronger and must be more heat-stable than consumer margarines formulated for table use. Some commercial flavor compounds available contain butyric acid and/or lactones to improve the buttery flavor in baked products. Most baker's margarines are also formulated with higher salt levels, usually 3.0% versus the 1.5 to 2.0% used in consumer products.

2.4 Preservatives

Preservatives are also optional ingredients permitted by the margarine standard of identity. The meaning of preservative is an additive to protect against spoilage or deterioration, but there are several categories of possible deterioration. Margarine preservatives fall into three categories: antimicrobial, antioxidants, and metal scavengers. The standards list sorbic acid and benzoic acid and their sodium and calcium salts and allow use levels of 0.1% individually or 0.2% in combination. These compounds protect margarine against microbial spoilage. Benzoic acid is more active against bacterial action while sorbic acid gives better protection against yeast and mold. Salt is also a preservative although it is usually added for flavor enhancement. The salt level in most margarines ranges from 1.5 to 3.0%. Since the moisture level in margarine usually ranges from about 16 to 19%, the salt concentration in the aqueous phase is 8 to 19%. In most cases, the concentration of salt in the water phase will provide sufficient antimicrobial activity, but without other preservatives or acidulents, mold can develop.

Propyl, octyl, and dodecyl gallates, BHT, BHA, ascorbyl palmitate, and ascorbyl stearate are permitted individually or in combination at a maximum level of 0.2%. These materials are antioxidants for the fats and oils component. Antioxidants may be necessary for the oxidative stability of products formulated with meat fats but are not usually added to vegetable oil margarines. Vegetable oils contain natural antioxidants, tocopherols, that survive processing to provide adequate oxidative stability for most margarine products.

Lecithin, both ascorbyl palmitate and stearate; isopropyl citrate; and calcium disodium EDTA all act as antioxidant synergists. A synergist performs two important functions: (1) it increases the antioxidant effectiveness, and (2) it ties up or chelates the trace metals that are oxidative catalyst. EDTA is also effective as an agent to retard oxidative bleaching of the carotenoid colorants used in margarine [16].

2.5 Vitamins and Colors

Fortification of all margarine products with vitamin A is mandatory; it must contain not less than 15,000 international units per pound. The use of vitamin D is optional, but when added, it must be at a minimum level of 1,500 international units per pound of finished margarine. Vitamin E addition is specially excluded under the standards of identity. However, the natural antioxidants in vegetable oils, tocopherols, are major sources for vitamin E, and variable amounts survive in processed margarines [17]. The mandatory vitamin A level for margarine is usually attained by the addition of beta-carotene for colored margarines, with vitamin A esters used to adjust for the required potency. The colorless vitamin A esters are used for all the uncolored margarines. Natural extracts containing carotenoids like annatto, carrot oil, and palm oil have also been used to color margarines. Apocarotenal is a synthetic pigment that is used primarily as a color intensifier for beta-carotene.

2.6 Margarine Oils and Fats

The physical and functional aspects of a margarine product are primarily dependent upon the characteristics of the major ingredient—the margarine oil, or marbase. Margarine consistency, flavor, and emulsion stability depend upon crystallized fat. In the United States hydrogenation is the preferred process utilized to change the solids/liquid relationship of margarine basestocks. A direct relationship exists between the fat solids content and the structure, consistency, and plasticity of the finished margarine. SFI values at 50, 70, and 92°F (10.0, 21.1, and 33.3°C) are utilized by most margarine manufacturers in the United States for margarine consistency control. The SFI values are indicative of the crystallization tendencies and the finished product quality, as shown in Table 10.4.

TABLE 10.4. *Consumer Margarine Solids Fat Index Effect upon Product Characteristics.*

Solids Fat Index		Characteristic Affected	Solids/Liquid Relationship Influence
Temperature	% Solids		
10.0°C/50°F	10 to 28	Spreadability	Optional range for spreadability at refrigerator temperatures: Too high = Firm, brittle, nonspreadable
		Printability	Low SFI = Colder chilling temp. required High SFI = Channeling possible with cooling
21.1°C/70°F	5 to 18	Consistency	Body and resistance to oil separation Too high = Brittle, firm, poor spreadability Too low = soft, soupy, oil separation
33.3°C/92°F	3.5 max.	Mouth feel	Quality consumer table grade margarine Melts quickly with a cooling sensation Too high = Lingering pasty, greasy, waxy sensation due to coating of the palate

CONSUMER MARGARINE OIL FORMULATIONS

Consumer margarines are formulated by blending two or more base-stocks with different degrees of hardness. This permits the margarines to be spreadable directly out of the refrigerator and to maintain a solid consistency at room temperature. Hydrogenated cottonseed oil was a component of most all-vegetable oil margarines to induce a beta-prime crystal habit to prevent graininess for quite some time. However, with the development of more spreadable margarines, the use of multiple basestocks and uniformly low cold storage temperatures has reduced the transition of 100% soybean and corn oil margarines to the beta crystal form, which causes sandiness.

The hydrogenated vegetable oil basestocks best suited for table grade products have steep SFI slopes to provide the desired eating, melting, and nonoiling physical characteristics along with machinability. The soybean oil basestocks previously presented in Table 4.7 in Chapter 4 included products that may be used to produce margarine oils as well as shortenings.

Margarines have been able to conform to the increased health consciousness of the U.S. consumer to increase the polyunsaturated fatty acids and liquid oils levels. Although polyunsaturate level is less emphasized now, apparently it is still perceived as healthy by the consumer. Therefore, it is still advantageous to have a high polyunsaturate-to-saturates ratio contributed by a high liquid oil level.

Table 10.5 identifies the compositions and SFI values for seven table grade marbases; three are for stick products, two are soft-tub types, and two are for liquid or squeezable margarines. The soft stick marbase probably represents the softest stick-type margarine that can be packaged successfully. The present-day packaging equipment will deposit wrap and carton product that is softer, but the margarine could not withstand the normal abuse after packaging in storage and distribution. This product was introduced after the soft-tub products in an effort to take advantage of the soft concept in the old familiar stick package.

The all hydrogenated stick table grade product represents the type that was the main consumer margarine product for quite some time. It is still the most preferred table fat product for baking and some cooking due to better oxidative stability. This product also prints well due to the high 50°F (10.0°C) SFI, especially in equipment that doesn't deposit the margarine into preformed quarter's parchment. The high liquid oil stick marbase represents the majority of the stick margarine production. The high liquid oil in the formula provides a marketing claim but reduces the finished margarine's oxidative stability and therefore shelf life. The surface of the margarine develops a darker color from oxidation, which is quite noticeable when the surface layer is scraped away during use.

Soft-tub marbase compositions are somewhat like a compound shortening formulation: a blend of a soft and a hard basestock. The hard basestock cannot be as saturated as the low iodine value hard fats used for shortenings and should have a steep SFI slope for good eating characteristics. Two different soft-tub compositions are shown in Table 10.5. The all hydrogenated product provides the oxidative stability and the firmest margarine consistency. Slight consistency differences can be the difference between a soft-tub margarine with a picture perfect surface appearance or excessive "lid slosh." The all hydrogenated marbase has a better chance of retaining a smooth surface because it should set quicker than the product formulated with liquid oil.

Liquid margarine has been marketed for quite some time, but it has never achieved significant consumer acceptance. Food processors have accepted and utilized this product for specialty applications more so than consumers. Liquid margarine has been prepared using both beta and beta-prime type hard fats. Beta-prime hard fats have been found more suitable

TABLE 10.5. Table Grade Marbases Composition and Solids Fat Index.

	Stick Marbases			Tub Marbases		Liquid Marbases	
	Soft Stick	All Hydro	High Liquid Oil	All Hydro	High Liquid Oil	CSO Hard Fat	SBO Hard Fat
Basestock							
109 IV H-SBO		42		82	32		
85 IV H-SBO							
74 IV H-SBO		20	25				
66 IV H-SBO	50	38	15				
60 IV H-SBO				18	18		
63-T H-SBO							4
60-T H-CSO						5	
Liquid soybean oil	50		60		50	95	96
Solids Fat Index, %							
at 50°F/10.0°C	22.0±	28.5±	30.0±	12.0±	11.0±	7.0±	3.0±
at 70°F/21.1°C	13.5±	17.5±	17.0±	7.5±	5.5±	6.0±	2.5±
at 92°F/33.3°C	2.0±	3.0±	1.5±	3.0±	0.7±	5.5±	2.0±

for the product packaged without a tempering or crystallization process stage. Product prepared with beta crystal forming hard fats requires tempering of the supercooled mixture at an elevated temperature for a period of time under agitation to develop and stabilize the beta crystal. This liquid margarine process resembles the liquid shortening process closely. The beta crystal formulation and procedure produces a more fluid, less viscous product with better suspension stability than the beta-prime product but costs more to produce [18]. The beta-prime hard fat direct process provides a better mouth feel and flavor but requires constant refrigeration to avoid separation.

INDUSTRIAL MARGARINE OIL FORMULATIONS

Foodservice and food processor margarines and spreads are considered industrial products. These products are formulated or packaged for more specific applications than the consumer products. The most popular foodservice margarine is the stick margarine formulation packaged in 1-pound solids used for food preparation. Individual serving or portion control soft spread products are popular foodservice dining room products. Food processor products are formulated for more specific uses than either the consumer or foodservice margarine or spread products. The major difference that affects the desired functionality is the marbase composition.

One type of food processor margarine for foodservice and food processor application is simply table grade margarine, which utilizes the stick margarine oil, filled into a larger container. For food processor roll-in applications, the product is plasticized like a shortening instead of the margarine print procedure. Baker's margarine is designed to have a wide plastic range with good creaming properties like the standard all-purpose shortening. In fact, the marbase for baker's margarine can be the same composition as the all-purpose shortening described in Chapter 6: a 80 to 85 IV basestock with a fully hydrogenated beta-prime crystal forming hard fat. Special roll-in margarines for Danish and other pastries are prepared with fat blends similar to the anhydrous product reviewed in Chapter 6. Margarine formulations provide a buttery flavor and color to the finished product, as well as moisture to produce steam during baking to help expand the dough layers and improve flakiness. Several other specialty products also rely upon the marbase formulation for functionality and the margarine process to provide flavor, color, and moisture for functionality.

3. SPREAD FORMULATIONS

All products resembling margarine that contain less than 80%, but more than 40%, fat are required to be labeled as spread. However, these products must conform with the margarine standard in all respects, except for the fat content and that safe and suitable ingredients not provided for in the standard may be added to improve functional characteristics so that the spreads are not inferior to margarine. Soft-tub and stick spreads containing 40 to 75% fat are usually formulated from the same fat and oil blends as those used for regular consumer margarines. The other ingredients utilized are also basically the same with the following exceptions [6,19]:

- Milk protein acts as an oil-in-water emulsifier. Consequently, the use of milk, casein, or caseinates can result in a phase reversion. Therefore, most spread formulations are milk- and milk protein–free.
- Emulsifier levels are increased slightly to improve the physical characteristics of the emulsion and its stability, typically 0.4 to 0.6% alpha monoglyceride levels with mono- and diglyceride use. Soft mono- and diglycerides are essential in protein-free spreads. Mono- and diglyceride and polyglycerol emulsifier systems have been found effective in spreads containing significant quantities of protein.
- Lecithin use in low-fat spreads may decrease the emulsion stability and increase the tendency to oil off; however, it also functions to slow the emulsion breakdown in the mouth. Therefore, the use of lecithin and the level of use must be evaluated for each formulation.
- Gelatin, pectin, carrageenans, agar, xanthan gum, starch alginates or methylcellulose derivatives are gelling or thickening agents used in some spreads to improve the body.
- The higher emulsifier levels used for spread can produce tighter emulsions, and the gelling or thickening agents can affect the rate and order that flavors are perceived. The flavor content and types must be defined to produce oral responses similar to the high-fat products.
- Preservatives are more important in spreads than in regular margarines because of the higher moisture content.
- The light reflection of a spread is different from that of regular margarine because of the increased number of water droplets present. Therefore, it is necessary to add about twice the amount of color used in normal margarine to obtain the same color intensity.

4. MARGARINE AND SPREAD PREPARATION

Margarine and spread processing resembles shortening preparation, except that an emulsion is prepared by mixing the water-soluble and fat-soluble ingredients before chilling and crystallization. The basic steps for margarine and spread preparation are (1) marbase blending, (2) lipid-phase preparation, (3) aqueous-phase preparation, (4) emulsion preparation, (5) chilling and crystallization, (6) packaging, (7) tempering, (8) storage, and (9) distribution. The same procedures work well with both margarine and spread products containing from 52 to 80% fat. However, as the fat level decreases below 52%, strict process control for temperature and addition rates must be employed to guard against the formation of a mixture of W/O and O/W emulsions or an all O/W emulsion [19].

The lipid phase is prepared by heating the marbase to a temperature of at least 10°F (5.6°C) above the melting point before adding the oil-soluble ingredients. These ingredients include the emulsifiers, vitamins, color, and flavor if it is oil-soluble. The aqueous phase is prepared separately. When milk proteins are utilized, the milk or whey product is reconstituted if liquid skim milk is not used, pasteurized, and cooled until required. The water-soluble ingredients are added to the required amount of pasteurized skim milk, whey, or water in the case of protein-free products. The ingredients include salt, preservatives, water-soluble flavors, and any other water-soluble materials. The oil and aqueous phases are blended together with high-shear agitation to form a water-in-oil emulsion. Continuous agitation is necessary to maintain this emulsion; without agitation, the water- or milk-phase droplets immediately begin to coalesce and settle out. After preparation, the emulsion is transferred to an agitated holding tank that supplies the chilling units.

An alternative continuous emulsion preparation process consists of proportioning pump systems capable of metering individual ingredients of the aqueous phase together and also the oil-phase components. In-line static mixers are utilized to blend the separate phases, which are then mixed in-line and emulsified with another static mixer. After in-line blending, the loose emulsion is continuously fed to the crystallization system.

The solidification process for the various margarine and spread types all employ a scraped surface heat exchanger for rapid chilling, but the other steps are somewhat different than for shortening and the other margarine types. The solidification or plasticization process for the four basic table-spread types are as follows.

4.1 Stick Margarine or Spread

The temperature of the stick margarine emulsion is adjusted and maintained for most table grade products that melt below body temperature at 100 to 105°F (37.8 to 40.6°C) before pumping to the scraped surface heat exchanger. If the temperature is allowed to cool below the melting point of the marbase, precrystalization structures may be formed that affect the consistency of the finished product. It is rapidly chilled to 40 to 45°F (4.4 to 7.2°C) in less than 30 seconds. Stick margarine requires a stiffer consistency than shortening, which is accomplished with the use of a quiescent tube immediately after the chilling unit. This is a warm-water-jacketed cylinder that can contain baffles or perforated plates to prevent the product from channeling through the center of the cylinder. The length of this tube may have to be varied to increase or decrease crystallization time, depending upon the product formulation. The super-cooled mixture passes directly to a quiescent resting or aging tube for molded print forming equipment. For filled print equipment, a small blender may be utilized prior to the resting tube to achieve the proper consistency for packaging and a slightly softer finished product. A remelt line is necessary because in all closed filler systems some overfeeding must be maintained for adequate product to the filler for weight control. The excess is pumped to a remelt tank and then reintroduced into the product line.

Two types of stick margarine forming and wrapping equipment are in use in the United States, i.e., molded and filled print. The molded print system initially used an open hopper into which the product was forced from an aging tube through a perforated plate in the form of noodles. The margarine noodles were screw fed into a forming head and then discharged into the parchment paper wrapping chamber and finally cartoned. Closed molded stick systems now use a crystallization chamber instead of the aging tube and open hopper arrangement, which fills the mold cavity by line pressure. The filled print system accepts margarine from the quiescent tube with a semifluid consistency. It is filled into a cavity prelined with the parchment or foil interwrap. The interwrap is then folded before ejecting from the mold into the cartoning equipment.

4.2 Soft-Tub Margarine or Spread

The oil blends for soft-tub margarines or spreads are formulated with lower solids to liquid ratios than the stick-type products to produce a spreadable product directly out of the refrigerator or freezer. Crystallization technique contributes to the desirable consistency as well, but the products are too soft to print into sticks. Therefore, packaging in plastic tubs or cups with snap-on lids is utilized.

To fill the container properly, the soft margarine or spread consistency must be semi-fluid like shortening. Therefore, the crystallization process resembles shortening plasticization more closely than stick margarine processing. The temperature of a typical soft margarine emulsion would be adjusted and maintained at 95 to 105°F (35.0 to 40.6°C) before pumping with a high-pressure pump to the scraped surface heat exchanger. Creaming gas or nitrogen, added to further improve spreadability, is injected at the suction side of the pump at 8.0% for the most spreadable product, and lower levels for a firmer product.

Margarine blends are rapidly chilled to an exit temperature of 48 to 52°F (8.9 to 11.1°C). Spread fill temperatures are higher than margarine products because the emulsion is more viscous. If the fill temperature is too low, the product will mound in the bowl for excessive lid coverage, and the product may become dry and crumbly. The supercooled margarine or spread mixture then passes through a worker unit to dissipate the heat of crystallization. The shaft in the worker unit revolves at about 35 to 50 rpm with approximately 3 minutes residence time. Worked product is then delivered to the filler where it is forced through an extrusion valve at pressures in the range of 300 to 400 psi. Either a rotary or straight line filler may be used to fill the tubs with margarine. The excess product necessary for a uniform supply to the filler is transferred to a remelt tank and eventually reenters the solidification system.

4.2 Whipped Tub Margarine or Spread

The same equipment used to prepare, crystallize, and package regular soft-tub margarine or spread can be utilized for whipped tub products. The difference during crystallization is the addition of 33% nitrogen gas by volume for a 50% overrun. The nitrogen is injected in-line through a flow meter into the suction side of the pump. Larger tubs, required for the increased volume, necessitate change parts for the filling, lidding, and packaging equipment.

4.3 Liquid Margarine

Both retail and commercial liquid margarines can be crystallized with the same equipment and process flow used for soft-tub products, depending upon the formulation and suspension stability requirements. Liquid margarine formulations normally consist of a liquid vegetable oil stabilized with either a beta or beta-prime forming hard fat. Beta-prime stabilized liquid margarine can be prepared using the same rapid crystallization process used to prepare soft-tub margarines omitting the addition of creaming gas. These finished products require refrigerated storage for suspension stability.

Liquid margarines formulated with beta forming hard fats and some processed with beta-prime formulations incorporate a crystallization or tempering step to increase fluidity and suspension stability. This crystallization step consists of a holding period in an agitated jacketed vessel to dissipate the heat of crystallization. The product may be filled into containers after the product temperature has stabilized. Some products with higher solids to liquid ratios are further processed to stabilize the fluid suspension with either homogenization or a second pass through the scraped surface heat exchanger before filling. The additional processing for more stable crystallization will increase fluidity and suspension stability, but the increased production costs may not be justifiable.

4.4 Industrial Margarines or Spreads

Foodservice and food processor margarines and spreads may be either duplicates of the retail products in larger packages or designed for a specific use, either product- or process-related. Among the specific use margarines, puff paste or danish pastry applications are the most difficult with regard to crystallization. The characteristic features of a roll-in margarine are plasticity and firmness. Plasticity is necessary because the margarines should remain as unbroken layers during repeated folding and rolling operations. Firmness is equally important because soft and oily margarine is partly absorbed by the dough, destroying its role as a barrier between the dough layers. As with shortenings, the ultimate polymorphic form for roll-in margarines is determined by the triglyceride composition, but the rate at which the most stable form is reached can be influenced by mechanical and thermal energy. Therefore, the customary crystallization process for roll-in and/or baker's margarines is a duplicate of the shortening plasticization process depicted in Chart 2.10 in Chapter 2, except that margarines containing water or milk in emulsion form are normally not aerated. The aqueous phase of a margarine emulsion has the same effect as gas incorporation on appearance and performance. Food processor margarine products are usually packaged in 50-pound corrugated fiberboard cartons, 5-gallon plastic pails, 55-gallon drums, or special packaging designed for each specific use.

5. REFERENCES

1. Riepma, S. F. 1970. *The Story of Margarine,* Washington, D.C.: Public Affairs Press, pp. 5, 109, 148, and 152.
2. Melnick, D. 1968. "Margarine Products," *J. Home Economics,* 60(10): 793–797.
3. Massiello, F. J. 1978. "Changing Trends in Consumer Margarines," *J. Am. Oil Chem. Soc.,* 55(2): 262–265.
4. Schwitzer, M. K. 1956. *Margarine and Other Food Fats,* New York, NY: Interscience Publishers, Inc., p. 64.
5. Stuyvenberg, J. H. 1969. *Margarine: An Economic, Social and Scientific History 1869– 1969,* Liverpool, England: University Press, p. 233.

6. Chrysam, M. M. 1996. "Margarines and Spreads," in *Bailey's Industrial Oil and Fat Products, Volume 3*, Fifth Edition, Y. H. Hui, ed. New York, NY: A Wiley-Interscience Publication, pp. 67–68 & 92–96.

7. Miksta, S. C. 1971. "Margarines 100 Years of Technological and Legal Progress," *J. Am. Oil Chem. Soc.*, 48(4): 169A–180A.

8. Anonymous, 1958. "The Role of Dietary Fat in the Human Health," in *Publication 575 National Academy of Sciences*, Washington, D.C.: Food and Nutrition Board, National Academy of Sciences, p. 56.

9. Anonymous, 1961. "Dietary Fat and Its Relation to Heart Attacks and Strokes," in *Circulation 23, American Heart Association*, p. 133–136.

10. Jolliffe, N. 1957. "Fats, cholesterol and coronary heart disease," *NY State J. Med.* 57: 2684–2691.

11. *Fats and Oils Situation*, April, 1971, Economic Research Service, USDA, p. 27.

12. *Fats and Oils Situation*, January, 1981, Economic Research Service, USDA, Table VIII.

13. *Oil Crops Yearbook*, October, 1996. Economic Research Service, USDA, pp. 57–58.

14. *Margarine Statistics Reports*, 1960 through 1995. National Association of Margarine Manufactures, Washington, D.C.

15. 21 CFR 101.56. 1994. in *Code of Federal Regulation*, Washington, D.C.: The Office of the Federal Register, National Archives and Records Administration.

16. Weiss, T. J. 1983. "Margarine," in *Food Oils and Their Uses*, Second Edition, Westport, CT: AVI Publishing Co., Inc., p. 190.

17. Slover, H. T. et al. 1985. "Lipids in Margarines and Margarine-like Foods," *J. Am. Oil Chem. Soc.*, 62(4): 781.

18. Weiderman, L. H. 1978. "Margarine and Margarine Oil, Formulation and Control," *J. Am. Oil Chem. Soc.*, 55(11): 828.

19. Moustafa, A. 1995. "Consumer and Industrial Margarines," in *Practical Handbook of Soybean Processing and Utilization*, D. R. Erickson, ed. Champaign, IL and St. Louis, MO: AOCS Press and United Soybean Board, pp. 352–353.

CHAPTER 11

Liquid Oils

1. INTRODUCTION

Cooking, salad, and high-stability oils are prepared from vegetable oils that are refined, bleached, and deodorized at a minimum but can also require modification by dewaxing, winterization, hydrogenation, and/or fractionation processes. All of these oil types have one common physical property—all are clear liquid oils at room temperature. The three types of liquid oils may be defined as

- cooking oil—an edible oil that is liquid and clear at room temperature that may be used for cooking. Cooking oils are used for pan frying, deep fat frying, and in packaged mixes or wherever a clear, liquid oil has application without refrigeration. Cooking oils may solidify at refrigerator temperatures.
- salad oil—an edible oil that will remain substantially liquid at refrigerator temperatures, about 40°F (4.4°C). The standard evaluation method for a salad oil is the cold test analysis. A salad oil sample that remains clear after 5½ hours while submerged in an ice bath meets the criterion for a salad oil. Salad oils may be used for cooking but are low in saturated fatty acids to prevent or delay clouding and graining during refrigeration.
- high-stability oil—an edible oil that is liquid at room temperature, with exceptional oxidative or flavor stability. High-stability oils are used for food preparation applications where the functional characteristics of a liquid oil are desired and a long shelf life is required.

459

1.1 Historical

Climate and availability influenced the eating habits of our ancestors. Inhabitants of central and northern Europe first derived their edible fats almost entirely from wild animals and later domestic animals. Consequently, their foods developed around the use of solid fats like butter, lard, and tallow. In the heavier populated areas of Asia, southern Europe, and Africa where it was impractical to dedicate land to livestock grazing, diets were developed with vegetable oils as the major fat source. All oil-bearing plants produce oils that are liquid at the prevailing temperatures where the plant grows. Even the tropical oils that are solids at room temperature in the cooler climates are liquid oils in the tropical climates where these plants flourish. North America's preference for animal fats was influenced by the early immigrants from northern Europe who adapted large sections of the country to raising domestic animals [1].

Fats and oils processors in the United States developed the differentiation between cooking and salad oils. The original intent was to retain fluidity and clarity for cottonseed oil during the colder months; oil fluidity and clarity were not problems at the warmer summer temperatures. Domestic ice boxes and, later, refrigerators, along with large-scale commercialization of the mayonnaise and salad dressing industry, increased the need for a liquid oil that remained clear at reduced temperatures. Initially, salad oil was referred to as "winter oil," which indicated the fractionation process used to produce the cloud-free oil at cool temperatures. The winterization process evolved from practical experience at refineries when it was observed that storing cottonseed oil in outside tanks exposed to low temperatures during the winter months allowed the higher melting triglycerides to settle to the bottom of the tank, leaving a top layer of clear oil. The top layer was decanted, deodorized, packaged, and marketed as a salad oil. The topping process did not provide consistent results because it depended upon nature to provide the coolant, and as popularity increased, it became impractical to maintain the quantity of oil in storage tanks at the volume and time required. Therefore, mechanically refrigerated chilling and filtering systems were developed to simulate the outside storage and topping process. Salad oil terminology came about because of the winter oil's application for mayonnaise preparation. At the same time unwinterized oil, which had been known as "summer oil," became cooking oil [2,3].

Salad oil identification later changed from an oil processed by the winterization process to an oil that can meet the requirements of a winter oil, i.e., resist clouding for more than 5½ hours at 32°F (0°C). This revised salad oil definition encompassed the oils identified as natural winter oils. These oils that have fatty acid compositions that are high in unsaturates with corresponding low saturated fatty acid levels do not require winterization to remain clear and brilliant at refrigerator temperatures. Soybean oil is a natural winter oil and does not require fractionation to meet the cold test requirements. Corn, canola, safflower, and sunflower oils would be natural winter oils, except for waxes that cloud at cool temperatures. These oils are normally dewaxed with a winterization-type process to remove the waxes. Salad oils cannot be prepared from peanut oil because it gels when chilled to the extent that it cannot be separated. The tropical oils' saturated fatty acid levels are too high for consideration as salad oils.

Liquid oils are consumed as naturally (undeodorized) and artificially flavored oils and as neutral, deodorized products. Olive oil is almost always marketed in the undeodorized form. The natural flavor is an important asset that deodorization would destroy. Soybean, peanut, sunflower, and sesame oils are generally consumed in their crude form in India and China; however, the oil seeds are expressed at low temperatures without previous heat treatment. These cold-pressed methods produce low yields, but the oils have a relatively mild flavor and odor [1]. Even though olive oil is growing in popularity as a gourmet oil, consumers in the United States are accustomed to deodorized oils. Cottonseed oil, the only oil consumed in the United States for many years, is so strongly and unpleasantly flavored that deodorization is considered necessary to made it edible. The efficient methods used to separate oils from vegetable seeds involves a cooking, or heating, process before extraction. These methods do not change the stability or nutritive qualities of the oil, but they do develop a stronger flavor and odor that is objectionable. The higher oil yields compensate for the refining and deodorization expenses required to make the oils palatable. A market has developed for cold-pressed peanut, sesame, safflower, and sunflower oils because of implied nutritional considerations, but no reliable consumption data are available [4].

In general, a greater utilization of liquid oils has developed in areas that had been partial to solid fats than of plastic fats in oil-consuming countries. Chart 11.1 compares the edible fats and oils preference changes in the United States beginning in 1950 when liquid oil usage was at a much lower level than either shortening or solid fats, and the table spreads composed of butter and margarine [1,5,6]. Liquid oils usage more than tripled during this period. Cooking and salad oil usage surpassed table spreads in 1970 and has been close to the volume of shortenings since 1980. This trend has developed because of

- the global movement of people
- high polyunsaturates content of most oils
- saturated fatty acids implication in cardiovascular diseases

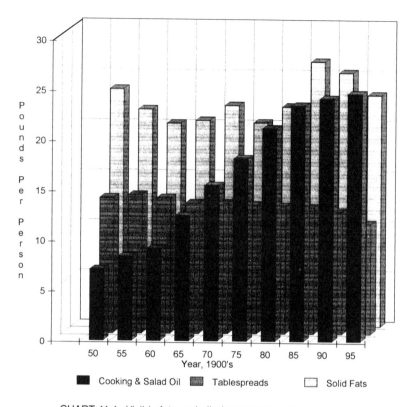

CHART 11.1. Visible fats and oils (per capita consumption, lb).

- process improvements for more bland oil flavors with improved oxidative stability
- finished product formulation and process development to accommodate liquid oils
- emulsifier development, which reduced the dependence of the crystalline properties of solid fats for functionality
- more convenient handling

Liquid oils are suitable and preferred for many types of food preparation where it was consumed shortly after preparation, frying with high absorption rates for rapid turnover, frozen and refrigerated food preparation, and other areas with protective packaging. These application restrictions still apply for cooking and salad oils, but high-stability oils have been developed for applications requiring a long-term oxidative stability. The primary characteristic of a high-stability oil is liquidity at room temperature with enhanced oxidative stability while retaining functional and nutritional properties. Two techniques for producing a high-stability oil are (1) hydrogenation followed by fractionation to separate the stearin or hard fraction from the olein fraction, which becomes the high-stability oil, or (2) the use of plant breeding techniques to produce oils low in polyunsaturates and high in monounsaturated fatty acids. Both of these techniques are used to produce high-stability liquid oils for use wherever stability and liquidity are the important functional requirements.

1.2 Cooking and Salad Oil Sources

The steady growth in the consumption of cooking and salad oils is evident from the USDA Economic Research Service Oil Crops Situation and Outlook Reports for domestic consumption of salad or cooking oils. Consumption and the source oils data for salad and cooking oils utilized since 1960 are summarized on Table 11.1 [6–8]. Deodorized cooking and salad oils are principally prepared from soybean, corn, cottonseed, and peanut oils. Cottonseed, corn, and peanut oils can be deodorized to bland neutral products that do not develop exceeding disagreeable flavors upon reversion or oxidation. Conversely, the characteristic oxidized flavor of soybean oil kept it from being an acceptable salad or cooking oil for quite some time.

Winterized cottonseed oil was the principal salad oil in the United States until the late 1950s. Soybean oil surpassed cottonseed oil as the principal vegetable oil in the United States soon after World War II; however, cottonseed oil remained the preferred salad oil source until the flavor and odor deficiencies of soybean oil were addressed. Soybean salad oil with a reduced linolenic fatty acid (C-18:3) was introduced to the U.S. market in the late 1950s/early 1960s. The linolenic fatty acid content was reduced with hydrogenation and subsequently winterized to remove the hard fraction created during hydrogenation. This specially processed soybean oil quickly replaced the more expensive cottonseed salad oil as the preferred food processor salad oil and was cautiously introduced to the retail consumer. It became the preferred retail bottled oil during the 1960s [9]. In the late 1970s with improvements in the processing of soybean oil, processors were able to produce a refined, bleached, and deodorized (RBD) soybean oil acceptable to industrial customers. This improved RBD soybean oil was introduced to the retail market as "all natural" and "light." The oil appearance was almost water white. This product quickly replaced the hydrogenated, winterized, and deodorized soybean salad oil sold in the retail grocery stores. Improved processing and packaging had provided a

TABLE 11.1. Salad and Cooking Oils: U.S. Source Oils.

	Year (million pounds)							
	1960	1965	1970	1975	1980	1985	1990	1995
Corn oil	247	239	247	281	350	515	636	429
Cottonseed oil	752	915	527	432	460	384	460	251
Olive oil	51	44	62	48	58	105	213	251
Peanut oil	28	53	140	100	148	110	139	W
Safflower oil		9						
Soybean oil	887	1,564	2,471	3,031	4,042	4,749	4,662	5,473
Unidentified				136	109	137	33	591
Total	1,966	2,824	3,464	4,028	5,167	6,000	6,143	6,744
Pounds per person	9.2	14.1	17.3	17.3	21.2	23.5	24.2	24.7

W = Withheld to avoid disclosing volume for individual companies.

more bland flavor stable oil than the soybean oil of the 1930s and 1940s, but the U.S. consumer had also become more accustomed to the soybean oil flavor and odor, which made it less objectionable.

Cottonseed oil has a unique flavor property that makes it a desirable frying media for snack foods. Potato chips fried in cottonseed cooking oil have a pleasant nutty character, while other frying oils produce chips with clean potato flavors that lack the nutty note [10]. This product usage for cottonseed cooking oil and the other industrial use of cottonseed salad oil probably accounts for most of the liquid cottonseed oil usage since cottonseed salad oil retail marketing appears to be limited to the gourmet shelves.

Corn oil is second only to soybean oil in salad and cooking oil volume, as well as for other products, and has wide consumer recognition even though it is produced as a by-product of corn sweeteners, starch, meal, and ethanol production. Corn oil has been regarded as exceptional in flavor and quality while obtaining a healthy oil image. Promotion of polyunsaturated oils for nutritional purposes during the 1950s and 1960s exploded the commercial use of corn oil in various foods, snack frying, and as a retail bottled oil. Corn oil contains 69% linoleic (C-18:2), about 13% saturated fatty acids, and only trace amounts of linolenic (C-18:3) fatty acid. The distinctive musty flavor and odor developed by corn oil has popular acceptance both for home use and for prepared food products like snack frying and mayonnaise or salad dressing preparation.

Peanut oil cannot be winterized to produce a salad oil because of the unfilterable gel produced with cooling. Therefore, it is marketed as a cooking oil both for industrial uses and retail. Peanut oil is considered a premium cooking and frying oil due to the pleasant flavor and cost. The oil separated from peanuts is the recipient of the pleasant peanut flavor, rather than the meal. Peanuts, and therefore the oil, are subject to mandatory price supports in the United States, which maintains the premium pricing. Nevertheless, peanut oil maintains a respectable cooking oil market for snack frying, foodservice frying, and as a retail cooking oil.

The increase in U.S. olive oil consumption during the past decade is due, in part, to promotions of the health benefits of Mediterranean dietary patterns and to market expansion from gourmet shops into supermarkets. The high monounsaturates of olive oil are being promoted by some nutritionists as a healthful oil. Olive oil, produced by crushing and pressing olives, has a high oleic fatty acid (C-18:1) content, ranging from 55 to 83%. The Mediterranean diet is based on the findings of a Harvard University Press study that concluded that the adult population of countries such as Greece and Italy live longer. This diet allows 25 to 35% of its calories to come from fat as long as that fat is olive oil. This study linked olive oil with lowering low-density lipoprotein cholesterol (problem) levels while maintaining high-density lipoprotein cholesterol (good) levels. However, these results have not been observed on a consistent basis and are therefore questioned by some researchers [11,12].

Since being approved as generally regarded as safe in 1985, canola oil has been offered as a salad oil to the industrial and retail consumers in the United States. Canola oil has been promoted with a healthy image, showcasing its low saturates content of 6% or less and high level of the monounsaturated fatty acid, oleic (C-18:1), approximately 60%. The USDA Economic Research Service has not included canola oil in the cooking and salad oil reported volumes, but J. Stanton, as a result of private communications with the USDA service, indicated that 700 million pounds were used in the 1991–1992 crop year [13]. Canola oil has been marketed to the retail consumer market by most of the major bottled oil suppliers both as branded and private label products.

Retail salad oil bottlers have attempted to create a market for sunflower oil on several occasions. In the early 1990s, the major branded retail salad oil producers introduced sunflower oil nationally; however, the volume generated has never been enough to report by the USDA in the *Fats and Oils Situation and Outlook Report*. Canola oil introduction as a retail bottled oil appeared to replace the last sunflower oil national marketing attempts. Then, sunflower became one of the components of the blended salad oils introduced in 1991, with a third less saturates than soybean oil. These oils were blends of soybean, sunflower, and canola oils to attain a 10% saturated fatty acid content [14]. Blended vegetable oils are still marketed, some with the same source oils, but the saturates are reduced to 5% by increasing the amount of canola oil in the blend. Other blends available are canola and corn oils or canola and soybean oils.

2. RETAIL CONSUMER OILS

None of the retail consumer oils currently available have been modified chemically to change the physical characteristics. Processing for most of the household bottled oils is merely refining, bleaching, and deodorization, except for winterization for cottonseed salad oil and dewaxing for canola, corn, sunflower, and safflower oils. Additionally, most of the household oils meet the requirements of a salad oil, except for peanut and olive oils. These consumer oils will become a semisolid at refrigerator temperatures.

The broad dietary shift from animal fats, which began in 1950, has favored liquid oil products. The U.S. consumer has become increasingly aware of the fats and oils role in coronary disease. As a result, consumers have replaced solid shortenings with liquid oils. The oils favored by consumers have tended to reflect the most recent study's findings. Initially, when it was thought that only polyunsaturated oils were useful in lowering serum cholesterol, corn oil began its rise in popularity. Reports that monounsaturates were equal to polyunsaturates in lowering serum cholesterol appeared to help peanut and olive oil sales. The Mediterranean diet helped to improve the sales of olive oil. Canola oil introductions capitalized on the low saturate level, i.e., 94% saturate-free. In 1991, an oil composed of soybean, sunflower, and canola oils blended to one-third the saturate level of soybean oil was introduced [14]. Blended retail oils are still marketed by the major branded suppliers but with different compositions. One blended oil is composed of soybean and canola oils for 7% saturates and another is a blend of corn and canola oils with 15% saturates.

Consumer retail oils are normally packaged in clear polyvinylchloride (PVC) containers in 16-, 24-, 32-, 38-, 48-, and 64-ounce bottles and 1-gallon jugs. Some different sizes are used for the gourmet oils such as olive oil, the cold pressed oils and specialty oils like avocado, grape, and so on. All retail oils were packaged in glass containers prior to the mid-1980s for product protection. Polyethylene containers had proven unsatisfactory for packaging oil due to oxygen permeability. However, age evaluations comparing glass to PVC bottles filled and sparged with nitrogen showed that this material was comparable to glass. No significant differences were found in sensory evaluations of the oils aged in the long-term ambient temperature tests regardless of the packaging material used for bottling. These results helped convince the edible oil industry to convert most consumer retail oil packaging to the translucent plastic bottles. Nitrogen blanketing with reduced headspace oxygen to less than 2% has also been proven to substantially increase the shelf life of the liquid oils [15].

Liquid oils are sensitive to light, which catalyzes oxidative reactions. It has been proven that liquid oil packaged in dark or amber glass extends the oxidative stability of the oil [16]. However, retail consumers have definitely shown a preference for clear containers. Fats and oils processors that have attempted to improve the oxidative protection for retail oil by tinting the bottles to amber have suffered reduced sales that have increased substantially when the container was changed back to clear glass [17].

3. INDUSTRIAL SALAD OIL APPLICATIONS

Salad oils are the major ingredient in most dressing products, which include mayonnaise, spoonable salad dressing, and pourable salad dressings. Salad oils contribute lubricity and a creamy, pleasant mouth feel to dressings of all types. Other functions of salad oils include the solubilization of oil-soluble flavors and contribution of an overall balanced flavor profile.

The predominant salad oils in the United States have been cottonseed, corn, and soybean oil. Cottonseed oil requires winterization to remove the hard fractions to remain clear at refrigerator temperatures. Corn oil must be dewaxed to remove waxes that cloud at room temperature. Soybean oil is a natural winter oil and only requires minimal processing, refining, bleaching, and deodorization before use as a salad oil. Therefore, soybean oil has the advantage of requiring the least processing, and being the least-cost source oil, it has become the dominate industrial salad oil. Canola oil has generated interest because it is lower in saturates and only slightly more expensive than soybean oil but has not shown any significant gains currently.

Salad oils are required for dressings, sauces, and other food products prepared or stored at cool temperatures that require a liquid oil. The major salad oil applications are salad dressings, i.e., mayonnaise, spoonable salad dressing, and pourable and separating salad dressings. These products all require high levels of salad oil in the formula, i.e., 30 to 80%. Salad oils are also used for other prepared sauce-type products that may be refrigerated, causing the emulsion to break and develop an oil layer. Other uses are when a liquid oil is desirable for application or appearance.

3.1 Mayonnaise

Mayonnaise, the most prominent savory dressing with a high oil content, is an oil-in-water emulsion. It is used both in the home and in the restaurant for preparing sandwiches, salads, and similar cold dishes. An emulsion is a mixture of two liquids, one being dispersed as globules in the other. The liquids must be immiscible in each other, incapable of forming a uniform mixture. The liquid that is broken into globules is termed the dispersed phase, while the liquid surrounding the globules is known as the continuous phase. In mayonnaise, salad oil is the dispersed phase, vinegar and water are the continuous phase, and egg yolk is the emulsifying agent. Salt and sugar are dissolved in the water and vinegar, while the spices are held in suspension in the water phase. Oil separation occurs in mayonnaise when the egg yolk films are broken, allowing the oil droplets to run together. The emulsion is most often broken by shaking or jarring, heat, evaporation, or freezing.

The French are responsible for introducing mayonnaise to America during the 19th century. It had been prepared in France back to days of Louis XIV as a condiment. Mayonnaise was limited to home or restaurant preparation because the dressing would separate after standing for short periods. It was first manufactured commercially during the early part of the 20th century [18], and a U.S. Standard of Identity was established for mayonnaise in 1928 [19]. These standards have been revised from time to time, but the most recent requirements are [20]

- oil—at least 65% of one or a blend of two or more edible vegetable oils. The oil may contain oxystearin, lecithin, or polyglycerol ester to inhibit crystal formation.
- acidifying ingredient—one or both of the following: (1) vinegar diluted with water to an acidity, calculated as acetic acid, of not less than 2½% acidity by weight or vinegar mixed with citric acid, provided the citric acid does not account for more than 25% of the vinegar calculated as acetic acid; and (2) lemon, lime, or both juices either fresh, dried, frozen, canned, or concentrated diluted with water to an acidity calculated as acetic acid of not less than 2.5% by weight of the product.
- egg yolk containing ingredient—liquid, frozen, or dried egg yolks; liquid, frozen, or dried whole eggs; or either, mixed with liquid or frozen egg whites. No minimum requirement is stated.
- optional ingredients—mayonnaise may also contain salt; nutritive carbohydrate sweeteners (which would include sugar, dextrose, corn syrup, honey, other syrups); monosodium glutamate; and any spice, except saffron or turmeric, or natural flavoring, provided egg yolk color is not simulated. Sequestrants, including, but not limited to, calcium disodium and/or disodium EDTA, may be added to preserve color and/or flavor.

MAYONNAISE INGREDIENTS

Each mayonnaise ingredient has a specific function to achieve the desired characteristics. Mayonnaise must have a pleasing and desirable flavor, consistency, and appearance that is acceptable to a majority of the consumers; resist separation caused by abuse and cool temperatures; and extend keeping qualities or oxidative stability. The proportions of oil and egg are balanced to obtain body, viscosity, and texture within the limitations of formulation, mixing procedures, and equipment. Salt, vinegar, and spices provide flavor; however, the emulsion type also affects flavor perception. A tight emulsion masks flavors to provide a mild flavor sensation, while a loose emulsion will release the flavors quickly to emphasize tartness, sweetness, or saltiness with a different intensity.

Salad Oil

Mayonnaise must contain 65% liquid oil to conform to the U.S. Standard of Identity: however, the producers use higher levels to obtain the desired consistency. Most mayonnaise is prepared with the oil content in the 75 to

82% range. Usually, the viscosity of the finished mayonnaise increases as the oil content is elevated. Oil levels of 80 to 84% make a thick, heavy-bodied mayonnaise preferred for foodservice use because it does not soak into bread or soften and flow over salads. Oil levels in excess of 84% will overload the system to cause emulsion stability problems [21].

Since salad oil is the major ingredient, mayonnaise quality is heavily dependent upon the oil quality. The most commonly used salad oil is RBD soybean oil, which has the advantage of requiring the least amount of processing, and it is usually the most economical source oil even before processing. Other salad oils, including winterized cottonseed oil dewaxed corn, safflower, and sunflower oils; RBD canola oil; or partially hydrogenated and winterized soybean and canola oils, may be used to produce mayonnaise. Oils high in saturated fatty acids or oils that solidify at refrigerator temperatures are seldom used because they cause the emulsion to break at refrigerator temperatures. Table 11.2 compares the quality parameters for the salad oils that are available for use in the United States. Each analytical result indicates the suitability of the salad oil for a particular effect upon the finished mayonnaise:

- Lovibond red color—Salad oil colors are relatively light and do not impart any appreciable color to mayonnaise; however, a dark-colored oil of 3.0 Lovibond red is acceptable because a yellow color is desired in the finished product.
- free fatty acid—High free fatty acid levels are an indication of the process quality control. Well-deodorized oils have a free fatty acid level of less than 0.05%, which should be maintained until the salad oil is either heat abused or catalyzed with moisture or mineral acid. Salad oil use with a high free fatty acid will effect a tighter emulsion because of a higher alpha monoglyceride content.
- peroxide value—Oxidation is a major cause of poor oil flavors. Peroxide value is one of the most widely used chemical tests for the determination of fats and oils quality; it has a good correlation with organoleptic flavor evaluations. However, a peroxide determination does not provide a full and unqualified evaluation of oil quality because of the peroxide transitory nature and their breakdown to nonperoxide materials. Therefore, high peroxide values usually mean poor flavor ratings, but a low peroxide value is not always an indication of a good flavor.
- flavor—Organoleptic or taste evaluations will always be the final assessment of oil quality. Flavor reversion is most prevalent with oils high in polyunsaturates, which describes all of the salad oils used for mayonnaise. Flavor testing of the salad oils provides a preview of the oil flavor that the consumer will experience when the product is con-

TABLE 11.2. Salad Oil: Typical Quality Assessment Characteristics.

	Processing						
	Refined, Bleached, and Deodorized		Winterized and Deodorized	Refined, Bleached, Dewaxed, and Deodorized			Hydrogenated Winterized and Deodorized
Oil Source:	Soybean	Canola	Cottonseed	Corn	Sunflower	Safflower	Soybean
Lovibond red color	1.5	1.5	3.0	3.0	1.5	1.5	1.5
Free fatty acid, % max.	0.05	0.05	0.05	0.05	0.05	0.05	0.05
Peroxide value, meq/kg	1.0	1.0	1.0	1.0	1.0	1.0	1.0
Flavor	Bland	Bland	Bland	Sl Corn	Bland	Bland	Bland
Iodine value	130 ± 5	118 ± 8	108 ± 8	125 ± 3	135 ± 5	145 ± 8	110 ± 3
AOM stability, hours min.	8	18	15	18	11	10	15
Cold test, hours min.	20	10	10	20	20	50	10

sumed and can identify off-flavors that the analytical evaluations may not uncover.

- iodine value—The iodine value is a simple and rapidly determined chemical constant that measures the unsaturation of a fat or oil. The higher the iodine value, the more unsaturated the oil. Therefore, it is an indicator of the oxidative stability of the salad oil.
- AOM stability—The oxidative stability of the salad oil will determine whether the mayonnaise will have an acceptable flavor for the desired shelf life. AOM is a measure of the oxidative stability of the salad oil. The longer that an oil can survive the AOM abuse before reaching the 100 peroxide endpoint, the longer it will retain an acceptable flavor.
- cold test—A salad oil must resist clouding at refrigerator temperatures to prevent the mayonnaise emulsion from breaking and allowing the oil to separate. Obviously, longer cold test results indicate that the emulsion stability of the finished product will be maintained for a longer period than an oil with a low cold test.

Vinegar

When mayonnaise was made on the table by the host just before it was placed on the salad, keeping qualities were not important. Vinegar was used for its flavor. Today, the preservative protection of vinegar against microbial spoilage is as important in mayonnaise as its flavor. The tart flavor and preserving characteristics of vinegar are obtained from the acetic acid it contains. Distilled vinegar is usually available to the dressing processors at 50 or 100 grain strength, with each 10 grains being equal to 1% acetic acid. The higher strength vinegar may be diluted with water to attain the concentration or acetic acid equivalent desired. All distilled vinegars do not have equivalent flavors. The flavor can vary between producers because of oxidation of the alcohol, and cider, malt, and wine vinegar have unique flavors that are distinguishable in the finished mayonnaise. Vinegar may also contain high quantities of trace metals that can be detrimental to the oxidative stability of the oil in the finished mayonnaise.

Eggs

One of the most important ingredients in mayonnaise is eggs. Egg yolk contributes the surface active agents lecithin and cholesterol, and the egg protein aids in emulsification by forming a solid gel structure when coagulated by the acid component. Almost all mayonnaise manufacturers use frozen and salted egg yolks [18]. Eggs also contribute to the flavor and color of the finished mayonnaise. Much of the creaminess and richness of mayonnaise is obtained from the combination of oil and eggs in the formulation.

Salt

In mayonnaise, salt enhances flavor, it is a preservative, and it stabilizes the emulsion. The amount of salt required for preservation will vary with the amount of moisture and egg yolk in the mayonnaise formulation. The usual range of salt is 1.2 to 1.8%. In most cases, flavor will be the limiting factor of the amount of salt used; for the other functions, a greater danger exists with the use of too little salt than too much.

Sugar

Many hotel and restaurant chefs do not use sugar in their mayonnaise dressings prepared fresh just prior to serving. However, most commercial mayonnaise does contain sugar to provide a slight sweet flavor to balance the tart flavor from the vinegar.

Mustard

Mustard is probably the most important spice used in mayonnaise. Mustard has been added to mayonnaise in flour or oil form. Some developers have considered mustard flour to contain emulsifying properties, while others dispute this theory. Mustard is used chiefly for flavor although mustard flour does add slightly to the color of the mayonnaise.

Pepper

White pepper has been preferable to black pepper because of the black specks resulting from the dark spice. White pepper is added for flavor.

Paprika

Oleoresin paprika is added to mayonnaise principally for color. It has a characteristic hot or mild flavor and odor, depending upon the source. The mild product is usually used in mayonnaise that is effective for color addition, but the paprika flavor is too mild to be noticeable at the levels used.

MAYONNAISE PROCESSING

Most emulsions tend to form with the major ingredient in the continuous phase and the minor ingredient in the dispersed phase. The mayonnaise emulsion is just the opposite, which makes it a difficult emulsion to prepare. A mayonnaise emulsion breaks when it reverses to the stable form where the oil becomes the continuous phase, and the aqueous phase is discontinuous although not dispersed.

Batch or continuous processing methods are used to produce mayonnaise. The classic batch system is to disperse the dry ingredients into beaten eggs along with one-quarter of the vinegar–water mixture. Then the oil and remaining water–vinegar mixture are added simultaneously in separate streams. The oil addition should be completed slightly before the water–

vinegar mixture. After all the ingredients are added, the emulsion is mixed 1 minute at low speed before filling.

Two-stage continuous mayonnaise production involves the preparation of a premix somewhat like the batch preparation sequence. The premix may be prepared as a batch or with metering pumps, leading to an inline mixer or preemulsifier. The premix emulsion is then homogenized with a high-speed mixer or a colloid mill. Other continuous systems are one-stage systems utilizing different sequences to produce smooth, stable emulsions.

Temperature of the oil and other materials during mixing influences the body, viscosity, and stability of the mayonnaise. If the salad oil and the other ingredients are too warm when incorporated into the emulsion, the finished mayonnaise will have a poorer stability and thinner consistency than those prepared with cool ingredients. The recommended ideal temperature varies from 40°F (4.4°C) to as high as 70°F (21.1°C). Emulsion failures are almost assured with finished mayonnaise temperatures above 75°F (23.9°C) [1,19,21].

3.2 Spoonable Salad Dressing

Spoonable salad dressing refers to a product similar in appearance to mayonnaise and somewhat similar in taste, which is often confused with the more costly mayonnaise. Spoonable salad dressing was developed as a low-cost alternative for mayonnaise during the depression years. Salad dressing is tart and tangy, while mayonnaise is relatively bland with a more subtle flavor. Mayonnaise and salad dressing are used interchangeably on sandwiches, salads, and as a condiment for various foods. The spoonable salad dressing market is divided by flavor preference of one product over the other, which is usually not quality- or cost-related.

U.S. government standards were defined for both salad dressing and mayonnaise in 1928 to relieve the confusion due to the lack of uniformity of the dressing products. The Standard of Identity was finally promulgated by the U.S. FDA and adopted August 12, 1950 [18,19]. Salad dressing is defined as an emulsified semisolid food containing [20]

- salad oil—at least 30% by weight of vegetable oil or a blend of oils. The vegetable oil may contain an optional crystal inhibitor, including, but not limited to, oxystearin, lecithin, or polyglycerol esters of fatty acids.
- starchy paste—a cooked or partially cooked starch paste prepared with a food starch, modified food starch, tapioca flour, wheat flour, rye flour, or any two or more of these. Water may be used to make the paste.
- acidifying ingredient—one or both of the following: (1) any vinegar, vinegar–water mixture, or vinegar–citric and/or malic acid mixture, providing the weight of the citric acid is not more than 25% of the

weight of the acetic acid in the vinegar; any blend of vinegars may be used and considered as vinegar; (2) lemon and/or lime juices in any appropriate form that may be diluted with water

- egg yolk containing ingredient—one or more of the following in an amount to be equivalent to at least 4% by weight of liquid egg yolks: liquid, frozen, or dried egg yolks; liquid, frozen, or dried whole eggs; or either, mixed with liquid or frozen egg whites
- optional ingredients—Salad dressing may also contain salt; nutritive carbohydrate sweeteners; monosodium glutamate; stabilizers and thickeners; and any spice, except saffron or turmeric, or flavoring, provided egg yolk color is not simulated. Sequestrants, including, but not limited to, EDTA, may be added to preserve color and/or flavor.
- protected atmosphere—Salad dressing may be mixed and packed in an atmosphere in which air is replaced in whole or in part by carbon dioxide or nitrogen.

Addition of a starch paste distinguishes salad dressing from mayonnaise; otherwise most of the ingredients are common with the two spoonable dressings. Salad dressing may vary in oil content from the lower quality products of 30 to 35% oil to the higher quality dressings of 40 to 50% oil. Liquid oils for salad dressings are selected using the same criteria as for mayonnaise. Oil modifies the mouth feel of the starch paste, making it smoother and richer. Oil content is not the major source of viscosity and body in salad dressing. This function is controlled by the type and amount of starch used. Starch paste must be increased when egg and oil levels are decreased, which also necessitates a change in the acidity and sugar content to provide stability and flavor balance. Improvements in flavor, body, texture, and stability are effected by increasing the oil content because this change brings salad dressing closer to the characteristics of mayonnaise.

Preparation of the starch paste to be mixed with the egg and oil components is the first step in the production of a salad dressing. This preparation is one of the most critical elements in the whole process. The ingredients in the starch pasted formulation must be balanced for the correct relationship to produce the desired salad dressing. Fermentation is the most serious form of spoilage encountered with salad dressings. A presence of a sufficient level of vinegar, salt, and sugar is necessary to avoid fermentation of the salad dressing. The acetic acid level should be 0.90 to 0.928% to provide a pH below 4.0 to destroy *Salmonella* and *Staphylococcus* [18]. The minimum amount of salt is 1.5% by weight. Excessive tartness resulting from the large amount of vinegar and the saltiness from the salt is overcome to a degree through the use of a high percentage of sweetener. It is not uncommon for a salad dressing to contain the equivalent of as much as 10% sugar. Salad dressings are more highly seasoned than mayonnaise, and the choice of spice forms is more critical. Starch paste is affected by proteins from spices and egg whites, which act as emulsifiers when coagulated by the acid in salad dressing. The proteins disperse with milling to modify the starch texture to produce a drier, less pasty dressing.

The procedure used for cooking the starch paste also has a great deal to do with the keeping quality of the finished salad dressing. The starch paste is frequently the problem source when careful control of cooking, cooling, and handling are not observed. Undercooking of the starch paste produces a thin unstable product with the potential to ferment, while overcooking leads to a combination of varying viscosity and instability. Many cooking and cooling procedures have been used, from batch steam-jacketed kettles and water-jacketed coolers, to continuous cooking and cooling of the paste. Cooking temperature depends upon the equipment, amount of agitation, starch type, and whether or not sugar and vinegar are a part of the paste. One constant is that the paste must be rapidly cooled to 90°F (32.2°C). Preparation of the salad dressing after the starch has been produced is essentially the same as for mayonnaise. The starch paste is blended with a modified mayonnaise base to produce the finished spoonable salad dressing. Packaging and handling of the salad dressing is also the same as for mayonnaise.

3.3 Pourable Salad Dressings

A huge variety of separating and emulsified pourable salad dressing is available on the grocery store shelves and from the foodservice distributor. This salad dressing category was introduced in limited flavors originally. French dressing was one of the very early varieties and has the distinction of being the only pourable salad dressing covered by a U.S. Standard of Identity. All other pourable salad dressings are not limited by a standard but are still required to utilize ingredients recognized as safe by the U.S. FDA. Most pourable salad dressings are formulated with relatively high vinegar, or acetic acid, and salt levels to prevent fermentation during distribution and use at ambient temperature.

Pourable salad dressings are produced in two different finished product forms: separating and emulsified. Separating salad dressing has a separate oil phase above an aqueous phase. In many cases, the two phases are prepared separately and filled into the container as two separate products. These products must be shaken before use and separate quickly after pouring. Some separating salad dressings use an emulsifier like polysorbate 60 to help retain the shaken emulsion slightly longer for dispensing a more uniform dressing. The most popular separating salad dressing variety is the "Italian" flavor.

Emulsified or one-phase pourable salad dressings can be homogenized or blended to maintain the creamy nonseparating consistency. Homogenization reduces the oil droplet size to produce the smooth, creamy dressing. Blended dressings are stabilized and thickened with gums. Green Goddess dressing, the first homogenized salad dressing in the category, was introduced in 1964. Most of the single-phase pourable salad dressings are now emulsified by homogenization rather than relying on gums to provide and hold the emulsion together. Other homogenized pourable salad dressings now include Ranch or Buttermilk, Creamy Italian, Caesar, Russian, and others.

FRENCH DRESSING

French dressing is the only pourable dressing covered by a U.S. Standard of Identity. To be marketed as French dressing, the product must meet the requirements of 21 CFR 169.115 [20]. French dressing is the separable liquid food dressing or the emulsified viscous fluid food dressing prepared from

- salad oil—not less than 35% of vegetable oil. The salad oils may contain crystal inhibitors, including, but not limited to, oxystearin, lecithin, or polyglycerol esters.
- acidifying ingredient—one or both of the following: (1) any vinegar, vinegar–water mixture, or vinegar–citric and/or malic acid mixture, providing the weight of the citric acid is not more than 25% of the weight of the acetic acid in the vinegar; any blend of vinegars may be used and considered as vinegar; (2) lemon and/or lime juices in any appropriate form, which may be diluted with water
- optional ingredients—French dressing may also contain salt, nutritive carbohydrate sweeteners, spices and/or natural flavorings, monosodium glutamate, tomato paste, tomato puree, catsup or sherry wine, eggs and ingredients derived from eggs, additives that impart the traditionally expected color, and stabilizers and thickeners. Sequestrants, including, but not limited to, calcium disodium EDTA and/or disodium EDTA, may be used to preserve color and/or flavor. Calcium carbonate, sodium hexametaphosphate, without restriction, or dioctyl sodium sulfosuccinate at 0.5% of the thickener or stabilizer weight maximum may be added as solubilizing agents.
- protected atmosphere—French dressing may be mixed and packed in an atmosphere in which air is replaced in whole or in part by carbon dioxide or nitrogen.

French dressings normally contain 55 to 65% salad oil, even though the standard only requires 35%, to achieve the preferred mouth feel. The gums and/or stabilizers used are more prominent at lower salad oil levels, which can produce a slimy character to the dressing instead of the creamy, pleasant mouth feel experienced with higher oil levels.

4. HIGH-STABILITY OILS

The primary characteristics of a high-stability oil is liquidity at ambient temperatures and resistance to oxidation. Most vegetable oils that are liquid at room temperature generally contain high polyunsaturated fatty acid levels. Polyunsaturated fatty acids have low melting points for liquidity but are also extremely susceptible to oxidation. Therefore, most liquid oils applications are limited to products where an extended shelf life is not a prerequisite. Two specialized techniques have been developed to enhance the stability of liquid oils while retaining functional and nutritional properties: (1) hydrogenation and fractionation to separate the stearin or hard fraction from the olein or soft fraction, which becomes the high-stability oil, and (2) the use of plant breeding techniques to produce liquid oils with low polyunsaturate and saturate levels but very high monounsaturated fatty acid levels. Table 11.3 compares three commercially available high-stability oils produced by these techniques. All of the oils have high-oleic fatty acid contents with relatively low saturated fatty acids, which provides high stability and liquidity.

4.1 Processed High-Stability Oils

Hydrogenation saturates the double bonds to convert linolenic (C-18:3) to linoleic (C-18:2) and linoleic to oleic (C-18:1) fatty acid. Preferably, the reaction could be stopped after converting all the polyunsaturated fatty acids to monounsaturates. However, the hydrogenation process is not selective enough to target only certain double bonds to saturate. Therefore, while saturating the polyunsaturates, significant levels of stearic fatty acid with a high melting point are also produced. Increased higher melting stearic fractions results in an oil that is more oxidative stable but no longer a clear liquid at ambient temperatures.

It is necessary to physically remove the high-melting hard fractions to regain the liquidity desired for high-stability liquid oils. Fractionation techniques involve separation of the solid and liquid triglycerides on the basis of melting point differences. Several fractionation methods are practiced: (a) winterization, a form of dry fractionation; (b) detergent fractionation; and (c) solvent fractionation. A liquid oil fractionation benefit is that the naturally occurring antioxidants, tocopherols, are concentrated in the olein or liquid fractions to further improve oxidation protection.

TABLE 11.3. High-Stability Liquid Oils.

	Modification Technique				
	Processed		Plant Breeding		
	Refined, Bleached, Hydrogenated, Winterized, and Deodorized	Refined, Bleached, Hydrogenated, Solvent Fractionated, and Deodorized	Refined, Bleached, and Deodorized		
Plant Seed:	Soybean	Soybean and Cottonseed	Safflower	Sunflower	Canola
Fatty acid composition, %					
Lauric C-12:0		1.0			
Myristic C-14:0		1.0	0.1		
Palmitic C-16:0	9.0	9.0	3.6	4.0	3.6
Palmitoleic C-16:1		1.0			0.6
Margaric C-17:0					0.3
Stearic C-18:0	3.8	5.0	5.2	4.0	4.7
Oleic C-18:1	63.0	78.0	81.5	80.0	85.7
Linoleic C-18:2	22.0	5.0	7.3	9.0	2.1
Linolenic C-18:3	1.0	trace	0.1	trace	0.1
Arachidic C-20:0			0.4	0.5	0.7
Gadoleic C-20:1					1.7
Behenic C-22:0			1.2	1.0	0.3
Erucic C-22:1					0.2
Oxidative stability rating*	3.5	0.6	1.7	1.9	1.2
Iodine value	92±	78±	82±	85±	78±

*1 = Best, 10 = Poorest (see Chapter 4, Section 4.1).

479

4.2 Plant Breeding

Plant breeding has been used for a number of years, primarily to increase grower yield and the quality of the crops. Advanced plant breeding technology also offers another means of producing high-stability oils. High-oleic fatty acid oil varieties have been developed for several different source oils, including canola, safflower, soybean, and sunflower [22]. Table 11.3 compares the high-stability oils fatty acid compositions for three oils commercially available. All of the high-stability oils have low saturated and polyunsaturated fatty acid levels with high monounsaturated levels. These oils are more like olive oil than the normal U.S. domestic oils. The high oleic fatty acid content leads to a significantly higher viscosity similar to olive oil. Plant breeders have been able to tailor the hybrid seed oils to specific end uses, in this case, high-stability liquid oils.

4.3 High-Stability Oil Applications

High-stability oils can be used wherever liquidity and oxidative stability influence the finished product quality or processing conditions. The monounsaturated oils function as

- frying oils—Frying stability is dependent upon the degree of unsaturation or the points where oxygen can attack the oil to initiate oxidation, which progresses to polymerization. The predominantly oleic fatty acid composition of the high-stability oils substantially increases their frying stability by limiting the opportunities for oxidation. Frying stability for the high-oleic oils is close to the selectively hydrogenated heavy-duty frying shortenings.
- protective barrier—When applied to the surface of food products, the high-stability oils protect the product from moisture and oxygen invasion, prevent clumping, and impart a glossy appearance. Specific applications include raisins and other fruits, breakfast cereals, nutmeats, snacks, croutons, bread crumbs, spices, and seasonings.
- carriers—Color, spices, flavors, and other additives may be blended in the high-stability oils to preserve the flavor and color or activity without fear of off-flavors from the carrier oil.
- pan release agents—Lubricants for baking pans, confectionery products, and other materials where liquidity and oxidative stability are important.
- food grade lubricant—Alternative to mineral oils for lubrication of equipment that may come in contact with the food product.
- compatibility—The high-stability oils are compatible with all types of fats and oils since crystal type is not a concern.

5. REFERENCES

1. Mattil, K. F. 1964. "Cooking and Salad Oils and Salad Dressings," in *Bailey's Industrial Oil and Fat Products*, Third Edition, D. Swern, ed. New York, NY: Interscience Publishers, a Division of John Wiley & Sons, pp 249–263.

2. Neumunz, G. M. 1978. "Old and New in Winterizing," *J. Am. Oil Chem. Soc.*, 55(5): 396A–398A.

3. Weiss, T. J. 1967. "Salad Oil Manufacture and Control," *J. Am. Oil Chem. Soc.*, 44(4): 146A–148A.

4. Krishmaurthy, R. G. and V. C. Witte, 1996. "Cooking Oils, Salad Oils, and Oil Based Dressings," in *Bailey's Industrial Oil and Fat Products, Vol. 3*, Fifth Edition, Y. H. Hui, ed. New York, NY: A Wiley-Interscience Publication, p. 195.

5. *The Fats and Oils Situation*, Economic Research Service, U.S. Dept. Agr. July, 1977, p. 17.

6. *Oil Crops Yearbook*, October, 1996. Economic Research Service, U.S. Dept. Agr., pp. 54 and 56.

7. *The Fats and Oils Situation and Outlook Report*, Economic Research Service, U.S. Dept. Agr. April, 1971, p. 27.

8. *The Fats and Oils Situation and Outlook Report*, Economic Research Service, U.S. Dept. Agr. January, 1981, Table IX.

9. Erickson, D. R. 1983. "Soybean Oil: Update on Number One," *J. Am. Oil Chem. Soc.*, 60(2): 356.

10. Gupta, M. K. 1993. "Cottonseed Oil for Frying: Benefits and Challenges," presented at *42nd Oilseed Processing Clinic*, March 8–9, 1993. New Orleans, LA. p. 19.

11. Haumann, B. F. 1996. "Mediterranean Product Consumed Worldwide," *INFORM*, 7(9): 890–903.

12. Kimbrell W. and D. A. Maki. 1995. "Moving to Mediterranean," *Food R&D*, 2(2): 1 and 6.

13. Stanton, J. 1993. "Canola in the United States," *Cereal Foods World*, 38(7): 483–485.

14. Anonymous. 1991. "P & G Test Markets a Blended Crisco Oil," *INFORM*, 2(4): 373.

15. Mounts, T. L. 1993. "Using Nitrogen to Stabilize Soybean Oil," *INFORM*, 4(12): 1377.

16. Warner, K. and Mounts, T. L. 1981. "Flavor and Oxidative Stability of Hydrogenated and Hydrogenated Soybean Oils, Efficacy of Plastic Packaging," *J. Am. Oil Chem. Soc.*, 61(3): 549.

17. Weiss, T. J. 1983. "Household Shortenings," in *Food Oils and Their Uses*, Second Edition, Westport, CT: AVI Publishing Co., Inc., p. 184.

18. Moustafa, A. 1995. "Salad Oil Mayonnaise, and Salad Dressing," in *Practical Handbook of Soybean Processing and Utilization*, D. R. Erickson, ed. Champaign, IL & St. Louis, MO: AOCS Press and United Soybean Board, pp. 314–338.

19. Finberg, A. J. 1955. "Advanced Techniques for Making—Mayonnaise and Salad Dressing," *Food Engineering*, 27(2): 83–91.

20. United States Standard of Identity, *Code of Federal Regulations*, 21 CFR 169.115, 21 CFR 169.140, and 21 CFR 169.150. 4-1-90 edition.

21. Weiss, T. J. 1983. "Mayonnaise and Salad Dressing" in *Food Oils and Their Uses*, Second Edition, Westport, CT: AVI Publishing Co., Inc., pp. 211–246.

22. Miller, K. L. 1993. "High-Stability Oils," *Cereal Foods World*, 38(7): 478–482.

CHAPTER 12

Quality Management

1. INTRODUCTION

Fats and oils processing involves a series of processes in which both physical and chemical changes are made to the raw material. The beginning ingredients, edible vegetable oils and rendered animal fats, are natural products with variable characteristics contributed by nature. Control of these processes requires consideration of the problems associated with (1) the properties of the raw materials, (2) large-scale material handling, (3) characteristics of each process or operation, (4) internal and external measurement methods and variability, and (5) identification of the customers' wants and needs. Control of these processes at satisfactory economic levels requires a quality staff and a system to maintain and improve quality [1].

Effective edible fats and oils quality control, at attractive economic levels, requires the cooperative activities of all concerned, i.e., technical service, product development, quality assurance, sales, plant operations, and quality control. The basic quality functions for each of these areas are

- Technical service interfaces with customer and product development to identify the product attributes necessary to provide the required functionality before the sale and to lead the resolution of quality complaints after the sale.
- Product development responsibilities are twofold: (1) develop products that can be produced consistently and most economically by plant operations to provide the required functionality identified by technical service and (2) identification of new products or processes that are functionally and/or economically superior to competitive products.

483

- Quality assurance is responsible for developing and implementing an integrated sequence of special controls for materials, processes, and products based on the quality aspects of the customer requirements, product design, and manufacturing process requirements to assure timely shipment of the product with the proper quality at a minimum cost.
- Sales interfaces with the customers on a routine bases and is usually the first aware of a new requirement or a product deficiency; therefore, sales has the responsibility for involving the quality support areas necessary to satisfy the customers' needs.
- Plant operations has the responsibility to produce and deliver products that consistently satisfy the customers' needs most economically.
- Quality control is the manufacturing activity responsible for implementing the integrated sequences of assessments of materials, processes, and products based on customer requirements, design specifications, and manufacturing process requirements to assure timely shipment of quality product at a minimum cost.

Quality management systems are networks of related procedures and practices designed to deliver a quality product to every customer. Quality management begins when the customers' requirements are determined and continues through product design, sales, manufacturing, product costing, delivery, and service for the customer after the sale. A successful new product introduction followed by continued quality production requires the cooperation of many individuals from all parts of the organization. All of the individual actions must dovetail with each other to produce products that meet or exceed the customers' requirements. Each process furnishes a supporting function that must be designed to operate in proper relation to the other processes to produce a functional quality product. Carefully planned and tested procedures are a prerequisite for assuring that the proper actions are performed in the time frame necessary to achieve and maintain customer satisfaction. Operating standards is a descriptive name for the network of procedures to produce quality products. Therefore, the operating standards are written instructions for each area to follow, which assure a quality product that meets the customers' wants and needs.

2. OPERATING STANDARDS

Operating standards should fulfill the first requisite of good quality management, that all operations performed and the quality requirements expected for each operation should be clearly defined and successfully communicated to all levels of management and operating personnel. Control is the main objective of the operating standards, but they also serve other purposes as well: (1) a basis for product costing, (2) a continuous record of products and process improvements, (3) a record of changes with specific reasons for the changes, (4) the basis for product and process audits, (5) and more.

Operating standards must cover all of the operational, technical, quality, and special requirements necessary to produce the desired product in the simplest, but most complete, format possible. Organization into three separate sections most successfully meets these criteria: (1) specifications, (2) procedures, and (3) methods.

3. SPECIFICATIONS

Specifications are integrated sequences of specific controls for materials, processes, and products based on the quality aspects of customer requirements, design performance, and manufacturing process capabilities to assure timely shipment of quality products at the lowest cost. Specifications provide specific guidelines for (a) the purchase of raw materials, processing aids, ingredients, and packaging supplies; (b) the processing of intermediates; and (c) formulation of the finished product. The finished product then becomes an ingredient or raw material for the food processors, who develop their own specifications. Thus, specifications define the performance and test requirements that must be demonstrated to assure functional compliance with the customers' requirements.

Properly designed specifications require logical thinking and program planning to produce simple, explicit, easily understood, complete instructions. The purpose of the specifications is to define what is needed to a large cross section of people who have to buy, supply, receive, process, package, evaluate, store, ship, and use the products. Each specification should define the characteristics of the product to the extent that possession of the document alone is sufficient to identify what is needed. Five different types of specifications are required for a rational system of controls for the complex processing of edible fats and oils into functional ingredients for prepared foods; ingredient, package, product, customer instructions, and summary specifications.

3.1 Specifications Format

Each specification should be identified with the name of the company issuing the specification, the date, and the common or usual name of the ingredient, package, or product. Most concerns have also adopted a specification numbering system to better identify the individual products for the user personnel and the various computer systems utilized. This and the other useful information common to all specifications should be presented utilizing a uniform format. Uniform presentation is best controlled by the adoption of a form that can be adapted for all of the specification types. This information should appear to be professionally designed to help convey the importance and official status to internal, as well as external, users of the documents.

Form 12.1 presents an efficient specification form that can effectively serve purchase and product specifications, as well as other types beneficial

COMPANY NAME AND LOGO		SPECIFICATION		
PRODUCT, MATERIAL OR OPERATION *(A)*		**EFFECTIVE DATE** *(C)*		**SPEC NUMBER** *(B)*
		SUPERSEDES: SPEC. NO.	**ISSUE** *(D)* **DATE**	**ISSUE NUMBER** *(B)*
DISTRIBUTION *(E)*	**REASON FOR CHANGE** *(F)*			
	ORIGINATOR *(G)*		**CUSTODIAN** *(I)*	
	DATE *(H)*		**DATE** *(J)*	

FORM 12.1. Universal specification format.

to the quality management effort. This form has three sections, two of which require information common to all specifications. The other section is the body of the specification where the definitions, formulations, quality requirements, and other characteristics that change with the product are presented in a logical sequence for each specification type. The content of the common sections is outlined here with references to the blocks within each section indicated in Form 12.1:

(A) Product description—defines the product with the common or usual name of the ingredient, the internal product description, or the package description. An indication of use of the material can also be included in this descriptive block, as well as other information that could help identify the product.

(B) Numerical identification—the specification number is a permanent number for the identified product. Each time a product change of any magnitude is made, a new issue number must be assigned, but the specification number remains constant. The identity relationship between the product and specification number cannot be changed unless the product is eliminated; then after 1 year of inactivity, the specification number can be reactivated for another product. However, even at this point, the next issue number for this specification should be assigned when it is reassigned. This time factor should allow all of the systems to be purged of this specification number and product relationship during this time period.

(C) Effective date—The effective date indicates to the operations and cost departments when this specification replaces any previous issues or the first date that it can be produced.

(D) Superseded information—The superseded specification number, issue number, and effective date document the previous product time of manufacture for information purposes, which might be needed for a problem or complaint that occurs.

(E) Distribution—this block identifies the copy distribution as determined for each individual specification. Distribution should be identified by job functions at specific locations, rather than personal names, to assure that all the necessary functions and locations are covered. A letter and number code for the distribution list saves space on the specification sheet, for example:
  • Letters indicate the location, i.e.,
    G = General office
    S = One plant location
    J = Another plant location

- Numbers indicate the job function, i.e.,

  1 = Plant manager   10 = Operations vice president
  2 = Plant QC manager   11 = Cost accounting manager
  3 = Plant QC laboratory   12 = QA manager
  4 = Plant processing manager   14 = R&D vice president

(F) Reason for change—the specification reason for change requirement helps to alert the users to the specific changes and the reason for each. This section lists all the changes for the user, rather than requiring a study of the document to identify them. A further aid to identify changes for the personnel utilizing the specification is a "c" indicator in the right-hand margin in line with the actual change.

(G) Originator—the signature of the originator, usually the author, of the particular specification indicating that the printed specification meets the customer's and/or product functionality requirements

(H) Originator date signed—the date the document was signed by the originator

(I) Custodian—It is essential that only one custodian authorize all specifications or at least all of a single type of specifications to minimize the risk that specifications published by multiple custodians would be different, causing confusion to the users. The custodian's signature also confirms that the requirements have been evaluated to confirm that operations should be able to comply with them.

(J) Custodian date signed—the date the document was signed by the custodian

3.2 Ingredient Specifications

Ingredient specifications are exacting standards required of the ingredients, including all quality characteristics, composition, legal, and other requirements that might affect the end use of the product. Active and prospective suppliers are provided copies of the purchase specifications and all revisions. These specifications are the base for communications between the supplier and the fats and oils processor serving several requirements for both parties:

- description of the user's needs
- basis for the ingredient costs
- legal and technical basis of the purchase contract
- user's standard for ingredient acceptability

The ingredient specification must be a composite of the user's needs and the supplier's ability to produce. It should define the ingredient in performance terms rather than how it should be produced. Ingredient manufacturers have more knowledge and experience making their product than the user. Therefore, the ingredient user should describe the requirements and let the manufacturer determine the best procedure to meet them.

Proper preparation of ingredient purchasing specifications necessitates consideration of several general factors regarding the manner to describe the product requirements. First, the objectives that the ingredient is expected to fulfill must be determined followed by the parameters that best describe them. The parameters must be controllable and measurable; it is self-defeating to specify factors that cannot be measured or controlled. Whenever possible, the test methods should be standard procedures accepted by the industry producing the ingredient. However, special evaluations or improved procedures that provide more accurate or timely results than the standard procedures should also be included.

The allowable variation for each specific limit must be determined, recognizing that variation exists for most all analytical and performance evaluations. Ingredient specification authors tend only to allow minimal tolerances; however, unnecessarily tight specification limits are usually self-defeating. A tight limit has no merit if it only reduces the ingredient availability or increases the cost. The allowed tolerances should be as wide as possible to perform satisfactorily but not so broad that the specification is ineffective.

The relative importance of the different parameters must be identified. Some are more critical to the acceptability of the ingredient and require a narrower range of acceptability. Others may only have a minor importance and do not need as much attention. One method of conveying the importance of the requirements is to separate them into critical, major, and minor categories.

The ingredient supplier should have the latitude to use the best procedure to produce the product; however, the complete composition must be known and specified. In most cases, the ingredient becomes a part of the user's finished product, requiring labeling, identification of potential hazards, or some other regulatory requirement.

All ingredient specifications should be as complete as possible. Although references to other sources of more detailed information such as analytical methods, government of industry standards, and religious requirements may be necessary, every ingredient specification should be as complete as possible to avoid the need to refer to other information sources for understanding [2].

Numerous resources are available for the ingredient specification preparation. Most times, the best source for ingredient information is the manufacturer, who usually has detailed product profiles for their products.

These usually include product specifications or typical analysis, which may or may not be acceptable in entirety as the ingredient purchase specification.

Other resources include the *Food Chemical Codex* and the *Code of Federal Regulations*. The *Codex* lists specifications for many food additives, and the *Code of Federal Regulations* is the source of the Standards of Identity for many foods and the standards for grades of many agricultural materials. Also, many trade organizations have developed standards for their products, and numerous technical societies have developed standard analytical methods for use with their product types, i.e., the American Oil Chemists Society, the American Association of Cereal Chemists, the Association of Analytical Chemists, and others. These methods are an important part of the ingredient specification. Additionally, reference books and publications on food products should not be overlooked as resources. Another valuable information source for preparing similar new ingredient specifications are the specifications already in existence [3].

INGREDIENT SPECIFICATIONS FORMAT

A uniform format can be followed for all of the ingredient specifications using Form 12.1. The adaptation of the form for ingredient specifications is shown in Form 12.2. The body section for all ingredient specifications should contain the following contents [3]:

(1) Regulatory statement—The first item on the specification sheet should be a general statement requiring compliance with current regulations of the appropriate governing agencies.

(2) Definition—A description of the ingredient, defining the product, physical form, and composition. It is desirable to specify that the ingredient meet the requirements of a standard like a U.S. Grade, a standard from *Food Chemicals Codex,* vegetable or animal origin, natural or artificial, any religious authorization requirements, and so on. In many cases the processing may be important; for example, should it be hydrated, pasteurized, dried, or treated in some specific manner? Some materials may be designated by function; for example, an additive may be specified as an antifoamer, antioxidant, flavor, colorant, enrichment, or other.

(3) Formula—If the ingredient is to be prepared by a specific quantitative formula, this should be outlined on the specification. This should be done only if the ingredient performance depends upon a specific composition and/or preparation procedure.

(4) Chemical and physical evaluations—This section is the heart of the ingredient specification. Chemical requirements include such parame-

COMPANY NAME AND LOGO	INGREDIENT SPECIFICATION		
PRODUCT, MATERIAL OR OPERATION	EFFECTIVE DATE *(C)*		SPEC. NUMBER *(B)*
(A)	SUPERSEDES SPEC. NO. *(D)*	ISSUE DATE	ISSUE NUMBER *(B)*

REGULATORY STATEMENT:

DEFINITION:

FORMULA:

CHEMICAL AND PHYSICAL EVALUATIONS:

MICROBIOLOGICAL REQUIREMENTS:

MANUFACTURING CONDITIONS: (OPTIONAL)

PACKAGING:

HANDLING AND SHIPPING:

APPROVED SUPPLIERS:

DISTRIBUTION	REASON FOR CHANGE *(F)*	
(E)	ORIGINATOR *(G)*	CUSTODIAN *(I)*
	DATE *(H)*	DATE *(J)*

FORM 12.2. Ingredient specification contents.

ters as moisture, acidity, heavy metals, pesticide residue, preservatives, emulsifier level, and so on. Physical characteristics include color, flavor, melting point, viscosity, stability, foreign material, and performance requirements. The evaluation methods should always be specified with the expected results and the allowed tolerances.

(5) Microbiological testing—All ingredients susceptible to the growth of harmful organisms should require certification of acceptable microbiological test results to ensure the receipt of safe, legal, uniform products.

(6) Manufacturing conditions—This section should include necessary procedures that are more unusual than those that are standard methods of operation. These may be important because, in many cases, the procedures are related to attributes that are difficult to measure upon receipt of the ingredient.

(7) Packaging—Generally, as a minimum, the ingredient specification should include the type of container in which the product is packed, the net weight, and identification information. The identification should include the name of the product, the producer, and a lot number for packaging date identification at the very least. Many other details may be included in the packaging section, such as
   • composition of the packaging material, strength, weight, thickness, and so on
   • unitizing, number per pallet, stretch wrapping, and so forth
   • no glass, straps, clips, staples, wires, or the like
   • specifications for bulk container sizes, sealing requirements for protection, cleanliness, and so on

(8) Handling and shipping—Any special handling or shipping requirements should be identified on the ingredient specification or a referral to another reference specification if lengthy and the requirements apply to many different products. These special instructions could include fumigation requirements; protection from heat, cold, or freezing; or the need for a certain arrival temperature.

(9) Approved suppliers—This section should be completed and updated when a new supplier is approved or an old supplier deleted for one reason or another. This listing keeps operations informed of the approved suppliers to contact in the case of a serious problem when purchasing and quality assurance personnel are unavailable. Normally, this section is deleted before the specification is given to new prospective suppliers.

3.3 Packaging Specifications

Packaging requirements must be considered as carefully and thoroughly as any other component of the finished product. Packaging selection is predicated on product protection, custom, end-use application, economics, and customer requirements. Edible fats and oils packaging materially affects the quality and cost of the finished products. In fact, packaging materials and the attendant labor expense represent the second most significant portion of the product cost after raw materials. Quality-wise, a product that has been meticulously processed to obtain optimum performance can be protected or lost by the selection of the packaging material [4].

The packaging specification, like ingredient specifications, must be a composite of the user's needs and the supplier's ability to produce. It should define the package in composition and performance terms, rather than how it should be fabricated. The package supplier should be allowed to determine the best procedure to meet the packaging requirements.

Preparation of packaging specifications requires the consideration of several general factors regarding the product package:

- the expectations for the packaging material
- the parameters that are controllable and measurable and that best describe the packaging
- the allowable package variation that will perform satisfactorily on the designated packaging equipment
- coordination of the packaging components to produce a complete package. Many of the edible fats and oils product packages require more than one component that must fit together. For example, the packaging bill of materials for a soft margarine could include (1) a plastic tub, (2) a plastic lid, (3) a printed fiberboard sleeve, (4) a printed shipping case, and (5) fiberboard case pads. All of these components must fit together snugly for product protection but easily for high-speed packaging. These requirements necessitate consideration of the allowed specification tolerances for all the components to ensure trouble-free performance.
- satisfaction of the applicable regulatory agency requirements for net weight, product protection, food grade materials, and construction utilization, plus any other regulations that may apply.

The functional design of a package can be formulated internally, with an outside designer, or with the supplier. Final decisions regarding the container must meet the anticipated need of the environment to which it will be subjected in normal and extreme conditions. The packaging specification preparation should rely upon the designers and the identified or expected customer requirements for determination of the critical, major, and minor requirements.

PACKAGING SPECIFICATION FORMAT

Uniformity of presentation is as important for packaging specifications as any of the other specification types. Form 12.1 can be easily adapted for packaging with the main section identifying all the requirements for each package component. The package specifications should be presented in a

uniform format for clarity and for rapid location of the individual requirements. The adaptation of the universal form is illustrated in Form 12.3,
with a more detailed review of the suggested contents below:

(1) Regulatory statement—Like the ingredient specifications, those for
    packaging are purchase specifications, which should always include a
    general statement requiring compliance with current regulations of
    the appropriate governing agencies.

(2) Definition—A description of the package component, defining the
    container part, its form, construction, materials used, fittings, intended

COMPANY NAME AND LOGO		PACKAGING SPECIFICATION	
PRODUCT, MATERIAL OR OPERATION **(A)**	EFFECTIVE DATE *(C)*		SPEC NUMBER **(B)**
	SUPERSEDES: SPEC. NO.	ISSUE *(D)* DATE	ISSUE NUMBER **(B)**

REGULATORY STATEMENT:

DEFINITION:

COMPOSITION:

SIZE:

COATING:

TARE WEIGHT:

QUALITY STANDARDS:

PRINTING REQUIREMENTS:

HANDLING AND SHIPPING:

APPROVED SUPPLIERS:

DISTRIBUTION *(E)*	REASON FOR CHANGE *(F)*	
	ORIGINATOR *(G)*	CUSTODIAN
	DATE *(H)*	DATE *(I)* *(J)*

FORM 12.3. Packaging specification contents.

use, style, and any other pertinent information to better define the packaging is required. It is desirable to specify that the container meet the requirements of any existing container standards, regulations, or recognized descriptor terms.

(3) Composition—Many different materials are used to package edible fats and oils finished products. The traditional containers are constructed of metal, glass, plastic, corrugated, and fiber-wound composite materials with many different coatings, liners, finishes, tints, and other protective products.

(4) Size—The container's size is given, which can include capacity, dimensions, volume, thickness, and any other measurements that specifically identify the container. The allowed tolerances for these measurements must be identified.

(5) Coating—Coatings might be applied to an exterior or interior surface to reduce friction, to protect the product, or to complete the package.

(6) Tare weight—The weight of the container should be as uniform as possible for weight control and for determining the gross weight of the finished product for transportation information.

(7) Quality standards—Application testing of packaging materials can include confirmations of the composition, container sizes, protective coatings, tare weights, or other evaluations such as compression testing, bursting strength, resistance to temperature changes or abuse, stress tests, and any specific testing for a particular attribute. All test requirements should be identified with the allowed tolerances and communicated to the suppliers through copies of the current specifications.

(8) Printing requirements—Many containers are preprinted with all or a portion of the product's label. The current governmental regulations for the specific product label must be transmitted to the supplier for compliance. Any changes in the product or the label requirements should initiate a change in the specification to update these instructions.

(9) Handling and shipping—This section instructs the supplier on the desired packaging of the packaging materials. For example, the instructions for pallets of empty salad oil bottles might be as follows: The cases of empty bottles are to be shipped with the cases right side up, except for the top layer of cases on the pallets. The cases on the top pallet layer are to be inverted with the top case flaps folded in a normal closure position, but not glued. The pallet load is to be either strapped or string tied.

(10) Approved suppliers—This section should be completed and updated when a new supplier is approved or an old supplier deleted for any reason. This listing keeps operations informed of the approved suppliers to contact in the case of a serious problem when purchasing and quality assurance personnel are unavailable. Normally, this section is deleted before the specification is given to new prospective suppliers.

3.4 Product Specifications

These standards provide the criteria for judging the suitability of the product for the intended use. Product specifications are detailed descriptions of the composition, process conditions, quality requirements, and special instructions for each product. The preferred fats and oils product specifications treat the product from each process as a finished product designated here as process control system. The purpose of the process control system is to identify product defects where and when they occur, so that corrective action can be taken immediately at the point in the process where changes can be made most effectively with a minimum of reprocessing, lost time, and lost product. A requirement of the process control system is that each product must satisfy all the specified limits before transferring to the next process. The control points established for each process must be preventive, participative to involve all functions, and practical to be understandable and enforceable.

Quality must be created during product design, identified by specification, and built into the product during production. Only when process capabilities and product tolerances are compatible with each other and the product functional requirements can a quality product be produced. Coordinated quality programs provide product production data to the quality assurance and product development groups so that product design is consistent with the current production capabilities.

The formulation of each product will vary, depending upon the product's performance requirements; not only the composition is dictated by the desired performance, but also the component's allowed tolerances. Normally, a fats and oils product is composed of two or more basestocks blended together to meet the finished products requirements. The allowed tolerances for the various components are critical to the degree of acceptability. A fats and oils product's analytical characteristics can be met many times by several different compositions; however, only the product complying with the design composition will perform properly. Therefore, it is important to carefully review the components and the tolerances specified to determine that the specification does not allow a substandard product.

The evaluations that control the product during processing must be carefully considered to achieve the best product and process control. Adequate control of each product at each stage of processing must be determined by the specification originator and confirmed by the custodian. The considerations required for adequate control are identified in the following checklist:

(1) Characteristics to measure—Identify the attributes that separate this product from the others.

(2) Measurement methods, equipment, or instrumentation—Determine the analytical and performance evaluations that measure the product attributes reliably and timely. All evaluations must be completed in time for necessary corrective action to be taken in the case of adverse findings.

(3) Performance standards required—Variability is assured in every process from two sources: (1) process inherent variability and (2) test method and equipment variability. Therefore, the prerequisites for realistic product specifications preparations are
   • Consider the accuracy of the test methods employed. The product tolerances specified can be no better than the established variation for the measurement results.
   • Determine the tolerances required by the process equipment to be utilized. Products that require tighter tolerances than the capacity of the process cannot be produced routinely. Delivering product within the tighter limits would require a sorting process with a predictable failure rate. An alternative consideration would be to upgrade the process or equipment to produce the desired quality product with maintenance, training, better control, new equipment, or a process change.

(4) Control points—Control measurement should be exercised at several points during each process: usually, evaluations during processing to determine that the process is in control and analysis of the finished product for each process to confirm that it remained in control.

PRODUCT SPECIFICATION FORMAT

Form 12.1 is also applicable to product specifications with the body section identifying all the requirements for each individual product and process. The product requirements should follow a uniform presentation in the identification section for clarity and to facilitate quick location of individual requirements. The adaptation of the universal form for product specifica-

COMPANY NAME AND LOGO	PRODUCT SPECIFICATION		
PRODUCT, MATERIAL OR OPERATION	EFFECTIVE DATE		SPEC. NUMBER
	SUPERSEDES SPEC. NO.	ISSUE DATE	ISSUE NUMBER

FORMULA:

PROCESSING:

ADDITIVES:

QUALITY STANDARDS:

SPECIAL INSTRUCTIONS:

NOTES:

DISTRIBUTION	REASON FOR CHANGE	
	ORIGINATOR	CUSTODIAN
	DATE	DATE

FORM 12.4. Product specification contents.

tions is shown in Form 12.4, and the contents in order of presentation are reviewed below:

(1) Formula—This section specifies the product's components, both by common name and specification number, with the target percentage and allowed tolerances.
(2) Processing—The process is identified with all specific conditions that are required to produce the desired product specified. For example, a hydrogenation process specification would identify the conditions for pressure, temperature, and catalyst type and concentration.

(3) Additives—Many products benefit from fat and nonfat additives. In most cases the addition levels are minute, but the performance differences are substantial. Additive examples are antioxidants, antifoamers, colorants, emulsifiers, flavors, antisticking agents, crystal inhibitors, and chelating agents.

(4) Quality standards—The physical, analytical, and performance requirements are listed in this section, indicating the required control point for each evaluation and specifying the target result with the allowed tolerance. It is not necessary to perform all of the analysis at each process control point, which can be indicated with an "X" denoting that the evaluation need not be performed at this point.

(5) Special instructions—Some products require slightly different handling, or the customer may have some specific requirements that must be satisfied to meet their needs adequately. These instructions should be presented in this section, or a directive should be written, indicating the location of more detailed instructions.

(6) Notes—This section is reserved for notes to explain in more detail any entry in the specification. These notes might clarify the addition point or sequence for an additive, show that two additives should be mixed together, indicate a potential problem, advise an allowed substitution or generally provide more detail when needed.

3.5 Customer Instruction Specifications

Edible fats and oils are ingredients for the preparation of other food products whether the consumer is a large food processor, a restaurant, or a homemaker. Many customers have specific requirements for code dating, analytical characteristics reporting, preshipment samples, packaging instructions, delivery temperatures, and so on that are different from the standard practice. These instructions, in most cases, are customer specific and too lengthy to incorporate into the product specification or list on each individual order and too important to rely on general correspondence to convey the requirements. Creation of a separate specification type for these instructions helps to convey the importance and provides a perpetual reference tool that can be easily updated when the customer requirements change.

CUSTOMER INSTRUCTIONS SPECIFICATION FORMAT

The specification form shown in Form 12.1 is adaptable to customer instructions with some revisions in the information section and, of course, the body or informational section. Customer instruction specification's numerical identification should be replaced with the customer's name, and the specific requirements are listed in the body or informational section. The requirements must be presented in a logical manner since a predetermined contents for all customers would be awkward and lengthy. Form 12.5 illustrates a typical customer instruction specification.

3.6 Summary Specifications

Summary Specifications are useful tools for several different purposes but basically serve two functions, either to compile specific information

---

COMPANY NAME AND LOGO	**CUSTOMER INSTRUCTIONS SPECIFICATION**	
PRODUCT, MATERIAL OR OPERATION **Shamrock Baking Company** **General Offices: Clover, Texas**	EFFECTIVE DATE *April l, 1996*	SPEC. NUMBER Shamrock - 1
	SUPERSEDES: ISSL **1** DATE **1/1/95** CUSTOMER: **Shamrock - 1**	ISSUE NUMBER 5

Certificate of Analysis Requirements:

A certificate of analysis showing the Shamrock product name, code number and date of shipment, and all analysis indicated with an (S) on the individual Product Specification must accompany each shipment. The COA should be addressed to the Quality Control Manager at the receiving plant.

Shamrock Product Identification:

Company Spec. No.	Shamrock Name	Shamrock Code Number
8003	Shamrock Cake & Icing	1102-3
8039	Shamrock Icing	1304-6
8040	Shamrock All Purpose	1001-2
8099	Shamrock Filling Shortening	1605-1
8341	Shamrock Salad Oil	1050-4

Package Identification Requirements:
All packaged products must have the Shamrock code number stenciled on each package on two sides, in 3 inch letters.

Shamrock Package Shipment Requirements:
No staples, nails, wire ties, or similar devices should be used as external closures
Three (3) code dates maximum per product
All pallatized products must be stretch wrapped

DISTRIBUTION G: 10-11-12-21-25 S: 1-2-3-4-19-32-40 J: 1-2-3-4-19-32-40-43	REASON FOR CHANGE Addition of Shamrock Filling Shortening Code 1605-1	
	ORIGINATOR /s/ O. K. Sample	CUSTODIAN /s/ Q. A. Wright
	DATE May 29, 1996	DATE May 29, 1996

FORM 12.5. Customer instructions specification.

from many individual specifications or to provide supplemental information for the individual specifications. A listing of the potential summary specification's titles probably describes the objectives of this specification type better than a lengthy discussion. Some selected typical summary specification titles are:

- summary of basestock limits and controls
- shortening plasticization conditions
- margarine manufacturing conditions
- flaking conditions
- packaged product's tempering and storage requirements
- label instructions for stenciled packages
- packaged products target weights
- product ingredient statements

The summary specifications are time-saving devices for operations and provide useful compiled information for other users, such as process and product development, cost accounting, transportation, and sales. In many cases, these specifications provide all the information necessary, rather than the need for a complete set of specifications for every user. These specifications also provide convenient comparisons of the products. Sometimes, development of similar products can result in radically different requirements. The summary specifications help identify these incidences before the newly developed product specification is issued.

Summary specifications require constant attention to keep the information current. Every new or revised product specification may also require a change in one or more of these specifications. However, a change in regulations or another attribute that affects most of the product specifications might be handled by a change in the summary specification, rather than changing all of the individual specifications.

SUMMARY SPECIFICATION FORMAT

The format utilized for these composite specifications would follow the same guidelines as the customer instructions specifications. However, it would be more practical to use specification numbers, rather than the names of the summaries only; otherwise, the customer instructions specification in Form 12.5 should serve as an example for the summary specifications. As with customer instructions, no set style for the information section is practical. The summarized information should be presented in a logical format, varying for each individual specification as the information dictates.

4. PROCEDURES

Every company has its own particular personality and individuality in its approach to business. Policy, determined by top management, establishes the broad principals that guide the company's actions. The various departments within the company must develop plans and specific procedures for the execution of the policy as it affects their area of responsibility. Thus, quality assurance must tailor their procedures to fit the company policy or mission statement.

Many persons in all parts of the organization must make decisions and take actions that directly affect product quality. All of these individual actions must dovetail with each other regarding purpose and timing to meet the company's quality goals. The dovetailing is accomplished by carefully planned procedures. The purpose of the procedures are to guide the actions of the company as related to quality management activities.

The quality programs must fit the current policy yet be flexible, with provisions for recognizing when a change is needed and to execute it on a timely basis. Quality procedures must have provisions for revision caused by changes in the system, obsolescence, and knowledge gained by use and experience, which may be incorporated into the individual procedures. The individual procedures should be reviewed annually, if warranted, to assure its success. Initiation of system does not automatically guarantee compliance to the procedures. Therefore, periodic audits of the procedures should be conducted to determine (1) if the procedure is functioning as it was designed and (2) if the procedure is still conceptually adequate.

Procedures establish the method of execution of a function, operation, or a process to meet the identified objectives. Quality assurance has the responsibility of assuring that the quality objectives and goals have been clearly and adequately defined for the entire organization, with special emphasis on product performance, protection, and legality.

4.1 Procedures Format

A uniform format should be utilized for procedures that can be easily distinguished from the specification format. Procedure should be identified with the company name, procedure title, effective date, issue number, and a page number. If the procedure is a part of the company's administrative procedure system, it may also have a procedure number assigned. Normally, procedures are not individually signed but are distributed with a signed cover letter that also identifies the procedure's copy list.

The actual procedure format usually is in outline form with the following components (see Form 12.6):

(1) Purpose—defines the objective of the procedure and a short explanation of the reason it is needed
(2) Scope—describes the guidelines pertaining to the procedure
(3) Procedure—a step-by-step procedure for the actions required and the responsible area, function, or personnel for each
(4) Forms—an explanation of any forms developed to assist in performing the procedure

COMPANY NAME AND LOGO **POLICY AND PROCEDURES**	**TITLE** **PROCEDURE NO.** **PAGE** **OF** **ISSUE DATE** **REVISION NO.**

PURPOSE: Defines the objective of the procedure and a short explanation of the reason it is needed.

SCOPE: Describes the guidelines pertaining to the procedure.

PROCEDURE: Step by step guidelines for the actions required and the responsible area, function, or personnel for each.

FORMS: An explanation of any forms developed to assist in performing the procedure.

REPORTS: The documentation necessary for the procedure, if any, with the required distribution.

FORM 12.6. Typical procedures format.

(5) Reports—the documentation necessary for the procedure, if any, with the required distribution

4.2 Quality Procedures

Most edible fats and oils organizations will require many of the same type procedures, depending upon the products produced. Those engaged in packaging products will require a more extensive quality procedures system than those shipping bulk tank products only. Eight fats and oils quality procedures are briefly described in the following paragraphs with the objectives, reason required, and the process.

COMPLAINT HANDLING PROCEDURE

In spite of functional quality control systems, fats and oils processors can still experience product complaints. Each quality complaint may involve some or all of six basic steps to resolve the problem to the customer's satisfaction while learning how to prevent a reoccurrence in the future:

(1) Replacement of product—Fats and oils product problems may interrupt the customer's production or cause the product produced to be off-quality. Shipment of good quality product as soon as possible to replace the questionable product should be the first consideration.

(2) Financial adjustment—Unless the problem is found to be customer-generated, reimbursement for complaint-generated costs should be the second essential step toward reestablishing customer relations.

(3) Restoration of customer goodwill—A product failure is an annoyance at best, and the irritation can easily grow out of bounds without proper followup. Most customers are most anxious to know what preventive measures have been taken to ensure that their current problem will not be repeated shortly.

(4) Evaluation of complaint product—The actual complaint product should be analyzed and the production records reviewed to identify the source of the problem.

(5) Prevention of a recurrence—Corrective actions identified from the complaint evaluations should be instituted on a priority basis. In many cases the underlying cause is broad enough to affect other products and/or customers, and measures must be taken to avoid a spread of the problem.

(6) Document the complaint and resolution—A written report has several benefits: closure for the complaint, source of information for future

problems, backup detail for potential process improvements, and informing all involved parties of the problem identification and resolution.

The organization machinery needed to process complaints should be no more elaborate than the complaint load justifies. However, the lack of a formal complaint procedure can easily result in excessive attention required to resolve a complaint and lost customers from slow or poor handling. It must be remembered that each individual customer complaint must be taken seriously, and corrective actions need to be implemented promptly to restore customer satisfaction.

A complaint form specifically designed for the user company and products can help simplify and speed up the complaint handling process. The form must be simple, explicit, easily understood, and contain complete instructions for completion on each form. The form should be a checklist of all the basic information necessary to reach a speedy conclusion for the problem. Form 12.7 shows a complaint form, which requires information felt necessary for a bulk tank shipment or packaged shortening, margarine, or oil complaint.

Complaints are only helpful as an indicator of the process of a quality control program but do specifically identify problem areas. A continuing analysis of complaints will indicate the effectiveness of the problem-solving process implemented to effect corrective actions. The complaint analysis should track quality complaints by problem categories and report the followup conclusions to management and the personnel involved.

PRODUCT RECALL PROCEDURE

It is inevitable that, occasionally, defective products will be produced. Further, it is possible that a defect could be of such a magnitude or seriousness that it constitutes a health hazard, with the attendant risk of injury, which results in legal liability for the processor. Other, less serious product defects represent lesser, but real, legal, regulatory, or financial risks to the processor. Should defective product leave the control of the producer, it is essential that a procedure exist to identify, retrieve, and dispose of the defective product before (1) consumers are exposed to any hazards; (2) federal, state, or local regulations are violated; or (3) any legal financial or regulatory penalties are levied.

The FDA Enforcement Policy defined *recall* as "a firm's removal or correction of a marketed product that the Food and Drug Administration considers to be in violation of the laws it administers and against which the agency would initiate legal action, e.g., seizure. Recall does not include a

COMPANY NAME AND LOGO	QUALITY COMPLAINT EVALUATION	COMPLAINT NO: DATE ENTERED:

CUSTOMER INFORMATION		PERSONNEL INVOLVED
CUSTOMER:		SALES:
ADDRESS:		TECH SERVICE:
CITY, STATE:		Q A MANAGER:
TELEPHONE:		Q C MANAGER:
CONTACT PERSON:		

PRODUCT INFORMATION		
SPECIFICATION	LABEL	PRODUCT NO.
PLANT	PACKAGE	ORDER NO.
SHIP DATE	CODE DATE	CUSTOMER PO NO.
TANK CAR NO.	TIME	ARRIVAL DATE
TRUCK NO.	CUBE NO.	QUANTITY

COMPLAINT DESCRIPTION   SAMPLE REQUESTED?

COMPLAINT EVALUATION

CORRECTIVE ACTION

COMPLAINT CLOSURE DATE COMPLETED Q C MANAGER Q A MANAGER	DISTRIBUTION

FORM 12.7. Quality complaint evaluation form.

market withdrawal or a stock recovery." The FDA assigns class designations to recalls to indicate the relative degree of health hazard for each product as follows:

- "Class I is a situation in which there is a reasonable probability that the use of, or exposure to, a violative product will cause serious adverse health consequences or death."

- "Class II is a situation in which use of, or exposure to, a violative product may cause temporary or medically reversible adverse health consequences or where the probability of serious adverse health consequences is remote."
- "Class III is a situation in which use of, or exposure to, a violative product is not likely to cause adverse health consequences."

Product recovered, but not designated as a recall, may be identified by two additional terms:

- *Market withdrawal* is the removal or correction of a distributed product, which involves a minor violation or no violation subject to legal action by the FDA. The withdrawal reason could be normal stock rotation, adjustments, repairs, and so forth.
- *Stock recovery* is the removal or correction of a product that has never left the producer's direct control. The product must be located on premises owned or under control of the producer, and no portion of the lot should have been released for sale or use.

Recalls can usually be divided into four phases:

(1) Defect discovery—The product defect is identified, significance determined, and the problem extent ascertained. The FDA is notified when the product defect has been identified to meet the recall criteria.
(2) Recall defective products—Identify the product lot involved and the location of all potentially defective product. Notify all locations to place the product "on hold" pending instructions for further disposition.
(3) Disposition—Determine the handling required for the defective product, i.e., destruction, rework, or other.
(4) Termination of recall action—A recall is terminated when the FDA agrees that all reasonable efforts have been made to remove or correct the problem.

The recall procedure implemented should be tested periodically for effectiveness. The primary concern for defective product resulting from an internal problem is the system's ability to reconcile production records with inventory and quantities shipped to identified locations for a particular lot number. Defects arising out of an external recall of a product utilized as an ingredient, processing aid, or packaging material are more demanding. Each shipment of these materials must have a lot number identification and records maintained to trace usage in the particular lots of finished products.

PRODUCT IDENTIFICATION PROCEDURES

The manner and depth of product identification varies within the edible fats and oils industry; nevertheless, the importance of adequate identification cannot be overemphasized. The main reasons for and the benefits derived from good product identification are

- traceability—Tracking of all the components and processing of the individual products with the identity of the ingredients, processing aids, and packaging materials preserved. Traceability is a vital requirement for an acceptable recall program.
- complaint evaluation—Exact product identification for both packaging date and specification recognition is mandatory for evaluation of a product complaint to determine the cause and corrective action to prevent future problems.
- stock rotation—The code date or packaging date is a useful tool for practicing first in/first out in warehousing.
- shelf life control—Edible fats and oils products are perishable because of changes in flavor, consistency, appearance, and so on. Code dating provides a means for controlling the age of product shipped and a pickup control from the customer.
- product audits—Product identification is necessary for product audits evaluating product movement and quality. The code date information identifies the age of the product and, coupled with the specification identification, should enable the auditor to retrieve the original processing data for a more complete picture of the product attributes and changes caused by age and distribution.
- inventory control—Product identification codes and code dates are routinely utilized for the inventory systems for production control and the financial area.

There are many approaches to product identification. In some cases, customers specify the product identification coding to be used on packaged product; some systems require a decoder to identify any information, while others use completely recognizable information; and the depth of some identification is very detailed, while others only have the bare minimum. In any case, for the reasons listed and processor protection, a complete product identification system should be developed and utilized for every product, both those shipped in packages and in bulk.

Many edible oil processors utilize a batch number system for product traceability. Beginning with the major raw materials, crude edible fats and oils are procured in lots that can be an identifiable batch number. The batch number identification system can be utilized throughout all of the processing, including the shipment of bulk products. Traceability maintenance requires documentation of the batch numbers and transmission to the next process tier of documentation. In this system a batch number is maintained until a change in the product has been made by blending or processing.

Every individual packaged product produced, as well as any outer shipping cases, should be marked with a code date to identify the date packed and other identifying codes to identify the product. This information should be applied to each container, as permanently as possible, in a location or on a part that should not be separated from the actual product in normal use.

Several types of code dates can and are utilized for food products, i.e.,

- Open code dates that depict the packaging date, "sell by" date, or "use by" date, usually identifying the month and year, are utilized for many retail products.
- Closed code dates utilize a system of letters and numerals that are not immediately recognized as a date. Typically, these systems utilize the Julian day of the year in some combination with the last numeral of the year and a letter designation for the producing plant. However, some processors utilize systems requiring a decoding sheet or device to identify the code date.
- Individual packaging times and case sequential coding are also practiced by some processors for improved traceability. This information allows the processor to identify the beginning, end, middle, or some other point during packaging in addition to any interruptions that may have occurred.

Product information codes are usually imprinted on the packages in the same location or general area as code dates. This information may be called a product code, which is usually the same as the UPC bar code. The product code will typically identify the product label, package size, and specification number. Some concerns will also print the product specification number on each package for additional identification.

Product liability in the case of a problem is minimized with each improvement in traceability. The more extensive systems providing case sequential coding and time packaged, as well as cross checking information for product identity, are valuable assets to pinpoint problem areas and sources. These aids can help to salvage the majority of a packaged product batch instead of losing the whole batch if this coding were not available.

SPECIFICATION CHANGE REQUISITION

Specifications should represent the best and latest known methods and procedures for each operation. This means that the specifications must be flexible and open to change when an improvement has been identified. Therefore, effective specifications are continually updated and revised to reflect changes caused by quality improvements, manufacturing changes, technology advances, cost reductions, formula revisions, package design modifications, government regulations, and probably the foremost reason, customer requests. Progress and change are usually partners; however, change does not always result in progress. Therefore, a definite process for changing a specification with appropriate checks and balances is necessary to avoid useless or unwarranted revisions.

Anyone using the specification system should be entitled to request a specification change. Quality assurance, as the custodian for the specification system, is the logical clearinghouse for these requests. Input should be solicited from any area that could be affected by the revision to help determine if the change should be approved. Some of the logical questions that each area should review for each request are

- Are the requested revisions realistic?
- Will a change in the process or equipment be required for this revision?
- Are the specified limits realistic?
- Will this change affect the functionality of the product?
- Will the change introduce any special handling requirements?
- Does the revision reflect the customer's specification or product requirements?
- Will the customer agree to the proposed change?
- How will product or delivery costs be affected?
- Does the request require more extensive study or development before passing judgment?

A specification change requisition form designed around the information needed for fats and oils products to determine the feasibility of the revision can help simplify and speed up the approval process. The form must be simple, explicit, and self-explanatory for completion and handling. Form 12.8 illustrates a specification change requisition form designed to fulfill these requirements.

NEW PRODUCT APPROVAL PROCEDURE

Operations produces products according to the guidelines provided by product development with assistance from quality assurance. Specifications are the guidelines that define the materials, processes, sampling plan, quantitative and qualitative measurements, and the acceptance or rejection criteria for each product. Quality assurance has the responsibility for preparing the specifications with direction from product development, marketing,

COMPANY NAME AND LOGO	SPECIFICATION CHANGE REQUISITION	SPEC. NUMBER ISSUE NUMBER REQUEST DATE
ORIGINATED BY:	RECOMMENDED BY:	

CHANGE OR ADDITION REQUESTED:

REASON FOR CHANGE:

SUBMITTED FOR CONSIDERATION TO:	AGREEMENT		REASON:
	YES	NO	
OPERATIONS:			
PRODUCT DEVELOPMENT:			
TECHNICAL SERVICE:			
SALES:			
OTHER:			

QUALITY ASSURANCE ACTION: APPROVED DISAPPROVED	REASON:

S/S Q A CUSTODIAN
Q A DIRECTOR

COMMENTS:

FORM 12.8. Specification change requisition form.

and technical service, followed by approval from operations that the guidelines are reasonable and achievable. Once written and approved, operations has a description of the quality constraints it must work within, and a product profile has emerged for marketing.

It is at the new product conception stage that maximum flexibility exists for using creative ingenuity and technical talents to design quality products that can be produced as error-free as possible. Quality and safety are best built into a product during the development stages where, previously, performance has been the major concern for many products. Thus, quality assurance should be involved early in the development stage to identify the potential product constraints and help design systems to avoid them. Some of the potential quality and safety constrains for edible fats and oils products are

- product protection—sanitation and health hazard control
- quality control—product measurement and tolerances
- legal—federal, state, and local regulations
- economic—product worth versus cost to produce
- raw material—quality, type, and availability of crude edible fats and oils
- customer needs and demands—In addition to performance, other requirements like delivery temperature, packaging, product limits, and quantities could be major constraints.
- labeling—Packaged product may have to conform to certain nutritional requirements or meet another limitation.
- other influences—any other limitations that may arise

A new product approval procedure should have definite requirements to assure that the product performs as expected and can be effectively produced by operations. A three-hurdle procedure to determine if product development can pass the new product production responsibilities to operations involves the following requirements:

(1) Experimental specification—A product specification with complete composition, processing, quality standards, and all other pertinent requirements but limited to a one-batch plant test, which is the complete responsibility of product development. The actual production is supervised by product development personnel and the disposition arranged with sales marketing for successful product or with operations for substandard product to rework or for disposal.

(2) Probational specification—After a successful plant experimental production and customer approval, the product enters the probational period. Operations produces the product, but product development retains responsibility and is on call for any problems that may be encountered.

(3) After the successful production of 5 to 10 batches of the probationary product, the specification is finalized and becomes the full responsibility of operations.

The three-step procedure to obtain an approved product specification provides product development with an opportunity to plant test a laboratory and/or pilot plant developed product and demonstrate that it performs as designed. It also provides plant scale data for evaluation of the quality standards and tolerance levels.

WEIGHT CONTROL PROGRAM

The purpose of a weight control program is to assure that the average fill weight for each packaged product is as close to the declared net weight as possible. From a practical standpoint, it should be recognized that, because of variation in product and filling equipment, it is impossible to have all packages weigh the exact specified amount. Overweighs are expensive for the producer but underweighs are unfair to the customer. Therefore, most weight control programs are designed to prevent the shipment of unreasonable short or overweight packages and to conform to the applicable federal, state, and municipal regulations.

Control charts have been established as the preferred statistical method for control of packaging line weighing and filling systems. The primary purpose of control charting is to show trends in weight toward minimum and maximum limits. While the control chart indicates when the established limits have been exceeded, its main purpose is to provide the means of anticipating and correcting whatever causes may be responsible for defective weights. Prevention of weights outside the acceptable range, rather than just discovery and correction of unsatisfactory lots, is the fundamental principal involved. Some other benefits for control chart statistical weight control systems are

- reduction of operator bias in the recording and use of the weight data
- more concise adjustment criteria potential for the packaging equipment operation
- a better understanding of each packaging line through the continuous collection and analysis of the data
- the ability to isolate and correct definable causes of process variation

QUALITY COST SYSTEM

Product quality has two readily identifiable stages: design and conformance. Quality of design is the degree to which the product satisfies the customer. Quality of conformance is the degree to which a product conforms to the specification. Cost-wise, the quality of design expenses are considered unavoidable because they are necessary to produce the product. Generally, the unavoidable quality of design costs are

- sales and marketing costs involved with determining the customer's quality requirements
- product development costs to create the product and process, which meets the identified quality requirements

- product development and quality assurance costs of translating the product requirements into specification form, which meets the requirements of operations
- operations costs needed to secure the process required to meet the product specification.
- the standard costs identified for the product, which usually include the raw materials, ingredients, processing aids, processing, direct labor, maintenance, depreciation, overhead, laboratory control analysis, and product shrink

Quality of conformance improvements or avoidable cost reduction creates an inverse relationship between quality and costs. As the quality of a product improves because of decreases in variation, there should be a reduction in manufacturing costs. This cost reduction is a result of reducing the level of poor quality product. Evaluating the cost of poor quality involves more than simply determining the cost of production and correction or disposal of bad product. The four areas contributing to poor quality costs are

- internal failure—costs attributed to the production of defective product not shipped to a customer
- external failure—costs attributed to the production of defective product that is shipped to the customer
- appraisal—costs related to monitoring the process and determining the condition of the process
- prevention—costs related to minimizing the level of failure and appraisal costs

Some quality of conformance costs can be avoided at a minimal expense, such as centering the result of a capable process, relaxation of an exceptionally tight tolerance, or rectifying a specification misunderstanding. Other quality costs are avoidable at a substantial, but still economic, investment, such as changes in the process, new equipment, or more extensive process controls. Still other quality costs are avoidable only at an investment cost that costs more than the cure can produce [5].

A major quality question for any company is whether to enter into a program for reducing loss caused by defects and, if so, in which areas. Computation of an accurate cost of quality will provide the data to determine if the quality of conformance costs are a major problem and, if so, the most lucrative areas to attack first. If expenses are considered minor, then the question is how to maintain this level of quality. Routine cost of quality updates would alert management of trends, either towards improvement or a decrease in the level of quality. A decreasing level of quality from any indicator should naturally trigger consideration of corrective actions.

A distinction between avoidable and unavoidable costs for the cost of quality evaluation is not always clear-cut. One suggestion for separation of the unavoidable from the avoidable costs is to assume that operations performed only occasionally are avoidable. Both the prevention and appraisal activities can have a multiple nature, while external and internal failures require little consideration other than collecting accurate cost data. A partial listing of an edible fats and oils operation's potential avoidable costs are listed by the four quality of conformance areas below:

(1) External failure costs:
- returned product—freight charges, extra handling, lost packaging supplies, product rework expense, rush order to replace, and so forth
- complaints—destroyed product, expense to evaluate complaints, customer expense, the returned product expense, lost volume, lost customer, customer goodwill, and so on

(2) Internal failure costs:
- reprocessing—processing in excess of the standard for all processes from refining through warehousing
- reblending—additional blends past the first attempt to meet specified limits
- heel disposal—costs to utilize excess product prepared
- rework or regrade—reprocessed product due to performance failures
- scrap product—product that cannot be salvaged due to contamination with foreign material
- package problem—repackaging or rework due to faulty packaging or packaging error
- overage product—rework and reprocessing costs for packaged products in excess of the product's shippable life
- replacement production—product production to replace unacceptable product
- product giveaway—packaged product weights over target and margarine or spread products with excess fat contents

(3) Appraisal costs:
- reinspection—inspection requirements in excess of the specification requirements
- resampling—product resampled because of suspected error with the first sample
- troubleshooting—to determine the cause of a complaint or internal failure
- reanalysis—product reanalyzed because of suspected error with the first analytical result

(4) Preventive costs:
- diagnosis—determination of problem solution and expense

- corrective actions—expense for equipment and/or procedure required
- retraining—additional or new training of personnel

LABEL APPROVAL PROCEDURE

Various agencies of the federal, state, and local governments have some authority over the labeling of food products, such as Food and Drug Administration (FDA), U.S. Department of Agriculture (USDA), Federal Trade Commission (FTC), U.S. Treasury Department (BATF), and the individual state and local governments. Some regulations, like the Fair Packaging and Labeling Act, apply only to packages and labels that are intended to be displayed to consumers when sold at retail and shipped in interstate. Other regulations specify special requirements for selected food products; for example, fats and oils labeling requirements are different for a shortening processor than those specified for a prepared food product. A shortening label must label each source oil actually in the product, in the order of predominance. A prepared food product utilizing the same shortening as an ingredient can list source oils not present in the product if they may sometimes be used for the shortening. These examples are just a few of the regulations involved in determining the acceptability of a label for a food product. The complexity of the regulations, along with the continuing changes, necessitates a procedure for a specialist to review all new and changed labels for conformance.

Marketing normally designs product labels and packages to attract the attention of the customer. Quality assurance is the logical area to assume responsibility for assuring the legality or conformance to regulations for all packaging and labels. These reviews must begin with the initial design and continue with the proofs from the printers and, finally, with representative samples of the production printing. The procedure developed must provide a means for ensuring the evaluations in a timely manner, documentation of the results, and a formal approval notification to the necessary personnel in marketing, purchasing, and operations.

5. METHODS

Quality is not as simple as it seems. In the final analysis, it must relate to the end use of the product. There would be little merit in a clean, light colored oil for shortening production if it could not be used to produce a bland flavored product with a predictable physical behavior and keepability, nor would a light colored oil be suitable for retail salad oil if it clouds and has an oxidized flavor when bottled or something similar. It is essential that edible fats and oils laboratories have the facilities and capabilities to certify that a product meets all the requirements for which it was intended, as well as all federal, state, and industrial standards required to achieve its merit of excellence.

Processing has a tremendous affect upon oil quality. Some changes that occur, such as elimination of some of the odoriferous components or their precursors and reduction of color, are deliberate and beneficial to fats and oils quality. Other changes, such as isomerization and polymerization, are coincidental and/or unwanted. Monitoring and assessing oil quality at any given processing step is of utmost importance. Proper process control allows more efficient production of oils of superior quality.

Control of the edible fats and oils processes and products quality is heavily dependent upon the measurement methods employed in the plant laboratories. These measurements are the basis for the acceptance and cost of raw materials, processing decisions, and release of the outgoing product. Additionally, the laboratory evaluations are the means for improvement in product uniformity through constant assessment of the effects of process variables, equipment behavior, and the processing procedures affects. Therefore, accurate, precise, reliable laboratory results are essential for quality production.

Development of analytical procedures for fats and oils was one of the reasons for the organization of the Society of Cotton Products Analysts, whose successor is the American Oil Chemists' Society. It was recognized that uniform procedures were necessary for commodity trading, quality control in processing, storage stability, and nutritional labeling of food products and as a means of describing the functionality of a product. Development of analytical methods remains a major effort of the American Oil Chemists' Society with those that have wide use and applicability published in the AOCS *Official and Tentative Methods*. Old, less satisfactory or unused methods are dropped, and new ones are added by the Uniform Methods Committee based on suggestions from members after testing and study by one of the technical committees. These methods have become the accepted edible fats and oils analytical procedures throughout the world [6].

The laboratory methods employed should generally be the ones universally accepted in the industry, both by the suppliers of the raw materials and the purchasers of the finished goods. However, new products, techniques, or equipment may compel a laboratory to modify the methods for their use. In this case the accepted methods provide a "benchmark" by which the new or modified methods and instruments may be evaluated. Additionally, various methods from other recognized associations or even in-house–developed evaluation procedures may have to be adopted for certain processes or products to meet the requirements for some specialized products. Therefore, evaluation methods specific to the available equipment and facilities should be developed for use in the quality control laboratories.

5.1 Methods Format

Laboratory methods are written instructions for the measurement of chemical, physical, and performance aspects of raw materials, ingredients, by-products, and the in-process and finished fats and oils products. The methods utilized must be the most accurate and precise evaluations available, precision indicating the reproducibility of a result. Clear, concise instructions that follow a logical sequence to the results are necessary for the required accuracy and precision.

A definite style should be adopted for all internal laboratory methods to present the procedures in a manner that clearly communicates the requirements to the user analyst. A uniform format for laboratory methods, like all other procedures, should be observed to improve the understandability of the requirements. Form 12.9 presents a form that would help to maintain method uniformity. Use of this form with the following preparation guidelines should better maintain the style adopted:

- title—a specific, descriptive title or name
- division—The methods may be grouped into divisions for easier reference, such as (1) chemical analysis, (2) physical analysis, (3) organoleptic analysis, (4) instrumental analysis, (5) performance evaluations, (6) by-product analysis, (7) environmental analysis, and (8) miscellaneous analysis.
- effective date—identifies the date this method was adopted or the date of the last revision
- issue number—the number of changes for an individual method evident by the issue number. The next consecutive number is assigned with each change in a method. The issue number allows a rapid audit check to assure that the most recent method edition is in use.
- superseded information—identification of the previous issue number and date of issue for the identified method for reference or referral, if required
- method number—a numerical identification of the method to assist in filing and retrieval. The method number and the method title must have a permanent relationship.
- page number—Laboratory methods are rarely limited to one page. The method page number related to the total number of pages assures that all of the method is available.

COMPANY NAME AND LOGO	LABORATORY METHODS		
TITLE:	EFFECTIVE DATE:	ISSUE NUMBER:	METHOD NUMBER:
DIVISION:	SUPERSEDES: METHOD: ISSUE: DATE:		PAGE NUMBER: OF

DEFINITION - A short description of the method.

SCOPE - Defines the application of the particular method.

APPARATUS - Lists the equipment necessary to perform the particular method.

REAGENTS - Defines the reagents required and the proper preparation or method for preparation.

PROCEDURE - List the step by step procedures to perform the evaluation.

CALCULATIONS - Shows the proper calculations with explanations to determine the proper results.

REFERENCE - Refers to the method source or other pertinent information.

NOTES - Information of particular importance to the method performance.

PRECISION - Identification of the precision and accuracy of the method.

DISTRIBUTION:	REASON FOR CHANGE:	
	ORIGINATOR:	CUSTODIAN:
	DATE:	DATE:

FORM 12.9. Laboratory methods format and guidelines.

- definition—a short description of the method, the value of the result, and any specific information that may provide a better understanding of the method
- scope—defines the application of the method and any limitations
- apparatus—a listing of the laboratory equipment required for this analysis, including the manufacturer's name, model numbers, and any other identification available for purchasing the desired apparatus
- reagents—all reagents used in the method listed with sources of supply or referral to a preparation method in the case of special solutions or

standards. The grade and purity should be identified by the designation utilized by the laboratories' suppliers.

- procedure—clear, detailed instructions for performing the evaluation that follow a logical sequence to the end result
- calculations—identification of the required calculations with an explanation of all factors. Prepared worksheets for involved determinations and/or calculations are helpful for recording observations, performing the calculation, and as a record for review if necessary.
- references—provide background information and assist in referral to the original work, if necessary
- notes—information of particular importance to the method, such as safety considerations, potential problems, sensitivity, detection level, results reporting, and substitutions
- precision—identification of the precision and accuracy of each method, when available
- distribution—identification of the copy distribution for each method. Distribution by job function at specific locations, rather the personal names, assures that all the necessary functions and locations are covered. A distribution letter and number code, similar to the proposal for specification distribution, saves space on the methods sheets.
- reason for change—the methods reason for change requirement alerting the users to the specific changes and the reason for each change. A further aid to identify changes is a "c" indicator in the right-hand margin in line with the actual change.
- originator—The signature of the originator, usually the method developer, indicating that the method has been reviewed for accuracy
- custodian—authorization by a custodian to minimize the risk of losing uniformity among the laboratory methods
- custodian and originator's date—the dates that the method was reviewed and signed

6. PROCESS CONTROL

Edible fats and oils processes are more difficult to control than the physical operations performed in most mechanical industries. Three major evaluation methods are available for the inspection and testing of products to control the quality of the products produced:

(1) Screening—inspection of every unit and defects screened out, sometimes called 100% inspection

(2) Lot-by-lot inspection—examination of a relatively small number of samples to judge the acceptability of the entire lot

(3) Process control—concerned with all causes of defective product, be it operator, operation, equipment or raw material. The purpose of process control is to discover defective product where and when it occurs, so that corrective action can be taken immediately.

Most of the edible fats and oils controls are chemical in nature, involving analysis of raw materials, analysis during processing, and analysis of outgoing products. Two of these areas are controlled by screening-type evaluations to determine conformance to specification, which automatically determines the acceptability of the lot or shipment. The first of the screening controlled areas is well established with vegetable oils trading. These raw materials are purchased on the basis of specific analytical limits. Trading rules establish the standards of quality for the various types of each source oil. Analysis of "official samples" is the basis for settlement adjustments for many of the oils traded.

Inspection and testing of the outgoing products, performed to assure that the customer's requirements have been met, are screening-type evaluations for bulk shipments but lot-by-lot evaluations for packaged products. These analyses are surveillance or after-the-fact controls. This evaluation point is at a go–no-go location; either the product is acceptable for shipment, or it must be rejected for reprocessing, downgraded, or some other remedy.

Process control is the most important function for product acceptability. If each step in the process is right and the flow or sequence proper, the desired quality will be in the finished product. Process control requires planning to establish the relative importance of each quality characteristic, points in the process to measure, and the methods and procedures for the measurement. The elements of a process quality control system are

(1) Control point determination—Establishment of control points are necessary to assure that all of the requirements are met in all areas. The quality of the finished product may be predicted by measurements obtained at correctly established control points in the process beginning with the raw material and concluding with shipment of the product. The evaluation points must be preventive, participative, and practical or before, not after, the fact, must involve all processes, understandable and enforceable.

(2) Process capability—The degree of inherent variability in each process must be established to determine if predetermined quality levels can

be satisfied and at what confidence level with the existing equipment and procedures. Product requirements outside the process capabilities will generate unacceptable product through the "sorting" process, which rejects product that does not meet all of the specified limits.

(3) Control sampling—Every sample point specified must be established carefully to ensure that the data collected will have a real value. Meaningless results from insignificant sample points can mask significant problems and divert energies to the wrong direction. The normal sample points for most semicontinuous edible fats and oils processes utilized are

- start up—analysis of samples to assure that the process is in control
- stream—analysis of samples after predetermined time periods to assure that the process remains in control
- finished batch—analysis of a designed batch or lot to assure that the process remained in control to meet all of the finished process limits

(4) In-process analysis—The interrelationships and links with upstream and downstream operations must be assessed for each process and each analysis. The analysis specified must have been considered in light of the process requirements to provide the finished product's performance. The expected variation from each analytical method must also be considered when the tolerances for the limits are established.

(5) Finished product acceptance—Conformance to specification is determined by the inspection and testing of the product after the final process. The analysis at this point should be minimal to ensure that changes have not occurred after the in-process control limits were satisfied.

Edible fats and oils process control development requires skill in the art of processing, as well as technical competence. The technical competence required is not only a knowledge of the process conditions and effects, but also a file of past performance records, which indicate the product changes to expect as the process conditions are varied. Further, it is necessary to know how these changes affect overall product quality as measured by a complete analytical profile.

6.1 Edible Crude Fats and Oils Process Control

Crude fats and oils incoming analyses are performed for several specific reasons: (1) for cost adjustment, (2) to identify required process treatments, (3) to assess the quality of the receipts, (4) for safety, and (5) for vendor rating. Most of the vegetable oils are purchased on the basis of specific trading rules that identify certain quality limits and adjustments in costs for deviations from those limits. Four associations with trading rules for specific source oils are as follows:

Association	Source Oils
National Cottonseed Producers Association	Cottonseed oil
	Peanut oil
	Sunflower oil
National Soybean Processors Association	Soybean oil
Canadian Oil Processors Association	Canola oil
National Institute of Oilseed Products	Canola oil
	Coconut oil
	Corn oil
	Cottonseed oil
	High-oleic safflower oil
	High-oleic sunflower oil
	Palm oil
	Palm kernel oil
	Peanut oil
	Safflower oil
	Sesame oil
	Soybean oil

The trading rules are specific for each source oil, but most specify limits for refining loss, refined color, refined and bleached color, free fatty acid, moisture and volatile matter, and flashpoint. Normally, cost adjustments are made on results outside the trading rule's specified limits. Flashpoint evaluations at receipt are safety precautions to ensure that the oil does not contain any solvent remaining from the extraction process; the solvents used are explosive.

Free fatty acid content determines the caustic treat required for most refined oils. Both refined and bleach colors can determine the refining process requirements and the bleach treatments to achieve the desired finished product color requirements, such as high color cottonseed oil that requires additional refining to achieve the desired red color or soybean oil with high chlorophyll that will probably require a higher bleaching media treatment. The crude analyses also provide the data for a vendor quality rating. Results summaries readily indicate good and problem suppliers.

Other helpful raw material analyses sometimes required on a skip lot basis, at certain periods of the year, or as the result of a particular problem are

- iodine value—Changes in the unsaturation level of the fat or oil receipt caused by new crop, growing season, feed, or other reasons can affect the finished product formulation or intermediate processing. Early knowledge of these changes will allow planning to compensate for the changes.

- fatty acid composition—Suspected changes in the raw materials can be identified more precisely with this analysis than iodine value. Nutritional labeling and other specialty product performance requirements may preclude the use of some receipts or cause a blend of the same source oil from different suppliers to meet the requirements. Purity of a particular fat and oil can also be determined rapidly with fatty acid compositions.
- trace metals—Determination of trace metals in vegetable oils, usually present in parts per million quantities, has always been a formidable problem. However, studies have shown that calcium and magnesium levels above 100 ppm in crude oils can cause problems after apparent adequate caustic refining. Calcium and magnesium depress the hydration of the phospholipids, leaving some in the refined oil. Residual phosphorus in refined and bleached oils, over 1.0 ppm, can poison hydrogenation catalysts and/or cause off-flavors [7]. Periodic trace metal analysis can effectively monitor crude oil receipts' quality.

Even with the crude oil analysis, more than one sample point exists. Samples from three separate control points can require analysis: (1) the official sample obtained by the seller at loading, (2) a destination sample obtained at receipt, and (3) samples during storage. The official sample analyses are the trading rule basis for settlement. The destination sample need only minimal testing to verify the official sample results unless it becomes the official sample when the seller fails to obtain one on loading. Sample analysis from storage should be performed to monitor the oil quality bimonthly. Again, only minimal testing is necessary—probably free fatty acid, moisture, and at times a bleach color.

Rendered animal fats have no association trading rules, which means that the incoming quality requirements must be established with the supplier, if uniformity is expected. Meat fats are reasonably uniform raw materials. Lard and tallow consistency varies somewhat from season to season, but these variations usually present no real or serious operating problems; however, the quality characteristics can vary, depending upon the method of rendering utilized by the supplier. The incoming analysis for the meat fat receipts should include the following:

- organoleptic—The flavor and odor should be characteristic of the product, that is, not sour or not a strong boar odor, which escalates with heating.
- moisture—A high moisture level (above 0.20%) may be a contributor to an off-flavor or a high free fatty acid caused by hydrolysis.

- impurities—Fine impurities in meat fat can participate during deodorization, causing very dark colors. Determination by an effective filtration method or a heating test simulating deodorizer conditions alerts operations that this product must be specially filtered before use.
- color—A high red color (above 1.5 Lovibond Red) may indicate that the product has been heat abused in rendering and require special bleaching. A green tallow, due to high chlorophyll content, requires bleaching with an acid-activated media.

6.2 Refining Process Control

The impurities in vegetable oils consist of phospholipids, metal complexes (notably iron, calcium, and magnesium), free fatty acid, peroxides, and their breakdown products: meal, waxes, moisture, dirt, and pigments. These impurities are present in true solution, as well as colloidal suspension, and their removal is necessary to achieve the finished oil quality standards for flavor, appearance, consistency, and stability required by end-use product applications [8].

The primary processing system used to purify crude vegetable oils in the United States is a combination of degumming and caustic centrifugal refining. As an option, crude oils can be degummed before refining by a water treatment, followed by centrifugation to remove the hydrated gums for lecithin production. Crude or degummed oils are treated with sodium hydroxide to saponify by impurities for removal by a primary centrifuge as soapstock. The refined oil is water-washed to remove soap traces and again centrifuged to remove the hydrated soap. Refined, water-washed oil is finally vacuum-dried to remove the traces of moisture remaining.

Refined oil quality standards must be established for each source oil; however, the control points are basically the same for all oils:

- before refining—Prior to refining, each batch of crude oil should have been analyzed for free fatty acid, neutral oil or cup loss, and bleach color. The caustic treat is calculated based on the amount required to neutralize the free fatty acid and a predetermined excess to insure removal of other impurities, such as phosphatides and color bodies. Usually, concentrated sodium hydroxide is diluted with water to obtain the desired Baumé or concentration. Analyses for percent NaOH and degree Bé should confirm that the desired dilution has been attained, normally, 20 to 50 degree Bé for cottonseed oil; 16 to 24 degree Bé for soybean, sunflower, peanut, and corn oils; and 12 degrees Bé for palm, palm kernel, and coconut oils. Some refiners use pH to ensure that the reaction mixture contains the proper caustic treat before the primary centrifuge; pH targets will range from 9.8 for lauric oils to 10.8 for cottonseed oil [9].

- primary performance—Soap analysis should be performed on the oil exiting the primary centrifuge to maintain a soap level compatible with the water-wash capability, usually 300 ppm maximum. High soap contents above 500 ppm cannot be corrected with back pressure adjustments, indicating that the machine needs cleaning. Some refiners utilize a spin test to provide a quick estimate of moisture, which provides an indication of the soap content at this point. This evaluation consists of spinning the oil sample for 1.5 minutes on a high-speed centrifuge using a calibrated 10-cc tube. Hourly checks are recommended to ensure that the product is consistent [10]. The soapstock should be monitored for neutral oil content at least every 4 hours. Neutral oil in soapstock should be less than 18% on a dry basis.

- water-wash performance—The soapy water solution from the secondary centrifuge should be composited over a shift and analyzed for neutral oil content. The neutral oil content should be less than 0.05%.

- dried oil evaluation—Final refined oil control samples are generally obtained downstream from the vacuum drier. At this point, moisture should not exceed 0.1%, and in most cases 0.05% free fatty acid and 50 ppm soap content maximum are specified limits for all oils. Phosphatide containing source oils such as soybean and canola are also controlled by the residual gum level. Refined oil gums precipitate when treated with acetone for visual quantity estimates or more specific determinations with spectrophotometer evaluations. Another qualitative method utilized is the acid heat break test; heat 60 to 150 ml of refined oil with three drops of concentrated hydrochloric acid to 550°F (288°C) and observe visual appearance. Any darkening or dark residue indicates incomplete refining. With other oils, like cottonseed oil, completeness of refining is controlled by comparing the laboratory-refined and plant-refined oil color results.

- as refined finished oil—Every batch should be analyzed for color, free fatty acid, moisture, soap, phosphorous, impurities, and other evaluations specific for the source oil or the individual operation. The color, free fatty acid, moisture, and soap results should match the production results and meet the specification limits. A plant-refined oil bleach color significantly lower than the laboratory results indicates unnecessary overrefining. Phosphorous content must have been reduced to less than 30 ppm if standard prebleaching is utilized. If any of these analyses indicate a problem, steps must be taken to correct it in prebleaching or consider re-refining.

The process control quality standards for refining for a typical crude soybean oil could be specified as shown in Table 12.1.

Traditionally, the method used to refine meat fats has been what is now identified as physical refining. The impurities in meat fat consist of proteinaceous material from the rendering process and free fatty acid. The proteinaceous material must be removed by filtration with low levels of bleaching clay and/or diatomaceous earth or, alternatively, by water-washing followed by bleaching. After clarification, the clean dry meat fats can be further processed, i.e., deodorized or hydrogenated and deodorized [11]. Typical quality standards for a tallow physical refining process control specification is presented in Table 12.2.

Prebleaching of vegetable oils, performed after refining, is normally done for color reduction, and color analysis is the usual method for determining bleaching earth amounts. Another more recent practice is the adjustment

TABLE 12.1. Typical Refined Soybean Oil Specification Limits.

QUALITY STANDARDS:	To Refining	Primary Centrifuge	Water Wash	Vacuum Dried	As Refined
Free fatty acid, % max.	1.0	0.05	0.05	0.05	0.05
Neutral oil loss					
Crude oil, % max.	2.5	X	X	X	X
Soapstock, % max.	X	18.0	0.05	X	X
Soap, ppm max.	X	300	X	50	50
Moisture, % max.	X	X	X	0.1	0.1
Heat break test	X	X	X	neg.	X
Phosphorous, max.		X	X	X	30
Visible impurities	X	X	X	none	none
Bleach color, max.	3.5	X	X	X	3.5
Chlorophyll, ppm max.	X	X	X	X	30.0

TABLE 12.2. Typical Tallow Physical Refining Control Limits.

QUALITY STANDARDS:	To Filter	During Filtration	As Filtered
Lovibond red color, max.	1.0	1.0	1.0
Free fatty acid, % max.	0.4	0.4	0.4
Moisture, % max.	0.2	0.1	0.1
Impurities test	neg.	neg.	neg.
High heat color rise	0.5	0.5	0.5

of bleaching earth levels to achieve a zero peroxide value to ensure the removal of peroxides and secondary oxidation products. The process also cleans up traces of soap and phosphatides remaining from the refining process. Altogether, absorptive bleaching removes color bodies, peroxides, secondary oxidation products, soap traces, and phosphatides. The final step in bleaching is filtering to remove the bleaching earth with the absorbed impurities from the oil.

On startup, the oil should be recirculated through the filter until the desired bleach oil color and peroxide value is achieved, the soap is removed, a clean filterable impurities test is done, and an acceptable heat stress color is determined. A typical heat stress color test consists of heating the oil sample to 320°F (160°C) over a 4- to 6-minute period and immediately determining the Lovibond red color for comparison to the oil color before heating. A heat stress color increase indicates incomplete removal of the phosphatides. The color and filterable impurities evaluations should be repeated hourly to ensure that a filtration problem has not developed and that the oil quality is consistent. Vegetable oils subject to high chlorophyll contents, like soybean and canola oils, will be bleached to a chlorophyll content, rather than the usual red color endpoints; limits of 6 to 30 ppm are utilized on the basis of the finished product requirements.

Every batch of bleached vegetable oil should be analyzed for color, soap, filterable impurities, free fatty acid, and peroxide value. The finished oil analysis should be consistent with the in-process results. The free fatty acid content should be 0.1 to 0.2% higher than the refined oil, especially when acid-activated bleaching earths are utilized. Phosphorous results for vegetable oils, like soybean and canola oils, must be lower than 5 ppm if the oil is to be successfully hydrogenated or deodorized. For canola and corn oils, sulfur levels should be determined to assure that it has been reduced to less than 5 ppm. Sulfur adversely affects hydrogenation reaction rates and deodorized oil flavor.

Periodically, or if the bleaching media addition rate is questioned, the amount of bleaching earth in the oil to filtration should be determined. This can be done by filtering the bleaching earth out of a known sample quantity, washing the filtrate with a solvent, drying the cake, and weighing to determine the percent treat. This periodic evaluation will ensure that the bleaching earth addition is accurate. Excessive bleaching earth usage results in high oil losses without benefit, and insufficient earth usage reduces product quality. The spent bleaching earth should occasionally be evaluated for oil content. This is to ensure that the correct filter blowing and steaming practices are being observed. Excessive filter blowing will reduce the spent earth oil content to less than 30%, which can present a fire hazard, and the final oil steamed from the earth can be of poor enough

TABLE 12.3. Typical Bleached Soybean Oil Specification Limits.

QUALITY STANDARDS:	To Prebleaching	From Prebleach Filter	As Prebleached
Lovibond red color, max.	X	6.0	6.0
Heat stress color rise, max.	X	none	none
Peroxide value, meq/kg max.	X	0.1	0.1
Soap, ppm max.	50.0	0	0
Impurities test	X	neg.	neg.
Chlorophyll, ppm max.	30.0	6.0	6.0
Moisture, % max.	X	0.1	0.1
Free fatty acid, % max.	0.05	X	0.07
Phosphorous, ppm max.	X	X	5.0
Cold test, hours min.	X	X	25.0
Bleaching earth, %*	X	A	X
Spent earth oil, %**	X	X	30.0

*Analyze bimonthly from stream to filter.
**Analyze bimonthly.
A = Analyze.
X = Do not analyze.

quality to affect the entire batch. Insufficient filter blowing results in a high oil loss to the spent earth. Contamination is a concern with bleaching systems utilized for different source oils. This problem can be identified by changes in the iodine value for most oils, or if the source oil is a natural winter oil, a cold test will point out contamination [12].

Typical quality standards for a bleached soybean oil could be specified as seen in Table 12.3.

The bleaching system can be used as the clarification or filtering step to remove the impurities from meat fat as previously reviewed for refining process control. Meat fats require little bleaching unless heat-abused or green-colored tallow due to a high chlorophyll content. A green color is more noticeable in tallow due to the usual lack of red and yellow coloration. The chlorophyll is readily removed with acid-activated bleaching earth.

6.4 Hydrogenation Process Control

Edible fats and oils are hydrogenated for two reasons. First, flavor stability is improved by reducing the number of double bonds or sites for oxygen addition. Second, the physical characteristics are changed so that the product has more utility. Edible fats and oils products, such as shortenings, margarines, frying fats, coating fats, and other specialty products, all have unique performance characteristics resulting from hydrogenation.

Hydrogenation involves the chemical addition of hydrogen to unsaturated fatty acids. The reaction is carried out by mixing heated oil and hydrogen gas in the presence of a catalyst. Hydrogenation can be continued until all of the double bonds are saturated or until only partially hydrogenated. In practice, the conditions utilized for the hydrogenation process allow certain reactions to proceed at a faster rate than others and provide different degrees of selectivity or the preferential reduction of the unsaturated fatty acids in the oil. A huge variety of products can be produced with the hydrogenation process, depending upon the conditions used and degree of saturation. The fatty acid composition and resultant characteristics of the hydrogenated products depend upon the controllable factors: temperature, pressure, agitation, catalyst activity, catalyst type, and catalyst concentration. Changes in the reaction conditions affect the selectivity of the hydrogenated basestock. Selectivity affects the slope of the solids fat index (SFI) curve; steep SFI slopes are produced with selective hydrogenation conditions, while flat SFI slopes are the result of nonselective hydrogenation. The hydrogenation controllable conditions must be identified for process control specifications: catalyst type, nickel percent, temperature, and pressure. The other controllable factor, agitation, is usually fixed in plant converters so that the speed or pitch cannot be varied.

The oil to hydrogenation must be of suitable quality for effective hydrogenation. In general, the requirements should be the same as those for prebleached oils: less than 0.07 free fatty acid, less than 1.0 ppm soap, 0.1% maximum moisture, a low color, 5.0 ppm maximum phosphorous, and 10.0 meq/kg maximum peroxide value. All of these impurities can act as catalyst poisons. Phosphorous in the form of phosphatides mainly affects selectivity to produce a higher degree of saturation with a decrease in *trans* isomers. Moisture inactivates the catalyst and may promote free fatty acid formation by hydrolysis. Soap reacts with the nickel catalyst to form nickel soaps, which reduces the available nickel proportionally. The oil analysis should be rechecked to hydrogenation if it has been in storage after prebleaching for any length of time.

Hydrogen gas used for hydrogenation should be 99+% pure. Gaseous impurities, especially sulfur compounds such as hydrogen sulfide (H_2S), sulfur oxide (SO_2), and carbon disulfide (CS_2), are very injurious to catalyst. Impurities such as carbon monoxide (CO), carbon dioxide (CO_2), methane (CH_4), and nitrogen (N_2) do not affect catalyst activity [13].

Normally, a basestock system of hydrogenated products with varying degrees of hardness and selectivity is produced for subsequent blending with other basestocks or unhardened oils to produce the desired finished product. Consistency of most fats and oils finished products is controlled using standard AOCS analytical methods such as solids fat index, iodine value, and melting points. However, time restraints during the hydrogenation process require more rapid controls. The sampling method for hydrogenation control most generally consists of drawing samples for analysis at short intervals while approaching the desired endpoint. Three different control evaluations are used to determine the endpoints based on the hardness of the basestock:

- soft basestocks—Refractive index (RI) of oils is related to the degree of saturation; for given oil the refractive index is almost linearly related to the iodine value analysis, except it can be affected by factors such as free fatty acid, oxidation, and polymerization. RI alone is a satisfactory control for hydrogenation reactions to produce basestocks with a 90 iodine value or higher. Batch to batch, SFI results have been relatively uniform with RI controls for the soft basestocks when the starting oil iodine values are consistent.
- intermediate hardness basestocks—Oils hardened to iodine values less than 90 produce increasing amounts of *trans* isomers, which have a marked effect upon SFI results. Consequently, it is essential that the intermediate basestocks be controlled with, not only a refractive index, but also a melting point. Generally, it is the practice to hydrogenate to a refractive index as an advisory limit, but the actual control is the Mettler dropping point. AOCS method Cc 18-80 measures the temperature at which an oil sample becomes fluid enough to flow after conditioning in a freezer for 15 minutes.
- hardstocks—Hydrogenation of low-IV hardstocks is far less critical than any other type of hydrogenated product. These oils are saturated to iodine values of less than 10 where *trans* isomers are eliminated and melting points are usually in excess of 130°F (54°C). Refractive index is rarely used to control hardstocks due to the high melting points. Quick titer evaluations are good control evaluations for these products. The test consists of dipping a titer thermometer into the hardened oil sample and rotating it between the fingers until the fat clouds on the thermometer bulb. Each oil type and iodine value will have a different result, which makes the quick titer accurate enough for control of this hydrogenation process [14].

Successful hydrogenation endpoint analysis may be used to release product to the next process, postbleach, and further to storage or blending, especially if the basestock is produced routinely and statistical evaluations indicate that the process is in control. The practice of holding all hydrogenated batches separate until all the consistency analyses have been completed must be balanced against the cost of lost production and time. A complete SFI profile can require 6 hours from the time that the analysis

TABLE 12.4. *Typical Hydrogenated Soybean Oil Basestock Quality Limits.*

QUALITY STANDARDS:	To Converter	From Converter	From Catalyst Filter
Moisture, % max.	0.10	X	X
Free fatty acid, % max.	0.07	X	0.10
Lovibond red color, max.	A	X	*
Peroxide value, meq/kg max.	10.0	X	12.0
Filterable impurities, max.	none	X	none
Refractive index @ 46°C	A	1.45520 ± 0.00030	
Mettler dropping point, °C	X	30.0 ± 1.0	30.0 ± 1.0
Solids fat index @ 10.0°C	X	18.0 ± 3.0	X
@ 21.1°C	X	7.5 ± 3.0	X
@ 26.7°C	X	2.5 ± 1.0	X
Iodine value	X	89.0 ± 2.0	89.0 ± 2.0
Qualitative nickel	X	X	negative
HYDROGENATION CONDITIONS:			
Gassing temperature, °F		325 min.	
Hydrogenation temperature, °F		400 ± 10	
Pressure, psig		20 ± 2	
Catalyst, % nickel		0.02	
Catalyst type		new	

* Lighter than to converter.
A = Analyze.

is started. Nevertheless, the consistency analysis should be determined on each batch or some frequency for high-volume basestocks to assure that the relationship between the quick controls and the more time-consuming analysis for solids fat index, iodine value, melting points, fatty acid compositions, and so on.

Before the filtered product is released to postbleach, the oil should be analyzed for color, free fatty acid, and nickel content. A qualitative evaluation should be adequate at this point, such as ammonium sulfide added to a sample obtained after the catalyst filter. If a black precipitate forms, nickel is present. This test has been found sensitive to approximately 2 ppm nickel; more accurate determinations require atomic absorption or other more precise, but time-consuming, analyses.

The filtered catalyst should be tested on a periodic basis to ensure that the proper filter blowing practices are being observed. The nickel content of the filtrate affects the salability of this waste product to reclaimers, which may be the best means of disposal.

The hydrogenated oil color should be lighter than the oil to hydrogenation due to heat bleaching. A darker color could indicate incomplete filtration or too high a drop temperature from the pressurized vessel into the atmosphere. Filterable impurities analysis should differentiate between the two causes for the dark color. Free fatty acid content should be only slightly higher than the bleached starting oil unless an abnormally long reaction time was necessary. A high free fatty acid could indicate that the oil to the converter was wet. A high moisture could alter the solids fat index/melting point/iodine value relationships by poisoning the catalyst.

Typical quality standards for a hydrogenated soybean oil basestock process control specification are illustrated in Table 12.4.

6.5 Postbleach Process Control

The primary function of the postbleaching process is to insure that trace nickel contents, as well as peroxides and secondary oxidation products, are removed. Color reduction is secondary and should never be the deciding factor for postbleaching hydrogenated oil. However, if heat bleaching is experienced during hydrogenation, a color decrease during deodorization should not be expected. Then the oil must be bleached to a color below the color limit of the finished product, taking into consideration additives that may increase color results, such as emulsifiers, lecithin, and so on.

On startup, the oil should be recirculated through the filter until the desired peroxide value and Lovibond red color is achieved with a clean filterable impurities test. Qualitative nickel should also be determined if a positive result was determined after the catalyst filter. These analyses should be repeated hourly during filtering to ensure that a filtration problem has not developed and that the oil quality is consistent.

Every batch of postbleached oil should be analyzed for Lovibond color, peroxide value, filterable impurities, moisture, free fatty acid, and a consistency control to verify that contamination has not occurred. The results for the finished batch of postbleached basestock should be consistent with the bleaching in process results and the hydrogenation consistency controls. One exception is that free fatty acid should be 0.1 to 0.2% higher than the hydrogenated oil results. Citric or phosphoric acid is added to the oil before postbleaching to chelate any traces of nickel remaining after catalyst filtration. Consistency evaluations to determine contamination have to be the same analysis performed on the hydrogenated oil after the endpoint had been attained—refractive index, iodine value, Mettler dropping point, or quick titer. Contamination is indicated by a change in the consistency evaluations outside the recognized method variations.

Typical postbleach quality standards for process control for the same soybean oil basestock illustrated for the hydrogenation process are presented in Table 12.5.

TABLE 12.5. Typical Postbleach Soybean Oil Basestock Quality Limits.

QUALITY STANDARDS:	From Postbleach Filter	As Postbleached
Lovibond red color, max.	1.0	1.0
Peroxide value, meq/kg max.	0.0	0.1
Filterable impurities, min.	none	none
Moisture, max.	0.10	0.10
Qualitative nickel, min	negative	negative
Free fatty acid, % max.	X	0.12
Mettler dropping point, °C	X	30.0 ±1.0

6.6 Winterization Process Control

Oils that solidify or cloud at temperatures above 40°F (4.4°C) must be winterized or dewaxed to qualify as a salad oil. Winterized is the term used to identify the removal of the high melting point fraction or stearin that solidifies at low temperatures from certain vegetable oils, notably cottonseed oil and partially hydrogenated soybean oil. Soybean oil is a natural winter oil with a high resistance to clouding, but hydrogenation to improve flavor stability destroys this characteristic. After hydrogenation, soybean oil must be winterized to remove the high melting point fractions to regain the resistance to cloud at low temperatures.

The process of winterization involves a partial crystallization followed by separation of the solids from the liquid portion. The liquid fraction is used for salad oils; the solid fraction or stearin can be utilized in some shortening or margarine formulations. In the winterization process, the oil is cooled to a predetermined temperature for a predetermined period of time prior to separation of the liquid from the solid fractions. In practice, several different processes are utilized to winterize edible oils. The design of the winterization process, the rate of cooling of the oil, the temperature of crystallization, and the agitation of the oil are crucial and play a significant role both in separation of solid fats as distinct crystals and in helping separate them from the liquid oil. Parameters for these conditions must be specified separately for each product for process control.

Determination of the extent of the winterization of an oil is a simple procedure. The universally accepted method is the cold test (AOCS Method Cc 11-53), which measures the ability of an oil to resist fat crystallization. It is defined as the time in hours for an oil to become cloudy at 32°F (0°C). It is obvious that the cold test results are obtained after the fact; therefore, winterized oil is usually held separately until the testing period has exceeded the specified limits. Oils that fail the test must

TABLE 12.6. Typical Cottonseed Salad Oil Quality Limits.

QUALITY STANDARDS:	To Chill Tanks	From Filter	As Winterized and Filtered
Lovibond red color, max.	3.5	X	3.5
Free fatty acid, % max.	0.10	X	0.10
Filterable impurities, max.	none	none	none
Moisture, % max.	0.10	0.10	0.10
Peroxide value, meq/kg max.	2.0	X	4.0
Cold test, hours min.	X	10.0	10.0
Iodine value	109.0 ± 3.0	110.0 ± 3.0	110.0 ± 3.0

be rewinterized; those that pass are transferred to storage until deodorized.

The other salad oil control analyses should be consistent with the starting oil analytical results, except for the iodine value results. Winterized oils should experience an increase due to the hard fraction removal. Typical winterization quality standards for process control of cottonseed salad oil are shown in Table 12.6.

6.7 Dewaxing Process Control

Some vegetable oils, including sunflower, corn, and canola oils, contain waxes from the seed shell, which can produce a cloudy appearance. Waxes are high melting point esters of fatty alcohols and fatty acids with low solubility in oils. The quantity of wax in oil varies from a few hundred ppm to over 2,000 ppm. To provide an oil with adequate cold temperature clarity, the wax content must be reduced below 10 ppm.

A traditional dewaxing process, normally performed after prebleaching, consists of slowly cooling the oil under controlled conditions to crystallize the wax, which is then removed by filtration. Another process utilizes a wetting agent and centrifuges. An aqueous solution containing a wetting agent is metered into the oil stream, which is cooled in a heat exchanger before entering the crystallizers. The aqueous solution is then centrifuged out of the oil carrying the waxes with it. This aqueous/wax mixture is then passed through a heat exchanger and recentrifuged to recover the wetting agent from the waxes. These and other dewaxing processes are in use for processing oils containing waxes to prevent clouding on the grocery store shelf.

Two laboratory evaluations are utilized to determine if an oil has been properly dewaxed: cold test and chill test. Cold test evaluations should

TABLE 12.7. Typical Dewaxed Corn Oil Quality Limits.

QUALITY STANDARDS:	To Chill Tanks	From Filter	Dewaxed Oil
Lovibond red color, max.	5.0	5.0	5.0
Free fatty acid, % max.	0.10	X	0.10
Mositure, % max.	0.10	0.10	0.10
Filterable impurities, max.	none	none	none
Iodine value	124.0 ± 4.0	X	124.0 ± 4.0
Chill test, hours min.	X	24.0	24.0
Cold test, hours min.	X	5.5	5.5

conform to AOCS Method Cc 11-53. Some oils can have adequate cold test results but still develop a slight cloud or wisp of a cloud at room temperature. The cloud or wisp can be detected with a chill test that is performed by drying a portion of the oil sample by heating to 266°F (130°C), allowing it to cool to room temperature and filling a 4-ounce bottle. A second 4-ounce bottle should be filled with a portion of the sample that has not been heated. After 24 hours at 70°F (21.1°C), examine the samples for cloudiness. A cloud or wisp, probably on the bottom of the sample, in the heated sample indicates that the oil has not been dewaxed properly. Cloudiness in the unheated sample indicates moisture cloud. Cloudiness in both samples should initiate retesting to assure that the heated sample has been properly dried.

Removal of waxes from the oils does not change the physical characteristics of the oil. The iodine value for the starting oil and the dewaxed oil should be the same. Also, the quality type analytical results should not change appreciably either. A change in these results should initiate an investigation to determine the cause and the corrective actions necessary with as much concern as a low chill or cold test result generates. Typical dewaxed quality standards for corn salad oil process control are presented in Table 12.7.

6.8 Blending Process Control

The shortening, margarine base oil, specialty products, frying fats, and even some salad oils performance and other physical characteristics are finalized during the blending process. The formulation with hydrogenated basestocks and/or refined, bleached oils provided by product development, is blended to meet the analytical consistency controls. The finished product requirements determine which analytical characteristics are essential for each product. These requirements frequently include physical analysis, such as, solids fat index, iodine value, melting point, and fatty acid composition.

At this point, the product assumes a final identity for traceability. Contamination with other products will be a major concern for all further transfers and processing. An accurate, timely control to identify contamination possibilities is the Mettler dropping point. A result is possible within 30 minutes or less that has a usual accuracy of ±0.7°C within most laboratories for products with dropping points above 33.0°C (91.4°F) [15]. Determination of the Mettler dropping point after each movement for process is a reliable control for assurance that the contamination has not accrued. In cases where a change in the dropping point is experienced, the result many times points to the contaminating product.

Determination of the Lovibond red color and the peroxide value will indicate if the basestocks have been abused in storage to require bleaching before deodorization to eliminate secondary oxidation products and to achieve the final specified color limit. Some heat bleaching during deodorization may be expected; however, the heating during hydrogenation probably destroyed most of the unstable carotenoids in the basestocks. Typical blend process control limits for the composition and quality standards for an all-purpose emulsified cake and icing shortening utilizing the soybean oil basestock illustrated previously for the hydrogenation and post bleach processes are presented in Table 12.8.

TABLE 12.8. *Typical Blending Composition and Quality Standards*

COMPOSITION: Ingredients:	Spec. No.	Percent	Tolerance
HB 89 I.V. SBO Base	6689	90.0	± 1.0
HB 60-T CSO Hardstock	6401	10.0	± 1.0
TO BLENDING:		100.0	
QUALITY STANDARDS:		As Blended	
Solids fat index, %			
@ 10.0°C		25.0 ± 2.5	
@ 21.1°C		18.5 ± 2.0	
@ 26.7°C		18.0 ± 2.0	
@ 33.3°C		15.0 ± 1.5	
@ 40.0°C		11.0 ± 1.5	
Mettler droping point, °C		49.0 ± 2.0	
Lovibond red color, max.		1.5	
Filterable impurities, min.		none	
Peroxide value, meq/kg max.		2.0	
Free fatty acid, % max.		0.12	

6.9 Deodorization Process Control

Deodorization is the last major processing step in processing of edible fats and oils. It is responsible for removing both the undesirable flavors and odors occurring in natural fats and oils and those created by prior processes. This process establishes the oil characteristics of bland flavor and odor that are expected by the customers for salad oils, shortenings, frying fats, and other specialty performance products. Fats and oils deodorization improves flavor and increases the oxidation stability by nearly complete removal of free fatty acids and other volatile odor and flavor materials, partial removal of tocopherols, and thermal destruction of peroxides. The thermal treatment also heat bleaches the oil by destruction of the carotenoids that are unstable at deodorization temperatures.

Deodorization is primarily a high-temperature, high-vacuum, steam distillation process. The usual deodorization process path is deaeration, heating, steam stripping, and finally cooling of the oil, all with zero exposure to air. Careful attention to all of the processing steps is necessary to produce a quality deodorized fats and oils product. Deodorization cannot produce an acceptable product unless the feedstock has been properly processed and protected prior to steam distillation, which reduces free fatty acid content and eliminates offensive flavors and odors while heat bleaching the oil if the unstable carotenoids have not been eliminated previously. Deodorization will not destroy or remove secondary oxidation products as many processors mistakenly attempt. Experience has shown that flavor and odor removal correlates well with the reduction of free fatty acid content of the oil. A feedstock oil with a 0.10% free fatty acid will be rendered odorless and bland flavored when the free fatty acid has been reduced to 0.01 to 0.03%, assuming a zero peroxide value.

The deodorization process sampling points are (1) to the deodorizer if the blend or prebleached oil has been stored or waiting for deodorization more than 24 hours; (2) stream samples after the deodorizer filter; (3) finished deodorizer batch analysis; (4) after additives, like emulsifiers, antioxidants, and so on; and (5) after each 24-hour storage period. The blending process analytical analyses, with the exception of Mettler dropping point results, can be used as the "To Deodorizer" requirements if the analyses are not more than 12 to 18 hours old. The dropping point analysis should be performed each time an oil is moved to insure that no contamination has occurred. Stream sample after the filter should be obtained at short intervals, after every deodorizer charge or cycle, and during startup until the process is in control; that is, all the required quality analysis or flavor, color, and impurities results are in limits. Thereafter, the process stream should be analyzed either hourly or every 2 hours, depending upon operations confidence in the process.

TABLE 12.9. Typical Deodorization Composition and Quality Standards.

COMPOSITION: Ingredients:	Spec. No.	Percent	Tolerance	Total Percent
Undeod. Shortening Mix	7003	100.00	±0.0	94.1
AS DEODORIZED: Additives:				
43% Mono- and diglyceride	9501	6.25*	±0.5	5.9
FINISHED PRODUCT:		100.00		100.00

QUALITY STANDARDS:	To Deodorizer	From Filter	As Deodorizer	With Additives	After Storage
Lovibond red color, max.	1.2	1.2	1.2	1.5	1.5
Free fatty acid, % max.	0.12	0.03	0.03	0.15	0.15
Peroxide value, meq/kg max.	2.0	0.0	0.0	0.5	0.5
Moisture, % max.	0.10	0.10	0.10	0.10	0.10
Flavor, min.	X	bland	bland	bland	bland
Filterable impurities, max.	none	none	none	none	none
Mettler dropping point, °C	49.0 ± 2.0	X	49.0 ± 2.0	47.0 ± 2.0	47.0 ± 2.0
Solids fat index					
@ 10.0°C	X	X	X	24.0 ± 2.5	X
@ 26.7°C	X	X	X	17.0 ± 2.0	X
@ 40.0°C	X	X	X	10.0 ± 1.5	X
Alpha monoglyceride, %	X	X	X	2.8 ± 0.3	X

* Based on 100% deodorized base.

The finished batch, or "as deodorized" evaluations, repeat the quality analysis to assure that the process remained in control all during the process and physical or consistency analysis to assure that no changes have occurred due to contamination or isomerization. The consistency analyses are those that are required for the finished product, allowing for any additives that will effect changes when added. After these results are confirmed to meet the limits, the specified additives are incorporated, mixed for a minimum of 20 minutes, and the product sampled for the appropriate analysis.

If deodorized product is stored for 24 hours or more, it should be reevaluated daily for color, free fatty acid, peroxide value, and flavor. This daily testing will indicate reversion problems. The specified limits for these requirements must be met before the product can be transferred to packaging or shipped in tank trucks or railcars.

Typical deodoration process control limits for an all-purpose emulsified cake and icing shortening utilizing the blend previously illustrated are presented in Table 12.9.

6.10 Bulk Shipment Process Control

A high volume of edible fats and oils is shipped in tank trucks, railcars, or bulk containers primarily to food processors. These products must be handled properly in loading and transit, as well as the customers' tanks, to ensure the quality integrity at the time of use.

Before loading, it is important that the transportation equipment is of the proper type or size, in good mechanical condition, and properly cleaned. The bulk containers should be cleaned and inspected thoroughly before loading. The tanks must be completely clean and dry. Truck outlet valves should be inspected, as well as hoses and pumps, for cleanliness and moisture and evaluation of the protection available during shipment. On tank cars the foot valves must be operable and internal heating coils pressure tested for leaks. All of these inspections prior to loading should be documented to help assure that the inspections are actually performed and to record the condition of the transport at loading.

All bulk shipments in tank trucks, railcars, or containers should be core or zone sampled following AOCS Method C1-47 [16]. The sample should be evaluated for all the quality or organoleptic analyses, as well as Mettler dropping point and filterable impurities. These results determine the shipability of the car, truck, or container after loading. After quality release, the tanks and/or containers are sealed, preferably at all possible entry points, to protect against tampering during transit. The seals utilized should have to be destroyed to gain entry into the tank or container.

TABLE 12.10. *Typical Bulk Shipment Composition and Quality Standards.*

COMPOSITION: Ingredients:	Spec. No.	Percent	Tolerance
All-purpose emulsified shortening	8003	100.00	± 0.0
BULK SHIPMENT:		100.00	
QUALITY STANDARDS:		As Loaded	
Lovibond red color, max.		1.5	
Free fatty acid, % max.		0.15	
Peroxide value, meq/kg max.		0.5	
Moisture, % max.		0.05	
Flavor, min.		bland	
Filterable impurities, max.		none	
Mettler dropping point, °C		47.0 ± 2.0	
Alpha monoglycerides, %		2.8 ± 0.3	

A portion of the car, truck, or container sample should be retained. The retention time should be for a period at least three times longer than the customer could possibly have the product in storage; longer times approximating the shelf life of the customer's product may be required in some cases.

Typical bulk shipment release quality standards for the all-purpose emulsified cake and icing shortening illustrated for blending and deodorization process control are presented in Table 12.10.

6.11 Shortening Packaging Process Control

The physical form of shortenings is important for the proper handling and performance in prepared food products. Many food applications depend upon the physical properties peculiar to each packaged product, such as softness, firmness, oiliness, creaming properties, melting behavior, surface activity, workability, solubility, aeration potential, pourability, and others. For plastic shortenings, consistency is important from the standpoint of usage and performance in bakery and related products. For liquid shortenings, consistency is important for handling characteristics and adequate suspension of the additives to perform properly [17].

Shortening consistency is controlled by two dominating factors: (1) composition of the fat blend and (2) processing conditions used to crystallize, package, and temper the products. Experience has shown a definite correlation between the physical characteristics of the product blend and the consistency of the packaged shortening. Therefore, it is important to develop and specify crystallization conditions for each product for uniform functionality. The plasticization conditions that affect consistency are

- pre-cooler temperature—The heated shortening product is cooled to just above the melting point, 10 to 15°F (6 to 8°C), to reduce the heat load on the chilling unit, which provides the maximum cooling capacity. Precrystallization caused by cooling below the crystallization point must be avoided at this point.
- creaming gas—Nitrogen is injected into the inlet side of the chiller in precisely controlled quantities, normally 13.0 ± 1.0 by volume for standard plasticized shortenings, to provide a white creamy appearance and increase the shortening's workability.
- chiller unit pressure—A pressure control value should maintain a constant pressure of approximately 350 to 375 psig at the discharge of the pump to the precooler. Pressure requirements will vary with equipment design.
- chiller unit outlet temperature—The finished shortening consistency becomes softer as the temperature is decreased, while firmer, more brittle shortenings develop with higher temperatures. The chilling unit outlet temperature tolerance should be controlled to ±1.0°F if possible and no higher than ±2.0°F.
- chiller throughput—The product flow rate can be adjusted to control outlet temperature but should be within the design limits of the equipment.
- worker unit pressure—An internal pressure of 300 psig minimum on both the worker unit and the chilling unit should be maintained by an extrusion value after the worker. This pressure insures thorough air dispersion while breaking up the crystal aggregates.
- worker exit temperature—A temperature rise of 10 to 15°F (6 to 8°C) from the worker unit should be expected; heat of crystallization dissipation is the cause of the temperature rise.
- filler pressure—Extrusion values are utilized at the filler in most systems to deliver a homogenous smooth product. The usual operating pressure for most fillers is 300 to 400 psig.
- tempering temperature—The primary purpose of tempering is to condition the solidified shortening so that it will withstand wide temperature variations in subsequent storage and still have a uniform consistency when brought back to room temperature (70 to 75°F, or 21.1 to 23.9°C). In practice, holding at 85°F (29°C) for 24 to 72 hours or until a stable crystal form is reached is typical.

- storage temperature—The usual shortening storage temperature is 70 ± 2°F unless the customer has other specific requirements.

Shortening packaging process control points are (1) to chilling, (2) first and last piece line samples, and (3) packaged after tempering. The initial packaged product evaluations, or the "to chilling" sample, are performed to substantiate that contamination has not occurred during the transfer from the deodorizer and that the product has not been heat abused to affect color or flavor. Mettler dropping point or another physical characteristics analysis, like iodine value, solids fat index, or even fatty acid composition, could be utilized for this control, but the dropping point result has the advantage of a short elapsed time to perform and good reproducibility. The analyses determining quality acceptability are color, free fatty acid, peroxide value, flavor, moisture, and impurities. Specific limits should always be specified for each analysis, but changes from the previous results before movement should also be evaluated. Substantial changes, even if the results are still within the specified limits, should be cause for concern. Evaluation of changes can uncover problems that could develop into substantial problems after packaging.

First and last piece line samples are evaluated for contamination at startup and during the packaging operation to detect any deteriorative effects upon the product by the chilling, working, or filling process. The usual analysis to provide these controls are the same as the "to chiller" sample, with special attention to any changes in results. Potential problems at this point could be a result of the previous product packaged or impurities from the transfer lines, the equipment, any additives, creaming gas incorporation, packaging materials, or the atmosphere. Additionally, all shortening products should be in-line screened for impurities after chilling and should pass effective metal detection after packaging.

The final packaging product controls are performed after tempering for evaluation of the product characteristics the customer will experience. Product consistency changes are not complete when the product is packaged; the crystallization process continues even after tempering and shipping but at a substantially reduced rate. Consistency evaluations such as penetrations, appearance, and texture ratings, as well as performance evaluations, like the cake and icing tests for the illustrated product, should not be relied upon until after the tempering process; evaluation prior to this point would not be representative of the product as received or used by the customer.

Typical packaging process control quality standards for the all-purpose emulsified shortening, previously illustrated at the blend and deodorization process stages, with color and flavor added at packaging are presented in Table 12.11.

TABLE 12.11. Typical Shortening Packaging Process Control Standards.

COMPOSITION: Ingredients:	Spec. No.	Percent	Tolerance	Total Percent
All-purpose emul. shtg.	8003	100.00	± 0.0	99.97502
TO PACKAGING: Additives:		100.00		
22% Beta-carotene	1681	0.009*	± 0.0	0.00899
Flavor stock	1563	0.016*	± 0.0	0.01599
AS PACKAGED:				100.00000

QUALITY STANDARDS:	From Supply	After Additive	Line Samples	As Packaged
Lovibond red color, max.	1.5	X	X	X
Free fatty acid, % max.	0.15	0.17	0.17	0.17
Peroxide value, meq/kg max.	0.5	0.5	0.5	0.5
Moisture, % max.	0.05	0.05	0.05	0.05
Flavor, min.	bland	buttery	buttery	buttery

TABLE 12.11. (continued).

Filterable impurities, min.	none	none	none	none
Mettler dropping point, °C	47.0 ± 2.0	X	47.0 ± 2.0	47.0 ± 2.0
Alpha monoglyceride, %	2.8 ± 0.3	match	X	2.8 ± 0.3
Color standard, min.	X	X	match	match
ASTM penetration @ 80°F	X	X	X	150 ± 30
Appearance rating, min.	X	X	8.0	8.0
Texture rating, min.	X	X	X	8.0
140% sugar cake rating, min.	X	X	X	8.0
Icing volume, sp/gr. min.	X	X	X	0.80

CRYSTALLIZATION CONDITIONS:

Pre-cooler product temperature, °F	57.0 ± 2.0
Creaming gas, % nitrogen	13.0 ± 2.0
Chilling unit pressure, psig min.	350
Chilling unit outlet temperature, °F	70.0 ± 1.0
Chilling unit through put, lb/hour	33000 ± 300
Worker unit outlet temperature, °F	85.0 ± 1.0
Worker unit pressure, psig	300
Filler pressure, psig	375 ± 25
Tempering temperature, °F	85.0 ± 2.0
Tempering time, hours	48 ± 8
Storage temperature, °F	70.0 ± 2.0

*Based on 100% finished shortening.

6.12 Liquid Shortening Packaging Process Control

Liquid opaque shortenings have been developed for food products where fluidity at room temperature and below is important. The major uses for these products are deep fat frying, pan and grill frying, bread, cake, and nondairy applications.

Liquid shortenings are flowable suspensions of solid fat in liquid oil. The viscosity and suspension stability are controlled by two basic factors: (1) composition of the shortening and (2) the crystallization technique utilized. Hard fats with a low iodine value and beta crystal habit are used to seed the crystallization. The hard fat level can vary from as low as 1.0% to higher levels, as required to produce the desired finished product viscosity. Equally important for liquid shortening fluid stability are the processing conditions. These conditions must be conducive to the production of stable beta crystals in a concentration where the viscosity is low enough to effect easy pumping but high enough to prolong suspension stability.

Additives contribute heavily to the functionality in almost all liquid shortenings. The major additives utilized are: (1) antifoaming agents for frying applications; (2) emulsifiers for bread, cake, and nondairy applications; (3) color and flavor where these characteristics are desired; and (4) lecithin for flow properties in cakes and antisticking properties for pan and grill applications.

Liquid shortening crystallization, tempering, and packaging control points are (1) to crystallization, (2) as crystallized, (3) after additives, (4) line samples, and (5) as packaged. Initially, samples are drawn from the storage or supply vessel to determine that contamination has not occurred during the transfer from the deodorizer and that the quality or organoleptic characteristics have not been abused to adversely affect flavor and stability. The analysis most likely used to assure that contamination had not occurred would be iodine value. Mettler dropping point analysis could be utilized by maintaining the Mettler furnace in a freezer at $-20.0 \pm 2.0°C$ $(-4.0 \pm 3.6°F)$ to facilitate the analysis of relatively low melting points. The quality analyses are those normally evaluated for every deodorized oil: Lovibond red color, free fatty acid, moisture, impurities, and flavor. A control limit should be assigned for each analytical evaluation, but any significant change from the previous evaluations should be investigated for cause. The deaeration procedure will probably be performed in the same vessel as the supply tank for crystallization. One of the most destructive agents for liquid shortening suspension stability and viscosity is air. The deaeration process is simply heating without agitation to eliminate any air incorporated after deodorization. Evaluation of the air content in the oil should be determined at each control point by obtaining the density of the product and comparing it with the density of the product without air—laboratory deaerated.

After the product has been crystallized and tempered to stabilize the crystal, it should be sampled for quality analysis, air content, and either iodine value or Mettler dropping point to determine if contamination has occurred. After crystallization, the additives such as antifoamer, color, flavor, or lecithin are incorporated if required. If an antifoamer, methyl silicone, is the only additive, identification analysis cannot be performed on a timely basis, and actual weighing of the material and mechanical slurring to disperse the material must be relied upon for control. However, a tracer can be employed by dispersing the antifoamer in an antioxidant mixture that can be analyzed on a timely basis. Color, flavor, and lecithin additions at this point can also be measured with analytical controls.

During packaging, the line samples should be evaluated for air content and the other control analysis to confirm that abuse or contamination has not occurred. The as packed analysis are after the fact evaluations to confirm that the liquid shortening has the proper viscosity and/or pourability after crystal stabilization, the quality characteristics have not been abused and the product meets all of the customer identified or required parameters.

Potential liquid frying shortening crystallization and packaging process control specified limits are illustrated in Table 12.12.

6.13 Liquid Oils Packaging Process Control

Liquid oils in general are suitable for all cooking and some bakery products where lubrication is the only performance characteristic required. In addition to household uses, liquid oils are used for frying snack foods and some other products consumed shortly after frying. Salad oils usage is still required for mayonnaise and salad dressings.

There is a significant difference between cooking and salad oils. The practical definition of a salad oil is that it remains substantially liquid at refrigerator temperatures, 40°F (4.4°C). The standard method for evaluation of a salad oil is the cold test, AOCS Method Cc 11-53, which measures the resistance to crystallization of a sample immersed in an ice bath at 32°F (0°C). If it remains clear after 5.5 hours, it meets the criteria of a salad oil. A natural salad oil or a well winterized oil will remain clear for periods longer than 5.5 hours, usually 10 to 15 hours minimum.

A practical definition for a cooking oil is that it is normally clear at room temperature but will cloud or solidify at refrigerator temperatures. Cooking oils can be used either in their natural state or after processing, depending on the source oil, local taste, custom, and so on. Some of the source oils utilized as cooking oils are cottonseed, peanut, corn, sunflower, and olive

TABLE 12.12. Typical Liquid Frying Shortening Process Control Standards.

COMPOSITION: Ingredients:	Spec. No.	Percent	Tolerance	Total Percent
Liquid shortening	8875	100.00	±0.0	99.98750
TO CRYSTALLIZATION: Additives:		100.00		
THBQ antioxidant mixture	1527	0.01250*	±0.0	0.01249
Methyl silicone	1601	0.00001*	±0.0	0.00001
AS PACKAGED:				100.00000

QUALITY STANDARDS:	Crystallization		After Additives	Line Samples	As Packaged
	To	As			
Lovibond red color, min.	1.0	1.0	1.0	1.0	1.0
Free fatty acid, % max.	0.04	0.04	0.05	0.05	0.05
Peroxide value, meq/kg max.	0.5	0.5	0.5	0.5	0.5
Moisture, % max.	0.05	0.05	0.05	0.05	0.05
Filterable impurities, max.	none	none	none	none	none

TABLE 12.12. (continued).

QUALITY STANDARDS:	Crystallization		After Additives	Line Samples	As Packaged
	To	As			
Flavor, min.	bland	bland	bland	bland	bland
Mettler dropping point, °C	28.0 ± 1.0	28.0 ± 1.0	28.0 ± 1.0	28.0 ± 1.0	28.0 ± 1.0
Air content, % max.	none	1.0	1.0	1.0	1.0
Qualitative TBHQ, min.	X	X	positive	positive	positive
Viscosity, @ 85 ± 2°F	X	500 min.	500 min.	500 min.	X
Viscosity, @ 70 ± 2°F	X	X	X	X	6000 max.
Iodine value	X	X	X	X	102 ± 3

CRYSTALLIZATION and TEMPERING CONDITIONS:

			After Additives		
Pre-cooler temperature, °F			100.0 ± 2.0		
Creaming gas, %			none		
Chiller unit temperature, °F			70.0 ± 1.0		
Chiller unit pressure, psig min.			350		
Throughput, lb/hour			33000 ± 300		
Worker unit outlet temperature, °F			80.0 ± 2.0		
Worker pressure, psig min.			300		
Tempering vessel:					
Agitation, hours			2 ± 1		
Temperature, °F			85.0 ± 2.0		
Storage temperature, °F			70.0 ± 2.0		

*Based on 100% deodorized base.

oils. These oils will become semisolid or at least cloud at refrigerator temperatures unless winterized, dewaxed, or fractionated. However, it is somewhat impractical to remove the solid fractions from peanut and olive oils. Both these oils have distinctive flavors, which are preferred by many consumers over the bland flavors of most processed oils. Peanut oil is utilized both deodorized and undeodorized, while olive oil is rarely processed past extraction and filtration.

Liquid oils for household use are normally packaged in 16-, 24-, 32-, 48-, and 64-ounce and 1-gallon plastic containers, as well as some odd sizes for specialty oils like olive, grape, and others. One- and five-gallon plastic containers and 55-gallon closed head metal drums are the usual packages utilized for food service and food processor users. Typical oil filling operations involve the following sequence for household, food service and food processor packaging:

(1) Transfer the oil from the process department to the oil filling department supply vessel.
(2) Pump the oil through a heat exchanger to the appropriate filler.
(3) Containers are fed to the filler and filled with oil to the specified weight.
(4) The containers are capped and labeled prior to packaging in corrugated cases.
(5) The case is sealed, labeled, and printed with traceability information.
(6) The filled plastic containers are exposed to metal detection before transferring to storage or shipment. Metal drums must be inspected through the two openings in the top with a light before filling for foreign material such as rust, moisture, insects, or dirt.

The process control points for liquid oil filling systems are (1) supply tank, (2) line samples, and (3) as packaged. Initially, samples are obtained from the supply tank prior to the start of the filling process. Analysis of this sample is to confirm that the liquid oil product has not been contaminated or the flavor degraded during transfer or from the previous product in the supply tank. Line samples, normally obtained at startup and near to the batch completion, are evaluated for product changes caused by the packaging operation or some other problem, such as a mis-pumping into the supply tank, addition of an additive to the wrong product, a leaking tank coil, product filtration problems, and others. The as packed sample is a final evaluation of the product to reconfirm the previous samples and determine any customer specific analytical results not previously performed. The process control quality limits for a typical cottonseed salad oil are presented in Table 12.13.

TABLE 12.13. Typical Cottonseed Salad Oil Filling Process Control Standards.

COMPOSITION: Ingredients:	Spec. No.	Percent	Tolerance	Total Percent
RBW CSO salad oil	8301	100.00	± 0.0	100.00
TO OIL FILLING:		100.00		
AS PACKAGED:				100.00

QUALITY STANDARDS:	Supply Tank	Line Samples	As Packaged
Lovibond red color, max.	3.5	3.5	3.5
Free fatty acid, % max.	0.05	0.05	0.05
Peroxide value, meq/kg max.	0.2	0.2	0.5
Moisture, % max.	0.05	0.05	0.05
Filterable impurities, max.	none	none	none
Flavor, min.	bland	bland	bland
Cold test, hours min.	15.0	X	15.0
Iodine value	110.0 ± 3.0	110.0 ± 3.0	110.0 ± 3.0
AOM stability, hours min.	X	X	15.0
Qualitative TBHQ	negative	X	negative

6.14 Shortening Flake Process Control Standards

Fat or shortening flake describes the higher melting edible oil products solidified into a thin flake form for ease in handling for remelting or for a specific function in a food product. Chill rolls are utilized to produce both low-iodine value, hard fat flakes and slightly lower melting point specialty use products. Solidification of the flake products is accomplished when the roll picks up a coating of melted product from a supply trough, which solidifies rapidly and is removed by a scraper blade. Hard fat flake products are normally packaged in 50-pound kraft bags and the specialty products in fiberboard cases many times sized for the individual food processor's batch requirements. A typical chill roll operation probably has the following processing sequence to produce and package flakes:

(1) Transfer the deodorized flake oil product from the process department to a chill roll supply tank.
(2) Add and disperse any additive required by the finished flake product profile.

(3) The product in liquid form is transferred through a heat exchanger to adjust temperature and filters to the chill roll feed trough.

(4) A thin coating of the product is picked up by the rotating chill roll from the feed trough.

(5) The solidified product is scraped off the roll after one revolution.

(6) The flakes are packaged, labeled, identified for traceability and exposed to metal detection before storage to dissipate heat before shipping.

The process control points for the flaking operation are (1) supply tank, (2) after any additives, (3) as flaked, and (4) as shipped. Initially, as with any transferred product, a sample obtained from the supply tank is analyzed for assurance that the product has not been contaminated or abused during or after the transfer from the previous department. Sampling after any additives have been dispersed in the product acknowledges that the desired change has been effected. In some cases, especially for colorants, adjustments may be required to achieve the desired appearance because flaked sampling can indicate any abuse to the product during processing by moisture, impurities, heat, or metals that cause an off-flavor or color. The as-shipped sample confirms the as-packed sample and identifies any changes in color due to crystallization or free fatty acid due to hydrolysis. The process control quality control limits for a low-iodine value soybean oil hard fat are presented in Table 12.14.

6.15 Margarine and Spread Process Control Standards

Consumer margarines and spreads are used primarily for table spreads and cooking. Industrial margarines are utilized for many different specialized products such as danish, cakes icings, puff pastry and others. Two standards of identity regulate U.S. margarine production: vegetable oil margarines are regulated by the FDA, and animal fat or animal–vegetable margarines are subject to USDA inspection regulations. The margarine standards require 80% fat minimum, specify only safe and suitable ingredients usage, allow one or more listed optional ingredients, and require not less than 15,000 international units per pound of vitamin A. Products with less than 80% fat resembling margarine must be referred to as a spread unless the requirements for diet or reduced-calorie product are satisfied [18].

Margarine and spread processing consists of the following basic operations for a batch system:

(1) The margarine or spread oil bases are blended and processed to meet the product specification requirements. The physical properties of the

oil base contribute heavily to each product's eating characteristics, consistency, and plasticity.

(2) Liquid milk or whey, when used, is usually prepared by reconstituting dry milk or whey solids, pasteurized, cooled, and held in sanitary tanks until use.

(3) The water-soluble ingredients are weighed or metered into a sanitary tank and mixed thoroughly.

(4) The oil-soluble ingredients are metered or weighed into a sanitary tank and mixed thoroughly.

(5) The oil and water mixtures are brought together to form an emulsion.

(6) The emulsion is quick-chilled with a swept surface heat exchanger. For stick margarines and others where a firm narrow plastic range is desired, the product is allowed to solidify without agitation, that is, no working unit.

TABLE 12.14. *Typical Soybean Oil Hard Flake Process Control Standards.*

COMPOSITION: Ingredients:	Spec. No.	Percent	Tolerance	Total Percent
HB 63-T SBO Hardstock	8421	100.00	± 0.0	100.00
TO FLAKING:		100.00		
AS FLAKED:				100.00

QUALITY STANDARDS:	Supply Tank	As Flaked	As Shipped
Lovibond red color, max.	1.5	1.5	1.5
Free fatty acid, % max.	0.05	0.05	0.10
Peroxide value, meq/kg max.	0.3	0.3	0.5
Moisture, % max.	0.05	0.20	0.20
Filterable impurities, min.	none	none	none
Iodine value, max.	5.0	5.0	5.0
Flake condition	X	dry	dry

FLAKING CONDITIONS:	
Feed tank temperature, °F	160 ± 5
Oil to roll temperature, °F	200 ± 10
Chill roll inlet coolant temperature, °F	80 max.
Chill roll speed, rpm	Set to obtain cool, dry flakes
Oil trough level	1/2 to 3/4
In-package flake temperature, °F	110 max

TABLE 12.15. Typical Table Grade Margarine Process Control Standards.

COMPOSITION:

Ingredients:	Spec. No.	Percent
Pasteurized milk	9203	4.41666
Potable water	1000	13.31188
Granulated salt	1701	2.00000
Sorbic acid	1501	0.05000
Margarine base oil	8102	79.60000
Lecithin	1309	0.20000
Mono- and diglycerides	9507	0.40000
Flavor stock	1436	0.01500
Vitamin A–carotene mix	1575	0.00646
EMULSION TO CRYSTALLIZATION:		100.00000

QUALITY STANDARDS:	Base Oil	Pasteurized Milk	Emulsion	Line Samples	As Packed
Free fatty acid, % max.	0.05	X	X	X	X
Peroxide value, meq/kg max.	0.5	X	X	X	X
Flavor, min.	bland	sweet and clean	X	salty and buttery	buttery
Mettler dropping point, °C	36.0 ± 1.0	X	X	X	X
Moisture, %	0.05 max.	X	17.0 ± 0.3	17.0 ± 0.3	17.0 ± 0.3
Quevenne lactometer @ 60°F	X	35.0 ± 0.5	X	X	X
Total acidity (lactic), % max.	X	0.15	X	X	X

554

TABLE 12.15. (continued).

QUALITY STANDARDS:	Base Oil	Pasteurized Milk	Emulsion	Line Samples	As Packed
Appearance rating, min.	X	X	X	match standard	match standard
Fat, % min.	X	X	X	80.0	80.0
Salt, %	X	X	X	2.0 ± 0.1	2.0 ± 0.1
Curd, %	X	X	X	0.4 ± 0.1	0.4 ± 0.1
Vitamin A, U/lb min.	X	X	X	X	15,000
50°F penetration, mm/10	X	X	X	X	50 ± 15
Standard plate count/g, max.	X	X	X	X	10,000
Yeast and mold/g, max.	X	X	X	X	100
Coliforms/g, max.	X	X	X	X	10
PROCESSING CONDITIONS:					
Potable water temperature, °F			45.0 ± 5.0		
Pasteurized milk temperature, °F			45.0 ± 5.0		
Margarine base oil temperature, °F			130 ± 5.0		
Emulsion temperature, °F			105 ± 5.0		
Scraped wall heat exchanger exit temperature, °F			47.0 ± 1.0		
Worker unit			none		
Creaming gas, % nitrogen			none		
Tempering temperature, °F			45.0 ± 5.0		
Tempering time, hours			24 min.		
Storage temperature, °F			45.0 ± 5.0		
Shipping temperature, °F			45.0 ± 5.0		

(7) Tub spreads, margarines, and other products where spreadability or plasticity is desired utilize a working unit similar those used for shortening processing.

(8) Margarine or spread products are packaged and transferred to refrigerated storage in most cases, but some industrial margarines or spreads require holding at a specific temperature to stabilize the fat crystal.

The batch process sampling points for margarine and spread operations are (1) base oil supply tank, (2) pasteurized milk storage tank, (3) product emulsion, (4) line samples, and (5) as packed. The first two process control points may be required for the process control or the two ingredients, instead of as a part of the margarine or spread requirements. Both the oil base and the reconstituted milk could be utilized for several different finished margarine or spread formulations. Each emulsion should be evaluated for moisture level for a timely indicator and allow correction if required of the composition. The proper moisture level indicates that fat, salt, and curd levels are also at the desired concentrations. Line sample evaluations confirm that the color and flavor levels meet the specified limits and should verify the emulsion sample results for composition. Portions of the line samples should be composited for a shift analysis of the moisture, fat, salt, and curd contents. As-packed samples, preferably obtained shortly after startup and near the completion of each packaging shift, should be reevaluated for appearance, flavor, penetrations, and consistency after 24 hours to allow the fat crystal to stabilize. The flavor of this sample will be notably different from the samples flavored during packaging because of crystallization of the fat portion, which somewhat masks the flavor and salt perception. Portions of the as-packed samples must also be evaluated for microorganisms. A typical table grade margarine emulsion, crystallization, packaging, and storage process control specification is presented in Table 12.15.

7. REFERENCES

1. Juran, J. M. 1962. *Quality Control Handbook*. New York, NY: McGraw-Hill Book Co. p. 23-3.
2. Wintermantel, J. 1982. "Ingredients Specification Writing," *Cereal Foods World*, 27:65.
3. Wintermantel, J. 1982. "Ingredients Specification Writing, Part III," *Cereal Foods World*, 27:235.
4. Leo, D. A. 1985. *Bailey's Industrial Oil and Fat Products,* New York, NY: John Wiley & Sons. pp. 311–312.
5. Juran, J. M. 1962. *Quality Control Handbook*. New York, NY: McGraw-Hill Book Co. pp. 1-30–1-46.

6. Dutton, H. J. 1978. "Analysis of Fats and Oils," *J. Am. Oil Chem. Soc.*, 55:806.

7. Flider, F. J. and F. T. Orthoefer. 1981. "Metals in Soybean Oil," *J. Am. Oil Chem. Soc.*, 58:271.

8. Wiedermann, L. H. 1981. "Degumming, Refining and Bleaching Soybean Oil," *J. Am. Oil Chem. Soc.*, 58:159–160.

9. Carr, R. A. 1978. "Refining and Degumming Systems for Edible Fats and Oils," *J. Am. Oil Chem. Soc.*, 55:765–771.

10. Roden, A. and G. Ullyot. 1984. "Quality Control in Edible Oil Processing," *J. Am. Oil Chem. Soc.*, 61:1109–1110.

11. Latondress, E. G. 1985. "Refining, Bleaching and Hydrogenating Meat Fats," *J. Am. Oil Chem. Soc.*, 62:812.

12. Roden, A. and G. Ullyot. 1984. "Quality Control in Edible Oil Processing," *J. Am. Oil Chem. Soc.*, 61:1110.

13. Puri, P. S. 1980. "Hydrogenation of Oils and Fats," *J. Am. Oil Chem. Soc.*, 57:852A.

14. O'Brien, R. D. 1987. "Formulation—Single Feedstock Situation," in *Hydrogenation: Proceedings of an AOCS Colloquium*, Robert Hastert, ed., Champaign, IL: American Oil Chemists' Society, p. 156.

15. "Method Cc 18-80," in *The Official Methods and Recommended Practices of the American Oil Chemists' Society*, 1994, 4th ed. Champaign, IL: American Oil Chemists' Society.

16. "Method C 1-47," in *The Official Methods and Recommended Practices of the American Oil Chemists' Society*, 1994, 4th ed. Champaign, IL: American Oil Chemists' Society.

17. Bell, R. J. 1992. "Shortening and Margarine Products," in *Introduction to Fats and Oil Technology*, P. J. Wan, ed. Champaign, IL: American Oil Chemists' Society, pp. 187–188.

18. Chrysam, M. M. 1985. "Table Spreads and Shortenings," in *Bailey's Industrial Oil and Fat Products, Vol. 3.* T. H. Applewhite, ed. New York, NY: John Wiley & Sons, pp. 57–60.

CHAPTER 13

Troubleshooting

1. INTRODUCTION

Troubleshooting can be defined as the process used to determine why something is not performing properly and for correction of the problem. As with many practical skills, troubleshooting is an art, as well as an analytical or scientific process. Troubleshooting begins when it is recognized that a problem exists and it is determined that it needs to be solved. A basic analytical process for solving problems is outlined in Chart 13.1. The first step is to define the problem. This is a description of the problem that includes the symptoms and any other pertinent information. The next step is to gather information about the problem. This includes questioning the involved parties for more detailed information, viewing the physical systems, analyzing the product involved, and gathering information regarding the processes involved. The third step is to narrow the scope to the process or processes causing the problem. After the problem has been narrowed down, a corrective action for the problem evolves. The next to last step is to test the identified corrective action. If successful, implement the changes necessary on a temporary basis until proven before making the changes permanent. If unsuccessful, the problem-solving process must revert back to an intermediate step to redefine the corrective actions required [1].

There are many different troubleshooting styles and methods. Each has advantages and disadvantages. The approach chosen is, most times, personal preference, convenience, or, many times, dictated by the problem. Many times more than one approach or a combination of methods will be necessary to solve the problem. Good troubleshooters always use several

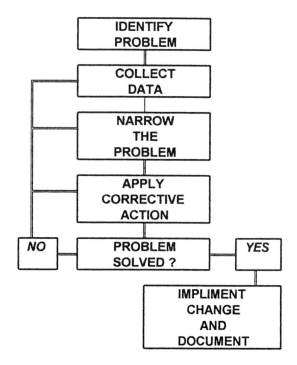

CHART 13.1. Troubleshooting Process.

approaches that are applied based on the problem symptoms. Some of the approaches applicable to edible fats and oils problem solving are

(1) Experience—the most common and usually the simplest, most expedient method or approach. An experienced technician can recognize many of the problems and the corrective action necessary immediately because it or something similar has happened previously. Also, if there are a number of possible solutions, experience can be used to identify the best area to attack first. This selection can be based on the highest probability, the most expedient, the least risk, the easiest, and so on. Experience is primarily on-the-job learned skills; usually, the more different job positions an individual has successfully performed, the more experience gained. However, experience can also be gained with hands-on laboratory and classroom training, but some actual on-the-job experience is normally necessary to reinforce the training. Good process and performance records are also a form of experience that can be effectively utilized for problem solving.

(2) When, what, and where—The three Ws' troubleshooting technique is a simple question and answer technique. It utilizes questions similar to the following:
- What is the problem? and What is not the problem?
- Where is the problem? and Where isn't the problem?
- When did the problem happen? and When didn't the problem occur?
- What has changed? and What hasn't changed?

This technique uses observation, experience, performance records, and testing to narrow down the problem to a workable size.

(3) Trapping a problem—Setting a trap is often the only effective method for catching the cause of a reoccurring problem. Systems employing trending of the archived data, along with possible additional monitoring points, can be employed to identify the problem area and help determine the necessary corrective actions.

There are probably as many troubleshooting approaches as there are individuals solving problems. Most of these approaches encompass the techniques described, at least in part. A common tread with all problem-solving techniques is identification of the problem and implementing corrective actions. These actions are almost always performed on a more timely basis and easier when experience is available. The remainder of this chapter is a compilation of edible fats and oils problems with corrective actions drawn from experience for the individual processes and product types.

2. PROCESSING TROUBLESHOOTING

2.1 Crude Oil Receipts

PROBLEM: MOLDY OIL SEED UTILIZATION

Possible Causes and Corrective Actions
(1) Aflatoxin—Moldy peanuts and other oil seeds, nuts, or fruits are routinely used as oil stock. When aflatoxic peanuts are crushed, only a small portion of the toxin is retained in the oil. Filtration of the oil lowers the aflatoxin level, and conventional refining and bleaching effectively remove any residual aflatoxin [2].

PROBLEM: DAMAGED SOYBEAN EFFECTS

Possible Causes and Corrective Actions
(1) Frost damage—Immature soybean plants are killed by severe frosts, which terminates development of the bean. The effects of frost damage on soybean oil a month prior to maturity are [3]

- increased fatty acids—Levels as high as 2.2% free fatty acid were determined with crude oil from frost damaged beans.
- excessive refining loss—Crude oil from immature beans can increase the refining loss substantially. Studies indicated more than a three-fold increase for frost damaged beans.
- higher green color—Crude oils from frost damaged beans are green in color because of the presence of porphyrins, identified by sharp absorption peaks in the 640–680 μ range, which are related to chlorophyll. The green color must be removed prior to caustic refining or in prebleaching before exposure to oxidation, which can set the color and becomes extremely difficult to remove. At prebleach, acid-activated clays have been found more effective than the neutral clays, and higher levels must be used as the green color intensity increases. As a general rule, soybean oil should be bleached to a green color limit rather than a red color limit.
- poor flavor stability—Increased bleaching requirements can decrease the flavor stability of the soybean oil. Some severely frost damaged oils have had to be declared inedible because of flavor stability and color problems.
- high saturated fatty acid content—Iodine values of oil from beans frost damaged a month before maturity have been found with over 7.0 IVs lower than oil from mature soybeans.

(2) Field and storage damage—Soybeans exposed to rain or damp weather in the field after maturity or stored at high moisture levels in excess of 13 to 14% result in decomposition of the oil fraction. The effects of the decomposition are [4]

- refining loss—Increased refining losses are attributed to two factors: (1) high free fatty acid content of the crude oil and (2) alteration and degradation of phospholipids. Prolonged moisture exposure activates a lipase within the beans, which results in hydrolysis. Degraded phospholipids become nonhydratable and act like emulsifying agents to entrain higher levels of neutral oil in the soapstock. Field- and storage-damaged soybeans have experienced oil losses in refining over 4.0%, as opposed to the normal 1.0 to 1.5% loss with undamaged beans.
- polymerization—Heat damage lowers polyunsaturated fatty acids by polymerizing linoleic and linolenic fatty acids. These effects are determined to flavor and oxidative stability of deodorized oils.
- trace metals—Two- to tenfold increases in iron content have been experienced with damaged soybeans from two sources: (1) that naturally occurring in the bean and (2) mainly that picked up from processing equipment. Metal inactivators, usually citric or phosphoric acid, must be utilized to reduce iron to levels that do not impair flavor and oxidative stability.

- nonhydratable phospholipids—Oils from field- and storage-damaged soybeans contained nonhydratable phospholipids that could not be removed easily by degumming or refining and ultimately affected the initial flavor and flavor stability of the finished salad oils produced.

PROBLEM: HIGH-COLOR COTTONSEED OIL

Possible Causes and Corrective Actions

(1) Off-grade—Off-grade or early season crude cottonseed oil production has required a higher level of acid pretreatment to obtain a lower bleach color and reduce refining losses. The amount of acid pretreatment also depends on where the cottonseed was grown and the local seasonal conditions. Wet humid climates produce higher colored oils. Phosphoric acid is used more commonly to pretreat crude oils because of its less corrosive properties and better commercial availability. The acid pretreatment level usually varies between 100 and 500 ppm by weight of the oil, depending upon the crude quality. However, some off-grade crude cottonseed oils have required pretreatment, with as much as 1,800 ppm phosphoric acid [5].

PROBLEM: IMMATURE CANOLA SEED

Possible Causes and Corrective Actions

(1) High chlorophyll—Immature canola or rapeseed with high chlorophyll levels can lead to a bluish green color in the oil, which is difficult to remove with refining. The maximum level for Swedish crude canola oil has been established at 3.0 red, 125 yellow on the 1.25-inch cell of the British Standard Lovibond Color Method. This corresponds to a chlorophyll level of 30 ppm. Above this level, extra processing is required to produce an oil with satisfactory flavor and stability [6].

PROBLEM: INADEQUATE MEAT FAT RENDERING CONTROLS

Possible Causes and Corrective Actions

(1) Dark color—Heating or cooking too long at high temperature during rendering scorches the protein content, which discolors the meat fat. Bleaching will probably be necessary to meet the expected color limits.

(2) Flavor—Steam-rendered animal fats flavor and odor can resemble boiled meats.

(3) Sour—Excessive protein and moisture allowed to remain in the rendered fat can cause souring during transit. Sour meat fats may be the major reason for USDA inspector rejections and enforced downgrading to inedible product.

(4) Dissolved phospholipids—Fat darkening will result unless the phospholipids are removed by water washing.

(5) High free fatty acid—The amount of free fatty acid an animal fat contains is a good indication of whether the fats were properly handled before rendering. Meat tissues contain fat-splitting enzymes, which start to hydrolyze the fat to form free fatty acids as soon as the animal dies. Rendering must be performed as soon after the animals are slaughtered as possible for a minimum of free fatty acid development. Increased processing losses will be experienced to neutralize any excess free fatty acids present.

(6) Proteinaceous material—Colloidal protein will cause darkening, especially with steam distillation or deodorization. Small amounts of diatomaceous earth and/or bleaching clay followed by filtration can remove these materials.

PROBLEM: GREEN TALLOW APPEARANCE

Possible Causes and Corrective Actions

(1) High chlorophyll content—Tallow can have a green appearance caused by a high chlorophyll content resulting from feeding fodder or pasture grasses. Since tallow contains low levels of red and yellow coloration, the green color is accentuated and should be readily apparent upon receipt. Tallow chlorophyll can be removed with acid-activated clay bleaching.

PROBLEM: YELLOW COLORED PALM KERNEL OIL

Possible Causes and Corrective Actions

(1) Palm oil absorption—Palm kernel oils with a yellow color and a carotene level above the norm of 4.4 ppm may be because of absorption of palm oil by a thinner tenera shell or because of breakage and palm oil penetration of the kernel. Palm and palm kernel oils are both obtained from the fruit of the palm tree. The outer pulp contains palm oil, and the kernel in the fruit contains palm kernel oil. Since these oils differ considerably in characteristics and properties, mixing causes substantial changes in performance.

2.2 Crude Oil Storage

PROBLEM: FREE FATTY ACID INCREASE

Possible Causes and Corrective Actions

(1) Blowing with steam during unloading—Discontinue using steam to blow lines, limit time used, or blow to a catch-all tank.

(2) Excessive moisture in receipts—Analyze before unloading and schedule high-moisture lots to be refined quickly to minimize free fatty acid increase.

(3) High-storage temperature—Do not heat crude oil during storage, repair leaking valves allowing unintentional heating, and do not allow centrifugal pumps to run against closed valves.

(4) Steam coil leaks—Repair coils.

(5) Tank bottoms—Agitate tanks and clean periodically to remove any sludge buildup.

(6) High-pressure steam—The steam pressure used to heat oils at any stage should be limited to 10 psig. This provides a steam temperature of 240°F (115.6°C), which should minimize overheating and scorching of the oils. Some facilities use only hot water to avoid contacting the oils with a temperature over 208°F (98°C).

PROBLEM: COLOR DARKENS DURING STORAGE

Possible Causes and Corrective Actions

(1) Long storage time—Rotate usage; first in/first out. Storage time should be limited to 3 weeks maximum.

(2) High-temperature oil—Refine oil as soon as possible.

(3) Aeration—Shut off agitators when liquid level is too low, and do not allow the oil to free-fall into the tank; fill the tanks from the bottom.

(4) Carotene decomposition—Carotenes decompose with oxidation to form compounds that are difficult to remove during refining and bleaching. Crude oil studies have shown that TBHQ antioxidant protects carotene against decomposition. The TBHQ-protected oils require less stringent bleaching conditions and a lower bleaching clay treat to obtain the desired product characteristics [7].

PROBLEM: COTTONSEED OIL BLEACH COLOR

Possible Causes and Corrective Actions

(1) Crude storage—Good quality crude cottonseed oil can be successfully stored for 5 months in cool, but not cold, weather without any appreciable change in FFA, color, flavor, and refining loss, but the ease of bleaching is decreased appreciably [8]. Oxidation of the complex pigment system in cottonseed oil affects bleach color. Refining before extensive oxidation of the pigments occurs removes the source of the color reversion. Therefore, it is more desirable to store cottonseed oil as refined, rather than as crude.

PROBLEM: REDUCED OXIDATIVE STABILITY

Possible Causes and Corrective Actions

(1) Crude oil storage time—Contact with atmospheric oxygen will cause oxidative damage, which is ultimately reflected in the quality of the processed oil. Vegetable oils with the lightest colors and best oxidative stability are processed almost immediately after extraction or those immediately protected with nitrogen.

(2) Antioxidant protection—Studies have shown that TBHQ antioxidant substantially increases the oxidative stability of vegetable oils and has a stabilization effect upon the natural tocopherol antioxidants. The AOM stability of crude soybean oil has been doubled with the addition of 200 ppm TBHQ with a significant improvement in the tocopherol content of the deodorizer distillate [7].

PROBLEM: EXCESSIVE TANK BOTTOMS

Possible Causes and Corrective Actions

(1) Poor or no agitation—Install properly designed agitators and agitate oil, except when liquid level is below the agitators.

(2) Poor quality crude oil—Analyze before unloading and refine as soon as possible to minimize settling of impurities.

(3) Tank sweating—A combination of high humidity, warm oil, and cool temperatures can result in sweating inside storage tanks. This moisture dripping into the oil is an effective degumming system. The result is a buildup of unwanted solids in the bottom of the tank. Sweating can be controlled somewhat by vent systems that allow airflow into and out of each storage unit.

PROBLEM: CONTAMINATION

One of the most common problems in any tank farm is the unwanted mixing of oils by operator error, faulty equipment, or design problems [9].

Possible Causes and Corrective Actions

(1) Piping—Where economically possible, separate piping systems for each major oil handled or at least oils with critical physical properties should be provided.

(2) Tank size—The use of a few large tanks, rather than numerous small tanks can minimize the chances of contamination.

(3) Valves—All valves should be chosen so that it is easy to recognize when in either the open or closed position.

(4) Flow system—Automated systems are available that will not allow a

pumping to begin until the correct valves are open or shut and other specific criteria are met.

(5) Spill prevention—Ruptured tanks, broken lines, or human error have resulted in lost product and significant harm to the environment, which could have been contained. Each storage facility must be properly diked or a drainage system constructed that will contain any conceivable spill.

(6) Procedures—The secret to a good storage facility operation is good planning and adherence to basic rules and common sense. Written operating procedures provide a means of operator training and guidance, while checklists allow daily followup and planning.

PROBLEM: STORAGE TANKS

The number, size and construction material of the storage tanks are among the most crucial points to be considered when planning the operation of an existing or new oil processing facility. The choices depend upon the quality of the oil to be stored; i.e., crude oil has different requirements than intermediate or finished product.

Possible Causes and Corrective Actions

(1) Number and size—A one oil processing facility needs larger, but a lesser number of, tanks than a facility handling many different source oils. Where many oils must be stored, a series of larger and smaller tanks is required to decrease the danger of contamination or mixing the different oils.

(2) Design features—The normal crude oil storage tanks have some very common design features for optimum performance:
- Cylindrical tanks are preferred over rectangular tanks, which are generally more expensive and more difficult to clean.
- Welded mild steel construction is used.
- A high height to diameter ratio for maximum protection of the stored oils from air exposure is necessary.
- The feed pipe should run from the top to the bottom, with a vent close to the top bend.
- The tank bottom should have a slope to a sump with a drain line for cleaning.
- The tank outlet should be about 1.5 feet (0.5 meters) above the bottom of the tank and at the side to allow sludge settling.
- Hot water heating coils should enter from the top of the tank should be shaped like a vertical inverse-U at one side of the tank to provide space for the melted material to expand.
- Side entering agitation counteracts layering in large tanks.
- Outside insulation is used where necessary.

- A plastic coating or lining is used to protect against fatty acids, if necessary—to be determined by the source oil to stored.

PROBLEM: INVENTORY MEASUREMENT ERRORS

Possible Causes and Corrective Actions

(1) Gauging—Six factors are involved in product weight determinations in any gauge measurement procedure in which an error in any one factor will create a discrepancy [10]:

- accurate precalibration of all the tanks involved—Charts supplied for the individual tanks must be assumed to be correct; however, some sources of error exist:

    —Tank bottoms may sag or buckle with varying loads of oil. The tanks should be checked periodically with weighed oil to verify to calibrations.

    —Tanks heated to keep a product liquid will expand to provide a greater volume than calibration at lower temperatures. This increase in volume can be calculated from the known coefficient of expansion of steel.

    —Irregular shapes and internal structures of the tanks may have affected calibrations to make the charts only approximate.

- proper sampling of the oil product for analysis—The sample obtained must represent accurately the entire amount of the oil product in the tank. If moisture or solid fractions are present, these materials must be included accurately in their proper amount by zone sampling.

- accurate temperature measurement of the product in the tank—The number of readings needed to obtain a true average temperature may vary greatly, depending upon the size and depth of the tank availability of agitation and heating coils. The importance of temperature is illustrated by the fact that an error of 1°F will result in a weight difference of approximately 1,300 pounds in a 400,000-gallon tank.

- accurate measurement of the depth or outage—Measurements should always be made at the same marked place on the tank opening with a high-quality steel tape, free from kinks and bends with clear readable figures and readings to the nearest one-sixteenth inch. Measurements should be repeated until successive results are identical. Two problems experienced in obtaining satisfactory tape readings are oil motion from heating and/or agitation and entrained air causing foam.

- accurate specific gravity analysis of the sample are mandatory for accurate calculations

- proper calculations based on the tank calibration and the data obtained.
(2) Metering—Under ideal conditions of pump pressure and oil temperature, metering is possible. However, ideal conditions are rarely met, particularly for weight determinations of bulk shipments or inventories where accuracy is required.
(3) Weighing—Actual weighing of the oil is the most desirable procedure. Weighing is accomplished with many different types of scales, i.e., tank, railcar, truck, and even drum scales. These scales must be calibrated frequently to assure accuracy.

2.3 Degumming

PROBLEM: DARK COLOR GUMS

Possible Causes and Corrective Actions
(1) Acid degumming—Phosphoric acid degummed soybean oil improves the refined oil yield but provides darker gums. The darker gums produce inferior lecithin products; the lecithin produced is darker in color than lecithin from water degumming processes. However, most soybean oil gums are not processed for lecithin recovery, and the separated gums are mixed with the soapstock obtained with caustic refining.

PROBLEM: LOW YIELD

Possible Causes and Corrective Actions
(1) Excess water—It is important to add only the amount of water necessary to precipate the gums, since any water excess causes unnecessary oil losses through hydrolysis. The normal water level for degumming is approximately 2.0%.

2.4 Caustic Oil Refining—Primary Centrifuge

PROBLEM: HIGH FREE FATTY ACID (OVER 0.05%)

Possible Causes and Corrective Actions
(1) Caustic treat too low—Check the caustic strength and correct if found low, or recheck the free fatty acid of the crude oil and recalculate the treat required. The theoretical quantity of caustic is based on the ratio of molecular weights of NaOH to oleic fatty acid. Most oils are refined

with 0.10 to 0.13% excess caustic. Exceptions are the lauric oils and palm oil, which require a minimal excess of approximately 0.02%, and cottonseed oil, which is refined for color reduction, requires a higher excess of approximately 0.16%.

(2) Inadequate mixing of caustic with the crude oil—Assure that the high-speed, in-line mixer and the vertical mixers are functioning properly.

PROBLEM: HIGH RED COLOR

Possible Causes and Corrective Actions

(1) Low caustic treat—When a high refined color is experienced on the startup samples, reconfirm the caustic treat and increase slightly, if needed, to obtain acceptable lab bleach color.

(2) Poor quality crude oil—The final step in most crude oil production is final solvent stripping. Temperatures in this step may reach 250°F (121°C). The longer an oil remains at the elevated temperatures, the harder it is to remove the color bodies in caustic refining. This treatment can cause color fixation. An increase in caustic treat may be necessary to obtain an acceptable refined and lab bleach color.

(3) Insufficient mixing—Confirm that the mixers are operating properly and sufficient caustic–oil mixing is being delivered.

(4) Oxidized tocopherols—alpha-Tocopherol oxidizes to a very colorful chroman-quinone. Determination of the cause of oxidation and corrective maintenance or procedures are necessary to prevent a reoccurrence of the problem. The high red color should be removed with vacuum bleaching.

(5) Cottonseed oil storage—It has been determined that cottonseed oil should be refined as soon as possible. Oxidation of the complex pigment system adversely affects the bleach color. Refining before extensive oxidation of the pigments occurs removes the source of the color reversion.

PROBLEM: HIGH GREEN COLOR

Possible Causes and Corrective Actions

(1) Chlorophyll content—High chlorophyll contents in soybean and canola oils are usually caused by crop damage by exposure to wet weather in the field, early frosts, or high moisture storage. Chlorophyll is better removed before alkali treatment since it tends to be stabilized by alkali and heat.

PROBLEM: POSITIVE ACID HEAT BREAK

Possible Causes and Corrective Actions

(1) Low caustic treat—Recalculate the caustic treat and increase if necessary. Use the minimum caustic treat to obtain a negative acid heat break.

(2) Uneven caustic flow—Adjust caustic flow each time the crude flow is changed and determine that the caustic pump is operating properly.

(3) Inadequate caustic and crude oil mixing—Ensure that the mixers are operating properly.

PROBLEM: HIGH SOAP CONTENT

Possible Causes and Corrective Actions

(1) High caustic treat—Check caustic treat and reduce if the acid heat break and lab bleach color are not adversely affected. The NaOH treat selected for the crude oil to be refined will vary with the free fatty acid content of the oil and the level of excess over theoretical determined for each source oil from previous experience. The theoretical quantity of caustic is based on the ratio of molecular weights of NaOH to oleic fatty acid. Most oils are refined with 0.10 to 0.13% excess. Exceptions are lauric and palm oils, which require a minimal excess of approximately 0.02%, and cottonseed oil, which is refined for color reduction, requires a higher excess of approximately 0.16%.

(2) Dirty centrifuge bowl—High soap levels, 500 ppm or higher, that cannot be corrected with back pressure adjustments indicate that the machine needs cleaning.

(3) Crude oil flow rate too high—Reduce the flow rate but make sure that the caustic treat is adjusted accordingly.

(4) Separator bowl rpm too low—Check rpm and correct if found low.

(5) Improper back pressure—Adjust back pressure to reduce the residual soap level in the oil without increasing the oil content of the soapstock. The back pressure usually falls in the range of 50 to 100 psig. Careful back pressure control is required to obtain a minimal soap content in the refined oil while achieving minimum neutral oil loss in the soapstock. The compromise adjustments usually border around 300 ppm soap, with 18% neutral oil in the soapstock on a dry weight basis.

PROBLEM: EXCESS NEUTRAL OIL IN SOAPSTOCK

Possible Causes and Corrective Actions

(1) High centrifuge back pressure—Increased centrifuge back pressure re-

duces the soap content in the oil phase but increases the neutral oil lost in the soapstock. Conversely, reducing the back pressure decreases the neutral oil loss in the soap phase, but can increase the soap in the refined oil to a level beyond the capacity of the water washing step. The primary back pressure must be carefully adjusted to the best compromise, usually about 300 ppm soap with 18% maximum neutral oil in the soapstock.

(2) Caustic addition temperature—Too high an oil temperature during the caustic addition can increase the saponification rate of the neutral oil and reduce the yield of the refined oil. The temperature of the crude oil at the point of caustic addition should be no higher than 100°F (37.8°C) with a more preferred limit of 90°F (32.2°C).

(3) Primary feed temperature—Most soft oil caustic mixtures are heated to 135 to 165°F (57 to 74°C) before delivery to the primary centrifuge to provide the thermal shock necessary to break the oil–caustic–soap emulsion. The best emulsion temperatures for each oil must be determined by experimentation or past experience.

(4) Centrifuge breakover—A breakover occurs when the interface suddenly moves toward the bowl periphery and no longer contains the column of oil under the top disc. Large amounts of oil are discharged through the soap outlet in a partial breakover. When a total breakover occurs, the large oil flow quickly forces all the soap from the bowl and all the oil follows through the soap outlet port. Breakovers can happen when the centrifuge bowl is clogged, the refined oil outlet back pressure is too high, or the refined oil temperature or caustic concentration drops considerably. Breakovers can usually be corrected by quickly opening the back pressure valve and readjusting until the soapstock and refined oil are discharging properly. The bowl may have to be reprimed or back-flushed to correct large breakovers [11].

(5) Caustic–crude oil contact time—After the caustic solution addition, the mixture must be blended to insure adequate contact with the free fatty acids, phosphatides, and color pigments. Caustic reacts with the free fatty acids to form soapstock, while hydrolyzing phosphatides and removing unsaponifiable material from the crude oil. Sufficient mixing is required, but it must not create a stable emulsion and allow proper soap conditioning. Experimentation may be required to determine the optimum conditions using past experience as guidelines.

(6) Solvent-extracted oils—Many times, solvent-extracted oils contain high levels of phosphatides and other thick, gummy, sticky materials that form deposits in storage tanks. These materials are difficult to handle, especially as oxidation and polymerization develop. The tank residues

can trap high levels of neutral oil to cause high oil losses. The gums emulsify the neutral oil, which is lost to soapstock during refining. Further, splitting the soapstock with acid is more difficult when high gum levels are present. Degumming of solvent-extracted oil prior to caustic refining has become customary to avoid these problems by some processors, especially in Europe [12].

PROBLEM: OVERREFINING

Possible Causes and Corrective Actions
(1) Low refined oil bleach color—A plant refined oil bleach color significantly lower than the laboratory results indicates unnecessary overrefining.

PROBLEM: POOR FINISHED OIL OXIDATIVE STABILITY

Possible Causes and Corrective Actions
(1) Oxidized refined oil—It is very important that oxidation be prevented during all stages of edible fats and oils processing because of the detrimental effects upon the finished, deodorized oils. For example, it has been estimated that a peroxide value increase of only 1.2 meq/kg for neutralized soybean oil after or during refining may reduce the oxidative stability of the deodorized oil by 50% [12]. Oxidation during processing can be minimized by avoiding contact with air, elimination of prooxidant metallic contaminates, nitrogen protection at all stages of processing, and/or the addition of antioxidants at the initial processing stages.

2.5 Oil Refining—Water Wash

PROBLEM: HIGH SOAP CONTENT

Possible Causes and Corrective Actions
(1) Low level of wash water—The amount of water required for water washing is 10 to 20% by weight. Generally, the best results are obtained at the 15% level. Determine wash water level and increase if below 15% of the crude oil flow rate.
(2) Inadequate mixing of wash water and refined oil—The water–oil combination is mixed to obtain intimate contact for maximum soap transfer from the oil to the water phase.
(3) Hard water—Hard water with calcium and magnesium deposits should not be used for the water washing operation. If the water hardness exceeds 20 ppm, a water softener or condensate from the steam system should be used.

(4) Dirty separator—The time duration between cleanings cannot be predicted; the water wash bowl must be cleaned on demand. The water wash centrifuge probably needs cleaning when the residual soap for the refined oil exiting the vacuum dryer exceeds 50 ppm if the soap from the primary is less than 500 ppm.

(5) High soap content from the primary—The water wash operation can only handle incoming soap levels of 500 ppm or less.

(6) Water addition temperature—The wash water temperature is important for efficient separation in the centrifuge. Wash water temperature should be 10 to 15°F (5 to 8°C) warmer than the oil to prevent emulsions.

(7) Oil temperature—Refined oil from the primary centrifuge should be heated as necessary to 190°F (88°C). A low temperature to the water wash centrifuge can cause breakover.

2.6 Oil Refining—Vacuum Drying

PROBLEM: HIGH MOISTURE

Possible Causes and Corrective Actions

(1) Low vacuum—A typical dryer operates at 70 cm Hg and should be equipped with a high-level alarm and automatic shutdown capability.

(2) Low water wash temperature—Normal water wash temperature is 185°F (85°C).

(3) High moisture to the dryer—Correct the problem in the water wash operation.

(4) Poor oil distribution in the dryer—Clean the plugged nozzles.

2.7 Prebleaching

PROBLEM: SHORT FILTRATION CYCLE

Possible Causes and Corrective Actions

(1) Inadequate body feed—The flow rate during precoating should be the same as during filtration. A filter with too slow a precoat rate or with an uneven hydraulic flow will produce an uneven precoat that results in blinded screens and short cycles.

(2) Too high flow rate—High flow rates cause packing of the solids to blind the screens.

(3) Too low flow rate—Low flow rates allow the solids to settle in the filter shell instead of precoating the screens.

(4) Blinded screens—Screens blinded with soap, uneven flow, or for another reason provide a reduced surface area that causes short cycles.

(5) Filter too small—A solids load too great for the filter utilized will necessitate frequent process interruptions to clean the filter.

PROBLEM: SLOW FILTER RATE

Possible Causes and Corrective Actions
(1) Full filter—Clean the filter.
(2) High moisture oil (over 0.1%)—Correct upstream problem at vacuum dryer or water wash.
(3) High soap—The soap content of vacuum-dried oil to prebleaching should not exceed 50 ppm. Higher soap levels require excessive bleaching earth usage or allow soap to remain in the oil, which can blind the filters.
(4) High phosphatides—Correct an upstream problem that was probably indicated by a negative acid heat break evaluation and caused by an improper caustic treat or the operation of the mixers. The phosphorus level in refined oil to bleaching should not exceed 6.0 ppm. Higher levels will require higher bleaching earth usage levels and increase the risk of higher phosphorus in the bleached oil. Bleached oil should not exceed 1.0 ppm phosphorus.
(5) High oil flow rate—The flow rate through the filter must not exceed the design limit. High flow rates cause premature blinding of the filter and reduce efficiency. A flow rate controller should be required at the exit of the filter.

PROBLEM: HIGH IMPURITIES IN FILTERED OIL

Possible Causes and Corrective Actions
(1) Inadequate precoating of the filter—Recirculate until the impurities improve; if impurities persist, stop and precoat the filter properly.
(2) Holes in the filter screens—Repair or replace the screens.
(3) Improper filter assembly after cleaning—Reassemble properly.
(4) Worn seals or gaskets—Replace defective seals and gaskets.
(5) Early delivery—During the startup cycle of the bleaching filters, care must be taken to return the initial cloudy oil to the bleaching vessel instead of contaminating the bleached oil storage.

PROBLEM: HIGH SOAP LEVEL IN BLEACHED OIL

Possible Causes and Corrective Actions
(1) High incoming oil soap content—Correct upstream problem, probably at the primary or water wash processes. The soap content of the in-

:uum-dried oil should not exceed 50 ppm. Higher bleaching ls are required to adsorb the excess soap.

:oating—Precoat filter properly to avoid channeling.

ate recirculation through the filter—Proper contact time be-
ne oil and the bleaching earth is essential.

(4) Blea⌣ ng earth addition—Ensure that the bleaching earth feeder is operating properly.

PROBLEM: POOR BLEACH COLOR

Possible Causes and Corrective Actions

(1) Low bleaching earth level—The amount of bleaching earth to be used should be the minimum amount needed to effect removal of the impurities as measured by peroxide reduction with a zero value as the target. Normally, 0.3 to 0.5% activated bleaching earth is sufficient, depending on the quality of the oil and press effect opportunities. The clay addition should be guided by performance and not a mandatory addition level.

(2) Short contact time—Recirculate through the filter and determine the effect. Usually, 15 to 20 minutes contact time between the oil and the bleaching earth is adequate to drive off the moisture in the bleaching earth and complete the bleaching reaction with atmospheric bleaching. In vacuum bleaching the greatest color reduction is realized within the first 3 minutes of contact because the air and moisture is removed by the vacuum. Extending the contact time does not harm the vacuum-bleached oil because oxidation is prevented by the vacuum.

(3) Low bleaching temperature—Bleaching temperatures can be increased but should not exceed 110°C (230°F). Heating should take place after the addition of the bleaching earth. Higher oil temperature reduces oil viscosity, which improves the adsorptivity of the bleaching earth; however, no significant improvement is experienced above 110°C. Problems with increased free fatty acid development, oxidation, and color fixing can result with temperatures exceeding 110°C.

(4) Bleaching earth addition temperature—Addition of the bleaching media to hot oil vaporizes the moisture in the clay too soon. The moisture release allows the lattice structure in the clay to collapse, which reduces the effective surface area before it has an opportunity to adsorb the color pigments and the secondary oxidation products.

(5) Agitation—Bleaching vessel agitation should be sufficient to achieve good contact of the earth with the oil without incorporating air.

(6) Inadequate refining—Check upstream problems: caustic treat or mixing in refining.

(7) Color set in the incoming crude oil—Re-refine or segregate for use where higher color is acceptable.

(8) Green color removal—High chlorophyll in soybean and canola oils caused by crop damage may be easier to remove before alkali treatment since it tends to be stabilized by alkali and heat.

(9) Cottonseed oil crude storage—Good quality crude cottonseed oil can be stored for up to 5 months at cool temperatures with no appreciable changes in FFA, refining loss, color, or flavor; however, bleach color noticeably increases [8]. Oxidation of the complex pigment system in cottonseed oil affects bleach color. Refining before extensive oxidation of the pigments occurs removes the source of the color reversion. Therefore, it is more desirable to store cottonseed oil as refined, rather than crude.

PROBLEM: HIGH OIL LOSS

Possible Causes and Corrective Actions

(1) Excessive bleaching earth—Activated bleaching earths can retain as much as 70% of their weight in oil, which is not recovered. Increased bleaching earth usage improves prebleaching effectiveness up to a certain point. After that, additional bleaching earth usage only increases oil losses and free fatty acid. The amount of bleaching earth required must be determined by the prebleached oil quality, i.e., acceptable red or green color, complete soap removal, and a zero peroxide value.

2.8 Hydrogenation

PROBLEM: SLOW REACTION

Possible Causes and Corrective Actions

(1) Poor catalyst activity—Increase the catalyst level. The poor activity may be due to catalyst poisons or a lower nickel content in a different lot of catalysts.

(2) Old catalyst—Add fresh catalyst. Catalysts that have been in storage for long periods may have deteriorated from abuse or contamination.

(3) Gas purity—The hydrogen gas used must be at least 99.0% pure. It should be dry and free of contaminant gases such as hydrogen sulfide (H_2S), sulfur oxide (SO_2), and carbon disulfide (CS_2). When hydrogen gas contains sulfur compounds, the sulfur combines with nickel, poisoning the catalyst and reducing activity and selectivity substantially. Impurities such as carbon monoxide (CO), carbon dioxide (CO_2), methane (CH_4), and nitrogen (N_2) do not affect catalyst activity [13].

(4) Not enough hydrogen gas—Short-term: stop the reaction and wait until sufficient hydrogen gas is available. Long-term: determine the cause of insufficient hydrogen gas and correct the problem.

(5) Poor agitation—Ensure that the agitators are turned on and operational. Internal inspection of the converter may be necessary to determine the condition of the agitator shaft and blades.

(6) Soap content—The oil soap content to hydrogenation should be 1.0 ppm maximum and must never exceed 20 ppm. Soap poisons the catalyst by reacting with the nickel to form nickel soaps that reduce the available nickel proportionally.

(7) Phosphatide content—Gums poison the catalyst and precipitate at hydrogenation temperatures, darkening the oil and clogging filters, making filtration almost impossible; the phosphorus content should not exceed 5 ppm.

(8) Sulfur poisoning—Particularly long induction periods occur in the presence of sulfur compounds. Reaction with sulfur inhibits the capacity of nickel to adsorb and dissociate hydrogen, reducing the total activity of the catalyst. As the ability of the nickel to hydrogenate is reduced, its tendency to promote isomerization is enhanced to produce large quantities of *trans* isomers. Wet milled corn oils can contain sulfur, which must be removed in prebleaching to prevent hydrogenation and flavor problems.

(9) Moisture content—Moisture inactivates the catalyst and can promote free fatty acid formation by hydrolysis. Analysis of the oil should confirm a 0.10% maximum moisture before the reaction is started. Hydrogenation feedstock oils with higher moisture levels should be dried before the catalyst is added.

(10) Cooling and heating coil leaks—Steam that leaks in the converter coils may be the cause of a high moisture content if the oil was dry from the supply vessel.

PROBLEM: WRONG MELTING POINT/IODINE VALUE RELATIONSHIP

Possible Causes and Corrective Actions

(1) Wrong catalyst used—Commercial nickel catalysts vary considerably in their inherent preferential selectivity. Also, used catalysts are less preferentially selective than new catalysts. High selectivity catalysts enable the reduction of linoleic fatty acid without producing high levels of stearic fatty acid for a product with good oxidative stability and a low melting point.

(2) New catalyst lot—The nickel content of a new lot of catalysts may be different enough from the previous lot to affect the activity; the new lot may require more or less total catalyst to achieve the desired physical requirements.

(3) Contamination with a previous batch—All precautions should be taken to ensure that no oil from the previous batch is left in the converter or drop tank to contaminate the next batch.

PROBLEM: CATALYST FILTRATION

Possible Causes and Corrective Actions

(1) Selective catalyst—Hydrogenation catalysts must exhibit both a high activity and a selectivity, while it is also essential that the catalysts can be filtered rapidly. These two objectives are somewhat incompatible because a very selective catalyst would have wide shallow pores, which implies small particles, which lead to filtration problems.

(2) Reused catalyst—Extensive reuse of catalysts can cause filtration problems because (1) free fatty acids in the feedstock react with the nickel to form nickel soaps that pass through the screens on most black presses and (2) the catalyst particle size will decrease with mechanical attrition, which will pass through most catalyst filters. Hydrogenated oils must be postbleached to insure the removal of trace quantities of nickel, which can adversely affect oxidative stability.

PROBLEM: BASESTOCK AOM STABILITY

Possible Causes and Corrective Actions

(1) Soft and hard basestock blending—Blending a slightly overhydrogenated batch with an underhydrogenated batch can probably better meet the desired physical characteristics like solids fat index (SFI) and melting point, but the AOM stability results will be poorer than if the two batches were hydrogenated to the correct endpoint. Very soft basestocks should not be produced to blend a very hard basestock into the desired limits. Hard basestocks should be hardened further to the next harder basestock or to almost complete hydrogenation to produce a low-iodine value hardstock.

(2) Nickel residue—Trace quantities of nickel passing through the catalyst filter will adversely affect the hydrogenated oil's oxidative stability. Stability studies have shown that 2.2 ppm nickel will reduce the AOM stability of lard by 50%. Hydrogenated oils must be postbleached to insure the removal of trace quantities of nickel remaining after catalyst filtration.

2.9 Postbleaching

PROBLEM: HIGH NICKEL CONTENT

Possible Causes and Corrective Actions
(1) Soluble or colloidal nickel—Correct the source of the colloidal nickel problem or insure that the hydrogenated oil is acid treated with 50 to 100 ppm citric acid and postbleached using bleaching earth.
(2) Short contact time between the oil and the bleaching earth—Recirculate through the press until the specified postbleach color and other analytical limits are met.
(3) Critic or phosphoric acid use—Hydrogenated oils require a special post-treatment to remove traces of nickel from the finished product. During hydrogenation, nickel soaps are formed and are sufficiently oil soluble to remain in the oil during filtration. Citric acid, at approximately 100 ppm, added prior to postbleaching, will chelate the remaining nickel and leave a residual nickel of less than 1.5 ppm. Industry results indicate that a residue of 0.1 to 1.5 ppm nickel may be present in deodorized oils without substantial harm to the oxidative stability [14].
(4) No bleaching earth used—Ensure that bleaching earth has been added to the oil.

PROBLEM: HIGH IMPURITIES IN THE FILTERED OIL

Possible Causes and Corrective Actions
(1) Inadequate precoating of the filter—Recirculate until the impurities improve; if the impurities persist, stop and precoat the filter properly.
(2) Holes in the filter screens, cloths, or papers—Repair or replace.
(3) Improper filter assembly—Reassemble properly.
(4) Worn seals or gaskets—Replace defective seals or gaskets.

PROBLEM: POOR BLEACH COLOR

Possible Causes and Corrective Actions
(1) Low bleaching earth level—The amount of bleaching earth to be used should be the minimum amount required to effect removal of the impurities as measured by peroxide reduction with zero as the target. The bleaching earth level should be guided by performance, not a mandatory addition level.
(2) Short contact time—Usually 15 to 20 minutes contact time between the oil and the bleaching earth is adequate to drive off the moisture in the bleaching earth and complete the bleaching reaction with atmospheric bleaching. In vacuum bleaching, the greatest color reduction is realized

within the first 3 minutes of contact because the air and moisture are removed by the vacuum. Extending the contact time does not harm the vacuum-bleached oil because oxidation is prevented by the vacuum.

(3) Low bleaching temperature—Bleaching temperatures can be increased but should not exceed 110°C (230°F). Heating of the oil should take place after the addition of the bleaching earth. Higher oil temperature reduces oil viscosity, which improves the adsorptivity of the bleaching earth; however, no significant improvements are experienced above 110°C. Problems with increased free fatty acid development, oxidation, and color fixing can result with temperatures exceeding 110°C.

(4) Inadequate refining—Check upstream problems: caustic treat or mixing in refining.

(5) Bleaching earth addition temperature—Addition of the bleaching media to hot oil vaporizes the moisture in the clay too soon. The moisture release allows the lattice structure in the clay to collapse, which reduces the effective surface area before it has an opportunity to adsorb the color pigments and the secondary oxidation products.

(6) Agitation—Bleaching vessel agitation should be sufficient to achieve good contact of the earth with the oil without incorporating air.

PROBLEM: HIGH PEROXIDE VALUE

Possible Causes and Corrective Actions

(1) Product skipped bleaching—Properly bleached oils have a zero peroxide value. High peroxide values for products out of bleaching indicate that the product has not been bleached with bleaching earth; it may have been pumped through the system without the addition of bleaching earth because the color met the required limits before bleaching. Postbleaching is mandatory after hydrogenation in many processes to ensure that any trace metals and secondary oxidation precursors are removed. A zero peroxide value out of postbleach indicates that the product has been bleached properly.

2.10 Winterization

PROBLEM: POOR COLD TEST

Possible Causes and Corrective Actions

(1) Filtered oil temperature too high—Cool the oil to the proper temperature before filtering (7°C or 44.6°F).

(2) Rapid cooling rate—Ensure that the oil is not cooled at too rapid a rate to produce the desired crystal; a temperature change from 40 to 7°C (104 to 44.6°F) over a 12-hour period is generally acceptable.

(3) Broken screen or torn cloth—Inspect and correct the problem.

(4) Leaking seals or gaskets—Inspect and correct the problem.

(5) Improper filter cooling after hot oil wash—Cool the filter to the desired temperature after each hot oil wash.

(6) Warm filter room—Keep the filter cool to prevent stearin from melting and remixing with the salad oil.

(7) Mixing low and high cold test oils—Low cold test products should be kept separate from salad oils with satisfactory analysis. Satisfactory results are not usually achieved when good and poor cold test products are blended. A laboratory prepared blend of the two products should be evaluated and found satisfactory prior to any blending of large batches.

PROBLEM: SLOW FILTRATION RATE

Possible Causes and Corrective Actions

(1) Cold oil—Check the oil temperature for adjustment of subsequent batches.

(2) Wet oil—Moisture levels above 0.1% will cause filtration problems. Trace and correct the moisture problem to speed up filtration on subsequent batches.

(3) High soap content (over 50 ppm)—Take corrective action in refining, water washing, and/or prebleaching to correct the problem for future production.

(4) High phosphatide level (over 4 ppm phosphorus)—Remedy the problem upstream in the refining process for future production.

(5) Overpressurizing the filter by the feed pump—Make sure that the back pressure control system is operational.

(6) Blinded screens—Externally wash blinded screens.

(7) Winterized oil crystal abuse—Do not pump or recirculate winterized oil any more than absolutely necessary before filtration; fractured crystals filter slowly.

(8) Low chill tank loading temperature—Improper graining results, which affects filtration rate results from cool chill tank loading temperatures. The chill tank supply oil temperature should be adjusted to meet the established limits before loading.

PROBLEM: POOR SALAD OIL YIELD

Possible Causes and Corrective Actions

(1) Long graining or chilling time—The solidified oil can entrap the liquid oil decreasing the salad oil yield.

(2) Hydrogenation conditions—The hydrogenation conditions for oil to be winterized for salad oil should produce the least possible saturates and *trans* isomers while reducing the linolenic fatty acid (C-18-3) content. Nonselective hydrogenation conditions of low temperature, high pressure, and a low catalyst concentration normally produce the desired low saturates and *trans* isomers.

(3) Low chilling temperatures—High viscosities caused by low crystallization temperatures inhibit the desired crystal growth. Filtration efficiency is dependent upon the crystal size and entrapped liquid oil within the crystal matrix.

2.11 Dewaxing

PROBLEM: POOR FILTRATION RATE

Possible Causes and Corrective Actions

(1) Sunflower oil temperature—Crystallization and maturation require gradually cooling the oil to 4 to 10°C (39.2 to 50°F) for stable beta crystals. However, the viscosity does not allow an acceptable filtration rate at this temperature. Heating the oil up to 12 to 14°C (53.6 to 57.2°F) increases the filtration rate approximately 60%.

(2) Gums—It is necessary to remove gums before dewaxing. The presence of gums in the oil influences the formation and stability of the formed crystals to diminish the filtration rate of the oil. Dewaxing should be performed after prebleaching and prior to deodorization.

(3) Diatomaceous earth addition—The addition of small amounts of diatomaceous earth to cold sunflower oil has been reported to increase the filtration rate by a factor of four [15].

PROBLEM: WAX CONTENT

Possible Causes and Corrective Actions

(1) Hybrid sunflower oil—Hybrid sunflowers that have replaced the open-pollinated varieties have improved plant uniformity, yields, and disease resistance. However, the higher oil content seeds with reduced hull content have increased wax levels in the hull, and the hulls are more difficult to separate from the kernels when processing to extract the oil. The waxes soluble in the oil at ambient temperature will cause an unsightly cloudy appearance at cooler temperatures. Dewaxing is required to maintain a clear sunflower oil.

(2) Hulled versus dehulled sunflower seeds—Sunflower oil wax content depends upon the oil content and the process technology. Generally, sunflower seeds with a 40% minimum oil content will have a 0.03 to

0.05% wax content from regular seeds and 0.008 to 0.015% wax content from dehulled seeds [15].

2.12 Esterification

PROBLEM: HIGH COLOR

Possible Causes and Corrective Actions

(1) Air contact—Fatty emulsifiers must be protected from air incorporation throughout the alcoholysis process to prevent oxidation of the unsaturated fatty acids or the alpha-tocopherols with attendant color development. Some potential air sources during fatty emulsifier production are (1) air leaks in fittings below the product level and in pumps and holes in the reaction vessel, heaters, and coolers; (2) improper deaeration of the reaction mixture; and (3) the stripping steam must be generated from deaerated water to be oxygen-free.

(2) Basestock hydrogenation catalyst—Emulsifier basestocks must be free of residual nickel, which catalyzes oxidation. Reused catalysts can cause filtration problems because of two situations. The first is that the free fatty acids in the hydrogenation feedstock react with the nickel catalyst and form nickel soaps that pass through filters. Second is the particle size reduction to sizes that will not be retained on the filter. Therefore, only new easily filtered catalysts should be utilized for emulsifier basestocks.

(3) Basestock postbleaching—Hydrogenated oils require a special postbleach treatment to remove traces of nickel from the finished product. Nickel soaps are formed during hydrogenation, which are oil-soluble and remain in the oil after filtration. Addition of 100 ppm citric acid prior to postbleaching will chelate the remaining nickel for more efficient filtration after bleaching. Additionally, the hydrogenated basestock must be bleached with bleaching earth, even when the analytical color limits indicate that bleaching is not required to obtain the specified color. A zero peroxide value indicating removal of secondary oxidation materials is also a prerequisite for emulsifier basestock postbleaching.

(4) Nitrogen protection—Relatively low levels of oxidation prior to alcoholysis can significantly impair the oxidative stability of the finished emulsifier. Quality products require a minimum of oxygen exposure before processing, which necessitates effective nitrogen blanketing of all emulsifier basestocks during storage or transportation, including the low-iodine value bases.

(5) Oxidized tocopherols—alpha-Tocopherol oxidizes to a very colorful chroman-quinone, which is probably catalyzed by excessive acidity of

the processed oil or emulsifier at some point during processing of the oils, basestocks, or emulsifiers.

(6) High discharge temperature—Proper cooling while protected by the vacuum before exposure to the atmosphere is mandatory for oxidative stability and color control. The product discharge temperature must be as low as possible while still maintaining the product as a fluid pumpable product.

(7) Alcoholysis catalyst selection—Less color development in mono- and diglycerides has been experienced with the use of calcium hydroxide, as opposed to sodium hydroxide.

(8) Filter aid—Only neutral filter aids should be utilized for emulsifier filtration. Modified filter aids can have deleterious effects upon the emulsifier color.

(9) Rust—Iron oxides are soluble in monoglycerides and many other emulsifiers. Rust from washed black iron surfaces and not sweetened with oil immediately will darken emulsifiers' color [17].

(10) High feedstock color—The feedstock color to the reaction should be analyzed to ensure that a color reversion has not occurred between the postbleach process and the glycerination feedstock tanks.

(11) Incomplete neutralization—The presence of catalysts will catalyze red color development. Neutralization should be confirmed with soap or pH analysis to ensure that the catalyst has been properly neutralized.

(12) Carbon steel reaction vessel—Ordinary carbon steel is not a suitable material for reaction vessels because it contaminates the product with iron soaps that produce dark colored product. Stainless steel equipment and vessels are preferred [28].

PROBLEM: LOW ALPHA MONOGLYCERIDE CONTENT

Possible Causes and Corrective Actions

(1) Reversible reaction—Esterification is a reversible reaction, especially in the presence of a catalyst at elevated temperatures. Reversion of monoglyceride occurs at the point of reaction completion before catalyst neutralization or after if neutralization is incomplete. During cooling prior to glycerol removal, the excess glycerine is less soluble in the fat phase and separates into the lower layer. This shifts the equilibrium somewhat to regenerate diglycerides and triglycerides. Therefore, catalyst neutralization and excess glycerine removal is a critical stage of glycerolysis, which must be performed with close and precise control [18].

(2) Excessive steam stripping—Most emulsifiers have a 1.0 or 1.5% free fatty acid limit to allow steam stripping to be suspended before the alpha monoglyceride content was reduced below the specified limits.

(3) Reaction temperature—The reaction should be performed within a temperature range of 175 to 250°C (347 to 482°F). Higher temperature reactions cause the formation of polyglycerols and polymerized glycerides, which affect both the monoglyceride yield and the functionality of the emulsifier product.

(4) Low glycerin to reaction level—It is necessary to maintain the prescribed glycerine level in the reaction mixture to obtain the specified alpha monoglyceride content. Two usual causes for a low glycerin level in the reaction have been (1) incorrect amount added and (2) glycerin flashed off before the reaction. An incorrect addition level can be caused by several different problems, i.e., calculation error, mis-pumping, pump failure, and so on. Glycerin can be flashed off or lost to entrainment because of high temperature, high vacuum, and/or stripping steam use during the drying process.

(5) Low catalyst level—The amount of reactive glycerin depends upon the level of catalyst used. Therefore, low catalyst level additions will result in low monoglyceride contents. A low catalyst level may be due to calculation error, a higher dilution than specified, faulty addition procedures or equipment, an excess of the other reactants, and so forth.

(6) Incomplete catalyst neutralization—Catalyst presence after the reaction enhances the potential reversion of the monoglyceride reaction. Catalyst neutralization should be monitored by soap or pH analysis or both. A high soap content or an alkaline pH indicates incomplete catalyst neutralization.

(7) Acid addition procedure—Phosphoric acid must be added slowly to allow for proper dispersion. Rapid addition of the neutralization acid, especially with poor agitation, will allow it to drop to the bottom of the reactor where it can be ineffective. Insufficient catalyst neutralization will reverse the reaction during stripping of the excess glycerine for a low active emulsifier content.

(8) Low reaction time or temperature—A low reaction temperature or time can provide an incomplete reaction for a low alpha monoglyceride or PGME content, depending upon the product being produced.

PROBLEM: IMPURITIES

Possible Causes and Corrective Actions

(1) Phosphate salts—Impurities identified as phosphate salts have been found in products shipped to customers shortly after production. Studies determined that the glycerolysis catalyst neutralization is time and temperature related; at 350°F (176.7°C) neutralization is completed

within 10 minutes, while 8 to 12 hours are required at 190°F (87.7°C). Emulsifiers neutralized at low temperatures may require a holding time before filtration or a second filtration to remove all of the phosphate salts.

(2) Early delivery—During the startup cycle of the phosphate salt filters, care must be exercised to recycle the initial cloudy emulsifier instead of contaminating the emulsifier storage vessel.

(3) Filter screen holes—Stream and representative batch samples must be evaluated for clarity to determine that the filters are operating satisfactorily. Positive impurities in any of these samples identify a problem that must be investigated and a solution determined before proceeding.

PROBLEM: HIGH FREE FATTY ACID

Possible Causes and Corrective Actions

(1) Excessive acid neutralization—Phosphoric acid is normally used to neutralize the alkaline catalyst. Excess phosphoric acid usage will analyze as free fatty acid and catalyze additional free fatty acid development.

(2) Insufficient stripping—Free fatty acids are removed by steam distillation after the reaction has been terminated. Insufficient steam distillation will not effectively reduce the free fatty acid level; however, overstripping can reduce the alpha monoglyceride content.

(3) Wet reactants—Addition of moisture to the reaction will cause hydrolysis, which results in free fatty acid development. A higher free fatty acid level may require additional stripping to meet the required limits, which may in turn lower the alpha monoglyceride content. Wet feedstocks or glycerin can be caused by moisture contamination from wet steam, steaming lines into the storage tanks, leaks in the heating or cooling coils, or the like. The source of the moisture must be identified and the reactants dried before use.

2.13 Blending

PROBLEM: AOM STABILITY

Possible Causes and Corrective Actions

(1) Soft basestock addition—Blending a mix with a soft basestock not specified will decrease the oxidative stability of the product. Blends should only be made with the basestocks specified to maintain the designed finished product's AOM stability.

(2) Exceeding composition tolerances—Blending with a specified basestock outside the specification tolerances can decrease the oxidative stability.

(3) Salad oil stearin usage—The natural antioxidants accumulate in the liquid fraction of a winterized or fractionated product. Therefore, the stearin fraction is less protected than the same hardness oils produced from regular source oils.

PROBLEM: GOVERNMENT REGULATIONS COMPLIANCE

Possible Causes and Corrective Actions

(1) Ingredient statements—FDA labeling regulations require an identification of each source oil in descending order of predominance. Blending changes outside the specified composition and tolerances can violate the ingredient statement requirements.

(2) Nutritional claims—Nutritional claims are made for edible fats and oils products based on the fatty acid compositions of the individual products. In most cases the product tolerances must be established based on the allowed rounding of the values. Blending outside the specification composition tolerance range or with unauthorized basestocks will result in product characteristics outside the nutritional claims. In this case there is no good or bad, just right and wrong.

PROBLEM: SOLIDS FAT INDEX COMPLIANCE

Possible Causes and Corrective Actions

(1) Basestock blending—Fats and oils mixes made with basestocks previously blended for SFI limits compliance will not have the same effect upon subsequent blends as hydrogenated basestocks produced in limits. Both the fatty acid composition and the triglyceride composition are changed with each blend.

(2) Sampling technique—Fats and oils blends should be agitated a minimum of 20 minutes at 10°F (5.6°C) above the expected melting point of the blend. Prior to sampling, the sample line and valve must be flushed with the current product to remove all previous product residue, and the sample container must be clean.

(3) Mis-pumping—Errors in pumping should be indicated first by a high or low melting point analysis. Material balance and a review of the product transfers should help identify the resultant blend composition for determination of corrective actions.

PROBLEM: POOR PERFORMANCE

Possible Causes and Corrective Actions

(1) Substitution basestocks—Substitution or addition of a nonspecified basestock may meet the analytical limits specified but change the prod-

uct's performance. Only specified basestocks should be used for blending. The product composition specified represents the product developed and approved by the customers. Product consistencies and performance are dependent upon compositions as much as and sometimes more than the analytical values. It may be possible to meet the specified analytical limits with several different basestock blends, but only one will perform properly.

2.14 Deodorization

PROBLEM: HIGH FREE FATTY ACID

Possible Causes and Corrective Actions

(1) Low vacuum—Identify and correct the vacuum problem. Refer to the poor deodorizer vacuum section for possible problems and solutions.

(2) Low oil temperature—Increase oil temperature.

(3) High oil flow rate—Increase residence time by reducing the flow rate.

(4) Poor steam distribution—Clean steam sparge ring.

(5) Too high or too low sparge steam pressure—Reduce if too high or increase with extreme caution if too low.

(6) Supply oil has high free fatty acid—Recirculate to attain the specified free fatty acid level, but identify upstream problem and correct for subsequent production.

(7) Excessive shell drain (semicontinuous deodorizer)—Drain more frequently until the reason for the increased shell drain is identified and corrected.

(8) High CO_2 content in nitrogen gas—CO_2 will analyze as free fatty acid. It can be determined that CO_2 is causing the high result by evacuating all the gases from the oil sample with a vacuum and retesting for free fatty acid content. A lower free fatty acid result would indicate CO_2 contamination of the nitrogen gas supply, probably at the in-house gas generating plant. Correction of the gas plant problem or purchase of liquid nitrogen would be two potential corrective actions.

(9) Chelating acid addition—Typical chelating acid addition levels are 50 to 100 ppm citric acid or 10 ppm phosphoric acid; 0.00227% citric acid analyzes as 0.01% free fatty acid [19]. The "as deodorized" free fatty acid level must be low enough to allow an increase from the addition of acid chelators.

(10) Antioxidant addition—Antioxidants will contribute to the apparent fatty acid content because the analytical method does not distinguish between acidity contributed by fatty acids or mineral acids. The free

fatty acid level of the deodorized oil must be low enough to allow for the increases contributed by antioxidants and acid chelates.

(11) Instrumentation accuracy—Preventive maintenance should be performed on a regular basis and instruments recalibrated whenever poor performance is suspected.

PROBLEM: DARK OIL

Possible Causes and Corrective Actions

(1) High soap content—Rebleach and redeodorize the immediate batch, but determine the cause of the problem upstream in refining or prebleaching to prevent future problems.

(2) High phosphatides content—Deodorizer feedstocks with a phosphatide content above 20 ppm will cause high deodorized oil colors. The phosphatides must be removed in refining and bleaching prior to the deodorization process. Some of the phosphatides and their associated metal complexes are not easily hydratable. These complexes require a phosphoric acid pretreatment for their removal in degumming or refining. The prebleach process removes the traces of soap and phosphatides remaining after the refining process. Oils that do reach the deodorizer with high phosphatide contents will have a darker color and a characteristic fishy odor after deodorization. These products must be bleached and then redeodorized to salvage the oils.

(3) Rework quality—Satisfactorily process and/or select any rework product to be used at the blending process. The immediate batch color problem may be correctable with rebleaching and redeodorization, or in extreme cases it may require blending at low levels with subsequent batches or sale as an off-quality product.

(4) Secondary oxidation—Identify the cause of the oil abuse, probably aeration and/or overheating, and correct. The immediate batch may be salvaged by rebleaching, with bleaching earth followed by redeodorization.

(5) Proteinaceous material—More effective filtering or water washing of meat fats is required to eliminate this problem. The immediate batch will require deodorization to char the protein material to facilitate removal in bleaching.

(6) High temperature air contact—Fats and oils must be protected from air incorporation throughout deodorization to prevent oxidation with attendant color development. Some of the potential air sources during deodorization are (1) air leaks in fittings below the oil level and in pumps and holes in the shell, heaters, and coolers; (2) the feedstock

must be deaerated; and (3) the stripping steam must be oxygen-free by generating from deaerated water.

(7) High discharge temperature—The oxidation rate for edible oils almost doubles with each 20°F temperature rise. Therefore, proper cooling of the oil before exiting the deodorizer is mandatory for oxidative stability evidenced by color, peroxide value, and flavor evaluations.

(8) Low deodorizer temperature—The thermal treatment that is a necessary part of the deodorization process also heat bleaches the oil by destruction of the carotenoids that are unstable at deodorization temperatures. The carotenoid pigments can be decomposed and removed at approximately 500°F or 260°C. The high side of the normal deodorization temperatures, 400 to 525°F (200 to 274°C), should deliver the heat bleach desired. However, the maximum deodorizer temperature that can be used must be limited due to the detrimental effect upon oil stability.

(9) Oxidized tocopherols—alpha-Tocopherol oxidizes to a very colorful chroman-quinone. The opportunities for oxidation must be evaluated and the cause eliminated for further production. The immediate product must be rebleached to remove the oxidized tocopherol and other secondary oxidation products.

(10) Crude oil storage—Storage of crude oils in contact with atmospheric oxygen can cause oxidative damage, which is ultimately reflected in the quality of the processed oil. Vegetable oils with the lightest colors and best oxidative stability are processed almost immediately or protected with nitrogen immediately after extraction.

PROBLEM: RAPID PEROXIDE VALUE INCREASE AFTER DEODORIZATION

Possible Causes and Corrective Actions

(1) Dirty laboratory glassware—All laboratory utensils must be rinsed with distilled water to remove all traces of soap or detergent.

(2) Light exposure—The deleterious effects of light exposure upon the flavor stability of edible oils is well known. Only limited exposure of the oil to sunlight or ultraviolet rays from florescent lightening will increase the peroxide value of the oil and impart off-flavors. Samples from any process should be protected from the light to ensure a representative analysis of the product. Clear glass or plastic containers should not be used to sample oils in process.

(3) Washed tanks—Process tanks must be clear water rinsed, neutralized, and sweetened with an oil rinse after washing.

(4) Chelating acid addition—30 to 50 ppm citric acid or 10 ppm phosphoric should be added in the deodorizer cooling tray to chelate metals. The absence of the acid addition may be due to a broken addition line, a clogged addition line, an empty acid supply tank, incorrect preparation of the acid solution, someone shutting off the system, or something similar.

(5) Deodorizer discharge temperature—Rapid oxidation of deodorized oils must be avoided by controlling the oil temperature in the cooling stage before exposure to the atmosphere. Liquid oils, high in polyunsaturates should be cooled from 100 to 120°F (38 to 49°C). Higher deodorizer discharge temperatures are necessary for higher melting products but should be maintained as low as possible. The speed of oxidation is doubled with each 27°F (15°C) increase in temperature within the 70 to 140°F (20 to 60°C) interval.

(6) Secondary oxidation—The immediate batch may be salvaged by rebleaching with bleaching earth followed by redeodorization; however, the source of the abuse should be identified to prevent a reoccurrence with subsequent production.

(7) Air incorporation—Repair pumps or other sources of air incorporation after deodorization.

(8) Polymerized oil buildup—Peroxide value increases can be caused by a buildup of polymerized or oxidized oil, which mixes with the oil in process. The corrective action is to wash the deodorizer. To prevent these problems from occurring, a regular wash-out schedule should be established. It should be standard procedure to wash the deodorizer each 6 months if it is not operated continuously 7 days a week or if the operation is subject to power failures. Otherwise, deodorizers should not be operated more than 12 months without a thorough wash-out [19].

(9) Prior oxidation—It is very important that oxidation be prevented during all stages of edible fats and oils processing due to the detrimental effects upon the finished deodorized oils. For example, it has been estimated that a peroxide value increase of only 1.2 meq/kg in neutralized soybean oil after refining may reduce the oxidative stability of the deodorized oil by 50% [10]. Oxidation during processing can be minimized by avoiding contact with air, elimination of prooxidant metallic contaminants, nitrogen protection at all stages of processing, and/or addition of antioxidants at the initial stages of processing.

PROBLEM: HIGH LOSS

Possible Causes and Corrective Actions

(1) Excessive blowing steam—A 15% increase in sparging steam will double the entrainment loss, and a 35% increase will be a fourfold loss;

therefore, the proper blowing steam level is essential to control deodorizer entrainment losses.

(2) Wet blowing steam—Ensure that the steam pressure to the deodorizer is within specified limits.

(3) Wet oil—Dry the wet oil before deodorization to control the entrainment loss and analyze the feedstock origin and eliminate the moisture source for future production.

(4) Charge or batch size—Ensure that the deodorizer charge or batch size does not exceed the capacity.

(5) High free fatty acid feedstock—Determine the upstream cause to avoid continued high losses.

(6) Leaking drop valves—Timely maintenance is necessary to prevent or repair leaking valves that cause erratic temperature recording and overflowed deodorizer trays.

(7) Leaking trays—Repair tray leaks should be identified by an increase in the deodorizer shell drain.

PROBLEM: CONTAMINATION WITH ANOTHER PRODUCT

Possible Causes and Corrective Actions

(1) Mixing in the measuring tank—Blow the supply lines into the measuring tank, followed by evacuation of this vessel with the deodorizer vacuum to prevent product heels.

(2) No empty trays between products (tray-type deodorizers)—One or two empty trays should separate products that are somewhat similar.

(3) No flush oil—Dissimilar products should be separated by flushing the system with a minimum of the first two trays of the new product.

(4) Nonrepresentative sample—The sample line and valve must be thoroughly rinsed with the product to be sampled to remove all congealed, oxidized, or gummy deposits; the sample container must be absolutely clean; all products must be liquid when sampled; and the sample must be adequately identified by product and source to avoid sample confusion.

PROBLEM: REVERTED OIL FLAVOR

After deodorization, certain oils develop objectionable flavors that may not be recognized as oxidative in nature. This flavor, representative of the original crude oil, has been described as reversion.

Possible Causes and Corrective Actions

(1) Air exposure—Oils containing linoleic and linolenic fatty acids are

subject to flavor changes with minimal exposure to air or oxygen of less than 1.0% [20].

(2) Crude versus refined storage—Most edible oils store best in the crude state. Degummed crude soybean oil oxidizes more rapidly than non-degummed crude oil. Apparently, the phospholipids in crude oil provide an antioxidant function above that of the natural antioxidants or tocopherols.

(3) Temperature control—The rate of oxidation or reversion is dependent upon temperature. Evidently, oxygen will diffuse into and react with the oil faster at higher temperatures. The oxidation rate approximately doubles with each 20°F (11.1°C) temperature increase; therefore, the 10°F above the melting point of the product rule should be enforced to control reversion.

(4) Nitrogen blanketing—Nitrogen blanketing of all tanks is an effective means of protecting edible fats and oils. Peroxide development and flavor instability can be virtually eliminated by keeping oxygen away from the oil at all stages of processing. Evaluations of soybean oil stored several months with and without nitrogen protection showed a decided more rapid flavor deterioration for the oil stored with air exposure.

(5) Phosphatide removal—Residual phosphatides will contribute to off-flavors and colors. Some of the phospholipids and their associated metal complexes are not easily hydratable with changes as a result of oil abuse. These complexes require phosphoric acid pretreatment for their removal in degumming or refining. The prebleach process cleans up traces of soap and phosphatides remaining from the refining process. Problems will develop with the deodorized oil if the feedstock has a phosphatide content above 20 ppm.

(6) Bleaching earth filtration—Complete removal of the bleaching media from the bleached oil is very important since residual earths can act as a prooxidant. The bleaching operation should be controlled with filterable impurities evaluations to insure that the bleached oil transferred to storage or the next process is free of any contaminates that could decrease the oxidative stability.

(7) Vacuum bleaching—The primary function of the bleaching process is to remove peroxides and secondary oxidation products. Secondarily, the process cleans up traces of soap and phosphatides remaining after the refining step and the adsorption of color pigments. Vacuum bleaching is more effective than atmospheric bleaching. It usually requires less bleaching earth, operates at lower temperatures, and minimizes oxidation by reducing exposure to air and providing an opportunity to cool the oil before returning to atmospheric conditions [21].

(8) High peroxide value to deodorization—Contrary to most beliefs, no significant oxidation breakdown products are removed by deodorization; their only opportunity for removal is the bleaching process. The thermal decomposition of peroxides is complete with deodorization, but the rate of peroxide formation in the oil during subsequent storage is increased and the flavor stability of the finished oil is compromised. Rebleaching before deodorization, when the peroxide value has been allowed to increase, is necessary for adequate flavor stability [21].

(9) Deodorizer temperatures—Normally, the deodorization temperatures used range from 400 to 525°F (200 to 274°C). The deodorization process is time/temperature dependent; lower temperatures require longer times, and higher temperatures, shorter times. Even though elevated temperatures have a favorable effect upon deodorization efficiency, excessive temperatures are detrimental to flavor stability. Twice as many tocopherols are stripped out of the oil at 525°F (275°C) as at 465°F (240°C); therefore, for flavor stability, the lowest practical deodorization temperature should be utilized.

(10) Reduced stripping steam flow—Inability to maintain acceptable oil quality when the deodorizer temperature, vacuum, and feed rate appear normal may be caused by a reduced or restricted stripping steam flow. It may be necessary to redrill the sparger holes if a boil out does not correct the problem.

(11) Avoid air contact—Fats and oils must be protected from air throughout deodorization. An oil reacts rapidly with oxygen at deodorizer temperatures with deteriorating effects upon flavors and oxidative stability. Potential air sources are
- air leaks in fittings below the oil level, in pumps, the shell, heaters, and coolers
- nondeaeration of the deodorizer feedstock
- stripping steam must be oxygen-free by generating from deaerated water.

(12) Chelating agent—Addition of citric acid (0.005 to 0.01% based on the weight of the oil) or phosphoric acid (0.001% based on the weight of the oil) before and after deodorization helps protect against oxidation and/or reversion. The acid inactivates the trace metals, particularly iron and copper, which may be present in the crude oil or picked up during processing. These acids decompose rapidly at temperatures above 300°F (150°C). The usual practice is to add the acid during the cooling stage in the deodorizer, during bleaching, and prior to caustic refining. Excessive amounts of phosphoric acid lead to the development of watermelon- or cucumber-type off-flavors, even with good oxidative stability results.

(13) Antioxidant replacement—Fats and oils resist oxidation until the antioxidants are destroyed during the induction period or the interval when oxidation proceeds at a slow rate. After the antioxidant ceases to function as a free radical terminator, there is a rapid increase in the rate of peroxide development. Replacement of the destroyed antioxidants with tocopherols or synthetic compounds will significantly improve the oxidative stability of the oil.

(14) Light exposure—The deleterious effects of light exposure upon the flavor stability of edible oils is well known. Only limited exposure of the oil to sunlight or ultraviolet rays from florescent lightening will increase the peroxide value of the oil and impart off-flavors. Samples from any process should be protected from the light to ensure a representative analysis of the product. Clear glass or plastic containers should not be used to sample oils in process.

(15) Blend stocks before deodorization—Shortenings, margarines, and some salad oils are blends of two or more basestocks or oils to achieve the desired product characteristics. Blending of these stocks prior to deodorization, rather than after, minimizes the handling and storage of the deodorized oil. It also allows process control to determine if the product has been abused inadvertently. This would allow the product to be bleached prior to deodorization for the removal of the secondary oxidation products.

(16) Boiler water treatment—Additives to treat boiler water can provide puzzling flavor results.

PROBLEM: POOR DEODORIZER VACUUM

Possible Causes and Corrective Actions

(1) Air leaks—Verification of air leaks can be accomplished with a drop test. This test is performed by closing the oil inlet and discharge line valves and then quickly shutting off the steam and water to the vacuum system. The vacuum may drop rapidly at this point, but it should stabilize at a value that must be above 20 inches for a valid test. Then the vacuum should be noted each hour to determine if the drop is within the limits identified by the deodorizer manufacturer. If the vacuum loss exceeds the design limits, tighten all piping connections, valve packing glands, and the equipment flanges. Continued vacuum loss after these measures indicates the need for leak detection using one or both of the following methods:
- soap detection test—Large leaks may be detected by painting all possible leak sources with a soapy water solution while the system is under 20 psig air pressure. The soapy water should form bubbles at the leak points. Mark the leaks, release the pressure, purge the sys-

tem with air, repair the leaks, restore the equipment and instruments, and then repeat the drop test to determine if the vacuum loss still exceeds the satisfactory rate. If so, smaller leaks can be identified with the following evaluation.

- ammonia/sulfur dioxide test—Prepare the system by closing all valves, blind the vacuum system tailpipes, disconnect the vacuum gauges, and admit ammonia gas until the system pressure reaches 5 psig and then further to 20 psig with compressed air. Examine the system by directing a small stream of sulfur dioxide gas around all the possible leak points. White fumes indicate the presence of a leak. Mark the leaks, release the pressure, purge the system, and then repair the leaks [19].

(2) High-temperature condenser water—When the vacuum system water is recirculated through a cooling tower, the average wet bulb temperature is a parameter used for the system design. Cooling towers will cool water to within a few degrees of the wet bulb temperature. Normally, the wet bulb and the corresponding water temperature chosen will not be exceeded 1.0% of the time in the summer months. However, the oil quality must be closely monitored during the warm summer days when the warm condenser water prevents operation at the desired vacuum.

(3) Other operational problems—Operational problems are the inability to maintain design vacuum. Two additional problems to those already reviewed are (1) low steam pressure or wet steam to the vacuum jets and (2) insufficient condenser water for one reason or another. Both of these conditions require maintenance or at least investigation to identify the source of the problem for corrective action and installation of preventive measures for the future.

(4) Vacuum maintenance—If operational problems are not the cause of the poor vacuum, the steam nozzles, steam chests, and throat of the boaster should be examined for wear. Replacement of the worn parts and adoption of a preventive maintenance program to regularly review the vacuum systems should help to relieve the poor vacuum problems.

PROBLEM: POOR STEAM EJECTOR OPERATION

Possible Causes and Corrective Actions [22]

(1) Low steam pressure—This may be caused by clogging of the steam strainers or orifice plates with pipe scale or sediment, improper operation of the steam pressure regulating valve, or low boiler pressure. A steam pressure gauge, for measuring the operating pressure, should be installed at a point close to the ejector steam inlet in order to determine the true operating pressure.

(2) Steam nozzles—In addition to the possibility of a nozzle clogging from pipe scale or dirt, a scale deposit might occur in the throats of the steam nozzles from impurities in the steam. When this occurs, it is necessary to clean out the nozzle with a proper size reamer or drill, taking care not to mar the internal surfaces.

(3) Insufficient cooling water—The water temperature entering and exiting the ejector condensers should be measured. If the temperature rise is not excessive, the cooling water supply is adequate and the source of the problem should be sought elsewhere.

(4) High back pressure at ejector discharge—The pressure at the final stage exit should be measured. Excessive pressure indicates that the piping should be changed to reduce the discharge pressure.

(5) Nozzle and diffuser wear—Operation of the ejectors with wet steam or corrosive gases or vapors will cause excessive wear and produce a rough wall surface. Periodic checks should be made to compare the throat diameter of the diffuser and nozzle with the original sizes to determine if replacement is required.

2.15 Antioxidant Addition

PROBLEM: ANTIOXIDANT ACTIVITY

Possible Causes and Corrective Actions [23–25]

(1) Incomplete dispersion—Dispersion problems may result from inadequate mixing equipment, poor mixing techniques, or shortcuts in procedures established for the antioxidant addition. Several acceptable addition methods are practiced, one of which is outlined below:

- The quantity of the batch must be large enough to allow adequate mechanical agitation; simply stated, the product must cover the agitator in the tank used for effective mixing.
- The temperature of the product must be maintained at least 10°F (5.6°C) above the melting point, but not too high, allowing oxidation before the antioxidant addition.
- Determine the proper antioxidant quantity for the individual product and weigh with a laboratory balance for each batch.
- Slurry the antioxidant in a portion of the product with vigorous mechanical agitation.
- Add the slurry to the product tank while under agitation at a temperature of 10°F (5.6°C) above the product's melting point.
- Continue agitation of the batch for a minimum of 20 minutes before sampling.

- Sample and obtain a positive qualitative antioxidant analysis before packaging or shipping in bulk quantities.

(2) Vessel agitation—Addition of a concentrated antioxidant mixture before agitation begins will result in a high level of the additive in a portion of the product while the major portion has little or no antioxidant present. This practice will produce serious stability deficiencies for most of the product and a very high concentration of antioxidant in the oil at the bottom of the vessel, which will result in performance problems. First and last piece product evaluations for packaged product should identify this problem. Tank truck or railcar products may indicate a negative antioxidant analysis for the as-shipped sample.

(3) Improper antioxidant levels—Maximum antioxidant levels are defined by the U.S. regulatory agencies. The allowed levels are generally recognized as the optimum for achieving good stability results. In general, successive additions of an antioxidant yields steadily diminishing returns. At higher concentrations an appreciable portion of the antioxidant is consumed by side reactions and thus does not function as a free radical terminator. In some cases decomposition of the antioxidant may produce substances with a prooxidant action.

(4) Antioxidant storage conditions—Proprietary antioxidant mixtures formulations are available from several suppliers. These mixtures vary in active ingredient contents and the solvent systems utilized to dissolve the antioxidant and keep them in solution until used. Separation or crystallization or other changes can occur if the products are handled improperly. Most antioxidant mixtures solidify at temperatures below 32°F (0°C), but the heating required to liquefy has no detrimental effect on antioxidant properties. Some deterioration occurs when the antioxidants are heated above 200°F (93.3°C) for extended periods.

(5) Incorrect antioxidants—Significant performance differences exist among the antioxidants in different source oils, processing, and handling conditions. Consideration for the selection of the antioxidant type or mixture formulation must be made during product development. Substitutions of one antioxidant for another cannot be allowed because of performance and especially labeling requirements. Each individual synthetic antioxidant must be identified by the prescribed nomenclature in the ingredient statement on the label.

(6) Addition stage—Natural antioxidants are removed from edible oils during processing, notably, refining, bleaching, and deodorization. Antioxidants can be added to oils at an early stage for protection, but it

should be recognized that deodorization will destroy the antioxidants. In this case, the antioxidants will have to be added again after deodorization for oxidative protection.

(7) Addition point—Antioxidants will not rejuvenate mistreated fats and oils products. Phenolic antioxidants function by inhibiting free radical formation, the initial step in oil oxidation. Therefore, it is essential that the addition point be as soon as reasonably possible after deodorization before the autoxidation reaction has a chance to start.

(8) Low antioxidant level—Analysis of the antioxidant level in the finished product will result in a lower level than actually added. The antioxidant level decreases as it functions to intercept oxygen before it can attack the fats and oils products.

(9) Induction period—The initial period of relatively slow oxidation is called the induction period. Antioxidants prolong the induction period but have no effect after the end of the induction period. Once oxidation has started, the rate is as rapid as experienced with unprotected fats or oils.

(10) Prior abuse—Edible fats and oils that have been allowed to oxidize significantly before deodorization will have a poor stability after deodorization. This is caused by the formation of oxidation products, which are odorless and tasteless and carry through into the deodorized product where they initiate and propagate further oxidation. Antioxidant addition to these abused oil products will have little or no stabilizing effect. Antioxidants cannot reverse or terminate the autoxidative process after it has passed the early stages.

(11) Tocopherols—The natural antioxidants, tocopherols, attain maximum effectiveness at comparatively low levels, roughly equivalent to their natural level of occurrence in vegetable oils. Tocopherol levels above the optimum concentration usually function as prooxidants.

2.16 Bulk Railroad Car or Truck Loading

PROBLEM: IMPURITIES

Possible Causes and Corrective Actions

(1) No analysis—Evaluate each product load before the shipment is released and refilter each load that does not meet established limits followed by upstream evaluations to determine the cause for the impurities.

(2) Poor sample—Zone sample all trucks and cars after loading is completed while the product is still liquid to identify problems before shipment that would result in a customer rejection.

(3) Dirty sampler or container—Clean all sampling apparatus before use.

(4) No in-line filter—Install in-line filtration on all loading lines.

(5) No filter element—Install proper element. The supervisor should ensure that all filters are being utilized properly.

(6) Filter not used—Eliminate by-passes around filters.

(7) Course filter element—Most loading filters should be 25 microns or less to effectively trap the impurities encountered.

(8) Holes or tears in the filter element—Replace the filter element and determine if the oil filtered should be refiltered unless this was the reason that the element failure was identified.

(9) Line blowing causes filter failure—Blow the line into a separate container or through another filter.

(10) Dirty filter shell—Impurities from the lid or the dirty filter element can fall off into the filter shell while the elements are being replaced, contaminating the next oil product filtered; therefore, the internals of the filter should be cleaned and inspected each time that the filter element is replaced.

(11) Dirty truck or car—All trucks and cars should be physically inspected just prior to loading and those found dirty should be cleaned before use.

(12) Open hatch—The loading hatch opening should be covered to prevent insects and other impurities from contaminating the oil product.

(13) Nonfiltered line from the in-line filter to the truck or car—Install a filter element or "sock" on the end of the loading line. Use a lint-free material for any "sock" arrangement to avoid contamination of the oil with the filter device.

(14) Loose or open hatches—Close all openings tightly and seal.

PROBLEM: RAIL CAR HANDLING

Possible Causes and Corrective Actions

(1) Cleaning—Tank cars should be cleaned and inspected thoroughly after return and before loading with another product. Physical inspection of the inside of the railcar is required to ensure that the tank is completely clean and dry and free of dirt, rust, and scale, as well as any possible infestation.

(2) Soap residue—All railcars washed with soap must be rinsed thoroughly to ensure that a soap residue will not contaminate the next product loaded. Evaluation of the rinse water to determine alkalinity will ensure that the railcar has been rinsed adequately.

(3) Broken coils—Railcars are equipped with internal and/or external coils for heating the product at destination. The internal coils have a ten-

dency to rupture during transit from "humping" by the railroad or after freezing if not properly drained of condensate in cold weather. Oil exiting from the coil outlet valve when uncapped is an indicator that the coil has been broken. Steaming through a broken coil contaminates the oil product with moisture and impurities from the heating coils. These railcars are equipped with two independent sets of coils, and a switch from the broken coil to the other set many times results in trouble-free heating for unloading. Each time a car is returned, it should be checked for heating coil leaks with air pressure and blown free of moisture as a freezing precaution.

(4) Insulated railcars—Insulated railcars have external heating coils, which eliminates the broken coil problem, but the insulation qualities can affect quality problems from residual heat. The products loaded do not lose the heat applied for loading sometimes for more than a week, depending upon the outside temperatures. Accelerated oxidation from the extended high temperature, exposure in a nonprotected atmosphere, and agitation from the car movement can result in products with high red colors, peroxide values, and off-flavors. Strict adherence to the rule of thumb to keep the fats no warmer than necessary to facilitate handling with pumps, which is usually 10°F (5.6°C) above the product melting point, or shipment by another means is necessary to deliver quality products under these circumstances.

(5) Impurities—Even with very conscientious inspections for cleanness, products have arrived at the customers' locations with unacceptable impurities in the product. One of the major causes of these impurities is the internal coils. Polymerized oil collected below the coils or on the coil supports has a tendency to flake off with movement during heating, contaminating the current product. Oil deliveries of bulk edible products should be filtered to collect all the impurities possible with this delivery mode.

(6) Aeration—The loading line should not allow the heated liquid product to cascade or fall through the air into the railcar. The desirable procedure is to fill the railcar from the bottom with subsurface entry to avoid splashing and excessive exposure of the oil, which accelerates oxidation.

(7) Dome hatch cover—The hatch opening allowing the loading line entry should be fitted with a temporary cover to prevent contamination with insects, rain, dust, and other airborne pollutants while product is being loaded. This cover will also protect the oil from direct sunlight.

(8) Unloading procedures—The purchaser of edible oils in railcars must use proper procedures to prepare the product for unloading to protect

the product quality. Two important product heating requirements are

- steam pressure—It is very easy to overheat or cause localized scorching of the oil; to avoid these problems the steam pressure used should be limited to 10 psig. This provides a steam temperature of 240°F (115.6°C).
- agitation—Agitation of the heated product should begin as soon as possible. Two methods of agitation are (a) recirculate the melted product by pumping from the bottom outlet back into the car through the dome, avoiding aeration by extending the line at least halfway to the bottom of the car, or (b) agitation with a portable agitator, which can be lowered through the dome to extend into the melted product.

(9) Car heels—All empty railcars returned must be visually inspected for product heels left by the customer. The heel might be due to incomplete melting, sloping of the unloading track, deliberate incomplete unloading, or some other reason. Removal of the heel may require squeegeeing manually to direct the residual oil into the outlet line or the addition of heated oil with agitation to melt the product for removal.

(10) Railcar service—Individual railcars should be kept in the same type service, that is deodorized, undeodorized, and/or crude oils. If possible, railcars should be dedicated to the same product service. Changing service from a deodorized oil to a crude oil does not present a real problem; however, the reverse can cause oxidation and impurities problems for several trips, even after caustic washing. This service change probably requires complete cleaning, including sand blasting, to avoid quality problems.

(11) Railcar inspections—A formal checklist identifies the critical areas to be inspected before and after each railcar is loaded for shipment. The checklist should include

- All interior surfaces must be thoroughly dry and free of all dirt, scale, and any possible infestation.
- Top hatch covers and gaskets are to be inspected to ensure cleanliness and must be free from infestation, undamaged, and will seal properly.
- Valves should be operative, leak-free, and close tightly.
- All exterior openings should be clean, capped, or otherwise protected from recontamination during transit.
- The top exterior must be clean and free from product, dirt, and other foreign material both prior to and after loading.
- All hatches and external openings must be closed and secured with registered seals.

2.17 Oil Filling and Packaging

PROBLEM: IMPURITIES

Possible Causes and Corrective Actions
(1) Dirty drums—Each drum should be inspected with a light to identify rust or other impurities before filling.
(2) Dirty smaller containers—Inspection of empty product containers at receipt followed by air cleaning at the filling line.
(3) Filter failure—Analysis of representative samples and the continual visual on-line operator inspection of clear containers should prevent excessive re-work. The filter elements should be changed when the filter pressure exceeds a certain predetermined level as a preventive measure.
(4) Dirty equipment—The filling equipment should be kept clean at all times to prevent contamination of the food product.

PROBLEM: WEIGHT CONTROL

Possible Causes and Corrective Actions
(1) Dirty scales and weights—All equipment should be kept clean and calibrated frequently to ensure the proper control; also, the quality check weigher should use separate equipment to identify line equipment problems.
(2) Oil temperature—Fluctuations in oil temperature will cause over- and underfilling, which should be controlled by an in-line heat exchanger to provide the optimum oil temperature.
(3) Fluctuating filler nozzle pressure—The filler oil supply should be from a clean, constant head, surge tank.
(4) Wrong tare weights—The container tare weights should be reestablished frequently to assure that the proper weight is being used. Individual containers must be tared before filling for accurate determination of fill weights.

PROBLEM: POOR FLAVOR

Possible Causes and Corrective Actions
(1) Overheating the oil—Temperature controllers should be used to control the storage tank temperatures and avoid overheating. The heating surface should never produce surface temperatures above 250°F (121°C). Exposure to these higher temperatures tend to scorch the oil products and deteriorate quality.

(2) Oil aeration—Identify the source of the aeration and repair or discontinue the poor practice. Potential sources are pumps, oil free-fall into the tank, agitation, blowing the lines with air, recirculation, and so on.

(3) Light exposure—Only limited exposure of the oil to sunlight or ultraviolet rays from florescent lightening will increase the peroxide value and impart off-flavors. Oils at all stages of processing should be protected from light. Clear glass or plastic containers should not be used to sample or store oils in process.

(4) Allowing the oil to stand in lines or unprotected surge tanks—Blow all lines with nitrogen and empty all surge tanks when not operating.

(5) Nonrepresentative sample—The sample line and valve must be flushed with the product to be sampled to remove all oxidized or gummy deposits, and the sample container must be absolutely clean.

(6) Poor filler cleaning practices—Clean the filler equipment frequently and thoroughly, ensuring that no soap or other residue is left on areas that could contact the product.

(7) No nitrogen protection—The oil should be nitrogen protected immediately after deodorization and continually through and into the package to prevent oxygen occultation. Dissolved oxygen in the oil will react with the oil, causing oxidation, which shortens the shelf life of the product.

(8) High head space oxygen content—The package head space should contain less than 2% oxygen to prevent excessive oxidation before purchase.

(9) Chelating agent omission—Citric or phosphoric acid should be added to deodorized oil before filtration to trap the trace metals, which accelerate oxidation.

(10) Dissolved oxygen—The liquid oil dissolved oxygen level should be 1.0% maximum. Higher levels indicate that the deodorized oil has not been nitrogen protected and aerated at some point before filling. Dissolved oxygen will accelerate oxidation, which causes off-flavors.

(11) Product heel—Fresh product should not be added to heels of previous lots. The older product can contain concentrations of oxidized products that will accelerate oxidation of the fresh product. Entire lots should be packaged to eliminate heels, or the remaining product must be recycled back to processing for adequate control.

PROBLEM: PACKAGING RATE

Possible Causes and Corrective Actions

(1) Oil temperature—The speed of filling and weight control are dependent largely upon the temperature of the oil. If the oil exceeds the

specified limits, the container may overflow or cause a slow down in filling speed to prevent overflow. If the oil is cold, the filling rate may be increased because there will be outage in the filled container and, consequently, less danger of oil spurting over the sides with rapid filling rates. An in-line heat exchanger should be utilized to control the fill temperature for the best performance.

2.18 Packaged Liquid Oils—Storage and Transportation

PROBLEM: CLOUDY OIL

Possible Causes and Solutions

(1) Storage temperature—Exposure of liquid oils to cold temperatures will cause the oils to cloud. The degree of solidification will depend upon the exposure time and the particular oil's resistance to cold temperatures. Oils that are not natural winter oils or that have not been winterized or dewaxed will cloud rapidly at refrigerator temperatures of 40°F (14.4°C) and below. Liquid oils should be stored in temperature-controlled warehouses in the winter months of the year.

(2) Shipping temperature—Liquid oils should never be shipped in mixed loads when the other products require refrigeration. Additionally, temperature controlled trucks and railcars are necessary to prevent liquid oil clouding in the winter months.

(3) Heating to clarify—Liquid oils may be clarified by controlled gentle heating if the solidification has not progressed too far. In some cases the oils will clarify when heated, but then a portion of the hard fraction will reappear when the oil is cooled to room temperature. In some cases, it may be necessary to heat the oil to quite high temperatures, 140°F (60°C), to melt the hard fraction that has developed.

(4) Wet oil—Water in amounts greater than 0.05 to 0.15% cause a turbid oil at refrigerator temperatures. Wet oils must be dried by deodorization or vacuum drying to avoid hydrolysis and clouding.

2.19 Shortening Packaging

PROBLEM: METTLER DROPPING POINT CHANGE
AFTER PUMPING

Possible Causes and Corrective Actions

(1) Pumping error—Blow air or nitrogen through the empty lines before actually pumping the product to ensure that it is lined up properly.

(2) Faulty valve allowing the product to deliver to more than one location—Location of the valve needing repair should be simplified by the Mettler dropping point results. The product with the change is probably in the tank with the faulty valve or at least a valve that lines up with this tank.

(3) Sample identification error—Determine the correct product identification by resampling, performing other fingerprint analyses, or checking transfer records. The sample tag should require both the product specification number and batch number, as well as the tank number for possible errors of this type.

(4) Nonrepresentative sample—The sample line and valve must be flushed with the product to be sampled to remove all previous product residue, and the sample container must be clean. If a poor sample is suspected, resampling is necessary.

(5) Cool tank temperature—The product temperature should be 10°F above the melting point before sampling to obtain a representative sample.

(6) Nonuniform product—Shortening, even if held 10°F (5.6°C) above the melting point, can stratify or separate to provide a sample that is nonrepresentative of the whole product. Tanks must be agitated for a minimum of 20 minutes before sampling to ensure product uniformity.

(7) Normal method variation—AOCS Method Cc 18-80 indicates that the expected variation within a laboratory for the Mettler dropping point result is 0.7°C.

(8) Emulsifier additions—Lecithin, mono- and diglycerides, and other emulsifier additions will change the Mettler dropping point result, which should be reflected in the product specification requirements.

PROBLEM: PEROXIDE VALUE INCREASE

Possible Causes and Corrective Actions

(1) Overheating—High temperature storage before packaging will accelerate oxidation. The rate of oxidation for edible oils at least doubles with each 20°F (11.1°C) increase in temperature. Storage tank holding temperatures should be adjusted to no higher than 10°F (5.6°C) above the melting point of the product. An abnormal peroxide value increase should be investigated thoroughly before packaging, even if the result is still within specified limits. It may be necessary to rebleach and redeodorize the product to restore the oxidative stability to an abused product.

(2) Chelating agent omission—Citric or phosphoric acid should be added to deodorized oil before filtration to remove trace metals that accelerate oxidation, as evidenced by the peroxide value increase.

(3) Aeration—Potential sources for air incorporation into the oil should be inspected and repaired or a poor practice identified and discontinued.

(4) Nonrepresentative sample—The sample line and valve must be flushed thoroughly with the product to be sampled to remove oxidized and polymerized residues from previous samplings. Additionally, the sample container must be free from any soap or detergent residues.

(5) Light struck sample—Only limited exposure of a shortening product to sunlight or ultraviolet rays from florescent lighting will increase peroxide value results and decrease flavor ratings. Samples determining the quality of product must be protected from light abuse to avoid inaccurate results. Transparent or open containers should not be used to obtain or store samples to be analyzed.

(6) Secondary oxidation—A rapid rise in peroxide value with an accompanied orange color may be due to secondary oxidation. The source of the abuse causing the problem was probably overheating, perhaps accompanied by aeration at some prior stage after bleaching. This product might be salvaged by rebleaching and redeodorization; however, it will probably never regain the oxidative stability of the oil before it was abused.

(7) Dirty glassware—A soap or detergent residue will give a false high peroxide value result. All laboratory glassware and washed sample containers should be rinsed thoroughly with distilled water to remove soap films. High peroxide value results should be rechecked, especially if several different products are also indicating high values.

(8) Laboratory reagent problem—Normally, all products tested from all process areas will have high peroxide value results if this problem exists. Replacing the reagent with a new lot in a single laboratory situation or performing duplicate analysis in another laboratory should determine if a reagent problem exists.

(9) Localized overheating—Good agitation while heating shortening in storage tanks prevents localized overheating. When steam coils are used to heat a storage tank without agitation, extreme localized heating of the oil near the coils occurs while the overall temperature rises slowly. Consequently, the damage to the oil near the coils affects the stability of all the oil in the tank.

(10) Heating surface temperature—The heating devices used should not provide surface heat above 250°F (121°C). Exposure to higher temperatures tend to scorch the fat and oil product and deteriorate quality.

PROBLEM: FREE FATTY ACID INCREASE

Possible Causes and Corrective Actions

(1) Nonrepresentative sample—The sample line and valve must be flushed with the product to be sampled to remove any traces of the previous product sampled, which may have had a high free fatty acid caused by emulsifiers or some other requirement.

(2) Antioxidant addition—Antioxidant mixtures additions will increase the product free fatty acid. Process specifications should recognize this increase and should allow a higher free fatty acid result after this addition. This will naturally require a lower free fatty acid from the deodorizer if the final limit cannot be increased because of performance or customer requirements.

(3) Colorant addition—Highly colored oils interfere with the visual endpoint determination for the free fatty acid analysis. Either a higher result must be allowed or the final analysis must be performed before the color addition.

(4) Emulsifier addition—The free fatty acid level of most emulsifiers far exceeds the 0.05% normally specified for a nonemulsified shortening. The addition level and effect should be a consideration when the product specified limits are being established. The free fatty acid increase can also be used as a quick indicator of the level of emulsifier added before the emulsifier analytical results are available.

(5) Identification error—A high free fatty acid result could be the result of a mix-up in samples or identification tags. All samples should be identified with the specification, batch, and tank number to help determine the correct product identification; however, any sample in doubt should be resampled and reanalyzed.

(6) Carbon dioxide in nitrogen gas—CO_2 will give a false high free fatty acid result. This problem can be identified by drawing a vacuum on the sample to remove all gases and retesting. If CO_2 contamination exists, the deaerated sample will test normal. The CO_2 source should be identified and the problem corrected.

(7) Wet oil—One cause of free fatty acid development is hydrolysis. Moisture in the oil will naturally accelerate this reaction; therefore, all fats and oil products should be "dry" when packaged.

(8) Pumping error—Either transferring the product to the wrong tank or inadvertent addition to another product through a faulty valve can provide a different free fatty acid result than desired. A lower free fatty acid should cause as much concern after a transfer as an increase; either change can indicate a problem that requires investigation.

(9) Excess chelating agent—Citric and phosphoric acid will catalyze free

fatty acid development when added in too high quantities: over 50 ppm citric and over 10 ppm phosphoric. The addition of the chelating agents after deodorization must be controlled to prevent catalyzation of the free fatty acid.

PROBLEM: COLOR INCREASE

Possible Causes and Corrective Actions

(1) Overheating—High-temperature storage before packaging will accelerate oxidation with an attendant color rise. The oxidation rate for edible oils at least doubles with each 27°F (15°C) increase in temperature in the 68 to 140°F (20 to 60°C) range. The storage tank's holding temperature should be adjusted to no higher than 10°F (5.6°C) above the melting point of the product. An increase in the product's color should be investigated thoroughly before packaging, even if the result is still within specified limits. It may be necessary to rebleach and redeodorize the product to restore the oxidative stability to an abused product.

(2) Colorant added—Color added to a shortening product, either on purpose or by mistake, will increase the color determination. If the product tested does not require an added color it probably was added to the wrong tank by mistake. Normally, colorants can be removed by the heat bleaching during deodorization, but some may require bleaching as well.

(3) Dark emulsifier added—Many emulsifiers are darker in color than a deodorized shortening product; therefore, the color requirements should be adjusted when the product limits are developed to allow for the resultant increase.

(4) Aeration—Oxidation caused by aeration of the fat and oil product is evidenced by a darker red color, as well as an increase in peroxide value and flavor degradation. Potential equipment failures and/or poor practices should be identified and repaired or discontinued.

(5) Secondary oxidation—The development of an orange color with a high peroxide value is an indication of oxidatively abused product. The source of the problem was probably overheating and/or aeration at some stage after bleaching. This product may be salvaged by rebleaching with a bleaching earth to remove the secondary oxidation materials. However, the oxidative stability of this product will probably never regain the original level of resistance.

(6) Identification error—Product specified color limits vary for the individual products based on the source oil, additives, and other considerations; therefore, an identification error can indicate that a product

is completely out of the specified limits when it isn't and in specified limits when it isn't as well. Each sample should be identified with the product specification and batch number, as well as the tank number sampled to reduce identification errors.

(7) Emulsified shortenings—Iron oxides are soluble in monoglycerides and many other emulsifiers. Rust from black iron equipment will be more readily absorbed by emulsified shortenings, which will affect the product's color and oxidative stability. Black iron surfaces should always have an oil film on the surface to prevent rust [17].

(8) Pumping errors—Product color may indicate a pumping error if the color change is considerably different. The oil pumper should check the line-up by blowing from the delivery point to the receiving tank before the product is pumped.

(9) Nonrepresentative sample—The sample line and valve must be flushed thoroughly with the product to be sampled before the sample is obtained in a clean container. The sample valve and line probably has a residue of the previous sample obtained, which could have oxidized and/or polymerized since the last sample was obtained, causing a nonrepresentative sample of the current product. Any indication of a sampling problem should be double-checked with another sample.

(10) Cloudy sample—The sample must be completely melted and clear to obtain a reproducible color determination. Evaluation of cloudy samples provides an invalid color result.

(11) Laboratory technician variation—Lovibond color determinations are made by matching the sample color visually with colored glasses by each individual technician and is a subjective evaluation at best unless the automatic tintometer is utilized. Some technicians have more difficulty determining color values than others, which can be the source of product color changes. All technicians should be tested to determine their ability to determine color differences. Since approximately 8% males and 0.4% females suffer from varying degrees of color blindness, there is a chance that some technicians will have defective color vision [26].

(12) Chelating agent omission—A metal scavenger should be added immediately after deodorization to remove trace metals, which accelerate the oxidation of fats and oils products. Color darkening and peroxide value increases are indications of oxidation.

(13) Localized overheating—Agitation is necessary while heating shortenings in storage tanks to prevent localized overheating. When steam coils are used to heat a tank without agitation, extreme localized heating of the oil near the coils occurs while the overall temperature rises slowly. However, the extreme damage to the oil near the coils ad-

versely affects the stability of all the oil in the tank, which probably results in a higher color and peroxide value.

PROBLEM: WET OIL

Possible Causes and Corrective Actions

(1) Leaking cooling coils—The cooling and heating coils in tanks should be checked for leaks by pressurizing on a regular basis and especially when high moistures are identified by laboratory analysis.

(2) Line steaming—Transfer lines allowed to set up with hard oils are frequently steamed to melt the solidified product. These lines must be flushed thoroughly with fresh product to remove any traces of moisture.

PROBLEM: SHORTENING CONSISTENCY

Possible Causes and Corrective Actions

(1) In-package heat rise—The temperature rise in the container, due to heat of crystallization, should not be more than 1 or 2°F (0.6 to 1.1°C). Increases above this level are indicative of substantial crystallization under static conditions and can cause (1) a firmer consistency than desired, (2) melted shortening in the center of the package, (3) low creaming gas or nitrogen content, (4) resolidified appearance, and (5) a dark color. The rpm and holding time in the worker unit should be evaluated to correct this discrepancy.

(2) Shortening picking—Too little working, or "picking," after chilling develops a shortening that is sloppy as it emerges from the filling spout, experiences a high in package temperature rise, sets up very rapidly, and becomes hard. Too much picking results in a shortening that fills sloppily, but sets up very slowly, with little temperature rise in the package. Ideal picking is normally between these two extremes, unless either of these conditions is desired for the individual product processed.

(3) Precrystallization—Sandy, lumpy, or riced plasticized shortenings or margarines are symptoms of precrystallization. Product crystallization prior to chilling can be caused by a low supply tank temperature, a low precooler temperature, or insufficient melting of the recirculated product.

(4) Postcrystallization—Consistency problems can be caused by postcrystallization immediately after the product has exited the chilling unit and filled due to insufficient cooling, working, or crystallization time. Excessive postcrystallization results in products that are too firm or too brittle, with poor spreadability or workability. The time required for

supercooled fat to transform from the alpha to beta-prime crystal form is the crystallization time. Different source oils crystallize at different rates because of the triglyceride composition. Crystallization time can be measured by observation of the lapsed time after cooling to the crystallization point and when crystals form. Product throughput is a major key to control postcrystallization, along with cooling temperatures and working.

PROBLEM: PLASTICIZED SHORTENING FAULTS

Possible Causes and Corrective Actions

(1) Streaking—Streaks in plasticized shortening can be caused by (1) chilling too low for the operating back pressure, (2) erratic creaming gas dispersion, (3) channeling that allows semi-liquid oil to pass through the chilling unit insulated from chilling properly, (4) mixing different chilling unit streams operating at different temperatures, (5) erratic chilling unit pressures, (6) low feedstock temperature allowing precrystallization, and (7) low pressure at the filler allowing a loss of creaming gas.

(2) Sandy—Small lumps about the size of grains of sand or larger can be caused by too cold chilling unit temperatures or precrystallization of the product prior to the chilling unit. This consistency fault is described as ricing when larger translucent hard particles develop.

(3) Ribby—This is caused by alternating very thin layers of hard and soft product, which has been described as feeling like a corduroy cloth surface. This condition can be caused by (1) too cold chilling unit temperature with low pressure, (2) mixing streams from two or more chilling units operating at different temperatures, or (3) too cold fill temperature causing excessive mounding in the package.

(4) Puffy—A soft plasticized shortening with large visible air cells or pockets that offer very little resistance when worked is caused by too high creaming gas content, too low back pressure to finely distribute the air cells, or a combination of the two conditions.

(5) Brittle—A brittle feel is contributed by (1) warm chilling unit temperatures, (2) either the absence or less than required working after chilling, (3) the product formulation has a narrow plastic range, or (4) low creaming gas content.

(6) Gray cast—Changes in the crystal structure brought about by high-temperature exposure may result in a gray cast.

(7) Oil separation—Free oil in a plasticized shortening is caused by (1) too warm chilling temperature, (2) precrystallization of the product before entering the chilling unit, (3) temperature abuse after pack-

aging, (4) formulation with insufficient hardstock or source oils for the wrong crystal habit, and/or (5) too low chilling unit and filler back pressure.

(8) Chalky—A dull white appearance is caused by too high a creaming gas content.

(9) Vaselated—A yellowish clear greasy appearance is typical of shortenings packaged without creaming gas—identified as a Vaseline-like or greasy appearance.

(10) Specks—Visible specks will appear in the finished shortening when the storage tank contains these impurities or the in-line filter is faulty. Also, heating of "frozen" or setup lines with a flame will polymerize and char the product to produce specks that can continue to contaminate products for extended periods of time.

(11) Rubbery lard—Tempering lard at 70 to 85°F (21.1 to 29.4°C) for 48 hours develops a plastic, rubbery consistency. Conversely, immediate refrigeration after packaging produces a loosely structured and brittle consistency.

(12) Too dark—The whiteness of a plasticized shortening is affected by the amount and degree of dispersion of the creaming gas. However, increasing the creaming gas content will not materially affect an oil that has a dark color before plasticization.

(13) Too soft—A product that is softer than expected can be caused by (1) a cool chill unit outlet temperature, (2) solids fat index results lower than specified, (3) the wrong composition, (4) contamination with another product, or (5) too much working after chilling.

(14) Too firm—A harder than expected plasticized shortening consistency can be caused by (1) a high chill unit outlet temperature, (2) solids fat index results higher than specified, (3) the wrong composition, (4) contamination with another product, or (5) insufficient working after chilling indicated by an in-package temperature rise of more than 2°F (1.1°C)

(15) Appearance—Blue plastic liners are normally used for shortenings to help make the plasticized product appear whiter. Different color liners alter the visual perception to make the shortening appear less white.

(16) Slack or overfilled containers—Faulty or erratic creaming gas control causes this condition if the product fill weights are adjusted properly.

(17) Poor aeration—Hydrophilic surfactants are dispersible in fats and oils but not soluble. These emulsifiers will separate from the liquid oils without good agitation and should be added just prior to plasticization.

Nonuniform dispersion will also result in an off-flavor for a portion of the plasticized shortening, i.e., bittersweet.

2.20 Liquid Opaque Shortenings

PROBLEM: SEPARATION

Possible Causes and Corrective Actions

(1) Aeration—Air incorporation promotes product separation and thickening. Storage studies indicate that an air content of less than 1.0% is necessary for a stable suspension. Sources of liquid shortening aeration during processing are (1) pumps sucking air, (2) agitation causing a vortex that whips air into the product, (3) nonpressurized filler, (4) product free-fall into a tank or packaging container, and (5) skipping the deaeration process.

(2) High temperature—Exposure of the product to temperatures above 95°F (35°C) will result in partial to complete melting of the suspended solids, causing separation. Complete melting and reprocessing are required to restore the product to the original opaque, creamy, pumpable product.

PROBLEM: THICK PRODUCT

Possible Causes and Corrective Actions

(1) Beta-prime hardstock—The small needle-shaped, beta-prime crystals pack together into dense fine grain structures that cause the product to thicken. The large, coarse, self-occluding beta crystals contribute the desired product consistency.

(2) Cold temperature—Liquid opaque shortenings will solidify at storage temperatures below 65°F (18°C) with a loss of fluidity. This solidification and fluidity loss can be reversed by controlled heating not to exceed the melting point of the product.

(3) High air content—An air content of less than 1.0% will cause liquid opaque shortenings to thicken and separate. Quality evaluations indicating a high air content should initiate immediate actions to identify the aeration source to eliminate a continuing recurrence with succeeding batches. The problem batch of liquid shortening will require remelting to deaerate, followed by recrystallization.

(4) Commingling—Liquid shortening contamination with another product normally results in a thick product. Pumping errors or pickup from previous product in tanks, lines, and processes are the usual causes for

commingling. Most of these products are firmer or have a beta-prime crystal that contributes to thicker liquid shortenings.

PROBLEM: THIN PRODUCT

Possible Causes and Corrective Actions

(1) Hardstock level—The hardstock level in opaque liquid shortening has a direct effect upon the consistency or viscosity of the crystallized product. The product becomes more viscous as the hardstock level is increased with all other variables kept constant. A uniform viscosity product should be produced batch after batch when the hardstock level is maintained within the tolerance levels determined acceptable by product design.

(2) Hardstock iodine value—The hardstock added to an opaque liquid shortening is the seed for crystallization and has an effect upon the final consistency of the liquid shortening. Lower iodine value hardstocks will induce more rapid and complete crystallization than higher iodine value products.

(3) Basestock hardness—The melting point and SFI results materially affect the consistency or viscosity of opaque liquid shortenings. As the basestock melting point increases, the product becomes more viscous or thicker.

(4) Temperature—Opaque liquid shortening viscosity or consistency is temperature dependent. Thin products are experienced at high temperatures, and thicker products result from cooler temperatures. Extended periods at high temperatures will cause product melting and separation for a very thin top layer.

(5) Separation—A portion of a separated product will be thin with the hard fraction settled to the bottom if temperature abuse is sufficient to completely break the product. Separation due to aeration usually begins in the center of the product or where a container shoulder traps the air that escapes from the product. It looks like a crack in the product. This crack or separated streak grows with time to eventually divide the product into liquid and hard fractions.

PROBLEM: POOR FLAVOR STABILITY

Possible Causes and Corrective Actions

(1) Liquid oil adjustment—Liquid shortening mixes that are firmer than specified should not be adjusted with any softer basestock than those specified in the composition. The flavor stability is dependent upon the weakest component in the composition. Therefore, adjustments with a

basestock containing higher polyunsaturates will cause a reduction in flavor stability.

(2) High precrystallization temperature—High-temperature storage before crystallization can promote oxidation to the point that packaged product will revert in flavor and develop a pinkish yellow color quickly. The oxidation rate for edible oils at least doubles with each 20°F (11.1°C) increase in temperature. An abnormal peroxide value increase or off-flavor indication from the crystallization supply tank should be investigated thoroughly before the product is packaged. Rebleaching followed by redeodorization will be required to restore the oxidative stability to an abused product.

(3) Aeration before crystallization—Liquid shortenings must be protected from air throughout the system, even though a deaeration step is part of the crystallization process. Liquid oils react quickly at elevated temperatures with atmospheric oxygen to revert in flavor. Some of the potential air sources are (1) air leaks in fittings, pumps, and flanges; (2) agitator use before being properly covered with product, creating a vortex; (3) blowing lines with air into the product; and (4) oil free-fall into the tank.

(4) Peroxide value increase—Analysis indicating a higher than expected peroxide value increase before packaging could be an indicator of secondary oxidation. If the oil product has been heat abused or improperly bleached, the oxidative stability may be substantially poorer than dictated by product requirements. Unexplained peroxide value increases indicate a reduced oxidative stability, which will be confirmed by accelerated stability evaluations, flavor reversion, and product discoloration.

(5) Localized overheating—Good agitation while heating in storage tanks before crystallization prevents localized overheating. When steam coils are used to heat a storage tank without agitation, extreme localized heating of the oil near the coils occurs while the overall temperature rises slowly. Consequently, the damage to the oil near the coils affects the stability of all the oil in the tank. Rebleaching followed by redeodorization may be required to reestablish the necessary oxidative stability.

PROBLEM: CRYSTALLIZATION PRODUCT AERATION

Possible Causes and Corrective Actions

(1) Deaeration process step—The first stage in many liquid opaque shortening crystallization processes is a deaeration step. This process should eliminate any entrapped air in the liquid oil before crystallization. An

air content analysis should be required for termination of this process stage because it may be the most likely shortcut taken to save time. Entrained air allowed to remain in the product can only increase from this point to potentially require a repeat of the crystallization process or cause customer displeasure with thick, viscous product.

(2) Agitation—Care must be exercised during the crystallization of liquid opaque shortenings to avoid whipping air into the product through improper agitation in the crystallization tank. Creation of whirlpools or vortexes that draw air into the product must be avoided.

(3) Product free-fall—Liquid opaque shortening products should not be allowed to cascade or fall through the air into crystallization and/or storage tanks or the packaging utilized. The most desirable procedure is to fill both the tanks and packages from the bottom with subsurface entry.

(4) Air content control—Some potential pourable liquid shortening processes incorporate a deaeration step just prior to packaging. Percent air analysis would evaluate the effectiveness of this process. This analysis is also applicable to determine the acceptability of other processes. Storage studies have indicated that an air content of less than 1.0% is necessary for a liquid opaque shortening's suspension stability.

2.21 Shortening Flakes

PROBLEM: FLAKE LUMPING

Possible Causes and Corrective Actions

(1) Heat of crystallization—Shortening flakes packaged immediately off of the chill roll can experience a heat of crystallization rise high enough to partially melt some of the product, causing lumping. Refrigerated air cooling of the flakes before packaging can effectively remove the heat of crystallization, which promotes product lumping.

(2) Low chill roll temperature—Low chill roll temperatures can result in "shock" chilling, which causes the oil film against the roll to rapidly solidify and pull away, providing an insulation from further cooling. In this condition the outside surface of the flake remains liquid, creating a wet flake that will fuse with others to create lumps in the package.

(3) Low oil to roll temperature—Milky or grainy appearing oil, indicating a low temperature, will not solidify properly on the roll. The flakes produced will appear wet or oily on one side, which will cause the flakes to fuse together in the package. The oil must be heated above the melting point and have a clear appearance, indicating the absence of solids when applied to the roll.

(4) Filling chute construction—The filling chute or any other enclosed passage for the flakes should be constructed of smooth material without corners or crevices where flakes could accumulate and fuse together. These lumps will fall into the package unnoticed until the customer finds them. Additionally, any enclosed areas should be accessible for cleaning on a regular basis.

(5) Fast roll speed—The chill roll should have a variable speed for production of the various shortening flake and chip products. However, the roll speed must be adjusted to produce dry flakes each time the product type is changed.

(6) Packaging—The usual packaging materials used with shortening flakes or chips are multiwall natural Kraft brown bags or corrugated cases with liners. Both packages can become good insulators to hold and cause further elevation of the product's temperature:

- bags—Several styles of Kraft bags are available with and without square bottoms. Some bag types stack together flatter and tighter than others with little or no free space between layers to allow the heat to dissipate. Partial melting from the nondissipated heat of crystallization and compression from the weight of the top layers on a pallet promotes flake clumping. The use of this package requires postcooling after flaking to remove the heat of crystallization to prevent fusion of the flakes.
- corrugated case and liner packaging—The case provides less insulation than the bag packaging. Some headspace will occur after filling the container because the chip products settle. The settled chips can fuse together if exposed to elevated temperatures for long periods. Therefore, it is desirable to store the case packaged products at cool or refrigerated temperatures.

(7) High-temperature exposure—High storage or shipping temperatures immediately after flaking will accelerate the heat of crystallization temperatures. The higher temperature exposure promotes surface melting and fusion of the flakes. Controlled temperature storage of 70°F (21.1°C) or less should be utilized for packaged flakes, with a holding period of 48 to 72 hours to stabilize the product temperature before shipment.

(8) Feed trough leak—The most common source of lumps in flaked products is a leak in the feed trough. This allows liquid to drip into the flake collector where it begins to solidify and acts as an adhesive, collecting other chips until a lump is formed, which breaks away to enter the package. Leaks need to be repaired when the roll is not operating. Attempts to repair a trough leak while operating by inserting some material or instrument into the leak could initiate a more serious problem. The temporary repair material can escape and find its way into

the finished product package. This material then becomes a foreign material in the flaked product, which could be more damaging than the original problem.

(9) Saturation level—Flake clumping or lumping problems become more severe as the melting point of the product increases. The flake release point from the chill roll is proportional to the fatty acid chain length and the degree of saturation, which control the solidification temperature. Therefore, flake products with the highest melting points will release automatically from the chill roll at a higher temperature than products that experience approximately the same heat of crystallization level, but the less hydrogenated product will attain a lower temperature after packaging as shown by the following selected data:

	Soybean Oil Flakes	
Iodine value	1.5	7.0
Fatty acid composition, %		
Stearic (C-18:0)	83.0	75.7
Oleic (C-18-1)	1.8	7.4
Product temperature, °F		
Initial in-package	110–115	100–102
30 minute in-package	140–142	130–133

PROBLEM: HIGH MOISTURE CONTENT

Possible Causes and Corrective Actions

(1) Heating coil leaks—Flaked products have melting points that require heating before chilling to remain liquid. The holding tank's steam heating coils will develop leaks eventually from wear and vibration. Each oil product should be sampled properly and analyzed before flaking to ensure that the product has not been contaminated with moisture or other foreign materials. Proper sampling technique requires agitation for 20 to 30 minutes before sampling to ensure a representative sample. High moisture analysis should prevent flaking of the product and initiate investigation to determine the cause and corrective action required.

(2) Line steaming—The melting point of flaked products necessitates that the transfer lines be heat traced to prevent the product from solidifying in elbows, low areas, and so on. When the heat tracing on the flake transfer lines fail or an operator has failed to clear the lines after a product transfer, the product remaining in the line will solidify and

prevent further transfer of product. One method used to open these lines is to steam them to melt the product causing the restriction. This process will contaminate the product holding tank with moisture. All of this product must be reprocessed to remove the excess moisture content.

(3) Wet roll—Moisture collects on the chill roll where and when it is not covered with a film of oil. The amount of moisture condensing on the roll is dependent upon the room humidity and the roll cooling temperature. This moisture will transfer to the surface of the flakes or chips produced. This moisture content cannot be eliminated but can be controlled somewhat by
  • dehumidification—Air conditioning and/or dehumidification units are required to control the room humidity to a reasonable level. These units would be more important in humid climates to control the product moisture content.
  • startup product—Product flaked on startup or after a down period will be more wet than during continuous operations. This product should be discarded until the startup excess moisture has been dissipated.

(4) Additives effects—Some of the preferred flavors for flaked products are water emulsions, which indicate a wet oil on analysis after the addition. This moisture presence must be recognized and allowed by specification. However, the level must be controlled in case another moisture source has also developed.

PROBLEM: FREE FATTY ACID INCREASE

Possible Causes and Corrective Actions

(1) High moisture—Condensation on the roll surface transfers to the flakes and chips. Moisture initiates the hydrolysis reaction where the chemical bonds that hold the triglycerides together are broken, creating free fatty acids, monoglycerides, diglycerides, and even glycerides if the hydrolysis is extensive. The expected free fatty acid increase from hydrolysis is usually from 0.05% or less before flaking to as high as 0.10% as shipped. The condensation moisture must be controlled by dehumidification and discarding startup product with high moisture to maintain acceptable free fatty acid levels.

(2) Lecithin addition—The addition of small quantities of lecithin to oils before flaking will emulsify some of the moisture from the roll to help reduce the free fatty acid development. Acceptance of the lecithin additive for this purpose has not been universally accepted. This addition could cause more problems than the benefit in some cases, such as

product darkening, smoking, and an off odor with high-temperature processing.

(3) Analysis end point—The free fatty acid analysis by AOCS Official Method Ca 5a-40 requires the titration of a solution of the oil sample in alcohol with an alkaline solution until a pink color develops. Colored flake products interfere with the identification of this color endpoint, which usually results in a higher free fatty acid. Special care must be exercised with these analyses to determine the true free fatty acid content.

PROBLEM: CHARTREUSE COLOR

Possible Causes and Corrective Actions

(1) Beta-carotene colored chips have shown a tendency to change color after flaking from the desired yellow to chartreuse color. The problem has been identified as an oxidation problem, promoted by the oxygen available in the moisture condensate molecules from the surface of the chill roll. The color change is executed by exposure of the colored chips to refrigerated temperatures immediately after flaking. Two feasible methods to prevent the color change have been identified:

- tempering—Holding the packaged chips at 85°F (30°C) for 24 hours prior to refrigeration. Apparently, this tempering stabilizes the beta-carotene colorant to resist the oxidative reaction.
- lecithin addition—An addition of 0.1% lecithin to the formulation binds the moisture enough to prevent a reaction with the beta-carotene colorant.

PROBLEM: METAL CONTAMINATION

Possible Causes and Corrective Actions

(1) Chill roll construction—Most older chill rolls were uncoated steel drums that rusted rapidly because of the condensate formed on the cold surface. The first products produced after a down period almost always contained visible quantities of rust from the roll surface. The product produced had to be physically rejected until it had visibly disappeared. The potential for rust contamination of the product flaked with these rolls is virtually assured with every production. Chrome plating of the older rolls and of new construction has eliminated rust as a contaminate from this source.

(2) Filter equipment—Rust can be developed from the moisture condensed inside the filter equipment as a result of the product heat, humidity, and cool room temperature. The rust can find entry into the

product by several avenues if it isn't removed from the filter shell periodically, or a better preventative measure would be to clean and dry the filter shell each time the filter cartridges are replaced. This metal contaminate visibly appears to be black specks which are not large enough to be detected by packaging line metal detectors. In-line magnets would collect the metal specks.

(3) Equipment maintenance—The liquid oil in the feed trough and the oil film on the chill roll are exposed while being solidified into flakes and chips. Therefore, this open process could allow the entry of maintenance items or nuts, bolts, or other equipment parts into the packaged product. Efficient metal detection of the packaged product is necessary to identify these metal contaminates before shipment to customers.

(4) Applicator warp—Several different application mechanisms are used to apply the heated liquid oil to chill rolls. Most, but not all, of these mechanisms can warp with use due to stress from heating and cooling, movement for flake size, heavy weight of the applicator supported only on the ends of the roll, and more. The amount of misalignment can cause the applicator to gouge the chill roll, causing metal contamination of the product. Constant observation of the application techniques subject to this problem or a change to an addition system that does not have this potential is necessary. In either case, metal detection of the packaged product is mandatory to insure that the products do not have metal contamination.

PROBLEM: COLOR INCREASE

Possible Causes and Corrective Actions

(1) Overheating—Heat accelerates the reaction of atmospheric oxygen with edible fats and oils. Heat abuse causes an increase in the color of flaked products by oxidation of tocopherols. For deodorized products, the speed of oxidation is doubled for each 20°F (1.1°C) increase in temperature. A holding temperature of 10°F (5.6°C) above the melting point for flaked products should be sufficient. Design of the transfer systems should also take into account the product requirements. For example, the transfer lines should be as short and straight as possible to minimize the amount of insulated heat traced lines where residual product can deteriorate.

(2) Colorant added—Several flaked products do require the addition of a colorant for a butter-like appearance. A mistaken addition to the wrong product or the use of a tank previously used for a colored product will result in a high red color. Yellow colors can be removed by redeodorization in most cases since the colors readily heat bleach.

(3) Identification error—The product sample must be identified by tank number, batch number, and product identity to help assure that the sample identification is correct. Different source oils, product types, additives, processing, and so on have an effect upon the flaked shortening's color requirements. A color result that is well within the specified limits for one product can be beyond the tolerance limits for another. Any suspicion of an error should be resolved by resampling and retesting before packaging.

(4) Pumping error—The product delivered to the indicated tank may show a major change in analytical characteristics, with color being an early indicator. The oil pumper must check the transfer lineup by first blowing air or nitrogen from the originating tank to the receiving tank for verification. It is also mandatory that the receiving department be notified, preferably in writing, of the product pumped and to which tank for identification and to avoid errors.

(5) Faulty valves—A partially open valve on a transfer line can allow contamination of another product even though the lineup precautions were taken and everything appeared satisfactory. Two indicators will identify this problem: (1) the product quantity delivered to the intended tank will be less than the originating tank, and (2) analysis of the tank with the faulty value will change. A visual color may be one of the first indicators of the commingled product if it is being packaged as the problem occurs.

(6) Nonrepresentative sample—The tank to be sampled must be agitated for 20 to 30 minutes, the product temperature must be 10 degrees above the melting point, the sample line must be flushed to clear the previous product sampled, and a clean sample container must be used to obtain a representative sample. Any indication that a nonrepresentative sample has been obtained should cause the product to be resampled.

(7) Cloudy sample—A visually cloudy product during sampling or analysis indicates that the product temperature is too low, and corrective actions are required. Cloudy product during sampling probably indicates that the tank temperature gauge or controller may be faulty, which requires maintenance. This product should not be sampled until a clear product is available. Likewise, the Lovibond color analysis will provide a high result if any cloudiness is allowed. Special precautions with high melting point products should be taken to assure that all of the product is melted. One fault that can easily occur with the analysis is to allow the color tube to stand on a cool surface to deaerate before reading the color. High melting products have a tendency to set up rapidly on the bottom of the tube, which gives a false high color reading.

(8) Trace metals—All products should be postbleached to help remove any trace metals present after hydrogenation and the catalyst filtering step. These trace metals are prooxidants, which cause product color rises through oxidation of the natural tocopherols even in the highly saturated products.

(9) Localized overheating—All storage tanks with heating devices should be equipped with a mechanical agitator. Power agitation will minimize damage from localized overheating while saving time and heating costs. If the agitation is temporarily out of service, the temperature differential between the product's melting point and the heating medium must be kept minimal to preserve the product quality. Damage to the oil near the coils due to localized overheating affects all of the product in the tank. The product may require rebleaching and redeodorization to regain most of the original product's color stability.

PROBLEM: FOREIGN MATERIAL

Possible Causes and Corrective Actions

(1) Unfiltered product transfer line—The product transfer line between the filter and the chill roll can be the cause of foreign material. The high melting point of most flaked products requires heat tracing of the lines or frequent thawing of setup lines at startup and anytime the product is not physically moving. Setup lines have to be thawed with heat from some source that can polymerize and char a portion of the oil in the lines. This material will appear as black specks in the flaked product. This material must be set aside for reprocessing to remove the black polymerized oil specks.

(2) Improper product package use—The product packages should *never* be used for any other use than for the product intended. Sanitation personnel have used the regular product containers for trash, which has inadvertently been shipped to a customer. Even the draw-off material on startup should not be filled into the regular product containers.

(3) Uncovered chill rolls—The chill rolls should have protective shields covering them to protect from airborne foreign materials. The shields must be designed to fit the chill rolls without hampering operation, maintenance, or restricting the cooling process.

(4) Open product containers—Product cases should not be stacked with the opening exposed to possible foreign material on the packing floor. Even in the most controlled environment, some foreign objects will find entry into the open cases. Erected cases must be covered or stored upside down before use. Also, the plastic liner should not be inserted until just before filling.

PROBLEM: IODINE VALUE CHANGE

Possible Cause and Corrective Actions

(1) Pumping error—AOCS Official Method Cd 1-25 for iodine value analysis is a good control to assure that the intended product has been pumped from one location to another. Any change outside the normal analytical variation (probably a 0.5% tolerance) indicates a change in the product delivered. This change could be the result of (1) the product pumped to the wrong tank, (2) a faulty valve allowing the commingling of two or more products, (3) product delivered on top of a heel from the previous product, or (4) a mis-pumping of another product into the designated product. Good control dictates that the products should be sampled and analyzed as close to the packaging time as possible to identify these problems.

(2) Sampling error—A representative sample requires that a predetermined protocol be established and observed. The following sampling protocol identifies the areas to be considered:

- product temperature—Flaked products have low iodine values with corresponding high melting points. The products must be held at 10°F above the typical melting point to remain clear and pumpable.
- agitation—The tank to be sampled should be agitated for at least 20 to 30 minutes prior to sampling. Agitation is necessary for a uniform temperature throughout the tank for a representative sample and product packaging. Product under the coils or adhering to the tank walls in a quiescent state will provide a nonrepresentative sample.
- flush sample line—Product from the previous product sampled will collect in the small-diameter sample lines. Flushing with the current product is necessary to clear the sample line to obtain a uniform product.
- clean sample container—Clean sample containers must be used for all samples to avoid contamination of the product with a previous product or another material.
- visual appearance—The sampler should observe the visual appearance of the sample while it is filling the sample container. A sample with visible hard fractions or a cloudy appearance indicates that the sample line has not been flushed or the tank temperature indicator is faulty requiring maintenance to properly identify the product temperature. The product should not be sampled until a clear product representative of the product in the tank can be secured.
- sample identification—The sample must be identified with the tank number, batch number, product identity, date, time and person obtaining the sample to assure that the sample is correct and provide information for followup if a discrepancy occurs.

(3) Additive changes—Some additives will change the iodin
which should be reflected on the product specification'
limits. Flaked product additives that can materially af'
value results are lecithin and other emulsifiers.

2.22 Packaged Shortening—Storage and Transportation

PROBLEM: ABSORBED FLAVORS

Possible Causes and Corrective Actions
(1) Most fats and oils products will absorb odors and flavors from other
foods and materials quite readily. Therefore, shortenings, oils, margarines, and other specialty fats and oils products should never be stored
or transported with aromatic foods like spices, garlic, onions, pickles,
fruits, smoked products, or any other materials such as ammonia, solvents, or petroleum products.

PROBLEM: OIL SEPARATION

Possible Causes and Corrective Actions
(1) Shipping temperature—The melting characteristics for the shortening
product must be considered when shipping temperature requirements
are established. Products with sharp melting points and steep solids fat
index curves require temperature-protected shipping, especially in the
summer months.
(2) Crystal structure—Beta-prime crystal habit shortenings are more heat
stable than beta crystal habit products and usually will withstand higher
temperatures if formulated with a sufficient level of hardstock.
(3) Stack height—Pressure from the weight of the cases stacked on top of
it can cause the bottom cases to have excessive oil leakage. The corrugated cases can act like an ink blotter to soak up the separated oil.
Small quantities of liquid oil will stain an entire case, and the stain will
migrate upward to the other cases. Stack height must be controlled for
these product formulations, especially in the summer months.
(4) Coated cases—In some cases, it is necessary to utilize wax or plastic
coatings for the cases to prevent shortening wicking problems caused
by separated oil.

2.23 Control Analytical Laboratory

PROBLEM: FAULTY ANALYSES

Possible Causes and Corrective Actions
(1) Technique—The work of a laboratory analyst is precision work. To be

efficient, their technique or working habits must also be precise. Because of the varied types of instruments and other equipment with which they work, there are many practices or tricks of the trade that must be learned that are too numerous to write into methods. These must be learned from literature, other laboratory workers, or experience. The essence of experimentation is technique or scientific skill, and the simplest procedure will give unreliable results if the technique is poor or incorrect. Listed below are some poor technique practices and an indication of some of the analyses affected:

- poor analytical balance technique—This is especially dangerous when small amounts are weighed for testing for analysis such as iodine value. If large weights are involved in the weighing of small amounts, the summation of the tolerances will cause a poor result for the evaluation. Other tests such as soapstock analysis and moisture content must be weighed quickly before moisture escapes in the dry atmosphere of the balance.

- misreading thermometers—Tilt of a thermometer, eye level while reading, divisions on a thermometer, faulty thermometers, agitation, time in a sample before reading, and type of thermometer are all factors to be considered in reading a thermometer. Proper reading of a thermometer is vital to the success of such tests as cloud points, melting points, congeal points, bleach, titer, quick titer, and smoke point.

- poor titration—Poor titrations due to improper agitation of sample, carelessness, or misreading a burette will give poor results on such tests as free fatty acid determinations, peroxide value, and caustic analyses.

- improper laboratory scale technique—Cleanliness, inaccurate weights, and other improper scale practices affect such tests as refining analysis, bleach, cloud point, peroxide value, smoke point, and others.

- use of shortcuts—Cutting short the time element on certain procedures or leaving out procedures that may seem unimportant can easily provide poor results for refining, iodine value, soapstock, bleach analyses, and others.

- improper baths—Wrong temperatures, improper agitation, and cutting short the duration in baths may give the wrong results for such analyses as refining, titer, melting points, cold test, AOM stability, and others.

- overheating—Overheating is dangerous to bleach and some moisture evaluations. Overheating of oils prior to color tests is a source of error. For example, if a sample of shortening is overheated while melting, the color reading will be higher than the true color of the sample. This can be easily be done by leaving the sample on the steam bath too long.

- care of equipment—Improper care of equipment affects most testing in the laboratory but especially refractive index, color readings, fatty acid composition, Mettler dropping point, and other analyses requiring laboratory equipment to perform the evaluations.
- misreading endpoints—Endpoints should be approached at the rate specified in the method, and the analyst must be able to recognize them. Improper recognition of an endpoint will give inaccurate results for such evaluations as most melting points, titrations, quick titer, cold test, and others.
- faulty comparative blanks—A blank is used in analyses such as iodine value and peroxide value. The blank requires as much care as the regular sample analysis to ensure accurate results.
- poor filtering or sample washing—The importance of sample filtering and washing techniques are reviewed and emphasized in most of the analytical chemistry literature. These techniques are essential in soapstock analysis and other evaluations with a somewhat similar procedure.
- improper sample drying—Trace amounts of moisture will produce abnormally low cold tests and adversely affect many other analytical evaluations.

(2) Sampling errors—In many analyses, the technician is required to use a small portion of a large sample for the evaluation. Extreme care must be excised to obtain a representative portion and when reducing a large portion to the very small amounts required for most analyses. If the oil to be sampled is a hardened oil, it should be completely liquid before weighing for the individual analysis.

(3) Mathematical errors—Misplacing a decimal point, transposing numbers, poor addition, inaccurate subtraction, or other simple mathematical errors will provide an inaccurate analytical result, even when good technique has been practiced throughout the testing procedure.

(4) Transcription errors—Excellent technique and execution of the evaluation can be lost in transcribing the data from a worksheet to the user department or the permanent record.

(5) Laboratory noise and confusion—The percent of errors in a laboratory can sometimes be traced to the amount of noise and confusion in the laboratory. Management, as well as the laboratory analyst, should keep any interference to a minimum. Noise and interference affect the work of each individual and all the others working in the same laboratory.

(6) Wild results—Seasoned analysts should be acquainted well enough with the history of the sample, the effect of various variables in the test, and the probable results to recognize a wild result and take the necessary

corrective actions, which may include an examination of the equipment, reagents, and sample identity, to determine any problems. If not acquainted with a history of the particular sample, the analyst must rely upon the specification, and any suspect result should be rechecked.

(7) Laboratory cleanup—Cleanliness and good housekeeping are integral parts of good laboratory technique. Minute amounts of impurities can spoil a sensitive test and, to a greater degree, provide erroneous results for any evaluation either qualitative or quantitative. Therefore, floors, tables, clothing, and so on must be kept clean, as well as the instruments, glassware, and other utensils used for the actual analysis.

(8) Glassware washing—The importance of absolutely clean containers for all laboratory evaluations cannot be overemphasized. Any film of soap, dirt, or oil on glassware or other laboratory containers will act as a prooxidant and seriously deteriorate the keeping quality of stored samples or the results of analysis. The procedures recommended for washing specialized laboratory glassware such as dilatometers, burettes, and pipettes is usually described in the individual methods. Regardless of the procedure used for washing, a visual inspection of all glassware should be made before the glassware is used, and any questionable glassware pieces should be rewashed.

2.24 Edible Fats and Oils—Organoleptic Evaluations

PROBLEM: OFF-FLAVOR

Possible Cause and Corrective Actions

(1) Flavor description—Experienced edible fats and oils tasters' description of the flavor type helps to identify the cause of an off-flavor. Descriptions of the flavor types encountered when tasting edible fats and oils are listed below:

- nutty—flavor reminiscent of fresh pecans; one of the least objectionable type flavors. When very intense, grades off into objectionable rubber-like flavor.
- rubbery—a flavor reminiscent of the odor of old rubber. Related to "nutty" in the description above; quite objectionable in slight to very objectionable in strong degree.
- monoglyceride—more of a "feel" rather than a flavor; when slight, characterized by a "rich" taste, or feel in the mouth, like cream. When very strong, it gives a bitter or puckering sensation and is one of the least objectionable type flavors when slight to moderate intensity.
- bitter—a flavor characteristic of products containing materials like polysorbate emulsifiers. It is not objectionable in slight degree but very objectionable when intensified.

- corn—a characteristic of corn oil, somewhat reminiscent of popcorn. It is not objectionable in slight degree.
- oxidized—a not quite fresh flavor, characteristic of oils exposed to air, or slightly old. It is sometimes "metallic" flavored to some observers and somewhat objectionable in slight degree, to moderately objectionable in strong degree. When very intense, it grades off into "rancid."
- rancid—a very disagreeable, sometimes sharp, biting, or nauseating flavor in very old or strongly oxidized fats and oils. It is related to "oxidized" in the above description and is very objectionable in any degree.
- sour—a sour-milky flavor like oil separated from fresh margarine. It is somewhat objectionable in slight to moderate degree. When very intense, it probably grades off into a very objectionable "fishy" or "fish oil" flavor.
- fishy—a flavor reminiscent of cod-liver oil, sometimes seen in heat mistreated soybean oil. It is related to "sour" described above.
- beany—a hay-like flavor, sometimes called "weedy" by some, most frequently seen in soybean oil products. It is somewhat objectionable in slight to quite objectionable in strong degree and probably grades off into a "painty" flavor when very intense.
- painty—a flavor reminiscent of the odor of linseed oil, or drying paint, after the solvent odor has disappeared. It is related to "beany" as described above. It is quite objectionable when in slight to very objectionable when in strong degree.
- raw—a flavor of not quite deodorized unhydrogenated oils, frequently observed in cottonseed salad oil and called "earthy" by some. It is somewhat objectionable in slight to moderately objectionable in strong degree. When very intense, it may grade off into "undeodorized" flavor.
- undeodorized—a flavor characteristic of the odor of hydrogenated oil prior to deodorization. It has an unpleasant aromatic or aldehydric flavor and is quite objectionable when in slight to very objectionable when in strong degree.
- musty—flavor reminiscent of the odor of a moldy or dark cellar or room. It is quite objectionable when in slight to very objectionable when in strong degree.
- watermelony—flavor reminiscent of biting into watermelon rind, or of cucumbers. It is quite objectionable when in slight to very objectionable when in strong degree.
- soapy—soap-like flavor, usually associated with the degradation of fats containing lauric acid. It is quite objectionable in only slight degrees.

- Dowtherm contaminated—caused by leakage in Dowtherm coils in the deodorizer. When very faint, it resembles "oxidized" flavor, but when moderate to strong is readily recognized as resembling the somewhat aromatic or phenolic odor of Dowtherm A. It is moderately objectionable when in slight to very objectionable when in strong degree.
- tallowy—flavor like beef fat from a roast after a day or so in the refrigerator, or like mutton fat. It is moderately objectionable when in strong degree.
- lardy—flavor like steam-rendered hog lard, maybe faintly skunky. It is moderately objectionable when in slight to quite objectionable when in strong degree.
- sour meat fat—very strong objectionable flavor, sometimes called "steep water" taste. This condition is usually caused by excessive protein and moisture in the rendered meat fat product.
- oxidized meat fat—rancid flavor caused by exposure to air, overheating and/or age. Mild characteristics are sometimes called "scorched."

(2) Oxidation characteristics—The oxidation course has two distinct phases. The initial period of relatively slow oxidation is called the induction period. Then after a certain critical amount of oxidation has occurred, the reaction enters a second phase characterized by a rapid rate of oxidation, which increases to many times greater than during the initial phase. The second phase start is when the fat or oil begins to smell or tasted oxidized.

(3) Flavor deterioration—Fats differ considerably in the way that their oxidation and accompanying flavor deterioration proceeds. The more highly saturated animal fats and hydrogenated oils have relatively little change in odor and flavor during the early phases of oxidation. Off-flavor development in these fats is both sudden and definite. However, relatively unsaturated oils, such as cottonseed or soybean oils, have a more gradual deterioration in flavor and odor [27].

(4) Light effect—Light has a deleterious effect upon the flavor stability of edible oils and fat-containing foods. Exposures of 0.5 to 1 hour produced significant changes in both flavor and peroxide values of laboratory control tests with refined, bleached, and deodorized (RBD) soybean, cottonseed, and safflower oils and two hydrogenated, winterized soybean oil products. The off-flavor was described as "light struck," which was described as grassy, or green, combined with a mouth sensation described as astringent. The oils' flavor and peroxide values were affected to the greatest degree in the first 30 to 60 minutes exposure, and from then on, the change is gradual. Light stability testing indicated

that the degradation of the oils was the same for all oils, even with measures such as hydrogenation, which improves the resistance to oxidation as related to temperature abuse. Containers that exclude light have effectively protected edible oils and fat-containing foods from off-flavors described as "light struck" [28].

(5) Absorbed oxygen—The amount of oxygen that must be absorbed to produce off-flavors varies considerably according to the composition of the fat or oil, the presence of natural or added antioxidants, the metal content (especially copper, iron, and heavy metals), and the product temperature.

(6) Thermal decomposition—In handling and storage of heated fats and oils, there is always a risk of thermal decomposition. Some products are more sensitive to heat than others and especially those with an additive. For example, lecithin addition to deodorized oils will increase the product's heat sensitivity. At temperatures above 160°F (70°C), lecithin will turn dark, smoke, and smell fishy. Other additives can also affect the product's resistance to deterioration. If heating is necessary, it should be done with either hot water or low pressure steam (<1.5 kg/cm^2).

(7) Absorbed odor and flavor—Fats and oils easily absorb odor and flavors from many different sources, for example, paint, solvents, laboratory chemicals, spices, gases, and fumes. Therefore, great care must be exercised when painting in or around the processing plant, for maintenance of equipment, and to avoid contamination with solvents and gases, and fats and oils packaged products should never be stored with other highly odoriferous materials or foods.

(8) Commingling of fats—Contamination of one source oil with another provides flavors and odors foreign to the initial product that are objectionable. For example, liquid domestic oils do not tolerate contamination with coconut, palm kernel, or other lauric oils. The short-chain fatty acids (C-4 to C-12) have a low taste threshold value when only slightly hydrolyzed in high moisture or high fat prepared food products.

3. SHORTENING APPLICATIONS

3.1 Baking Shortening

PROBLEM: PINK COLORED BAKED PRODUCTS

Possible Causes and Corrective Actions

(1) Antioxidant—Alkaline metal ions such as sodium or potassium in leavenings and dairy products used in baked products such as biscuits, pizza

crusts, and cakes can react with TBHQ and BHA antioxidants to cause a pink color. The color is not harmful, but it does present an undesirable appearance; therefore, these antioxidants should not be used in baking shortenings that will be used in products with baking leavening agents or dairy products.

PROBLEM: POOR CREAMING PROPERTIES

Possible Causes and Corrective Actions

(1) Too soft—Plastic shortenings extend into streaks and films in cake batters to lubricate large surfaces and entrap large quantities of air, which has a leavening effect. Shortenings with soft consistencies are not able to retain all of the incorporated air. Creaming volume evaluations show this deficiency with a decrease in the batter specific gravity determinations. The creaming volume performance test (see Method 8.1 in Chapter 3) measures air incorporation at 15 and 20 minutes mixing intervals of the first stage of an old-fashion pound cake formula. An increase in the batter specific gravity at the 20-minute determination indicates that the shortening is releasing the entrapped air. A soft consistency allowing this breakdown can be the result of

- steep solids fat index—Shortenings with wide plastic ranges provide optimum creaming properties. Wide plastic ranges are produced by combining a flat, soft basestock with a beta-prime crystal forming hardstock. A shortening with most of the solids fat index results between 15 and 25% has a wide plastic range.
- low solids fat index content—The percent solids in a compound shortening can be increased by increasing the hardstock level without a substantial change in the SFI slope.
- high creaming gas addition—In an effort to improve shortening whiteness at filling, some operators may increase the creaming gas content excessively. Shortenings with excessive creaming gas become puffy and soft with a resultant loss in creaming properties.
- low chill unit temperatures—Low chilling unit temperatures produce softer shortening consistencies. Penetration control values indicating a softer than desired shortening should trigger a change in the chilling unit outlet temperature to maintain the desired creaming properties.
- high working after chilling—High rpm, extended working, or a combination of the two conditions after chilling will increase the softening of the shortening consistency to potentially affect the creaming properties. Penetration evaluations should help identify a trend toward this condition, allowing corrective actions before a serious decrease in creaming properties is experienced.

- high shortening use temperature—Normally, the shortening's use temperature is 65 to 75°F (18 to 24°C). Higher use temperatures will soften the shortening and reduce the creaming properties. If a high use temperature is the norm for a particular operation, the shortening hardstock level can be increased to provide a specialty use shortening for optimum creaming properties in this situation.

(2) Too firm—Firm shortenings do not extend into streaks and films in cake batters easily, which reduces the air entrapment to diminish the leavening effect. Creaming volume evaluations will show high batter specific gravity results at both the 15- and 20-minute determinations. This condition should also be readily evident by consistency and penetration evaluations. Firm shortenings can be affected by

- too high solids fat index results—High SFI contents produce firm shortenings. An adjustment in the hardstock level to decrease the percent solids should automatically reduce the shortening's firm consistency. However, if the specified basestock is too hard, the required reduction in hardstock may be too high to maintain the shortening's heat resistance. In this situation, the basestock should not be used for this type of shortening. If this is a blending situation during processing, the out-of-limits basestock should be rehardened into another product.

- high chill unit temperature—Firm shortening consistencies can also result from high chilling unit temperatures. Penetration and physical consistency evaluations should identify this problem before the shortening is shipped, but this is an after-the-fact control. The proper crystallization conditions should be identified and specified for each product as developed. Then, penetration and physical consistency evaluations should identify slight drifts or trends that need attention, even though the current product still has satisfactory results.

- low working after chilling—Shortening consistency is dependent upon the amount of working after chilling. Lesser amounts of working will produce firmer shortenings that do not have optimum creaming properties in cake batters.

(3) Beta crystal habit—The large granular crystals of the beta polymorphic form produce products that are waxy and grainy with poor aeration potentials. Shortenings exhibiting a stable beta-prime crystal form have smooth consistencies that provide good aeration and excellent creaming properties. The source oils used in the composition of the shortenings determine the crystal habit of the shortening and hence the potential creaming properties. For example, soybean oil exhibits the beta polymorphic crystal form, while hydrogenated cottonseed oil is a beta-prime former. Therefore, soybean oil–based shortenings requiring a plastic range and creaming properties are formulated with 5 to 20% of a beta-

prime tending hard fat like cottonseed oil. The beta-prime hard fat must have a higher melting point than the soybean oil basestock in order for the entire shortening to crystallize in the stable beta-prime form.

(4) Emulsifier level—One of the functions contributed by an emulsifier addition to a baking shortening is aeration. Some emulsifiers are better aerators than others, and the levels required vary greatly with the different emulsifiers. The creaming properties contributed by the emulsifier or emulsifier system will be improved with increased levels up to a point. Too high an addition of the emulsifiers will deaerate the product rather than promote more air entrapment by producing a weak batter emulsion or cell structure in the finished product.

(5) Nonuniform consistency—Improper chilling, working, and tempering conditions can produce shortening and/or margarine with nonuniform consistencies, which could be described as streaking, sandy, ribby, puffy, separated, and petroleum jelly-like appearance. One of the important crystallization prerequisites contributing to these conditions is the pressure at the chill unit, worker units, and the filler. As a rule of thumb, the use of high pressures produces smoother, more uniform shortenings and margarine products.

(6) Hydrophilic surfactant—The more hydrophilic emulsifiers should not be a part of a conventional shortening system if the product is to be melted, because the incompatible emulsifier will separate from the liquid fat. These emulsifiers can be dispersed in plasticized shortenings if added to the heated fat with good agitation just prior to crystallization [29].

PROBLEM: LOW-VOLUME CAKES

Possible Causes and Corrective Actions

(1) Methyl silicone—This addition effectively retards foam development during frying to extend the useful life of frying shortenings. However, since cake batters are also foams, the unintentional addition to a cake shortening will also retard the cake volume. Methyl silicone is effective at levels of less than 1.0 parts per million, which obviously indicates the extreme handling care that must be exercised.

(2) High emulsifier level—An overemulsified cake will collapse from excessive aeration, which weakens the cell walls of the batter foam. This results in a low-volume cake or a dip in the center of each layer with a large coarse cell structure. The most obvious corrective action is to reduce the emulsifier level or use an emulsifier with less aeration capabilities.

(3) Low emulsifier level—Insufficient aeration is experienced with low emulsifier levels, which result in heavy, low-volume cakes. An increase in the emulsifier level should increase the batter aeration to produce a lighter cake with an increased volume.

PROBLEM: TUNNELS OR LARGE HOLES

Possible Causes and Corrective Actions

(1) Tunnels or large holes in a layer or pound cake are most times attributed to excessive bottom heat during baking. However, this condition can be aggravated by the use of an emulsifier system that aerates the batters to produce low batter specific gravities. The natural movement of the batter during baking is interrupted, which increases the incidences of tunnels or large holes in the baked cakes. A reduction in the emulsifier level, or the addition of 1.0 to 2.0% lecithin, will reduce the tunneling unless the cake formulation is too lean, which could require a complete reformulation.

PROBLEM: PREPARED MIX LEAVENING LOSS

Possible Causes and Corrective Actions

(1) Free fatty acid increase—Acidity increase of the flour or meal in a prepared mix is accelerated by shortening. A leavening change, or the loss of available carbon dioxide through bicarbonate decomposition, occurs, which is roughly proportional to the rate of free fatty acid increase. Emulsified shortenings tend to accelerate this reaction, but harder base fats and emulsifiers are less conducive to this problem.

(2) High moisture—Most leavening systems are activated by moisture. Shortenings with high moisture contents can reduce the shelf life of a leavened prepared mix considerably.

3.2 Frying Shortenings

PROBLEM: EXCESSIVE FOAMING

Possible Causes and Corrective Actions

(1) No antifoamer added—Analytical methods to determine the presence of methyl silicone have a poor reliability and require a lengthily elapsed time for the result; therefore, the use of a tracer so that the presence can be determined rather easily and is reliable is recommended. Mixing the antifoamer with a commercial antioxidant mixture provides a tracer and aids in the dispersion of the methyl silicone in oil.

(2) Poor antifoamer dispersion—Methyl silicone is only *dispersible* in oil—not soluble; therefore, the following procedure is presented, which has been successful in attaining good antioxidant dispersion when followed carefully. Agitate the methyl silicone with the antioxidant tracer and a small portion of the oil before addition to the product tank. Agitate the product tank for a minimum of 20 minutes before sampling to analyze for the tracer. During packaging, the supply tank should be continuously agitated until the oil level drops below the agitator. Last case analysis should be performed to ensure that the antifoamer did not settle out of the product during the period when the agitator was shut off.

(3) Soap contamination—Analysis of used and fresh frying shortening from a customer complaint should be performed. The presence of soap in the unused product indicates that the shortening was improperly refined and bleached or processed in equipment that had been washed and improperly rinsed or neutralized. The absence of soap in the unused product with a soap content in the used product indicates either a customer fryer washing procedure problem or introduction of a caustic material with the food fried. Deep fat fryers should always be rinsed thoroughly and neutralized with a weak acid like vinegar after boiling out with a cleaning material. Some foods that are deep fat fried have the potential for a caustic content due to the processing of the raw materials; for example, commercially prepared French fries are caustic rinsed to help remove the skin, and corn is steeped in a caustic solution to help remove the outer skin for the preparation of massa for tacos. Soap contamination of the frying fat accelerates foam development.

(4) Emulsifier presence—Emulsifiers allow the mixing of water and oil to form an emulsion that promotes foaming in a deep fat fryer. Emulsifiers could be introduced into a frying kettle by the commingling with an emulsified shortening before or during packaging, product mislabeling, or the use of the wrong shortening product for frying. Again, analysis of both unused and used product should identify the problem source quickly.

(5) Lauric and domestic oil blending—Immediate foaming in the deep fat fryer results with low quantities of lauric oil blends with a domestic oil or vice versa. The blends can occur during transportation, shortening processing, or in the customer's fryer. In any case, immediate foaming is the result. Fatty acid composition analysis of the used and unused samples will identify this problem by the lauric fatty acid (C-12:0) content.

(6) Wrong frying shortening used—(a) The use of an oil high in polyunsaturated in a low usage situation develops foam much sooner than a shortening or oil with a lower polyunsaturated level. (b) The use of a pan and grill frying shortening for deep fat frying will result in immediate foaming. Lecithin, the usual antisticking agent used in these products promotes immediate foaming in the deep fat fryer.

(7) Low turnover—Decreases in the quantity of food fried lessens the amount of frying shortening removed from the frying kettle by absorption into the fried foods. Therefore, less fresh shortening is needed to replenish the used shortening in the fryer. The lower replenishment quantity increases the polymerization level of the shortening in the fryer because it is exposed to heating for longer periods; polymerization contributes to foaming by holding the moisture captive. Forced oil turnover is practiced by some food operators to maintain the frying shortening quality at a constant level. In forced turnover, oil is removed from the fryer if the absorption is not enough to allow the addition of a prescribed quantity of fresh shortening.

(8) Overheating—High-temperature abuse of frying fats accelerates the polymerization rate of the frying media. Overheating could be caused by a faulty fryer thermostat or by deliberately frying at a high temperature. Polymerization contributes to foaming by holding the moisture captive.

(9) Food fats contamination—Almost all fried foods contain fats that leach into the frying fat. This fat transfer causes changes in the frying fat, which can be harmful; for example, a lauric oil transfer from a product previously fried in coconut oil will accelerate the foaming of a used shortening like it would a fresh shortening.

(10) Hot spots—Thin places on electric or gas tubes cause hot spots when heating. The higher temperature abuse in the hot spots starts the deterioration process. The deterioration products formed promote further deterioration to initiate a chain reaction. The results are polymerization evidenced by gum formation, leading to eventual higher absorption and foaming.

PROBLEM: GUMMING OR POLYMERIZATION

Possible Causes and Corrective Actions

(1) High-iodine value frying oil—A high iodine value indicates that the oil has a high polyunsaturated fatty acid level. Polyunsaturated fatty acids polymerize more rapidly than either monounsaturated or saturated fatty acids. These products can be used successfully to fry foods in a high

turnover situation but will polymerize excessively in frying situations with less than a 1:1 turnover (replacement of enough oil after 1 day's frying to again fill the fryer), causing a buildup of gummy material on the fryer walls, which is most noticeable at the oil level.

(2) Wrong shortening composition—Addition of an oil high in polyunsaturated fatty acids to a "heavy-duty" frying shortening will cause more kettle gumming because of the higher polymerization rate of the oil portion. Frying shortenings designated heavy-duty are usually all hydrogenated to reduce the polyunsaturated fatty acid content to a minimum to obtain maximum frying stability.

(3) Overheating—The gum buildup rate is proportionate to the temperature abuse; i.e., the higher the frying temperature, the quicker that gumming develops. Overheating can be caused by a faulty thermostat, deliberately frying at too high a temperature, hot spots caused by thin walls in the gas flue, uneven heating due to faulty electric coils, temperature recovery drifts too high, and other causes.

(4) Antifoamer omission—Methyl silicone inhibits polymerization, which is the cause of gumming. The omission of this additive will allow an early buildup of the gum deposits on the sides of the frying kettle and will eventually result in the excessive foaming of the frying fat. Mixing the antifoamer with an antioxidant mixture provides the possibility for a tracer analysis during packaging, which is not possible without the antioxidant. This analytical capability affords the shortening processor a control to help assure that the antifoamer has been added.

(5) Infrequent fryer cleaning—Some gum buildup is experienced, even with the most stable frying shortenings. These gum deposits are usually removed by the routine fryer boilouts; however, the buildup will become more noticeable with longer times between fryer cleanings. The following cleaning procedure has been used successfully to clean and neutralize deep fat fryers:
- Drain the frying shortening from the fryer.
- Fill the fryer with water to the usual frying fat level.
- Add a fryer cleaner at the manufacturer's recommended strength.
- Bring the cleaning solution to a boil, being careful that the water does not foam over the fryer sides. Turn the heat off when the foam starts to rise. Repeat boiling until all of the gummy deposits are loose.
- Scrub the top and sides of the fryer to remove deposits with a sink brush or scrub pad. Do not use a metal sponge or wire brush. These utensils will scratch the fryer surface, causing future cleaning problems, and may promote frying fat breakdown by exposing copper or brass to the hot oil. Drain part of the water and scrub the exposed sides of the fryer. Repeat this procedure until only a small amount

of water remains in the bottom of the fryer and scrub the bottom surface.

- Rinse the fryer with clear water.
- Refill the fryer with clear water.
- Add 1 pint of 50-grain vinegar to each 40 pounds of water in the fryer.
- Bring the vinegar water solution to a rolling boil.
- Turn the heat off after boiling. Splash the vinegar water over all surfaces where soapy water could have contacted the fryer.
- Drain and discard the vinegar water solution.
- Rinse the fryer with clear water.
- Dry the fryer surfaces thoroughly with paper towels. Paper towels should be used to avoid lint and possible soap residue from cloth towels.
- Fill the fryer with fresh shortening or used shortening that is still in satisfactory condition.

(6) Heating too rapidly—Plasticized shortening must be heated slowly at 200°F (93°C) until the voids or pockets are filled with melted fat and the creaming gas has been eliminated from the shortening. Unless the melting process is carried out with gentle heat, the portions of shortening in contact with the elements and fryer walls will be overheated. The high-temperature abuse will cause polymerization and gumming.

(7) Hot spots—Thin places on electric or gas tubes cause hot spots when heating. The higher temperature abuse in the hot spots starts the deterioration process. The deterioration products formed promote further deterioration to initiate a chain reaction. The results are polymerization evidenced by gum formation, leading to eventual higher absorption and foaming.

(8) Too strong vent draft—A strong vent draft will remove the protective steam blanket above the hot fat in the fryer, which results in an increased rate of oxidation. Fryer vent draft controls should be adjusted to retain the steam blanket as long as possible but still remove steam and cooking odors before they become excessive.

(9) Trace metals—Brass, copper, monel metal, and copper-bearing alloys are prooxidants that exert a marked catalytic effect to accelerate frying fat breakdown. Trace quantities of these prooxidants can be introduced by the food fried, the frying equipment, or contact of the frying shortening with these metals before addition to the frying kettle. The trace metals accelerate oxidation of the frying shortening, which is the combination of oxygen with the unsaturated fatty acids causing off-flavor and odor development with color darkening. Oxidation precedes polymerization of the unsaturated fatty acids in the frying shortening.

PROBLEM: DARKENING DURING FRYING

Possible Causes and Corrective Actions
(1) Oxidation—The combining of oxygen with the unsaturated fatty acids causes off-flavors and odors, as well as darkening. Control of the frying conditions can reduce the rate of oxidation but cannot eliminate it.

(2) Emulsifier presence—Frying shortenings should never contain emulsifiers that accelerate darkening, smoking, and other frying fat deterioration. The possible sources of the emulsifier could be from contamination with an emulsified shortening during processing, improper labeling of the shortening product, or an emulsified shortening being used by the frying operator either in error or on purpose. The possible contamination during processing should be eliminated by investigation of free fatty acid increases at any stage after deodorization. Another control is first and last package evaluation during the packaging operation; any real change in analytical characteristics should be immediately investigated. Control of the labeling process is the responsibility of the line operators and supervisors who must follow established quality practices in determining the proper product identification. The shortening processor should supply distributors and customers with product performance information regarding the proper use of the different products.

(3) Antioxidants—Some antioxidants, especially propyl gallate, accelerate the darkening of frying fats at frying temperatures. Propyl gallate should be avoided as a frying fat antioxidant, and the others should be used at the lowest possible levels necessary. Antioxidants protect frying shortenings from oxidation until heated to frying temperature. During processing, the antioxidants can be mixed with the antifoamer to act as a tracer for this minute effective additive. However, the minimum antioxidant quantities should be used to decrease the possibility of darkening in the frying kettle.

(4) High frying temperature—Frying fat darkening is accelerated with increases in frying temperatures. The foodservice frying temperature used for most foods is 325 to 375°F (163 to 191°C).

(5) Wrong shortening used—Pan and grill shortenings have been mistaken for deep fat frying shortenings, resulting in rapid darkening when heated to frying temperature. Most pan and grill products are yellow colored, which should alert the deep fat fryer operator that the wrong product has been added to the fryer. Pan and grill shortenings contain lecithin as an antisticking agent, which will darken at a relatively low temperature, smoke, and have an offensive odor.

(6) Faulty fryer thermostat—Overheating accelerates the darkening of frying shortenings, but early smoking should alert the frying operator of the problem.

(7) High egg product—Eggs yolks contain phospholipids, or lecithin, the same natural emulsifier used in pan and grill shortenings as an anti-sticking agent. Leaching of the phospholipids into the frying shortening will cause darkening, smoking, and an offensive odor when they decompose at frying temperatures.

(8) Inadequate filtering—Infrequent or no filtering leaves food particles in the fryer, which burn and carbonize to cause darkening of the frying shortening, as well as off-flavors.

(9) Poor cleaning practices—A soap residue from improper neutralization after cleaning accelerates frying fat breakdown, which is accompanied by darkening. Rinsing with a weak acid (vinegar) solution will neutralize the possibility of a soap residue.

(10) Hot spots—Areas of excessively high temperatures can be caused by thin places on gas flues or electric elements, allowing a higher heat transfer in an isolated spot, which accelerates frying fat darkening and breakdown. Inspection of the flues or elements during cleaning should help identify a possible problem before it is a major problem.

(11) Slow fat turnover—Frying fat is continually removed from the frying kettle by absorption into the food being fried. To maintain a constant quality of frying fat in the kettle, fresh fat must be added to replace the fat removed by the food. The rate at which fresh fat is added is designated as "fat turnover." Rapid fat turnover keeps the frying fat in good condition through frequent replenishment with fresh fat; however, a slow fat turnover exposes the fat to frying temperatures for longer periods, which results in a darker color fat due to oxidation and other frying fat deterioration. In a slow fat turnover situation, the frying fat must be discarded more frequently or measures taken to improve the fried food take-away. Forced turnover, or the replacement of a specific quantity of frying fat daily, has been successful for some operations. Another alternative has been the replacement of a large fryer with a smaller capacity fryer to improve turnover.

(12) Gum formation—Gum or polymerized fat deposited on the frying equipment tends to catalyze the formation of more gums, as well as contribute to foaming and a color increase of the fat. Regular cleaning of the frying equipment by boiling with a detergent solution and frequent wiping down of the equipment will help minimize this problem.

(13) Food preservatives—Prepared french-cut and other potato products will discolor or turn brown with exposure. Several preservatives are

used to prevent this product discoloration before frying. Too strong a concentration of these preservatives, excessive soaking in the preservative solution, or inadequate draining after treatment can accelerate frying fat deterioration.

(14) High-sugar foods—Sugars can transfer from foods to the frying fat. The sugars caramelize at frying temperatures, causing darkening and off-flavors to develop rapidly.

PROBLEM: EXCESSIVE FREE FATTY ACID DEVELOPMENT

Possible Causes and Corrective Actions

(1) Excessive chelating agent—Citric acid levels above 50 ppm and phosphoric acid levels above 10 ppm can catalyze free fatty acid development in the frying shortening during frying. The deodorizer addition systems should be calibrated routinely, and the acid/water solution should be prepared carefully.

(2) Emulsifier presence—The use of an emulsified shortening for frying will cause the breakdown of the fat. One indication of the breakdown will be a high free fatty acid level. Emulsifiers could be inadvertently added to the frying kettle by commingling of a frying and emulsified shortening before or during packaging, product mislabeling, or the use of the wrong shortening by the operator. Analysis of unused fresh shortening from the same lot and the used shortening should be performed to identify the problem source. Strengthening of the quality assurance requirements in the producing plant should help prevent a reoccurrence of the commingling problem, and reeducation of the operator on proper frying procedures should help retain a customer.

(3) High-moisture foods—Hydrolysis causes free fatty acid development; therefore, the more moisture introduced to the fryer, the higher the free fatty acid development. Food moisture levels can be controlled by drying soaked foods like french fries as well as possible before introduction into the fryer.

(4) Poor fryer cleaning practices—Soap residue in a fryer accelerates the free fatty acid development of the frying shortening. A weak acid solution (vinegar) rinse after cleaning will neutralize any soap residue remaining after cleaning, eliminating the possibility of a soap residue problem.

(5) Low turnover—Free fatty acid will be maintained at a level dependent upon the amount of fresh shortening added to the fryer daily. If enough fresh shortening is added to completely replenish the fryer every 3 days, the free fatty acid should be maintained at a satisfactory level for most fried foods. If the replenishment level drops below this point, the free fatty acid level will probably increase to a level where the flavor of the

fried food is affected. The frying shortening should be discarded at the first indication that the fried food flavor is beginning to change. The time before discarding will be longer in high turnover situations and shorter in low turnover situations.

(6) Drip back from exhaust vent—The material that collects on the exhaust vents is concentrated free fatty acids. If this material is allowed to drip back into the fryer, it will substantially increase the free fatty acid level, which results in excessive smoke and gives a bitter, acid flavor to the food fried. This material also collects on the underside of the lid of a pressure fryer. Pressure fryer lids should be wiped clean after each use to keep from contaminating the food fried.

(7) Abused fat addition—Some operators attempt to forgo the frying fat break-in period by adding used fat to fresh shortening after a complete change of frying shortening. This addition increases the free fatty acid content, which provides more food browning than 100% fresh shortening. Other operators attempt to improve their food costs with this practice by saving some of the normally discarded shortening. This practice has no sound economic basis because the degraded fat contains all of the ingredients required to bring about rapid breakdown of the fresh fat.

(8) Batter type—Free fatty acid development is increased with foods fried with batters containing egg yolks. Egg yolks contain phospholipids, which decompose at frying temperatures to cause darkening with an increase in free fatty acid. A positive correlation was found between the percentage of free fatty acid and color darkening in studies at Brigham Young University [30].

PROBLEM: EXCESSIVE SMOKING

Possible Causes and Corrective Actions

(1) High free fatty acid—The smoke point of frying shortening decreases in direct relation to the increases in free fatty acid; therefore, if the free fatty acid is controlled, smoking should not be a problem. See the free fatty acid section for causes and corrective actions.

(2) Emulsifiers present—The presence of an emulsifier in any shortening significantly lowers the smoke point. Excessive smoking shortly after heating to frying temperature is an indication that an emulsifier may be present. The causes and corrective actions for emulsifier presence in a frying shortening are covered in the free fatty acid problems section.

(3) Inadequate filtration—Breading, batters, small food pieces, and other materials that fall off the food fried are contaminants for the frying shortening, which char and smoke if allowed to remain in the fryer.

Frying shortenings should be filtered with a dependable filter at least once per day.

PROBLEM: HIGH FAT ABSORPTION INTO THE FRIED FOOD

Possible Causes and Corrective Actions

(1) Low frying temperature—Fried foods absorb fat at a rapid rate until the outside surface is sealed. The seal or crust is effected by a caramelization of the foods' sugars and starches by the heat transfer; therefore, the time required to achieve the seal materially affects the fat absorption rate. The frying temperature should be adjusted to seal the food as quickly as possible but still cook the inside of the product.

(2) Fryer load—Excessive food loads extend the heat recovery time, which causes a low frying temperature situation. Excessive fryer loads can increase the frying time to more than twice the normal length of time required to fry the food. The recommended deep fat frying load is a maximum of 1 pound of food for each 7 pounds of frying shortening.

(3) Shortening level—Frying with low shortening levels affects both the frying shortening and the food fried. The food fried can be grease soaked, soggy, or limp because of a slow temperature recovery from the increase in the fryer load to frying fat ratio. Frequent addition of fresh shortening to maintain the desired level reduces the breakdown rate by dilution.

(4) Frozen foods—All frozen foods should be thawed before frying to facilitate heat recovery and to be able to remove the excess moisture. The colder temperatures of the frozen foods will extend the time before the desired frying temperature is achieved, thereby increasing the fat absorption. Excessive moisture will increase the breakdown rate of the frying shortening through hydrolysis, which also increases the fat absorption rate.

(5) Foaming shortening—Foods fried in a foaming shortening are submerged in the foam layer rather than the liquid fat as usual. The foam layer contributes a much lower heat transfer than usual deep fat frying, which extends the frying time and increases the fat absorption. Foaming frying shortenings should be discarded and replaced with fresh product.

(6) Slow fryer recovery—The fryer temperature is lower than desired during the recovery time after the food is dropped into the frying fat. The lower frying temperature causes a higher fat absorption into the fried food, which increases with time; the longer the recovery rate, the higher the fat absorption. Recovery rates are a function of the fryer

thermostat. A routine check of the fryer recovery time and replacement of thermostats with slow recovery times will help assure that a uniform quality product is served.

(7) High moisture—Frying starts immediately when the food is immersed into the hot frying fat; an exchange of heat from the hot fat to the cold food begins. The temperature of the fat drops, and the temperature of the food rises. The moisture in the food starts to evaporate and the steam evolved causes a temporary bubbling on the fat surface as it escapes into the air. The steam emission temperature is only 212°F, even though the frying fat is at a higher temperature: 300+°F. As long as this process continues, the temperature of the food will not exceed 212°F. Thus, it is necessary to evaporate all of the moisture from the outer surface of the food before the sealing process begins. Therefore, foods with higher moisture levels extend the low-temperature frying time, increase the fryer temperature recovery time, and accelerate the hydrolysis rate. All foods should be drained properly to be as dry and free of excess moisture as possible before frying.

(8) Hot spots—Thin places on electric coils or gas tubes cause hot spots when heating. The higher temperature abuse in the hot spots starts the deterioration process. The deterioration products formed promote further deterioration to initiate a chain reaction. The results are polymerization evidenced by gum formation leading to eventual higher absorption and foaming.

(9) Batter type—Fat absorption is increased with foods utilizing egg and baking powder batters. Both egg yolk and baking powder increase the porosity of the batter to create more voids and increase the surface area for fat absorption.

PROBLEM: OFF-FLAVOR

Possible Causes and Corrective Actions

(1) Source oils—Each source oil has a characteristic flavor that it reverts to with abuse, some more objectionable than others. The flavor stability of each oil is dependent upon the level of polyunsaturated fatty acids in most cases. The unsaturated fatty acids combine with oxygen and revert to their original flavor profile. The intensity of the flavor reversion is dependent upon the amount of oxidation. Therefore, the selection of the source oil is an important consideration for the flavor of the finished product, especially if the oil has not been hydrogenated to reduce the unsaturated fatty acids. In some cases, the characteristic flavor of an oil is desirable; for example, a peanut oil flavor is supposedly characteristic for English fish and chips.

(2) Lauric oils—Coconut, palm kernel oil, and other high lauric fatty acid oils are quite resistant to oxidation, but more prone to hydrolysis than most other fats and oils. This characteristic creates a characteristic soapy flavor that is imparted to the food fried. If the oxidative stability of these oils is desired, the flavor of the fried food must be monitored closely to determine the replacement point for the frying oil; the normal indicators used by foodservice operators will not indicate when the frying oil should be discarded.

(3) Melting point—The melting point of the frying shortening will influence the fried food's mouth feel when eaten. A high melting point can mask the fried food's flavor, which can be desirable or undesirable. The effect of the melting point should be on the selection criteria when deciding upon a frying oil or shortening to use.

(4) Solids fat index—This analytical tool is a measurement of the percent solids at certain temperatures, usually 50, 70, 80, 92, and 104°F. The solids level at body temperature is an important indicator of the effect upon mouth feel or flavor of the fried food. High solids at 92°F will naturally mask the food flavors or delay the flavor impact. Low solids at 92°F will allow the flavor of the food to come through almost immediately. A steep solids fat index is desirable for most frying shortenings where frying stability and flavor are both desired; high solids at 50, 70, and 80°F that decrease rapidly to only 1 or 2% at 104°F are characteristic of a steep SFI.

(5) Multipurpose frying—Food flavor transfer from one fried product to another will occur if the same fryer is used for multipurpose frying. The flavor transfer can be minimized by frying potatoes between high-flavor foods. Potatoes have a high moisture content and provide a steam distillation effect to the frying fat to deodorize or carry off the flavor components.

(6) High free fatty acid—Both hydrolysis-caused fat splitting and the presence of other acidic materials will analyze as free fatty acid by the usual titration method, which contributes an offensive tart or acidic flavor to the fried food. The causes and corrective actions for the free fatty acid development are listed in the excessive free fatty acid section.

(7) Inadequate filtration—Food particles left in the heated frying fat will continue to caramelize and develop a burnt offensive flavor that will be transferred to the food fried in the same oil. The frying fat should be filtered at least daily with a filter aid that will trap the fine contaminants in the oil.

(8) Improper filter media—The filter aid or earth used with the mechan-

ical filter should be evaluated thoroughly before use. Some materials are very effective at removing the food particles and even reducing the free fatty acid level but impart an off-flavor to the frying fat that is carried through to the food fried.

(9) Inadequate skimming—The frying fat surface should be skimmed frequently to remove the small food pieces remaining in the fryer. These food particles will char and stick to fresh product being fried, giving off offensive flavors.

(10) Addition of used grill or pan frying fats—Some operators add the used fats accumulated from their grills or pan frying to the deep fat fryer, expecting to save on the fresh frying shortening additions. This practice speeds up the deterioration of the deep fat frying shortening and gives the food fried a strange or different flavor than the customer expects.

(11) Air exposure—Care must be exercised during the filtration process to avoid aeration of the hot oil. Oxidation rates are accelerated by the higher temperatures necessary to keep the fat liquid during filtering. The filter pump should be turned off as soon as the first sign of aeration is evident. Also, the oil should not be allowed to cascade into the fryer; the return line should be placed on the bottom of the fryer beneath the oil surface as soon as possible. Oxidation reverts the frying fats back to the source oil's original flavor.

(12) Copper or brass—Both of these metals are strong prooxidants that accelerate the oxidation process to revert the frying fat back to the original flavor, which immediately carries through to the fried food. Fryer coils, thermocouples, and other parts probably are copper or brass alloys coated with stainless steel. The fryer must be inspected routinely to assure that the stainless steel has not worn through with cleaning to expose these prooxidants to the frying fat. Also, fryer repairs with copper or brass should be avoided.

(13) Foaming fat—Foaming frying fat is evidence that it is completely broken down and should not be used to fry food for human consumption. Additionally, the flavor and mouth feel of any food fried would be offensive. The causes and corrective actions for frying fat foaming are presented in the excessive foaming section.

(14) High temperature—In addition to accelerating the breakdown of the frying shortening, the fried food will not be prepared properly with high temperatures. The exterior of the food will be done while the center is still raw. The usual foodservice frying temperature is 350 to 360°F, but large food pieces require a longer frying time, probably at a reduced frying temperature.

(15) Low temperature—Fat absorption is increased by low-temperature frying. It is necessary to absorb some fat in order to produce the desired character and eating qualities. It is the absorbed fat and the browning that impart the savor to fried foods, which no other form of cooking develops. But excessive absorption caused by a deteriorated fat or improper frying imparts an objectionable mouth feel and flavor. The potential causes for low-temperature frying are (a) faulty thermostat, (b) poor heat recovery, (c) operator error, or (d) poor frying procedures.

(16) Slack period heating—Keeping the fryer at frying temperature for 1 hour with no food fried will break down the frying fat more than several hours of actual frying. When not in use, the temperature of the fat should be lowered as much as possible to a point where the temperature will recover quickly when needed. Frying equipment with capacities in excess of the requirements increase the frying shortening degradation, especially during slack periods. The purchase of multiple fryers instead of one large unit would allow the operator of shut down complete fryers during slack periods, thereby reducing the frying media deterioration.

(17) Poor turnover—Frying turnover is the relationship between the amount of fat added over a period of a day to the total amount of fat present in the fryer. If the turnover is low, the fat deteriorates at a higher rate. If the turnover is rapid, the fat deterioration is lessened due to the fresh frying fat added to replenish the amount absorbed into the food fried. All frying fat deterioration negatively affects the fried food flavor and should be kept to a minimum.

(18) Improper cleaning practices—Minute residual traces of cleaning agents will catalyze frying fat deterioration. Through rinsing with a weak acid solution, like 1 pint of 50-grain vinegar to 5 gallons water, will neutralize any remaining alkali residue after cleaning.

(19) Food size—Equivalent size food pieces must be fried so that the entire batch can be fried uniformly. Frying small and large food pieces together in a fryer will necessitate removal of the food pieces at different times to achieve a uniform flavor and appearance.

PROBLEM: CHEMICAL ODOR

Possible Causes and Corrective Actions

(1) Antioxidants—Upon heating to frying temperature, an odor described as chemical may be attributed to the use of antioxidants. Many of the antioxidants like BHA have strong phenolic odors, which are more pro-

nounced because these compounds are volatilized with high-temperature heating.

PROBLEM: FOOD APPEARANCE

Possible Causes and Corrective Actions

(1) Burnt food particles—Burnt food particles suspended in the frying medium tend to impart a bitter taste, as well as an objectionable speckled appearance to the fried products, which increases with the quantity of burnt particles present. Daily or continuous filtering with a filter capable of removing the suspended food particles is necessary to preserve the quality of the frying fat.

(2) Abused fat addition—Some operators attempt to forgo the frying fat break-in period by adding used frying fat to fresh shortening after a complete change. This addition increases the free fatty acid content, which provides more food browning than all fresh shortening. Other operators attempt to improve their food costs with this practice by saving some of the normally discarded shortening. This practice has no sound economic or performance basis because the degraded fat contains all of the ingredients required to bring about rapid breakdown of the fresh fat. Additionally, the performance improvement is too limited to justify the more rapid degradation of the fresh shortening.

(3) Oily or shiny food—Foods fried in a frying oil that is liquid at room temperature will have a more oily and shiny appearance than foods fried in a fat that is solid at room temperature.

(4) Dry or dull appearance—Frying fats that are solid at room temperature impart a drier food appearance than liquid oils. The fried food appearance will become more dull as the melting point of the frying fat increases. The desired fried food appearance can be achieved by choosing the frying media that provides the desired performance.

(5) Food size—Uniform food pieces must be fried so that all of the food will be done at the same time. Frying small and large pieces together in a fryer will necessitate removal of the food pieces at different times to achieve a uniform flavor and appearance.

(6) High temperature—Deep fat frying at a high temperature will effect a seal or crust on the food surface quicker than lower temperatures to retain more moisture. However, the heat transfer time required to cook the food completely may produce a darker crust color than desired. The lowest frying temperature that produces food that is completely cooked and has a pleasing appearance must be determined for each operation. Normally, the desired restaurant frying temperature will fall between 325 and 360°F (163 and 182°C).

PROBLEM: IMMEDIATE FOAMING

Possible Causes and Corrective Actions

(1) Creaming gas—Plasticized shortenings have creaming gas or nitrogen incorporated for better handling properties. Upon heating in the fryer, the inert gas will be liberated. The deaeration process could be misinterpreted to be foam development, especially if the fresh shortening is being rapidly heated to frying temperature. The large clear bubbles will cease after the gas has been liberated, probably before the shortening reaches frying temperature. The foam caused by frying shortening deterioration is characterized by small, yellowish, bubbles that do not break or dissipate until the food is removed.

(2) Lauric and domestic oil blending—Immediate foaming in the deep fat fryer results with low quantities of lauric oils such as coconut or palm kernel oil blending with a domestic oil or vice versa. A blend of this nature can occur during fats and oils transportation, processing, or in the customer's fryers, but no matter how the blending occurred, immediate foaming is the result. This problem can be identified by fatty acid composition analysis of a sample of used or unused shortening by the C-12:0 or lauric fatty acid content.

4. MARGARINE AND SPREAD PRODUCTS

PROBLEM: HIGH BACTERIA DETERMINATIONS

Possible Causes and Corrective Actions

Bacteria are one-celled organisms, ranging in size from about 1/75,000 to 1/10,000 inch in diameter and from 1/25,000 to 1/5,000 inch in length. Bacteria reproduce by simple asexual cell division; one cell alone can reproduce by cell division. Bacteria need food, moisture, and heat to grow. They are not mobile and have to be transported from place to place, most often by hands, shoes, and clothes. Therefore, the best preventive measures are good manufacturing practices. The types of bacteria most margarine manufacturers check for are

(1) Coliform—Some members of the coliform group are found in the intestines of all warm-blooded animals. They are not generally considered pathogenic (disease producing) but rather as "fellow travelers" with intestinal pathogens. They do not survive pasteurization. When found in pasteurized product, the presence of coliform is suggestive of unsanitary conditions or practices during production, processing, or storage. Coliform testing measures the quality of sanitation procedures.

(2) Standard plate count—Total plate count is valuable as a sanitation indicator and for quality information. The bacteria that grow at the total plate count incubation temperature are known as mesophiles and include a wide variety of microorganisms. The media used is nutrient-rich and nonselective. Both pathogenic and nonpathogenic organisms may be present.

(3) Yeast and mold—Yeast and mold have very similar growth parameters. Both are able to survive extremes in conditions, such as pH, water activity, and high concentrations of sugar, that most bacteria cannot tolerate. Since yeast and mold can survive such conditions, they are important spoilage organisms in margarine and spread products. The presence of yeast and mold in these products indicates poor sanitation practices.

(4) Thermophile—The term *thermophile* is used to describe a group of microorganisms that grow in the 131° to 194°F range. These organisms are very heat resistant and can cause spoilage of product. They will grow into spores if held at elevated temperatures or if the product is improperly cooled.

(5) Pathogenic microorganisms—A pathogen is an organism that causes disease. The two types that are of interest for margarine and spread production are infections and intoxications. *Salmonella, E. coliform,* and *Listeria* are infectious organisms that can grow in margarine-type products. These live organisms are poisonous to humans and cause food poisoning when they are ingested. The other organism that poses a problem is *Staphylococcus.* Some strains of *Staph.* produce a toxin that is poisonous to humans when ingested. Pasteurization will kill the organism, but once the toxin has been formed, it remains active. This is the reason hand contamination of product after pasteurization can have very critical effects.

PROBLEM: MARGARINE PACKAGE PROBLEMS

Possible Causes and Corrective Actions
(1) Dented or crushed cartons, oil spots on the carton, and inadequate carton flap gluing are all operator adjustment problems that must be corrected upon occurrence to assure quality packaging.

PROBLEM: SOFT-TUB MARGARINE APPEARANCE PROBLEMS

Possible Causes and Corrective Actions
(1) Short fill can be caused by weight control problems or not enough nitrogen gas in soft-tub margarine production.

(2) Soft-tub margarine sloshed fill or excessive margarine on the lid indicates that the margarine fill temperature is too high, the line speed is too high, excessive rough handling by the case packers, or the wrong marbase utilized.

(3) Warm fill temperatures can also result in oil separation or a dull surface sheen when the product solidifies.

(4) A cheesy texture can occur from insufficient back pressure on the cooling system, too cold chilling temperature, or insufficient product throughput.

(5) Streaking is caused by the margarine channeling in the chilling unit. The streaked area represents the portion that received insufficient cooling.

(6) Grainy textures will result when a margarine transitions to the beta crystal with higher temperature storage. Margarines should always be stored at 45 ± 5°F (7.2 ± 2.8°C).

PROBLEM: FLAVOR AND MOUTH FEEL DEFECTS

Possible Causes and Corrective Actions

(1) Storage at high temperatures or for prolonged periods will allow the crystal habit to convert to the beta form, which has a higher melting character that will impart a waxy mouth feel. Margarine should be rotated on a first-in/first-out basis and held at refrigerated temperatures of 45 ± 5°F.

(2) Lack of flavor can be experienced because of (1) the use of a low level of flavoring material, (2) a loss of flavor with age, (3) beta crystal development, and (4) use of a marbase or margarine oil with too high a melting point or too flat an SFI curve.

(3) Oxidized oil will impart a reverted oil or rancid flavor to the margarine. The marbase used to prepare the margarine emulsion must have a good initial flavor and adequate oxidative stability for the product.

(4) Sour margarine flavors usually result because of the use of a poor quality milk.

(5) Plastic aroma and flavor may be imparted by the plastic container used to package soft-tub margarines.

(6) Fruity, artificial margarine flavors are generally due to poor warehousing practices. Margarine should never be exposed to aromas from fruits, fresh vegetables, or any products. Margarines will readily absorb flavors and odors.

(7) Old or aged margarine flavor is any change from the normal flavor sensation, which is usually a musty-type flavor with aging.

5. QUALITY MANAGEMENT

PROBLEM: SPECIFICATION PREPARATION MISTAKES [31,32]

Possible Causes and Corrective Actions

(1) Including all possible factors—Specifications frequently contain certain factors simply because they are known, not because they have any value. Unnecessary requirements confuse and sometimes contradict each other so that it becomes impossible to meet all of the specified limits. A good specification includes only those requirements that have a known significance.

(2) Duplicate evaluations—Sometimes the same property is specified to be measured by various methods instead of the one method that is most valuable and best suited to routine determinations. Unimportant details detract from the value of essential factors. Also, imposing unnecessary requirements upon a supplier must be avoided. A specification should not require any more evaluations than absolutely necessary.

(3) Not recognizing variability—The only property common to all ingredients is variability. A single value does not constitute a specification unless clearly shown to be a minimum or maximum. In all other cases, the property should be designated in terms of an acceptable range. This range should not be so narrow as to add cost or make compliance unduly difficult; however, it should only be as wide as end-use considerations can tolerate. Additionally, the variability resulting from the analytical method must not be forgotten.

(4) Zero tolerances—These are almost always unrealistic, especially with the extremely sensitive analytical capabilities available today. If a zero limit is specified, it must be in terms of a particular sampling plan and/or analytical procedure. However, even in the case of a simple particle size measurement, it is wise to allow at least a small tolerance for material that will remain on the most coarse screen or that will pass through the finest one.

(5) Significant figures observance—Significant figures used in specifications must not go beyond the accuracy of the analytical method employed or the requirements of the end use. For example, a maximum of 1% often is assumed to mean a maximum of 1.0%, or even 1.00%. This is not correct. A specified maximum of 1% really means that 1.499% is acceptable. Careful consideration must be given to the exact intended meaning when numbers are used in a specification.

(6) Coping a supplier's sales specification—This is the easiest way to prepare an ingredient or packaging specification. Unfortunately, it

does not always result in the best specification and in some cases may be very inadequate. Suppliers' specifications should always be carefully reviewed for compliance with the requirements of the finished product.

(7) Identification of specific analytical methods—The method of measurement for each factor specified must be identified. One of the most common reasons for disagreement between customers and suppliers concerning the acceptability of a product is that each is measuring compliance by a different procedure. This is a real problem when numerous standard methods are available for a given measurement, such as moisture, melting point, and so on.

(8) Special instructions omission—Special instructions required of the supplier should be included on each individual specification such as specific sampling, testing, and acceptance procedures, including pre-shipment samples, certificates of analysis, delivery temperature, and so forth. These special handling requirements prescribed by the customer should be clearly identified on each individual specification.

(9) No specification revision provision—There is no such thing as a final specification; if there were, it would mean that we have stopped learning. Sometimes, a new specification must be changed shortly after it has been issued, even after extensive original study has been conducted to make it as perfect as possible. Even then, inadequacies appear as the ingredient is produced or used. On the basis of experience and new technology, every specification will eventually need to be revised.

(10) Refusal to compromise or change—Getting exactly what is desired in an ingredient often is impossible. It may be necessary to work with what is available and build around it. Even after a specification has been prepared, the suppliers may find compliance with certain requirements impossible, and this situation will require compromise, usually without detriment to either party.

(11) Transferring a specification from one end use to another— An ingredient may be used in several products. Its function in each of these products may or may not be the same. Therefore, before an existing ingredient specification is considered adequate for a new use, it should be carefully reviewed to determine if the values and limits specified are appropriate for both products. Also, beware of reinstating an old specification since the product may no longer be available or the performance limits may have changed.

(12) Insufficient information—Realistically, the specification writer never has all the facts. The question is whether the proper use has been made of the available information. The writer must have some knowl-

edge of how an ingredient is made, the characteristic properties, and the intended end use. This information gathering requires communication with those who have the information, usually the manufacturer, product development, and operations.

(13) Specifying custom-made ingredients—Sometimes, a lack of knowledge of available ingredients or the required performance characteristics leads to specifications that identify an expensive custom-made product. A unique ingredient may be necessary on special occasions, but standard products readily available should be considered first. The developers and operations personnel initiating the requests for ingredients should consult with the specifications custodian early in their projects to avoid these problems. Likewise, the specification system should be friendly enough to avoid this type of confusion.

(14) Timely results—All analytical results specified should be completed in time for necessary corrective actions to be taken before the ingredient is used or consumed. Testing that is completed after the ingredient is used is worse than no testing at all, unless it is done as an audit with no intent of taking any action as a result of the findings.

(15) Incompatible analytical requirements—Many product analyses are interdependent; for example, iodine value is dependent upon the fatty acid composition, melting points are extensions of the solids fat index, and AOM stability is dependent upon the iodine value and/or fatty acid composition. Occasionally, the limits for a specification may be determined from a faulty laboratory analysis, an abnormal product, or just wishful expectations. These types of errors can make it impossible to qualify a supplier or contribute to excessive costs.

(16) Vague requirements—Specifications must be exacting standards for the product, including all quality characteristics, composition, legal, and other requirements. Vague requirements like variable for the tolerance or a recommendation rather than a directive become obstacles to a clear understanding of what is needed. All specification requirements should be specific, well defined, explicit standards.

(17) Specification clarity—Specifications are communication tools for all levels of management, suppliers, and the operations personnel. Every entry should be easy to understand and impossible to misunderstand.

(18) The umbrella method—Specifications are often prepared so that nobody is left out. Often, when preparing purchase specifications, the tendency is to include the weakest supplier, but when concerned with a production specification, the tendency is to include the ridged requirements desired by only one of the customers for that particular product. In both cases, the specification limits must be established utilizing factual data that identify the purchased product's minimum

requirements and the capabilities of the process to produce a particular product. Both purchased ingredients, process aids, and other adjuncts, as well as the internal processes, must be considered in light of the finished product to be produced.

(19) Tight limits—Specification limits set too tight just to gain or maintain a reputation for toughness, or hoping that production will meet them somehow, will result in a disregard for all specifications. Specification limits must be realistic for both the producer and the customer to prevent process interruptions or poor quality product when critical characteristics are ignored.

(20) The ostrich method—Specification limits cannot be changed to bring an out-of-control process into specification conformance. When a product's history has shown conformance to the design limits until a recent period, it indicates a serious process deficiency that should be addressed instead of changing the requirements to fit poor performance.

(21) Trivial versus important—The relative importance of the different parameters must be identified. Some characteristics are more critical to the acceptability of the product and may require a narrower range of acceptability. Others may only have a minor importance and do not need as much attention. One method of conveying the importance of the requirements is to separate them into critical, major, and minor categories.

(22) Incomplete reason for change—Each specification sheet should have a "REASON FOR CHANGE" section containing *all* of the changes to the current issue. This section lists all the changes for the user, rather than requiring a study of the document to identify them. A partial listing in this section could cause a change to be overlooked. A further aid to identify changes for the personnel utilizing the specification is a "c" indicator in the right-hand margin in line with the actual change.

(23) Packaging component coordination—Most of the edible fats and oils products packaging consists of multiple components. These components all have variations in the final dimensions, which could affect the performance during packing and the product protection. The components must have a tight fit for product protection but assemble easily for high-speed packaging. The allowed specification tolerances must be coordinated for all of the components to prevent "in limits" parts assembly problems.

(24) Processed product interchange—The process control specification system treats the products from each process as a finished product. Each of these finished products may be the base product for another single product or for many other finished products. The single base product

has only to meet the requirements for one product, but the multi-purpose base must conform to the requirement of all the succeeding products. The product designers and the specifications developers must coordinate the process components to determine if existing bases are satisfactory or a new basestock is required to adequately prepare the desired product. The specification developer must guard against the use of a basestock that will only meet the finished products requirements part of the time.

(25) Establishment of specification tolerances—The best specification limits and tolerances are arrived at by a determination of needs and capability. Specifications arrived at in this manner should assume the aspect of an industrial law in that they must be met and enforced. However, too few specification limits are based on performance data but, rather, are based on consensus of opinion, compromise, or mutual distrust. Historical performance data provide the means for determining process capability and finished product requirements. The observations that can be made from the historical data are

- The process will produce material that, for all measurable characteristics, has a central tendency: the average, median, or mean calculation of all the data points.
- The material will vary from the central tendency within certain limits, which can be determined by calculating the standard deviation.
- After the mean and standard deviation are identified, a probability can be established for the occurrence of material in any measurable distance from the mean or center point.
- Whenever product is produced outside the specification limits established in this manner, it is known that chance alone has not caused the deviation, but some real deficiency exists that requires corrective action.

New specifications should have temporary specified limits set on the basis of like products or pilot plant development work until a history is available to establish permanent limits.

(26) Abbreviations—Shortened words should be avoided unless the abbreviation is recognized universally, not just within a plant or company. Even some abbreviations in use for long periods can cause errors; for example, IV has been used to abbreviate iodine value for quite some time, but it could be mistaken for the roman numeral IV in some cases. Another illustration is the use of GLC to mean fatty acid composition. GLC is actually an abbreviation of the equipment used to provide a fatty acid composition and other analyses. Abbreviations can have too many different meanings to effectively communicate specification requirements.

PROBLEM: SPECIFICATION ENFORCEMENT

Possible Causes and Corrective Actions

(1) Specification and practice conflicts—The "black book" of the foreman or operator must not take precedence over the official specification. Specifications must be regarded as company law. No one should have the right to deviate from the established specification without first securing either a change authorization or a temporary authority to deviate.

(2) Wrong issue used—Replacement specification issues must replace the superseded issues promptly in all specification books to comply with product and customer changes as needed. Operations, quality control, costing, and others utilizing the specification network must devise a system to assure that the most current specification issues are available and being utilized. One effective system requires the return of the superseded copies to a central point for checkoff and disposal.

(3) Verbal changes—All specification changes must follow the designated specification change procedures. Verbal changes outside the established procedure almost always create problems, such as
- The change is never communicated to some of the people who need to know
- The change is never incorporated into the working system.
- Some people follow the change while others forget or ignore it.
- Verbal exchanges are easily misunderstood or forgotten, and the actual change implemented is different than desired.

With the speed of the communications systems today, the need for a verbal specification change should not exist. However, the personnel managing the system must be prepared to react on short notice at times to preserve customer goodwill.

(4) Deviations—Allowing deviations can have the effect of rewriting the specification. This creates a situation of having more than one source of the specification and in effect redesigns the product without authority. Once a specification has been issued, it should be regarded as an industrial law. No one should have the right to deviate from the specification without first securing a specification change or, in isolated cases, permission for the deviation from quality assurance.

PROBLEM: COMPLETE NEW SPECIFICATION

Possible Causes and Corrective Actions

(1) Auxiliary requirements—With the pressure to issue a new product specification, the auxiliary requirements have a tendency to be overlooked at

times. Utilization of a checklist enumerating all possible individual requirements helps to avoid these oversights. The following list should be a start for the development of a new specification checklist:

- Are any additional intermediate product specifications required? If so, have they been prepared and approved?
- Are purchase specifications available for all the ingredients, raw materials, and processing aids required with approved suppliers identified?
- Are the necessary packaging materials specifications prepared for this product with approved vendors identified?
- Are any new analytical, inspection, or performance evaluation methods required for this new specification? If so, are acceptable methods developed, procedures written, and plant testing complete?
- Have the summary specifications been updated to include this new specification's requirements?
- Does this product require certification from a governmental agency or a special interest group?

PROBLEM: COMPLAINT HANDLING

Possible Causes and Corrective Actions

(1) System—Complaints are the voices of the customers and deserve careful attention. The critical point is to have a system whereby all complaints, regardless of source, come to a central collection point where well-documented procedures for followthrough are understood and enforced. All complaints should be tabulated and receive continuing review by quality assurance and operations.

(2) Procedure distribution—The complaint handling procedure must be thoroughly documented and distributed so that all involved personnel understand their responsibilities for assuring product quality and reliability.

(3) Measure of quality—The number of complaints and returns are helpful as one indicator of the success of a quality control program but not sure evidence of customer satisfaction. The degree of reliability changes with the customer type:

- retail—Studies indicate that a very small portion of the dissatisfied retail customers will register a complaint; approximately 96% of the unhappy customers do not complain to the company. Most simply do not repurchase and tell their friends, neighbors, and relatives about their dissatisfaction.
- foodservice—A higher level of dissatisfied customers complain through their distributors; however, many of the complaints never reach the edible fats and oils processor.

- food processor—Direct sales of large product quantities to food processors result in a high level of the potential product complaints being registered with the supplier. Also, food processor complaints can be investigated more thoroughly to better identify the corrective action required to prevent a reoccurrence through direct contact with the customer.

(4) Results—An industrial customer probably understands that problems will occur. Therefore, he will remember your company, either for its inability to handle his problem or its outstanding support to him during a time of crisis. The impression caused can definitely affect future sales.

(5) Slow response—A speedy analysis is needed when your product is reported as the problem and you are not in agreement. Every effort should be made to identify the real problem to protect your product and turn this negative into a positive.

(6) Frustration—Passing the buck and voice mail are probably the two most frustrating experiences for a customer with a question or a complaint. This should be prevented by a policy that the first person involved must handle all the customer's inquiries until the situation has a successful conclusion. When voice mail is utilized, a procedure to return calls within a short period must be enforced.

(7) Points to consider—Personnel routinely handling complaints should have enough experience to answer questions regarding the product and a personality to fit the situation. The following are guidelines provided by persons routinely handling complaints at one company:
- Listen and be attentive to the customer.
- Keep your voice calm.
- Stay objective.
- Show a genuine willingness to help.
- Get all necessary information the first time.
- Stay with the customer; do not put him on hold or answer another call.
- Discuss a course of action with options.
- Commit to perform requested services on a specified date and meet that commitment.
- Ensure that your technical service representative understands the sensitivities of the situation when making a difficult call.
- Acknowledge the complaint in writing.
- Make a sincere effort to resolve the complaint.
- Make sure the customer has the proper information for any necessary followup calls.
- Keep the company's mission statement and principles of conduct in mind at all times.
- Remember, to the customer, you are the company.

PROBLEM: PRODUCT TRACEABILITY

The key component for any recall procedure is the ability to trace the use of specific lots of raw materials, ingredients, or processing aids to the finished product and then to account for all quantities produced.

Possible Causes and Corrective Actions

(1) Package product code dates—The code date should be as simple as possible and include the date packaged, batch or time period, and plant location. All packages and shipping containers must be coded with the same information. The actual code dates must be clearly discernible and placed in the same location, which can be easily located for rapid identification.

(2) Shipping records—Accurate product quantities by code date must be recorded on the shipping records for traceability. Two most common problems experienced are failing to record more than one code date per shipment and hidden code dates on the case panels facing the inside of the pallet, causing these code dates to be omitted.

(3) Processing records—Accurate batch and lot number recording for each component of a new process batch is mandatory for good traceability. Systems must be designed and implemented to preserve this information for the life of all the components.

(4) Universal product codes—UPC codes are used for almost all retail products, and many companies have integrated all products into this system. This system is used by many large distribution systems for inventory control.

(5) Sequential case coding—An additional sequential number case coding increases the degree of traceability to within a particular lot of product. This added feature allows the identification of product from any point during the packaging sequence. Suspected problems can be pinpointed by testing to determine the start or end of a particular difference or change in the product.

6. REFERENCES

1. Mostia, W. L. 1996. "Control for the Process Industries," *Control Magazine*, Putman Publishing Co., Chicago, IL. 9(2):65–69.

2. Parker, W. A. and D. Melnick. 1966. "Absence of Aflatoxin from Refined Vegetable Oils," *J. Am. Oil Chem. Soc.*, 43(11):635–638.

3. List, G. R. 1985. "Special Processing for Off Specification Oil," in *Handbook for Soy Oil Processing and Utilization*, D. Erickson et al., ed., Champaign, IL: The Am. Soybean Assoc. and Am. Oil Chem. Soc., pp. 355–377.

4. Robertson, J. A. et al. 1973. "Chemical Evaluation of Oil from Field and Storage—Damaged Soybeans," *J. Am. Oil Chem. Soc.,* 50(11): 443–445.

5. Hendrix, W. B. 1984. "Current Practice in Continuous Cottonseed Miscella Refining," *J. Am. Oil Chem. Soc.,* 61(8): 1369–1372.

6. Pritchard, J. R. 1983. "Oilseed Quality Requirements for Processing," *J. Am. Oil Chem. Soc.,* 60(2): 324–325.

7. Buck, D. F. 1981. "Antioxidants in Soya Oil," *J. Am. Oil Chem. Soc.,* 58(3): 275–278.

8. King, R. R. 1941. "Quality Changes in the Industrial Storage of Crude and Refined Cottonseed Oil," *Oil & Soap,* 18(1): 21.

9. Burkhalter, J. P. 1976. "Crude Oil Handling and Storage," *J. Am. Oil Chem. Soc.,* 53(6): 332–333.

10. Lathrap, C. A. 1965. "Determination of Weight of Bulk Oil Shipments," *J. Am. Oil Chem. Soc.,* 42(2): 155–158.

11. Carr, R. A. 1978. "Refining and Degumming Systems for Edible Fats and Oils," *J. Am. Oil Chem. Soc.,* 55(11):770.

12. Braake, B. 1976. "Degumming and Refining Practices in Europe," *J. Am. Oil Chem. Soc.,* 53(6):353.

13. Puri, P. S. 1980. "Hydrogenation of Oil and Fats," *J. Am. Oil Chem. Soc.,* 57(11): 852A.

14. Flider, F. J. and F. T. Orthoefer, 1981. "Metals in Soybean Oil," *J. Am. Oil Chem. Soc.,* 58(3): 181–182.

15. Morrison, W. H. et al. 1975. "Solvent Winterization of Sunflower Seed Oil," *J. Am. Oil Chem. Soc.,* 52(5): 148–150.

16. Sonntag, N. O. V. 1982. "Fat Splitting, Esterification, and Interesterification," in *Bailey's Industrial Oil and Fat Products, Volume 2,* Fourth Edition, Daniel Swern, ed. New York, NY: A Wiley-Interscience Publication, p. 134.

17. Weiss, T. J. 1983. "Shortening—Introduction," in *Food Oils and Their Uses,* Second Edition. Westport, CT: AVI Publishing Co., Inc., p. 125.

18. Sonntag, N. O. V. 1979. "Esterification and Interesterification," *J. Am. Oil Chem. Soc.,* 56(11): 752A.

19. Gavin, A. M. 1981. "Deodorization and Finished Oil Handling," *J. Am. Oil Chem. Soc.,* 58(3):181–182.

20. Dudrow, F. A. 1981. "Deodorization of Edible Oils," *J. Am. Oil Chem. Soc.,* 60(2): 274.

21. Weidermann, L. H. 1981. "Degumming, Refining and Bleaching Soybean Oil," *J. Am. Oil Chem. Soc.,* 58(3): 163–165.

22. White, F. B. 1956. "Deodorization," *J. Am. Oil Chem. Soc.,* 33(10): 505.

23. Sherwin, E. R. 1977. "Mistakes to Avoid in Applying Antioxidants to Vegetable Oils," *Food Engineering,* May, p. 83.

24. Sherwin, E. R. 1976. "Antioxidants in Vegetable Oils," *J. Am. Oil Chem. Soc.,* 53(6): 434–436.

25. Sherwin, E. R. 1978. "Oxidation and Antioxidants in Fat and Oil Processing," *J. Am. Oil Chem. Soc.,* 55(11): 809–814.

26. Belbin, A. A. 1993. "Color in Oils," *INFORM,* 4(6):652.

27. Sonntag, N. O. V. 1979. "Reactions in the Fatty Acid Chain," in *Bailey's Industrial Oil and Fat Products, Volume 1,* Fourth Edition, D. Swern, ed. New York, NY: A Wiley-Interscience Publication, pp. 144–145.

28. Moser, H. A. et al. 1965. "A Light Test to Measure Stability of Edible Oils," *J. Am. Oil Chem. Soc.,* 42(1):30–32.

29. Knightly, W. H. 1969. "The Use of Emulsifiers in Bakery Foods," in *Proceedings of the Forty-Fifth Annual Meeting of the American Society of Bakery Engineers,* March 3–5, 1969. Chicago, IL, p. 138.

30. Bennion, M. and R. L. Park. 1968. "Changes in Fats During Frying," *J. Am. Dietetic Assoc.,* 52(4): 308–312.

31. Wintermantel, J. 1982. "Ingredients Specification Writing, Part III," *Cereal Foods World,* 27(8): 372.

32. Manring, R. C. 1965. "Statistics for Better Spec's," *J. Am. Oil Chem. Soc.,* 42(12): 668A–708A.